DINOSAURS
AND PREHISTORIC LIFE

DK

Smithsonian

DINOSAURS
AND PREHISTORIC LIFE

Penguin Random House

FIRST EDITION

DK LONDON

Senior Art Editor Ina Stradins
Senior Editor Helen Fewster

Executive Editor Lori Cates Hand
US Editor Kayla Dugger

Managing Art Editor Michael Duffy
Managing Editor Angeles Gavira Guerrero

Jacket Designer Akiko Kato
Jacket Editor Emma Dawson

Jacket Design Development Manager Sophia MTT
Producer (Pre-production) Robert Dunn

Illustrators Andrew Kerr, Arran Lewis
Senior Producer Meskerem Berhane

Art Director Karen Self
Publisher Liz Wheeler

Design Director Phil Ormerod
Publishing Director Jonathan Metcalf

DK INDIA

Senior Art Editor Mahua Sharma
Senior Editor Dharini Ganesh

Art Editor Meenal Goel
Editor Aishvarya Misra

Jacket Designer Tanya Mehrotra
Picture Researcher Vishal Ghavri

DTP Designers Pawan Kumar, Rakesh Kumar, Vikram Singh
Senior Picture Researcher Surya Sankash Sarangi

Jackets Editorial Coordinator Priyanka Sharma

Managing Art Editor Sudakshina Basu
Managing Editor Rohan Sinha

Pre-production Manager Balwant Singh
Picture Research Manager Taiyaba Khatoon

Production Manager Pankaj Sharma
Managing Jackets Editor Saloni Singh

DK LONDON

Senior Art Editor Ina Stradins
Senior Editors Angeles Gavira Guerrero, Peter Frances

Project Art Editors Yen Mai Tsang, Francis Wong, Steve Woosnam-Savage
Project Editors Cressida Malins, Ruth O'Rourke, Gill Pitts, David Summers, Victoria Wiggins

Designers Sonia Barbate, Alison Gardner, Helen McTeer, Simon Murrell
Editors Jamie Ambrose, Daniel Gilpin, Salima Hirani, Tom Jackson, Nathan Joyce, Lara Maiklem, Emma Marriott, Claire Nottage, Miezan van Zyl

Design Assistants Riccie Janus, Rebecca Tennant
Editorial Assistants Elizabeth Munsey, Jaime Tenreiro

Production Editor Phil Sergeant
Researcher Graeme Lloyd

Creative Technical Support Adam Brackenbury, John Goldsmid
US Editor Jill Hamilton

Jacket Designer Duncan Turner
Indexer Hilary Bird

Illustrators Antbits Ltd (Richard Tibbitts), Dotnamestudios (Andrew Kerr), Francisco Gascó, Peter Minister, pixel-shack.com (Jon Hughes and Russel Gooday)
Production Rita Sinha

Cartographer Paul Eames
Picture Researcher Myriam Mégharbi

Special Photography Gary Ombler

Senior Managing Art Editor Philip Ormerod
Managing Editor Sarah Larter

Art Director Bryn Walls
Publishing Manager Liz Wheeler

Reference Publisher Jonathan Metcalf

DK INDIA

Art Director Shefali Upadhyay
Editor Garima Sharma

Designers Mahua Mandal, Neerja Rawat, Arijit Ganguly
DTP Designers Harish Aggarwal, Dheeraj Arora, Sunil Sharma, Preetam Singh, Jagtar Singh

Head of General Books Aparna Sharma

SMITHSONIAN

Established in 1846, the Smithsonian—the world's largest museum and research complex—includes 19 museums and galleries and the National Zoological Park. The total number of artifacts, works of art, and specimens in the Smithsonian's collections is estimated at 154 million, the bulk of which is contained in the National Museum of Natural History, which holds more than 126 million specimens and objects. The Smithsonian is a renowned research center, dedicated to public education, national service, and scholarship in the arts, sciences, and natural history.

Content previously published as *Prehistoric Life* in 2009

This American Edition, 2019
First American Edition, 2009
Published in the United States by DK Publishing
1450 Broadway, Suite 801, New York, NY 10018

Copyright © 2009, 2019 Dorling Kindersley Limited
DK, a Division of Penguin Random House LLC
19 20 21 22 23 10 9 8 7 6 5 4 3 2 1
001–299760–Sept/2019

A catalog record for this book
is available from the Library of Congress.
ISBN 978-1-4654-8249-5

DK books are available at special discounts when purchased in bulk for sales promotions, premiums, fund-raising, or educational use. For details, contact: DK Publishing Special Markets, 1450 Broadway, Suite 801, New York, NY 10018
SpecialSales@dk.com

Printed and bound in China

A WORLD OF IDEAS:
SEE ALL THERE IS TO KNOW

www.dk.com

Authors

YOUNG EARTH
Dr. Douglas Palmer—science writer and lecturer based in Cambridge, UK

MICROSCOPIC LIFE
Dr. Martin Brasier—Professor of Paleobiology, University of Oxford, UK

PLANTS
David Burnie, BSc—award-winning natural history writer

Chris Cleal—Head of Vegetation History, Amgueddfa Cymru (National Museum Wales)

Professor Sir Peter Crane, FRS—Director of the Royal Botanic Gardens, Kew until 2006; Professor at the Department of Geophysical Sciences, University of Chicago, USA

Professor Barry A. Thomas—Honorary Professor, Aberystwyth University, Wales

INVERTEBRATES
Dr. Caroline Buttler, Professor John C. W. Cope, and Dr. Robert M. Owens—Department of Natural Sciences, Amgueddfa Cymru (National Museum Wales), Cardiff, Wales

VERTEBRATES
Dr. Jason Anderson—vertebrate paleontologist and Assistant Professor of Veterinary Anatomy, University of Calgary, Canada

Dr. Roger Benson—paleontologist, University of Cambridge, UK

Stephen Brusatte—vertebrate paleontologist, University of Edinburgh, Scotland

Professor Jennifer A. Clack—Professor Emeritus of Vertebrate Paleontology, University Museum of Zoology, Cambridge, UK

Dr. Kim Dennis-Bryan—paleontologist, writer, and lecturer, Open University, London, UK

Dr. Christopher Duffin—paleontologist and teacher, London, UK

Dr. David Hone—vertebrate paleontologist, Beijing Institute of Vertebrate Paleontology and Paleoanthropology, Beijing, China

Dr. Zerina Johanson—curator of fossil fish, Natural History Museum, London, UK

Dr. Andrew Milner—vertebrate paleontologist and research associate, Natural History Museum, London, UK

Dr. Darren Naish—freelance researcher and author, University of Southampton

Dr. Katie Parsons—biologist and natural history writer, London, UK

Dr. Donald Prothero—Professor of Geology, Occidental College, and lecturer at California Institute of Technology, Los Angeles, USA

Professor Xu Xing—paleontologist, Beijing Institute of Vertebrate Paleontology and Paleoanthropology, Beijing, China

PERIOD INTRODUCTIONS
Dr. Ken McNamara—paleobiologist, Department of Earth Sciences, University of Cambridge, UK

HUMANS
Dr. Fiona Coward—Principal Academic in Archaeology and Anthropology, Bournemouth University, UK

GLOSSARY
Richard Beatty

Consultants

YOUNG EARTH
Dr. Simon Lamb—Associate Professor, School of Geography, Environment, and Earth Sciences, Victoria University of Wellington, New Zealand

Felicity Maxwell, BSc—freelance environmental consultant and science editor

MICROSCOPIC LIFE
Dr. Sean McMahon—Skłodowska-Curie Fellow, UK Centre for Astrobiology, University of Edinburgh, Scotland

PLANTS
Professor Sir Peter Crane, FRS—see above

Dr. Paul Kenrick—Research Scientist, Department of Earth Sciences, Natural History Museum, London, UK

INVERTEBRATES
Professor Euan N. K. Clarkson—Professor Emeritus of Paleontology, University of Edinburgh, Scotland

Dr. Caroline Buttler—see above

VERTEBRATES
Professor Michael J. Benton—Professor of Vertebrate Paleontology, University of Bristol, UK

Dr. Neil Brocklehurst—Research Fellow, Department of Earth Sciences, University of Oxford, UK

Dr. Stephen Brusatte—see above

Professor Robert K. Carr—Professor of Biology, Department of Natural Sciences, Concordia University Chicago, USA

Professor Jennifer Clack—see above

Dr. Gareth Dyke—paleontologist and lecturer, University of Debrecen, Hungary

Professor Christine Janis—mammalian paleontologist and lecturer, Brown University, Providence, USA

Dr. Sarah L. Shelley—Postdoctoral fellow in Mammals, Carnegie Museum of Natural History, Pittsburgh, USA

HUMANS
Dr. Fiona Coward—see above

Dr. Katerina Harvati—paleoanthropologist, Max Planck Institute, Germany

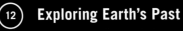

Scales and sizes

Most of the profiles in Life on Earth include a scale drawing to indicate the size (usually the maximum) of the animals or plants being described. In the plant entries, scale drawings are only included where there is reliable and easily accessible reference for reconstruction of the whole plant.

 1½in (4cm)

 7in (18cm)

 6ft (1.8m)

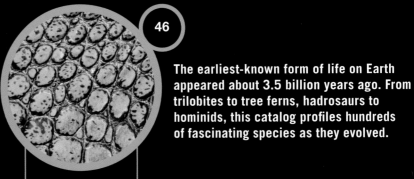

 (46)

The earliest-known form of life on Earth appeared about 3.5 billion years ago. From trilobites to tree ferns, hadrosaurs to hominids, this catalog profiles hundreds of fascinating species as they evolved.

contents

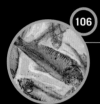

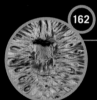
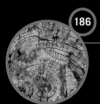

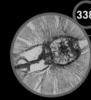

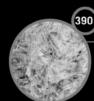

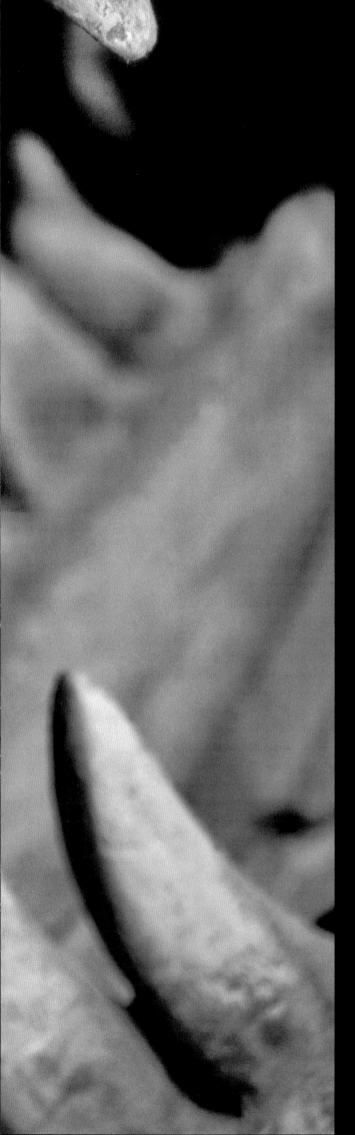

foreword

When Leonardo da Vinci picked up some fossil seashells from high in the Apennine Mountains of his native northern Italy at the dawn of the 16th century, he divined correctly that an ancient sea had once covered the area. So began the science of paleontology—the interpretation of the life of the ancient past. At a time when other philosophers explained such fossils as artificial "freaks of nature" or the lunchtime trash of Roman legionaries, Leonardo applied modern scientific methods of deduction.

Fossils can be beautiful and amazing. Anyone who has stood beneath the skeleton of a dinosaur or who has cracked open a rock to reveal the bright mother-of-pearl colors of a seashell that died over 100 million years ago will know the sense of wonder. That the dinosaur skeleton was once covered with muscle and skin and that the shellfish once lived on the seafloor, filtering food particles from the water, shows how fossils allow us access to different times.

By 1800, most naturalists accepted that the Earth was very ancient, although they could not measure its exact age; that many plants and animals found as fossils were actually extinct; and that there was some kind of succession of rocks and fossils through time. It took another two centuries to improve methods for dating rocks, to be able to read ancient environments from the sediments, and to understand the diets and habits of ancient life.

The modern paleobiologist makes observations in the field, but also does experiments in the chemistry laboratory and on supercomputers. Through trial and error, it is now possible to look beneath the surface and test ideas about the diet of a particular dinosaur or the scale and effects of a particular mass extinction. Paleobiology rests on the wonder of the fossils, but it is a powerful science where investigators piece together the clues like Sherlock Holmes to reveal the wonder of the life of the past.

This book is written by active professional paleontologists from all over the world, and it translates the current science into a rich narrative of the astonishing story of the evolution of life on Earth.

MICHAEL J. BENTON

YOUNG

Exploring Earth's Past

More than 200 years of scientific investigation have revealed a great deal about Earth's dynamic 4.5-billion-year history and development. Geological time has been measured and divided, and its environments and life reconstructed. Every year, new discoveries and new techniques of investigation generate new ideas about the planet's geological past, but there is still a great deal to be explored and explained by the relatively young science of geology.

The present and the past

The geological processes that transformed Earth from its origins over 4 billion years ago are recorded in rocks and minerals. Most of what we know about the planet's history has been obtained from the study of rocks and fossils found at or near the planet's surface. Additionally, some of these rocks preserve information about the environments in which living things evolved, along with their remains and traces. Reading this rock and fossil record, interpreting the information in it, and reconstructing a history of Earth and its living things have taken hundreds of years—and the task is still far from complete.

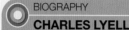

CHARLES LYELL

Trained as a barrister, the Scottish geologist Charles Lyell (1797–1875) used his skills of dealing with evidence and presenting arguments to examine the mass of emerging geological information in order to understand and reconstruct Earth's geological history. Lyell popularized an idea called the principle of uniformitarianism, which states that geological processes have occurred in the past at the same rate and intensity as they do today. This contradicted the prevailing view of an Earth shaped by a short history of catastrophic events, such as the biblical flood.

UNIFORMITARIANISM
A section of a modern sand dune (left) reveals a similar pattern of layers, called cross-bedding, to those in 200-million-year-old sandstones (above), suggesting formation by the same processes.

BROKEN RECORD
In the eighteenth century, Scottish farmer and naturalist James Hutton noticed "unconformities" or breaks in layers of gray shale and red sandstone at Siccar Point, Scotland (above). These gaps in the record, he reasoned, meant the Earth had a protracted history, with earlier episodes of mountain-building, erosion, and sedimentation.

Earth's structure

Different types of evidence suggest that Earth has a layered internal structure. By studying vibrations triggered by earthquakes, scientists have discovered that, like an onion, our planet has many layers. By combining this evidence with the study of volcanic rocks and meteorites, various layers with distinct compositions or physical properties can be recognized, from a hot, partly liquid core to a cold surface crust.

CORE TO CRUST
Earth's intensely hot iron and nickel core is divided into solid inner and liquid outer parts. Next is the mantle, made of dense silicate rock. Although almost completely solid, the hot, ductile rock flows. The outermost layer is the cool, thin crust with a variety of igneous, metamorphic, and sedimentary rocks.

outer core

inner core

mantle

crust

Rock and fossil records

Most rocks can be assigned to one of three types, depending on how they formed. Igneous rocks have cooled and crystallized from a molten state; they range from quick–cooled, fine-grained volcanic lava erupted at Earth's surface to coarse-grained rocks such as granites that have cooled slowly at depths of more than 3 miles (5km). Weathering and erosion of rocks generate sediment that is transported to areas such as inland basins or the oceans, where it is laid down as successive layers. Over time, with the application of pressure and heat, these layers are transformed into sedimentary rocks. Such rocks may contain fossils. Where either igneous or sedimentary rocks are subjected to extremely high temperatures and pressures, they are transformed or "morphed" into metamorphic rocks.

IGNEOUS ROCK
Melting of rocks deep in Earth's continental crust led to the formation of magma that slowly cooled and crystallized into this coarse-grained granite.

METAMORPHIC ROCK
When rocks are pushed deep into the crust, heat and pressure cause them to flow and recrystallize into metamorphic rocks such as this gneiss.

SEDIMENTARY ROCKS
The "layer-cake" appearance of Arizona's Painted Desert is actually a succession of sediment deposits laid down over 200 million years ago that contain both plant and dinosaur fossils.

REVEALED FOSSIL
Fossils can be used to date rocks. For example, the presence of the fossilized jawbone of a large, crocodilelike amphibian called *Rhinesuchus* in this rock layer in the Karoo region of South Africa shows that the rock is between 260 and 265 million years old.

Dating techniques

The first dating of Earth's rocks consisted of using common fossils to order recognized divisions of sedimentary layers. The types of fossils change over time because species evolve. Although there were many early attempts to calculate a chronological age for Earth and its formation, none were accurate until the phenomenon of radioactivity was discovered at the end of the nineteenth century. Knowledge of the fact that radioactive elements decay and have isotopes (see below) allowed scientists to date the formation of certain minerals within igneous rocks, from which it has become possible to establish an absolute chronology of Earth's history.

soil

volcanic ash ———————————————————— 1.5 million years ago

soil

volcanic ash ———————————————————— 1.75 million
soil ————————————————————————————— years ago

DATING FOSSILS
The radiometric dating of fossils depends upon dating the nearest appropriate igneous rocks, such as volcanic lava and ash deposited as part of the strata. Dating lava or ash above and below gives its minimum and maximum age.

RADIOMETRIC ROCK DATING
Radioactive elements such as uranium decay from the moment of their formation by shedding electrons (negatively charged particles present in each atom). This loss occurs over time at a regular rate, and results in the creation of a series of "daughter atoms" known as isotopes. By measuring the relative proportions of isotopes, the time that has passed since their formation can be calculated.

KEY
○ Uranium-235 atom
● Lead-207 atom

RATIO
32 uranium-235:
0 lead-207

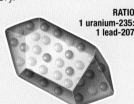

RATIO
1 uranium-235:
1 lead-207

RATIO
1 uranium-235:
3 lead-207

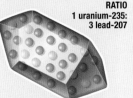

RATIO
1 uranium-235:
7 lead-207

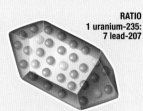

YEAR OF FORMATION
A mineral crystallizes from molten rock and contains radioactive uranium-235, which will decay to the isotope lead-207.

700 MILLION YEARS LATER
Fifty percent (half) of the uranium atoms are now atoms of lead-207. Thus, the "half-life" of uranium-235 is 700 million years.

1,400 MILLION YEARS LATER
By this time, another 50 percent of the uranium-235 atoms has decayed into lead 207. The ratio of uranium to lead atoms is now 1:3.

2,100 MILLION YEARS LATER
A geologist measures the ratio remaining in the rock (now 1:7) and dates the rock to three half-lives, that is, 2,100 million years old.

The Origin of the Earth

Born of gas and dust about 4.56 billion years ago, Earth has had a dramatic history. Scientists have attempted to reconstruct its earliest stages from evidence preserved in meteorites and Earth itself, as well as direct observations of distant stars and nebulae. But our knowledge of events in this remotest part of our planet's past remains incomplete.

The origin of the Solar System

The Solar System began to form about 4.56 billion years ago, when an immense cloud of gas and dust, the solar nebula, started to collapse under gravity. As it collapsed, the cloud flattened into an ever-faster spinning disk, with a bulging center that heated and condensed to form the Sun. The orbiting debris formed the four inner rocky planets. In the cooler outer disk, the four gas giants formed, then the small dwarf planets (including Pluto), and finally, a vast cloud of comets. Altogether, the Solar System extends about 3,700 billion miles (6,000 billion km) from the Sun.

dense, hot central region (protosun)

protoplanetary disk

planetesimals forming within rings

2 FORMATION OF THE PROTOSUN
Under the influence of gravity, the slowly rotating solar nebula began to contract and therefore spin faster. The cloud condensed into a disk with a dense, extremely hot, luminous center (the protosun) and a diffuse outer region (the protoplanetary disk).

3 RINGS AND PLANETESIMALS
The increasing speed of rotation condensed the icy gas and dust into rings within the protoplanetary disk. Colliding particles of dust and ice clumped together, and their increasing gravity attracted yet more material, forming planetesimals.

hot inner region of disk

Sun begins producing energy by nuclear fusion

accreting planetesimals

4 ROCKY PLANETS
Those planetesimals nearest the protosun consisted of the most heat-resistant and dense materials such as rock and iron. Attracted to each other by gravity, they collided and formed the four rocky planets (including Venus, below).

cooler outer part of disk

STAR-FORMING NEBULA
The first stars formed in the clouds of hydrogen and helium that dominated the early Universe. Their birth (above) and death (right) generated new elements such as carbon, oxygen, silicon, and iron.

SUPERNOVA EXPLOSION

The formation of the Earth

According to the most widely accepted theory of how the Solar System formed, known as the nebular hypothesis, the rocks and ice that shared the same orbit around the developing Sun coalesced under gravity, in a process called cold accretion. The largest bodies in each ring attracted the most material and formed planetesimals—loose collections of rock and ice with a uniform structure. As a planetesimal grew larger, its gravitational pull increased. It became more tightly held together, and it drew in rocks from its immediate surroundings with greater force, leading to a period of intense bombardment and growth. Earth, and the three other rocky planets of the inner Solar System, were formed in this way about 4.56 billion years ago.

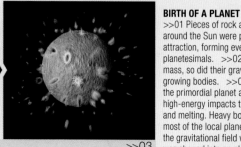

IMPACT CRATERS ON MERCURY
Mercury, the smallest of the rocky planets, has, like part of Earth's Moon, a highly cratered surface interspersed with dark lava fields. These impact craters all resulted from the same phase of intense meteorite bombardment as that suffered by Earth, and which lasted until about 3.5 billion years ago.

>>01   >>02   >>03

BIRTH OF A PLANET
>>01 Pieces of rock and ice sharing the same orbit around the Sun were pulled together by gravitational attraction, forming ever-larger clumps, known as planetesimals.   >>02 As they grew larger in size and mass, so did their gravitational field, forming steadily growing bodies.   >>03 Rocks were pulled toward the primordial planet at increasing speeds, producing high-energy impacts that generated localized heating and melting. Heavy bombardment continued until most of the local planetesimals were accreted and the gravitational field was so strong that the planet was shaped into an almost spherical spinning body.

1 THE SOLAR NEBULA FORMS
The solar nebula was initially a vast but dense cloud of cold gas and dust, several times larger than the present Solar System. The gas and dust is thought to have originated from the death of even older stars, and was effectively recycled.

THE NEBULAR HYPOTHESIS
The nebular hypothesis was developed by the French mathematician Pierre Laplace and the German philosopher Immanuel Kant in the late eighteenth century. The six stages shown here explain many of the known facts about the Solar System, such as why the orbits of most of the planets lie roughly in the same plane and why the planets all orbit the Sun in the same direction.

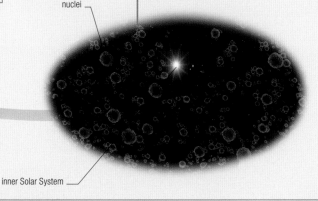

5 GAS GIANTS
In the cool outermost rings of the protoplanetary disk, beyond the asteroid belt, ice and gas could survive. Here, planetesimals made of rock and ice grew large enough to attract, and be enveloped by, deep clouds of gas (above). The four gas giants were formed, and shortly afterward the protosun became a fully fledged star.

gas giant forming

frozen cometary nuclei

6 REMAINING DEBRIS
Following the formation of the planets, some gas and other unaccreted material still remained in the protoplanetary disk. Most was blown away by radiation generated by nuclear fusion in the Sun. The remaining planetesimals formed the vast and distant Oort Cloud of comets at the edge of the Solar System.

inner Solar System

The Earth in the Solar System
Earth's size, orbit, and position in the Solar System have all promoted the evolution of life on this blue planet. The third planet from the Sun, Earth is the only one to lie in the habitable zone, where conditions are favorable to life. Earth's distance from the Sun and its mass, gravity, and internal heat have allowed the development and retention of an oxygen-rich atmosphere and abundant surface water. In contrast, Earth's small, lifeless Moon has little internal heat, an insubstantial atmosphere, and no surface water. Earth's orbit is elliptical, but its rotation and near-circular path mean that seasonal variation in exposure to solar radiation is not so extreme as to extinguish life.

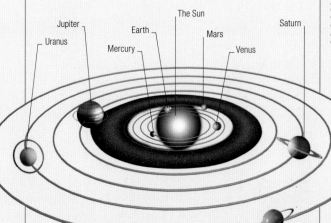

THE PLANETS	
Planet	**Average Distance from the Sun**
Mercury	36 million miles (57.9 million km)
Venus	67.2 million miles (108.2 million km)
Earth	93 million miles (149.6 million km)
Mars	141.6 million miles (227.9 million km)
Jupiter	483.6 million miles (778.3 million km)
Saturn	888 million miles (1.43 billion km)
Uranus	1.78 billion miles (2.87 billion km)
Neptune	2.8 billion miles (4.5 billion km)

The formation of the Moon
The earliest-known Moon rocks have been reliably dated at about 4.5 billion years old, indicating that Earth's satellite was formed not long after Earth itself. Most astronomers agree with the giant-impact theory, which proposes that the Moon originated when a massive asteroid collided with the young Earth and tore away a huge amount of its interior to create the Moon (see below). Continuous heavy meteorite bombardment over the following billion years left the Moon's rocky surface severely cratered. A period of volcanic activity then followed, and lava oozed out of cracks in the crust to fill low-lying craters. The lava solidified, forming the Moon's vast dark maria (seas), which are still visible from Earth today.

MOON ROCK
Over 838lb (380kg) of rock were recovered from the Moon by the Apollo missions. The rocks are mostly igneous, meaning that they formed from the solidification of cooling molten material (magma). Although closely related to Earth rocks, they are depleted in the volatile elements sodium and potassium.

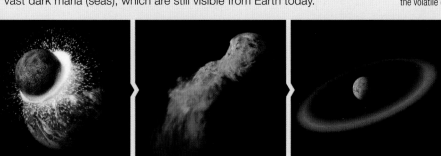

>>01 >>02 >>03 >>04

EARTH'S MOON
>>01 About 4.5 billion years ago, a Mars-sized asteroid collided with Earth and tore away a huge amount of silicate rock material from its interior. >>02 A vast cloud of gas and fragmented rock debris was ejected into space and rapidly cooled. >>03 Earth's gravitational pull held the ejecta in a circular orbit, forming a dense ring. >>04 The rocks collided, growing into ever larger clumps until eventually the Moon, a single satellite body over 2,100 miles (3,400km) wide, circled the Earth.

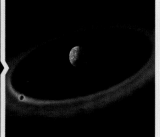

The First 500 Million Years

Earth's mass began to take shape slightly over 4.5 billion years ago. Within 50 million years, its core had formed and, in turn, generated a magnetic field. However, it was not until the meteorite bombardment had died down and the atmosphere and surface of the crust were relatively stable, about 3.8 billion years ago, that life had a good chance to evolve and thrive.

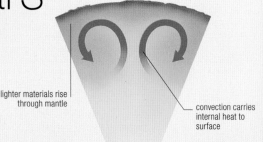

lighter materials rise through mantle

convection carries internal heat to surface

heavy materials sink to form dense core

Earth's core and mantle

Soon after it formed, most of Earth's mineral material separated from a uniform ball into the intensely hot metallic core and the cooler rocky mantle. The iron-nickel composition of the core is indicated by measurements of density, the chemistry of iron meteorites, and Earth's magnetic field. The magnetic field also reveals that part of the core must be liquid to circulate electrically conductive molten iron that generates magnetism. Analysis of earthquake waves shows the outer core is liquid, while the inner core is solid. As iron changes from solid to liquid at the boundary, energy is released, driving convection in the outer core. Within the mantle, gravity, which acts on differences in density between hot and cold rock, causes the mantle to flow in a pattern of convection. Colder, denser material sinks deep into the mantle, especially in subduction zones. This downward flow is balanced by the upward rise of hot and less dense mantle, either as plumes beneath hotspots, or upwelling beneath mid-ocean spreading ridges (see pp.20–21).

CONVECTION AND DIFFERENTIATION
The differentiation of Earth's mineral material into the metallic core and rocky mantle resulted in the establishment of the magnetic field and the transfer of heat energy away from the core, as hot rock of relatively low density rose up through the mantle.

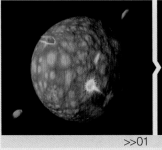  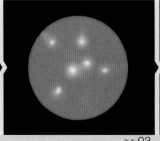 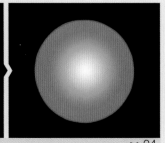

>>01   >>02   >>03   >>04

THE IRON CATASTROPHE
>>01 Around 4.56 billion years ago, gravity caused relatively cool nebular material to coalesce. >>02 As this continued, Earth contracted under the strength of its own gravity, heating up and weakening in the process.
>>03 Eventually, it was weak enough for dense iron and nickel to sink deep into the interior, starting the formation of the core.
>>04. The sinking and aggregation of this material released even more heat, possibly causing meltdown in the mantle.

IRON-OXIDE DEPOSITS AND OXYGEN IN THE ATMOSPHERE
Laid down in water 2.5 billion years ago, these deposits hold clues about when the atmosphere and oceans were oxygenated. Iron oxide is insoluble, so the iron must have been transported in a nonoxidized form. This could happen only if there was almost no oxygen in the atmosphere and ocean.

Earth's magnetism

With its two opposite poles, the Earth's magnetic field corresponds to that generated by a bar magnet, but it is formed by electrical currents generated by the fluid motion of the outer core. The mechanism may work like that of an electrical dynamo, which converts mechanical energy into electromagnetic energy. On average, the magnetic field switches its polarity or direction about every 500,000 years, but the last reversal was some 780,000 years ago. The axis of polarity is also aligned differently from Earth's axis of rotation (at present, they are separated by 11 degrees). The intensity of the field fluctuates, but it is sufficient for tiny, iron-rich particles or crystals within certain rocks formed at Earth's surface to line up, rather like compass needles. Because of this, some solidified lavas and other rocks provide a record of the field's polarity when they were originally formed. Measurement of these "fossil" or paleomagnetic fields has revealed a chronological history of Earth's polarity reversals.

EARTH'S MAGNETIC FIELD
Earth's magnetic field is doughnut-shaped (toroidal), like that of a huge bar magnet whose long axis is close to, but not aligned with, Earth's axis of rotation. The magnetic poles lie some way from the geographic poles.

geographic North Pole (where axis of rotation meets surface)

magnetic North Pole

lines of magnetic force

magnetic South Pole

geographic South Pole

MAGNETIC SHARK?
A sensory organ in this great white shark's snout detects weak electromagnetic fields, such as that produced by prey. We now know that sharks also navigate by sensing the Earth's magnetic field, positioning themselves relative to lines of magnetic force.

AURORAE
Luminous aurorae appear in the polar night skies when Earth's magnetic field traps charged particles carried from the Sun by the solar wind, producing a spectrum of colors.

The crust

Earth's outer crust is a thin layer that is typically about 5 miles (7km) thick beneath the oceans (oceanic crust), and 15–55 miles (25–80km) thick beneath the continents (continental crust). The oldest remaining fragments of existing crust date from about 4.0 billion years ago, but it is likely that primitive oceanic crust formed earlier than this, soon after Earth itself formed. As the oceanic crust formed, it was recycled back into Earth's interior (mantle) through the process of subduction (see p.20). In addition, deeper rocks were exposed at the surface by tectonic (crust-shifting) processes, resulting in weathering, erosion, and sedimentation. Eventually, these altered, water-rich rocks were dragged back down into the mantle, where they began to dehydrate and melt at the higher temperatures and pressures in Earth's interior, generating bodies of molten rock that—because they were less dense—rose up once again to the surface, resulting in volcanic activity and the formation of the continental crust (see p.19).

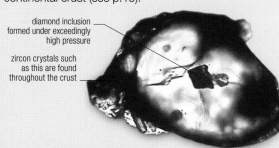

diamond inclusion formed under exceedingly high pressure

zircon crystals such as this are found throughout the crust

OLDEST TERRESTRIAL DIAMOND
These "microdiamonds" trapped inside a zircon crystal from the Jack Hills region of Western Australia are more than 4 billion years old—the oldest identified fragments of Earth's crust. It has been suggested that the diamonds crystallized under high pressure within 300 million years of Earth's formation. A typical zircon crystal is less than a millimeter across.

The formation of the atmosphere

Earth's present oxygen-rich atmosphere differs greatly from its original atmosphere, which consisted of the light gases hydrogen and helium and other volatile gases. However, in the latter stages of the Sun's formation, this first atmosphere was blasted away by a surge of the solar wind—the continuous stream of atomic particles given off by the Sun—only to be replaced by a second, more stable atmosphere as Earth continued to evolve and develop. Intense volcanic activity expelled vast amounts of volatile gases. Known as outgassing, this process released abundant nitrogen, carbon dioxide, and water vapor as well as ammonia, methane, and smaller amounts of other gases. The amount of oxygen in the atmosphere is believed to have slowly increased as microorganisms converted carbon dioxide to oxygen via photosynthesis. Clouds of water vapor condensed and precipitated, forming surface water and the first oceans.

ARCHAEA
Archaea are organisms similar to bacteria, yet most are anaerobic—they require no oxygen. *Pyrococcus furiosus*, shown here, thrives on sulfur in near-boiling sea water. Anaerobes such as these were probably among the first life forms on Earth.

CYANOBACTERIA
Earth's early seas became full of microscopic organisms similar to this blue-green algae or cyanobacteria. These organisms played an important part in oxygenating Earth's atmosphere.

solar wind

debris from space

hydrogen

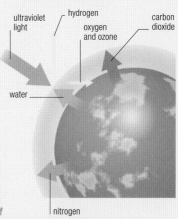

helium

FIRST ATMOSPHERE
During Earth's accretion within the solar nebula, the first atmosphere was formed from the light gases hydrogen and helium, only to be stripped away by an intense blast of solar wind.

ultraviolet light

hydrogen

oxygen and ozone

carbon dioxide

water

nitrogen

SECOND ATMOSPHERE
Earth's second atmosphere of water vapor, carbon dioxide, and nitrogen probably formed as a result of volcanic outgassing. Ultraviolet light split the water into hydrogen, oxygen, and ozone.

The formation of the oceans

Earth is unique among the planets of the Solar System in having abundant surface water that is being constantly recycled between its atmosphere and terrestrial water bodies such as seas, lakes, and oceans. Today, around two-thirds of Earth's surface is covered with seawater, and interactions between the oceans and atmosphere are vital for maintaining the planet's climate and life. The oceans probably started to form during the first 500 million years of Earth's history, when the planet first cooled sufficiently to allow water molecules to condense, fall onto the surface, and persist as free-standing water bodies. Zircon mineral grains laid down by water have been dated to over 4 billion years old, indicating that some surface water existed at that time. Some of Earth's oldest rocks are "pillow lavas" from western Greenland, many of which are up to 3.8 billion years old and were formed by the rapid cooling of lava in underwater volcanic eruptions. The early ocean waters reacted with carbon dioxide from the atmosphere to deposit calcium and magnesium carbonates as limestones. Weathering of rocks on the first continental landmasses also leached soluble salts into seawater. Australian limestone formations known as stromatolites, created by the biological activity of microscopic blue-green algae or cyanobacteria, indicate that fully saline oceans existed around 3.5 billion years ago.

300mph The speed at which tidal waves, known as tidal bores, crashed into early Earth's shores

LIVING REEF
Coral reefs are present-day biodiversity hot spots—the ocean's equivalent of tropical rain forests. The largest living structures on Earth, even the skeletons and shells of their inhabitants build up the seabed, altering the underwater environment both biologically and physically.

TIDAL FRICTION
The Moon's gravitational pull generated tides in Earth's first oceans, which in turn produced bores that ripped across coasts at speeds of 300 mph (480 kph). Such strong currents stripped, weathered, and eroded rock materials from the land, dumping the sediment at sea.

SCIENCE
CHANGING DAY LENGTH

Many living organisms record daily, monthly, and seasonal growth cycles by the changing rates of growth in their shells and skeletons. Coral, for example, deposits a new layer of limestone every day, and it is particularly influenced by lunar monthly growth cycles. By studying fossil corals from the early Devonian period (400 million years ago), such as the one shown here, we know from growth-line evidence that there were 410 days in a year during this part of Earth's history. Since Earth's orbit around the Sun has remained constant, the Devonian day must have been shorter—just 21 hours long—and Earth must have been spinning faster.

Early plate tectonics

Convection in the mantle, driven by gravity, has led to the formation of a number of tectonic plates, made up of Earth's cold, strong outer layer (called the lithosphere), which are in constant motion. As the early plates moved apart, there was an outpouring of new crustal material from the mantle. Since Earth's surface is not expanding, as much crust must return to the mantle as is newly generated, with the result that one plate is often forced below another along what are called subduction zones. Such plate movement began when Earth cooled sufficiently to form an outer crust over 4 billion years ago. The oldest rocks associated with this process are 3.8 billion years old, but this crustal recycling has destroyed all of the early oceanic crust and the oldest known existing oceanic crust beneath the present oceans is about 180 million years old.

THE CHANGING FACE OF EARTH
Using various lines of evidence, geologists can reconstruct the past distribution of Earth's oceans and continents. This map shows them as they probably looked about 650 million years ago. Before this date, not enough evidence has survived for accurate maps to be made.

South China
Arabia
North China
Australia
India
ANTARCTICA
PANTHALASSIC OCEAN
PANAFRICAN OCEAN
South Africa
Laurentia
Alaska
Siberia
West Africa
Greenland
Amazonia

EARLY TECTONICS
Early tectonic processes probably differed from those operating today. It is likely that during Earth's early history, convection in the mantle was more vigorous than it is now, leading to faster crustal movements, smaller plates, and more volcanic activity.

OPHIOLITE
An ophiolite is a part of the ocean floor that has been thrust above sea level by the action of plate tectonics—in this instance, when two plates collided. This brown rock formation in Oman, now weathered into rugged hills, is a rare and well-preserved example.

The first continents

Today, continents make up about one-third of Earth's surface but contain the oldest rocks on the planet, over 3.8 billion years old. Analysis of these rocks reveals even older zircon minerals that formed over 4 billion years ago. Geochemical investigation of the zircons and smaller fragments within them shows that they formed at relatively low pressures and temperatures in magma rich in water and silica at convergent plate boundaries, such as volcanic island arcs. This suggests that plate movement and subduction were active and liquid water and continental crust were present before 4 billion years ago. Subduction of the primitive crust rocks led to selective melting with increasing heat at depth. Preferential melting of silicate minerals with the lowest melting points and relatively lower density formed magmas that rose into the crust and solidified, forming granitic rock bodies near the surface. These initial island arcs, microcontinents, and their granitic bodies began to form the early continents as they converged and joined together.

BARBERTON GREENSTONE BELT

The 3.5-billion-year-old pillow lavas and sediments of South Africa's Barberton Greenstone Belt (left) preserve remnants of volcanic island-arc rocks involved in the early formation of continental crust. Magnesium-rich volcanic rocks, called komatiites, demonstrate the presence of liquid water at this time. The lava erupted on the early ocean floor and quenched to give rise to needle-shaped crystals (above). Eruption of this lava suggests the mantle was hotter, wetter, and more fluid than it is now.

THE DEVELOPMENT OF CONTINENTAL CRUST

It is likely that the first continental crust formed after a primitive crust had already developed and convection had started in the mantle. Continental crust is formed when rocks in the mantle melt and later solidify, in the process becoming differentiated from the mantle. The process was probably particularly rapid above sinking flows in the mantle and slower above rising flows, where the continuous supply of mantle rocks slowed the rate of differentiation.

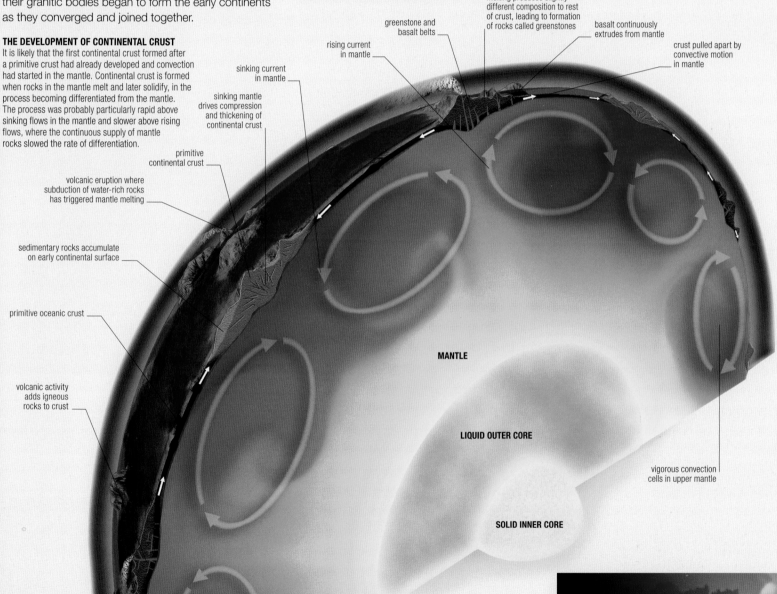

sinking current in mantle

sinking mantle drives compression and thickening of continental crust

primitive continental crust

volcanic eruption where subduction of water-rich rocks has triggered mantle melting

sedimentary rocks accumulate on early continental surface

primitive oceanic crust

volcanic activity adds igneous rocks to crust

greenstone and basalt belts

rising current in mantle

melting produces slightly different composition to rest of crust, leading to formation of rocks called greenstones

basalt continuously extrudes from mantle

crust pulled apart by convective motion in mantle

MANTLE

LIQUID OUTER CORE

vigorous convection cells in upper mantle

SOLID INNER CORE

LAVA

Volcanic activity has shaped Earth since its origins. Lava is the term for magma formed of molten rock that rises through the crust and erupts at the surface. It also refers to the solidified rock that forms when a molten lava flow cools.

Plate Tectonics

Earth's thin outer crust and upper mantle, down to a depth of about 60–200 miles (100–300km), are divided into continent-sized plates that jostle against one another. As the plates move, oceans are created and later disappear, and volcanoes and mountain chains are formed.

Divergent boundaries

Oceanic plates slide over the deeper parts of the mantle, driven by the force of gravity. As they do so, the mantle wells up, and the crust bulges, ruptures along weak points called faults, and eventually rifts apart. Pressure release allows the hot mantle to melt, forming magma that erupts as lava through the crest of a mid-ocean ridge. As this process continues, the splitting plate spreads apart or diverges, forming ridges and valleys on either side. As they slowly cool, the ridge flanks subside and their surface is smoothed out by the deposition of blankets of sediment.

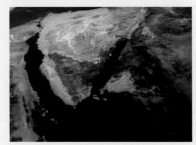

RED SEA
As Africa moves away from Arabia, the rifted valley between the two plates has been flooded by the Red Sea and will eventually become a wide ocean.

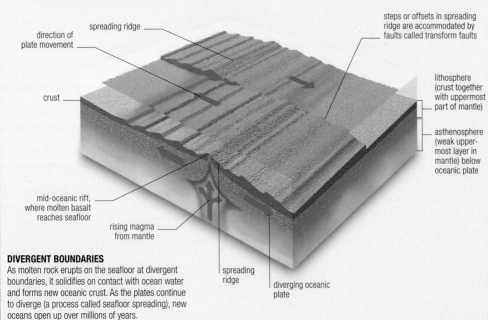

steps or offsets in spreading ridge are accommodated by faults called transform faults

spreading ridge

direction of plate movement

crust

lithosphere (crust together with uppermost part of mantle)

asthenosphere (weak uppermost layer in mantle) below oceanic plate

mid-oceanic rift, where molten basalt reaches seafloor

rising magma from mantle

spreading ridge

diverging oceanic plate

DIVERGENT BOUNDARIES
As molten rock erupts on the seafloor at divergent boundaries, it solidifies on contact with ocean water and forms new oceanic crust. As the plates continue to diverge (a process called seafloor spreading), new oceans open up over millions of years.

Convergent boundaries

New crust is created at spreading ridges, but since Earth is not expanding, divergence in one place results in convergence in another. On average, the crust is less dense than the mantle, and oceanic plates are denser than continental plates because they contain a thinner crust. As a result, where oceanic and continental plates collide, the heavier oceanic plate is overridden by the continental plate and descends (is subducted) into the mantle, melting and releasing magma, which erupts at the surface.

OCEANIC–OCEANIC
Where two oceanic plates meet, the older, cooler, and slightly denser plate is subducted and drawn deep into the mantle. Magma erupts at the surface to form chains of volcanic islands, like the one shown here.

OCEANIC–CONTINENTAL
Where plates of different densities converge, the denser plate is subducted below the other, descending hundreds of miles into the mantle. This usually generates earthquakes, as well as volcanic activity as water rising from the subducted slab leads to melting of the overlying mantle.

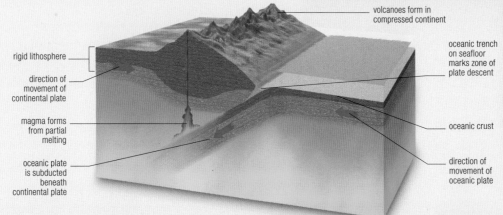

rigid lithosphere

direction of movement of continental plate

magma forms from partial melting

oceanic plate is subducted beneath continental plate

volcanoes form in compressed continent

oceanic trench on seafloor marks zone of plate descent

oceanic crust

direction of movement of oceanic plate

CONTINENTAL–CONTINENTAL
When two continental plates converge, neither is subducted. Instead, convergence leads to mountain formation as the crust folds and thrusts upward—which is how the Himalayas, above, were formed.

Volcanoes and earthquakes

Volcanoes and earthquakes are violent expressions of Earth's internal dynamic forces. The vast majority occur at plate boundaries, and they are intimately connected to plate interaction. Diverging plates stretch and break, generating shallow earthquakes and volcanic eruptions, most of which occur at spreading ridges in ocean depths and produce magma made up mainly of basalt. Converging plates, however, generate earthquakes as far down as 400 miles (700km). Magma rises through the crust, assimilating rock materials and changing composition as it goes before erupting explosively through surface volcanoes, some of which form volcanic islands.

OCEANIC VOLCANOES
Volcanoes can also rise from the ocean floor, away from plate boundaries above hot spots in the mantle. An example is Kilauea, which often erupts fluid lava.

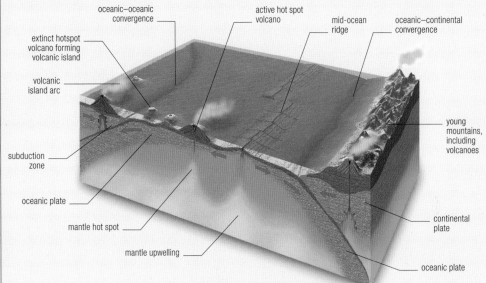

extinct hotspot volcano forming volcanic island

oceanic–oceanic convergence

active hot spot volcano

mid-ocean ridge

oceanic–continental convergence

volcanic island arc

subduction zone

oceanic plate

mantle hot spot

mantle upwelling

young mountains, including volcanoes

continental plate

oceanic plate

MID-ATLANTIC RIDGE, ICELAND
The typical volcanic rocks and structures of an active spreading ridge can be seen on Iceland, where the Mid-Atlantic Ridge has been lifted above sea level. During eruptions, lava flows from long fissures, such as the one seen here.

VOLCANOES AND PLATES
The vast majority of Earth's active volcanoes are situated along plate boundaries, and most of these lie within ocean-floor spreading zones. More evident are the volcanoes associated with subduction zones. But some eruptions also occur over hot spots in the middle of plates, where plumes of molten rock rise from deep in the mantle.

SAN ANDREAS FAULT
The continuing interaction between the Pacific and North American plates is most apparent at the 808-mile- (1,300km-) long San Andreas transform fault zone in coastal California. Sideways movement along the fault occurs at 1½in (35mm) per year and led to the catastrophic 1906 San Francisco earthquake.

Tectonics and life

Plate movement has had a significant impact on the evolution and distribution of life. Convergence brings different organisms together in competition, while divergence separates species groups, which then evolve in different conditions. An example is the supercontinent of Gondwana, formed around 542–488 million years ago. Evolving life forms spread throughout this enlarged land mass, leaving a record of themselves behind as fossils. Thus, fossils of the same species have been recovered in rocks from what are now widely separated continents. Only when Gondwana began to split apart, isolating these creatures from one another, did groups begin to evolve in different ways.

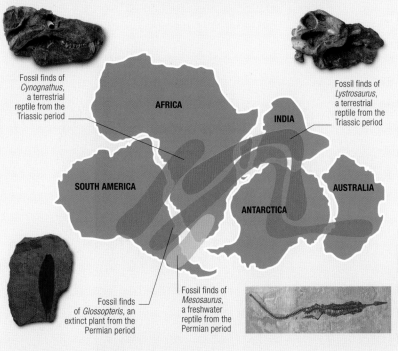

Fossil finds of *Cynognathus*, a terrestrial reptile from the Triassic period

AFRICA

INDIA

Fossil finds of *Lystrosaurus*, a terrestrial reptile from the Triassic period

SOUTH AMERICA

ANTARCTICA

AUSTRALIA

Fossil finds of *Glossopteris*, an extinct plant from the Permian period

Fossil finds of *Mesosaurus*, a freshwater reptile from the Permian period

WEGENER MAP
As long ago as 1915, German meteorologist Alfred Wegener collected geological and fossil evidence to show that the southern continents once joined together to form a supercontinent, but at the time he had no way of explaining how this had occurred.

WALLACE'S LINE
Distinctive Asian broadbills on the island of Borneo and Australian cockatoos on the island of Sulawesi, first noticed by Alfred Russel Wallace, were originally brought together by plate movements. The boundary is now called Wallace's line.

Changing Climates

Earth's climates have changed over time, encompassing cold ice ages and warmer greenhouse periods. Long-term changes are probably due to plate movements, volcanic and tectonic activity, and ocean circulation changes, together with changes in the composition of the atmosphere (greenhouse gases). Within ice ages, change is driven by cycles in Earth's orbit around the Sun, amplified by greenhouse-gas fluctuations.

Evidence in the rock record

Since the nineteenth century, evidence from rocks and fossils has pointed to repeated, often dramatic changes in past climates. Ancient coral-reef limestones, rain forest and tropical-swamp coal deposits, desert sandstones, and glacial deposits are all climatic indicators in the rock record. However, movement of tectonic plates across climate zones must be taken into account before a global change can be proven. The presence of successive glacial and interglacial deposits beyond the poles is one of the best indicators of climate change in the rock record. Glacial events have been identified in many stages of the geologic timescale (see pp.44–45); evidence for Precambrian ice ages has even been found in low latitudes, leading to the controversial idea that at times Earth may have been almost entirely covered in snow and ice.

DROPSTONE
Some 635 million years ago, a 3¾in- (8cm-) diameter rock dropped from melting sea ice into seabed sediment now exposed in northwest Namibia. Plate-tectonic reconstruction indicates Namibia was located at the Equator at the time—there were icebergs in low latitudes.

CARBONIFEROUS LIMESTONE CLIFFS
The limestone cliffs of the Gower Peninsula in Wales are actually deposits of tropical seas, as shown by their coral fossils. A tropical climate had not reached Wales, however; instead, plate tectonics had carried southern Britain into tropical latitudes.

GLACIATED LANDSCAPE
The U-shaped Lauterbrunnen Valley in Switzerland features almost vertical rock walls and hanging valleys. The flow of massive amounts of glacial ice created this valley, grinding and plucking rock fragments from the bedrock.

Evidence in the fossil record

Eighteenth-century discoveries of fossilized elephants in North America and Europe stirred debates about climate change and the biblical flood. By the next century, scientists realized these fossils were actually the remains of animals adapted to cold conditions. They had lived in an ice age that extended across northern Europe and North America—as had Neanderthals, our extinct human relatives. Finding fossilized tropical plants and animals at high latitudes and ancient glacial deposits in low latitudes also pointed to widespread past climate change, yet many (but not all) discrepancies were explained by the theory that plate tectonics had pushed continents through different climate zones at different times. Nonetheless, there is good evidence that global climates have repeatedly swung from cold to warm conditions throughout Earth's history.

FLORAL EVIDENCE FOR COLD CLIMATES
The mountain aven, *Dryas octopetala*, is a tundra and alpine plant. Its pollen is found throughout 12,000-year-old glacially related deposits in Europe.

FOSSIL FERN
About 100 million years ago, Antarctica was close to its present position, but without an ice cap. Dinosaurs roamed wooded polar landscapes and ate plants such as this now-extinct fern.

Oxygen isotopes

Isotopes are atoms with the same number of positively charged particles (protons) but different number of neutrons (neutral particles) in their nuclei. Oxygen is present in water as two isotopes: oxygen-16 and oxygen-18. Oxygen-16 is lighter than oxygen-18 and evaporates more quickly. During glacial periods, oxygen-16-enriched water vapor falls as ice and snow, depleting the ocean of this isotope and enriching it with oxygen-18 (see illustration below). Some marine organisms secrete calcium-carbonate skeletons derived from seawater, and the oxygen isotope content of these shells reflects the ratio of light to heavy isotopes in the water at the time, as well as the temperature of the water, both factors that are related to the climate. Thus, scientists can infer past climates from the ratio of oxygen-16 and oxygen-18 in fossil foraminiferan shells.

FORAMINIFERAN FOSSIL
Millimeter-sized, coiled, and chambered shells made of calcium carbonate are secreted by single-celled microorganisms called foraminiferans, which are common in ocean waters.

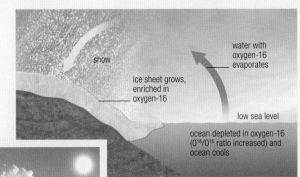

snow

water with oxygen-16 evaporates

ice sheet grows, enriched in oxygen-16

low sea level

ocean depleted in oxygen-16 (O^{18}/O^{16} ratio increased) and ocean cools

GLACIAL PERIOD
Preferential evaporation of the light oxygen-16 isotope in atmospheric water vapor and its precipitation as snow and ice lead to depletion of oxygen-16 relative to oxygen-18 in ocean waters during glacial phases.

INTERGLACIAL PERIOD
During interglacial phases, the melting of glaciers and ice sheets returns huge volumes of oxygen-16-rich fresh water to the oceans, thus altering the ratio of oxygen isotopes in sea water.

rain

ice sheet melts, returning oxygen-16 to ocean

high sea level

ocean enriched in oxygen-16 (O^{18}/O^{16} ratio decreased) and ocean warms

Orbital cycles

Over time, many factors control the amount of solar radiation that reaches Earth's surface and influences its climate, resulting in what are called Milankovich cycles (see panel, right). The shape of Earth's orbit varies from nearly circular to slightly elliptical, in cycles lasting 90,000–120,000 and about 413,000 years. In addition, the axis of rotation shifts between 21.8 and 24.4° over 41,000 years.

At its maximum tilt, summers receive more solar radiation and are warmer, while winters receive less radiation and are colder. The axis of rotation also wobbles (or precesses) due to gravitational forces exerted by the Sun and Moon as well as by Jupiter and Saturn. This results in a variation in solar radiation and length of the seasons in the Northern and Southern Hemispheres every 26,000 years.

BIOGRAPHY

MILUTIN MILANKOVICH

Serbian mathematician Milutin Milankovich (1879–1958) helped scientists understand the link between astronomical cycles, solar radiation, and climate change. Scottish scientist James Croll (1821–1890) first linked Earth's orbital eccentricity to solar variation, which influences surface temperature and climates. Milankovich built on this theory, with more accurate data about Earth's orbital variations. In 1920, he published an abstract on the curve of insolation (solar variation) at Earth's surface and a mathematical theory of periodic thermal effects of solar radiation, which established his international reputation. Milankovich went on to develop ideas about the motions of Earth's poles, linking them to the theory of glacial periods now known as Milankovich cycles.

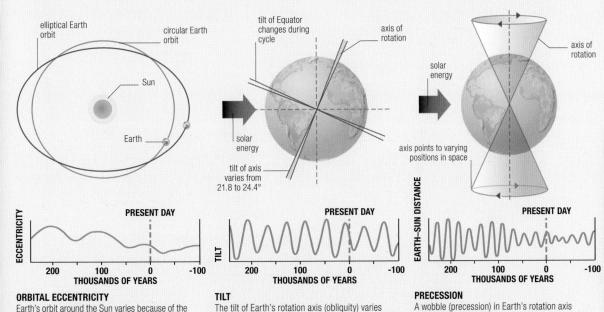

elliptical Earth orbit

circular Earth orbit

Sun

Earth

solar energy

tilt of axis varies from 21.8 to 24.4°

tilt of Equator changes during cycle

axis of rotation

solar energy

axis of rotation

solar energy

axis points to varying positions in space

ECCENTRICITY

PRESENT DAY

200   100   0   -100
THOUSANDS OF YEARS

TILT

PRESENT DAY

200   100   0   -100
THOUSANDS OF YEARS

EARTH–SUN DISTANCE

PRESENT DAY

200   100   0   -100
THOUSANDS OF YEARS

ORBITAL ECCENTRICITY
Earth's orbit around the Sun varies because of the gravitational influences of Jupiter and Saturn. The result is a variation in the amount of solar radiation that reaches Earth and in the length of the seasons.

TILT
The tilt of Earth's rotation axis (obliquity) varies periodically during a 41,000-year cycle, resulting in colder winters and warmer summers when the tilt reaches its maximum of 24.4°.

PRECESSION
A wobble (precession) in Earth's rotation axis relative to fixed stars causes the length and intensity of seasons to vary between Northern and Southern Hemispheres over a 26,000-year cycle.

Greenhouse gases in the atmosphere

Changes on Earth itself can affect the climate, especially the level of gases such as water vapor, carbon dioxide, and methane in the atmosphere. These "greenhouse gases" absorb infrared radiation emitted by Earth's surface, thereby controlling the way the atmosphere loses heat to space. As their level rises, the atmosphere tends to warm. Factors such as weathering of rocks, volcanic eruptions, and biological activity can affect greenhouse gas concentrations. Many types of rock react with CO_2 and water to form new minerals, removing CO_2 from the atmosphere. Volcanic eruptions can emit CO_2 derived from the Earth's mantle. As organisms grow, they incorporate carbon in their bodies by extracting CO_2 from the atmosphere; if they are buried when they die, the level of CO_2 in the atmosphere reduces. Burning this fossilized carbon returns CO_2 to the atmosphere.

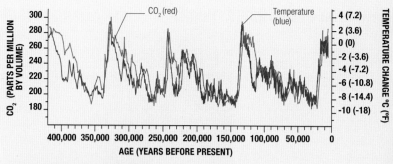

CARBON DIOXIDE AND TEMPERATURE
Changing levels of carbon dioxide can be measured in air bubbles trapped in ice. This graph, taken from an Antarctic ice core, shows the close link between carbon dioxide and temperature. Four glacial cycles can be seen on the graph.

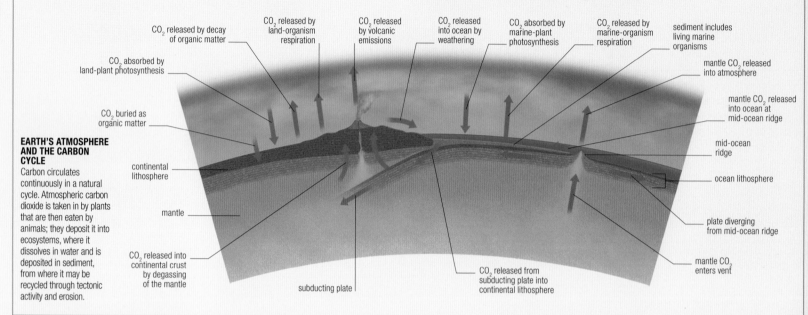

EARTH'S ATMOSPHERE AND THE CARBON CYCLE
Carbon circulates continuously in a natural cycle. Atmospheric carbon dioxide is taken in by plants that are then eaten by animals; they deposit it into ecosystems, where it dissolves in water and is deposited in sediment, from where it may be recycled through tectonic activity and erosion.

Changing geography

The global circulation of oceanic water redistributes heat and humidity to the atmosphere, affecting the climate. Tropical surface waters, heated by solar radiation, are carried toward the poles, warming the land and atmosphere in high latitudes. For example, coastal regions of northwest Europe are warmed by the Gulf Stream and warm air associated with it in winter. If the Gulf Stream were to cease or change direction, northwest Europe would have much colder winters. Since the circulation pattern is partly directed by the arrangement of ocean basins and continents, any change in that configuration affects regional climates. This means that plate movements can bring about climate change (for example, see illustration below).

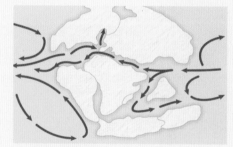

MESOZOIC CURRENTS
One hundred million years ago, tropical waters flowed east to west through the Tethys Ocean, which stretched from southeast Asia through what is now the Mediterranean to the Pacific between North and South America.

TODAY'S CIRCULATION
As a result of crustal plate movement, Africa has joined to Asia while North and South America are linked, leading to a much more fragmented global circulation.

RAIN, ALTITUDE, AND LEAF TYPES
The convergence of India with Asia 55 million years ago formed the Himalayas and elevated the Plateau of Tibet to the north. This uplift had a dramatic effect on the region's climate, creating the Southeast Asia monsoon (see illustration, right, which shows the flow of air in summer). Plants in the Himalayas have evolved different leaf shapes to suit the different moisture regimes at different altitudes. Their distribution in the fossil record can be used to date the rise of the Himalayas.

10,000–13,000ft (3,000–4,000m)
A dry, cold climate means plants found at this altitude have small, simple leaves

6,500–10,000ft (2,000–3,000m)
Plants have moderate-sized leaves with toothed margin where climate is cool and wet

less than 3,000ft (1,000m)
A hot, wet climate means plants have large leaves with a smooth margin and "drip tip"

DROWNED COAST
Eroded limestone in Ha Long Bay, Vietnam, shows one effect of the current warmer climate. Rising sea levels, caused by the melting of ice sheets and glaciers, have drowned this deeply eroded landscape, leaving a series of rocky islands.

Warm phases

"Greenhouse" phases with little or no polar ice and high sea levels have occurred throughout Earth's history, linked to significant increases in the concentration of greenhouse gases in the atmosphere. For example, fossil evidence shows that during the Cretaceous Period lush forests grew in Antarctica and northern Alaska, providing food for dinosaurs. Warm polar regions tend to promote warming at lower latitudes too, creating seasonally arid zones near the Equator, with huge, shallow seas.

HIGH AND DRY
Australia's marsupial animals, such as the wombat (above) evolved in a continent that was cooler and wetter than it is now. As a result, they are continually having to adapt to drier conditions.

MISSOULA FLOODS
Dry Falls in Washington was once the world's largest waterfall. In the last ice age, floodwaters 400ft (120m) deep poured over its edge, with a force about ten times greater than that of all the world's rivers combined.

FOSSIL MAMMOTH TOOTH
Found in glacial deposits throughout northern Eurasia and North America, fossil teeth of cold-adapted mammoths reflect the extent of frigid glacial climates.

THE LAST ICE AGE
Glacial erosion and deposition from the last glaciation have left traces on landscapes and sediments. Geologists use these to reconstruct the original extent of glaciers and ice sheets.

Cold phases

Repeatedly, Earth's climates have entered ice ages lasting millions of years at a time, when there were long-lasting ice sheets near the poles. They consist of alternating colder and warmer periods called glacials and interglacials. Geological evidence points to at least two major Precambrian ice ages, known as "snowball events." The most recent ice age began about 35 million years ago, with the last glacial event ending just 11,000 years ago. Fossils show how rapid climate change associated with ice ages had a drastic impact upon life, sea levels, and terrestrial environments on a global scale.

Life and Evolution

Genetic, biological, and fossil evidence shows that, for over 3.5 billion years, life has evolved, diversified, and changed from tiny, ocean-dwelling microbes to the millions of species that colonize all of Earth's habitable environments. This astonishing feat has been achieved by evolution.

What is life?

Life can be defined as a condition in which organic, animate matter is distinguished from inorganic, inanimate matter by its ability to renew its complex, highly ordered structure. This includes a capacity for change, growth, and reproduction, with functionality maintained until death, when its constituents disperse back into the environment. In order to maintain itself, life must be able to get energy and raw materials from the environment, as well as manufacture everything necessary for growth, repair, and replication. The only element that is known to form living structures is carbon, which can combine with itself and other elements—notably nitrogen, oxygen, and hydrogen—to form molecules of great diversity and complexity. Four main groups of organic carbon compounds are found in living organisms: carbohydrates and fats, which supply energy; proteins, built from amino acids, which form structural tissue; and nucleic acids, the basic components of genes.

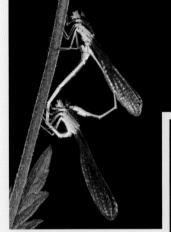

SEXUAL REPRODUCTION
Sexual reproduction, as seen in this image of two dragonflies mating, creates a new individual by the fusing of two sex cells or gametes—a male sperm and a female egg.

PHOTOSYNTHESIS
Plants use a pigment called chlorophyll to convert sunlight into food. Light energy captured by chlorophyll converts water, carbon dioxide, and minerals into "chemical energy" (sugars the plant uses as food) and oxygen, which benefits other life forms.

ASEXUAL REPRODUCTION
Some organisms, such as this ciliate protozoan, reproduce without sex cells (gametes). Here, the protozoan uses binary fission: it duplicates its DNA, then divides into two parts. Each new protozoan holds a copy of the "parent's" DNA.

LIFE IN THE PAST
Numerous species of plants and animals have evolved and become extinct, preserved only as fossils. The first dinosaurs evolved in the Triassic Period, and the group became extinct at the end of the Cretaceous.

What is evolution?

Evolution is a scientific theory that states that plants and animals change genetically over time, modifying and adapting over generations in response to the demands of Earth's changing environments. This biological process includes reproduction, diversification, and adaptation. Charles Darwin (see p.28), who formulated his theory of natural selection in 1837–1839, called the process "descent with modification": the descent from a common ancestor with the modification of biological characteristics over time. Natural selection is a key evolutionary process brought about by the survival of organisms best suited to their local environments: the "survival of the fittest."

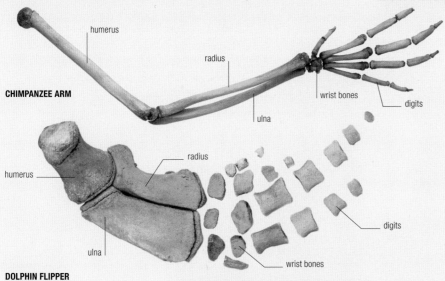

CHIMPANZEE ARM

humerus

radius

ulna

wrist bones

digits

humerus

radius

ulna

wrist bones

digits

DOLPHIN FLIPPER

ARMS AND FLIPPERS
Anatomy often reveals a shared ancestry between species. For instance, a chimpanzee's arm looks very different from a dolphin's flipper, but the sequence from the shoulder of a single humerus bone followed by a paired radius and ulna and a series of wrist and finger bones can be identified in both animals.

Genes and DNA

Almost every cell of every living organism, from single-celled amoebas to whales, has its own set of molecular instructions. The function of each cell is encoded in chromosomes: the threadlike part of a cell that carries hereditary information in the form of genes. Each human cell contains between 20,000 and 25,000 genes, each with its own set of instructions responsible for particular characteristics. The code is recorded mainly in the form of a molecule called DNA, which includes chemicals called bases arranged in pairs. A gene is encoded by a specific sequence of base pairs.

chromosome formed by one molecule of DNA

DNA molecule forms a spiral shape (a double helix) linked by four different bases

DOUBLE HELIX
The double-stranded coil of DNA can unravel into two separate strands, each of which acts as a template for a new piece of DNA, carrying information from one generation to another.

What drives evolution?

Evolution is fueled primarily by the processes of selection and competition. These act upon species, which respond by having offspring that contain inherited variations. Most species tend to produce more offspring than can survive. Natural selection favors and promotes the "survival of the fittest": in other words, those best adapted to the physical and biological environment, such as a particular climate or escaping a predator. These survivors select other, similarly adapted mates to produce offspring that survive in greater numbers to breed new generations.

PEPPERED MOTHS
Genetic variation is demonstrated in the pale and dark forms of the peppered moth. The latter survive where industrial pollution and blackened trees give them a camouflage advantage over their pale relatives.

SEXUAL RIVALRY
Male Hercules beetles are armed with giant, pincerlike horns with which they battle for sexual access to females. The horns are powerful enough to fracture the loser's tough exoskeleton (outer shell).

Adaptability

Without adaptability, life would never have moved out of the ocean. However, even adaptability is largely a result of cause and effect. Life cannot predict a future need. Instead, new traits are produced in individuals by genetic variation and mutation (when random changes in genetic code are introduced, for example, by errors in copying). If these are favorable, natural selection promotes their transmission to future generations. For instance, flight in pterosaurs and bats, based on a membrane stretched between or from fingers, may have originated as an adaptation to glide away from predators that themselves later became adapted to powered flight.

FLIGHT DEVELOPMENT
Adaptation for flight has been achieved many times by animals as diverse as extinct pterosaurs, such as the reptilian Jurassic-age *Dimorphodon* (above), and in mammals such as Daubenton's bat (right).

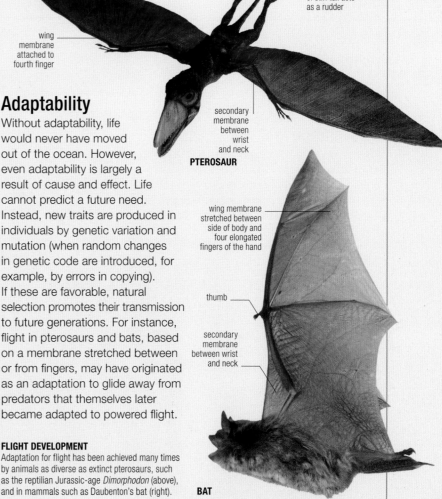

diamond-shaped flap of skin at end of stiff tail acts as a rudder

wing membrane attached to fourth finger

secondary membrane between wrist and neck

PTEROSAUR

wing membrane stretched between side of body and four elongated fingers of the hand

thumb

secondary membrane between wrist and neck

BAT

Speciation and extinction

The fossil record shows that, throughout Earth's history, species have evolved and died out. A species is a population of organisms that can interbreed, producing fertile offspring and sharing their genes in the process. Most species that have evolved are now extinct, but their genes survive in living descendants. Speciation is the process by which new species evolve from an ancestral species, and it can be brought about in a number of ways. For example, environmental change can fragment populations, leading to geographical and genetic isolation of different groups. Over time, genetic variations and adaptations to new circumstances alter the gene pool of these groups to such an extent that even if they did come together again, members of different groups would not be able to breed and produce fertile young as they would have evolved into separate species. Also, various regional and global environmental catastrophes have impacted upon life and resulted in mass extinctions. The largest of these, at the end of Permian times, caused 96 percent of all species on Earth to die out—and yet life recovered.

TAPIRS
Today, the Brazilian tapir (above) and Malaysian tapir (right) are both surviving descendants of an original piglike plant-eater, yet they are now geographically and genetically distinct. Such "living fossils" are relics of a much more widespread ancestral group of primitive mammals that lived 35 million years ago. They have survived because of their shy natures and retiring lifestyles in dense forests.

Coevolution

Where two or more species interact over a long period, the possibility of mutually beneficial adaptation or coevolution arises. For example, flowering plants develop a flower with colored petals to attract insects, which then cross-pollinate and fertilize the plants, so that the plants produce seeds. The form of different flower species have coevolved to accommodate specific insects and, less frequently, birds. In addition, the evolution of various attractive, nutritious, fleshy berries and fruits has led different birds and fruit-eating mammals to develop feeding habits that utilize such resources. The benefit to the plant is that its seeds pass through the gut of the consumer unharmed and are distributed in droppings of natural fertilizer.

BEES
Attracted by a flower, a bee is rewarded with nectar for transmitting pollen on its hairy body to another flower of the same species, which it will fertilize.

BILLS AND FLOWERS
Hummingbirds have coevolved with certain flowering plants. As flowers have developed deeper trumpet shapes, hummingbird bills and tongues have lengthened to reach more distant nectar sources. Viewed in detail, the shape of the birds' bills have also adapted to the shape of flowers of particular species.

BIOGRAPHY
CHARLES DARWIN
The son of a wealthy provincial English doctor, Charles Darwin (1809–1882) trained as a doctor, then as a clergyman, but he was mainly interested in natural history. The chance to join HMS Beagle's survey voyage in 1831–1836 gave him a unique opportunity to advance his knowledge, and his observations and specimen collections provided a wealth of ideas and questions about the nature of life, its origin, and history. In 1858, Alfred Russel Wallace's (see p.21) similar ideas prompted the copublication of an outline theory of evolution, but it was Darwin who wrote down and expanded their theory into a book, On the Origin of Species, in 1859.

Reversals in evolution

Animals as diverse as whales, snakes, and ostriches are all essentially tetrapods—four-limbed creatures—descended from a land-living ancestor that used its limbs for walking. Over time, they have all adapted to very different lifestyles and habitats, with radical modifications of their appendages and bodies. The whale's forelimbs have become paddles for swimming and steering, while the hind limbs have been lost altogether, although vestiges of the pelvic girdle remain. By losing all four limbs, snakes have adapted to a different mode of movement: throwing the body into S-shaped curves and using the friction of scales to move forward. Finally, the flightless ostrich is descended from a flying bird, which had forelimbs modified into wings. Such loss of original limbs is termed a reversal in evolution (although evolution is a process of constant adaptation rather than of steady progress).

LOSS OF FLIGHT
Having flown onto the islands of Réunion and Mauritius in the Indian Ocean, the dodo, a relative of the pigeon, underwent a reversal of evolution and lost its ability to fly. With no natural predators, this plant-eater thrived—until humans arrived in the 1590s, hunting it to extinction in less than 100 years.

RETURN TO THE SEA
Like all seals and walruses, the Australian sea lion belongs to a group of mammals that evolved on land. It has undergone an evolutionary reversal: over time, its limbs have been modified from ancestral legs used in walking into highly efficient flippers for fast swimming in order to catch its fish prey.

Macroevolution and microevolution

Genetic changes that lead to the creation of a new species are known as microevolution. Changes in higher-level taxonomic groups (such as families) and their patterns of evolution are termed macroevolution. For instance, the small-scale changes separating house-sparrow populations (see below) or the herring gull (*Larus argentatus argentatus*) from the lesser black-backed gull (*Larus fuscus graellsi*), which share a common ancestor, are microevolutionary. The changes that occurred between mammals such as sea lions and their remote, egg-laying ancestors, however, are macroevolutionary. Sea lions are aquatic mammals, but they give birth to live young on land and suckle them with milk. The platypus is a monotreme, the most ancient order of mammals; it has a ducklike bill and lays eggs yet it also suckles its young with milk.

SHELLED EGG
Primitive tetrapods laid unprotected eggs in water to be fertilized later, a process called external fertilization. Internal fertilization (in which sperm is transferred from the male directly into the female's body) and the laying of protected eggs allowed reproduction to occur outside water. These were major innovations that defined tetrapods such as crocodiles and birds.

PLATYPUS
The platypus is a curious primitive mammal that lays eggs. When the young hatch, however, the mother suckles them with milk—just like all modern mammals which give birth to live young.

HOUSE SPARROW SIZE
Since its nineteenth-century introduction to the United States, different populations of the house sparrow have evolved different traits, such as body size, as it spread across the continent.

KEY
AVERAGE MALE SPARROW SIZE
SMALLEST ──────────▶ LARGEST

Evolution through geological time

Over 99.99 percent of all species that ever lived are extinct because life has evolved for over 3.5 billion years, and species rarely survive beyond a few tens of millions of years. For 3 billion years, "life" meant microscopic marine organisms; it only moved into freshwater and onto land 470 million years ago, and not until 300 million years ago were plants able to colonize dry uplands. Whole groups of organisms have evolved and become extinct. Fossils reveal a pattern of rapid bursts of evolution, dispersal, and extinction, sometimes with waves of evolving populations replacing one another as well as gradual background evolutionary change. Evolutionary innovations have allowed new niches (the environment or ecosystem that an organism inhabits and interacts with) to be exploited; and extinction has emptied niches and allowed newly evolving groups to diversify into them. Additionally, climate change, changing sea levels, extensive volcanism, and plate movements have all affected evolution.

SCIENCE
PUNCTUATED EQUILIBRIUM

Darwin assumed that evolution required a great deal of time to accommodate all the macroevolutionary changes from the emergence of microbial life in the oceans. However, the American paleontologists Stephen Jay Gould (right) and Niles Eldredge claimed that rapid bursts of evolution are followed by inactive periods of little or no change—in other words, stasis. Called punctuated equilibrium, this idea has been difficult to prove, but detailed microevolutionary studies of large samples of continuously evolving related species such as trilobites reveal that it can happen.

Classification

With many millions of species estimated to be alive today, and many in danger of extinction, there is a need to make sense of the remarkable diversity of life on Earth. Beyond a mere key to identification, modern classification methods seek to reveal the evolutionary connections and the descent of all life from a common ancestor or ancestors.

Linnaean and Phylogenetic classification

Scientific classification of life was introduced by the 18th-century Swedish botanist Carl Linnaeus, who attempted to group like with like based on species in a hierarchy designed to illustrate a God-given order. Adopted by evolutionary biologists, it attempted to reflect natural groups and evolutionary relationships. Thus, feathered birds in the class Aves were

placed at the same level, or rank, as the class Reptilia. But fossils have shown that birds evolved from reptilian dinosaurs—their "class" nests within the "class" of reptiles. Phylogenetic classification tries to remove such contradictions by distinguishing a series of descendent groups with common ancestry that is based on unique shared characteristics.

LIVING AND EXTINCT FISH GROUPS

Fossils show that the Linnaean class Pisces is not a true taxon, but it is a useful informal grouping. This phylogenetic chart shows the evolutionary relationships between groups of fish. A group – such as Gnathostomata – branches off when its members evolve a unique, distinguishing characteristic, such as jaws.

KEY
- Extinct fish
- Living fish
- Gnathostomata (jawed fish)

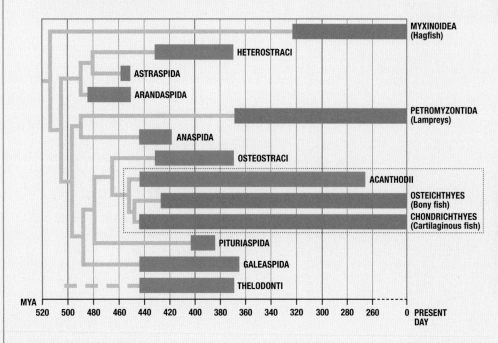

MYXINOIDEA (Hagfish)
HETEROSTRACI
ASTRASPIDA
ARANDASPIDA
PETROMYZONTIDA (Lampreys)
ANASPIDA
OSTEOSTRACI
ACANTHODII
OSTEICHTHYES (Bony fish)
CHONDRICHTHYES (Cartilaginous fish)
PITURIASPIDA
GALEASPIDA
THELODONTI

MYA 520 500 480 460 440 420 400 380 360 340 320 300 280 260 0 PRESENT DAY

LINNAEAN HIERARCHY

The Linnaean hierarchy begins with a biological species, identified by its Latin species and genus names; there can be one or more closely related species in a genus. Genera are grouped in ascending order into families, orders, classes, phyla, kingdoms, and domains with the aim of reflecting a meaningful relatedness in evolutionary terms.

DOMAIN Eukarya
The highest currently recognized taxonomic division of life, introduced in 1990, with three domains: Archaea, Bacteria, and Eukarya. Eukarya are distinguished by a membrane-bound cell nucleus.

KINGDOM Animalia
Today, at least 28 kingdoms are recognized, whereas scientists used to think there were just two: Animalia and Plantae. Animals are mainly multicelled eukaryotes that consume other organisms as primary food.

PHYLUM Chordata
Phyla are a major subdivision of the animal kingdom, made up of classes. Phylum Chordata comprises animals that at some point in their life cycle possess a notochord, which is the precursor of a backbone.

CLASS Mammalia
Classes are made up of orders and their respective subgroups. Mammals are distinguished from other chordates by having a single jaw bone (the dentary), hair, and mammary glands.

ORDER Cetacea
Each order contains one or more families and their subgroups. Cetaceans (whales and dolphins) are aquatic mammals that have lost their hind limbs and developed a fluked tail.

FAMILY Delphinidae
A family comprises one or more genera and their subgroups. Family Delphinidae includes all dolphins (a subgroup of toothed cetaceans) that have "beaked" jaws.

GENUS Delphinus
A genus is a subdivision of a family. Dolphins show considerable variation in body shape and color, but genetic analysis suggests there are just two or three species in the Delphinus genus.

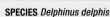

SPECIES Delphinus delphis
This group comprises similar individuals able to interbreed in the wild. Delphinus delphis, or the short-beaked common dolphin, is distinguished by its black-and-white color pattern and beak size.

Molecular clocks

The idea of a molecular clock is that evolutionary changes occur at regular time intervals. It is assumed that the rate of genetic change (mutation) in an organism's DNA has not changed over time, or at least can be averaged. The molecular genetic difference or "distance" between two species must be measured and their rate of genetic change estimated. The latter is derived from radiometric dating (see p.13) of the fossils of a known ancestral species. From these, it is possible to calculate when the two species diverged from their common ancestor.

50 MILLION YEARS AGO
Isolated sequence of 10 bases (see p.27) in common ancestor's DNA.

25 MILLION YEARS AGO
Descendent lineages have diverged. Each has had a single base mutation, so the lineages differ by two bases.

PRESENT DAY
Descendent lineages have diverged again by another base mutation. The lineages now differ by four bases.

CAATCGATCG

CAATTGATCG

CAATTTATCG

DIVERGING LINEAGES
The fossil record confirms that around 50 million years ago, an extinct primate was the common ancestor of both lemurs and galagos.

CAATTTATCT

CAATTTATTT

LEMUR AND GALAGO
These related tree-dwelling primates are now geographically and genetically distinct due to numerous genetic divergences.

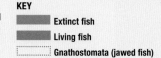

Classification of life

Taxonomic classification is still hotly debated in the scientific community, and no one definitive method is yet recognized globally. Problems with Linnaean classification have arisen during the last 100 years due to the knowledge that many organisms have similar anatomical characteristics as a result of parallel or convergent evolution. By contrast, phylogenetic classification or cladistics, developed by the German entomologist Willi Hennig in the 1950s and 1960s, identifies related groups of organisms descended from a common ancestor via the identification of new characteristics in their most likely evolutionary order. For the purposes of this book, all living things are grouped into three domains: bacteria, archaea, and eukaryotes. Protists and invertebrates are shown below within dotted lines because, although they are useful informal groupings, they are not true taxa (classification units).

Bacteria

DOMAIN Bacteria
KINGDOMS 10
SPECIES Several million

This major group of single-celled microorganisms lack a defined cell nucleus. They have a great abundance and diversity of form and have colonized virtually every habitat, from deep in Earth's crust to ocean-floor sediments and the living tissues of most organisms. Bacteria are essential to animal digestive systems, but they can also cause fatal diseases. There are millions of bacterial cells within a single pinch of soil, where they play a vital role in recycling nutrients.

Archaea

DOMAIN Archaea
KINGDOMS 3
SPECIES Several million

The microscopic and single-celled archaea have no defined cell nucleus or specialized structures. Their characterization as a group has been based on molecular data, and they have been intensely studied over recent years because of their remarkable tolerance to extreme environments and their various means of obtaining energy. The latter range from photosynthesis (converting sunlight into energy sources) through various modes of chemosynthesis: transforming metal ions, carbon, and hydrogen into a "food" source.

Eukaryotes

DOMAIN Eukarya
KINGDOMS At least 15
SPECIES 2 million

Most familiar living organisms are eukaryotes, ranging from the microscopic, single-celled amoeba to the giant redwood tree and the blue whale. This vast group of organisms all have common features in each body cell. The eukaryote cell is unique in having a complex structure that includes a distinct nucleus surrounded by a membrane. It also possesses mitochondria, structures that conduct the biochemical processes of respiration and energy production. Reproduction in eukaryotes involves the separation of duplicated chromosomes.

Protists

KINGDOMS At least 10
SPECIES More than 100,000

Single-celled protists are a diverse group of microorganisms that were viewed traditionally as a major group. However, details of their biology show that they have very little in common; some are plantlike, others animal-like, and yet others more like fungal slime molds, indicating that they have evolved from separate origins.

Red seaweeds

KINGDOM Rhodophyta
CLASSES 1 or more
SPECIES 5,500

The evolutionary relationships of red, brown, and green algae have long been problematic. Modern studies place them in very different groups. Red algae (rhodophytes) are one of the largest, most ancient eukaryote groups, with a fossil record extending back over 1.2 billion years.

Brown seaweeds

KINGDOM Phaeophyta
CLASSES 1
SPECIES 2,000

Mainly large, multicelled, cool-water organisms. Some grow up to 200ft (60m) and play an important role in the ecology of coastal waters; others are free-floating and form the unique habitats of tropical sargassum. The group is distinguished by genetic data and their chloroplasts, cell structures that conduct photosynthesis.

Plants

KINGDOM Plantae
DIVISIONS 6
SPECIES 283,000

The remarkable abundance and diversity of plants includes all the familiar forms, from green seaweed, mosses, and grasses to giant redwoods. Most plants obtain energy from sunlight through photosynthesis but may not have a single evolutionary origin because of fundamental differences in their modes of reproduction.

Fungi

KINGDOM Fungi
PHYLA 4
SPECIES 600,000

This diverse, abundant group has a fossil record extending back to Devonian times and includes microscopic single-celled and macroscopic multicelled organisms such as mushrooms, some with fruiting bodies many feet in size. Typically, they use spores for reproduction and have cell walls of chitin rather than cellulose.

Animals

KINGDOM Animalia
PHYLA About 30
SPECIES Over 1.5 million

The group name "animals" is deeply embedded in our language and common culture, with roots going back to classical times. The word's Latin root means "having breath" and has been used traditionally to distinguish plants from animals. However, since scientists have recognized that primitive life is much more complex than was previously thought, the dividing line between plants and animals has become blurred. As a result, modern usage of the name "animal" tends to restrict it to multicelled eukaryotes that must consume other organisms as food.

Invertebrates

PHYLA About 30
SPECIES Over 1 million

This group name is based on a negative attribute that differentiates all animals that are not vertebrates: they do not have a backbone. As such, it has no place in modern phylogenetic classification but is still a commonly used "catch-all" word for about 30 phyla, ranging from sponges and flatworms to arthropods and echinoderms.

Chordates

PHYLUM Chordata
SUBPHYLA 3
SPECIES 51,550

The chordates all have a distinct body plan—at least in their embryonic stages, if not always retained in adults. A stiffening rod, called a notochord, extends dorsally from front to back. Its fibrous tissue makes it flexible, and the paired muscle blocks on either side bend it into the sinuous, sideways curves that are associated with swimming. Above the notochord lies a dorsal hollow nerve cord, and below that lies the intestine. Most chordates also have gill slits and a tail at some developmental stage.

Lancelets

SUBPHYLUM Cephalochordata
CLASSES 1
SPECIES 50

Flattened, leaf-shaped lancelets grow to around 3in (8cm) long. There are no defined fins or head, and the front is marked by a cavity perforated by slits through which sea water is pumped for feeding and breathing. These filter-feeders burrow tail-first into the sea bed. Their poor fossil record extends to Cambrian times.

Tunicates (sea squirts and salps)

SUBPHYLUM Urochordata
CLASSES 4
SPECIES 2,000

Small, marine filter-feeding chordates that grow up to 6in (15cm) long, generally characterized by a notochord and tail in their larval, free-swimming stage. Typically, adults show a radical change in body form: the ascideans attach to the sea bed, the salps lose their tails and notochord but remain free-swimming, and the appendicularians remain free-swimming and retain both tail and notochord.

Vertebrates

SUBPHYLUM Vertebrata
CLASSES 8
SPECIES 49,500

In these animals, the notochord is surrounded by segmented skeletal elements to form a strengthened, articulated vertebral column. They have nervous regulation of the heart, muscles for eye movement, a sensory system of lateral lines, and two semicircular canals in the inner ear. Gnathostomata (jawed vertebrates), including sharks, bony fish, and all land vertebrates, are the most diverse and numerous group.

Mass Extinctions

In Charles Darwin's day there was plenty of evidence to suggest that life had changed radically over time, but little to indicate that evolution had been seriously disrupted by major extinction events that wiped out a huge percentage of all life on Earth. Yet, as we now know, the growth in the overall diversity of life has suffered several major setbacks, especially at the end of the Permian Period 251 million years ago, after which recovery took the best part of 30 million years.

Causes

There have been at least five mass extinctions in the last 540 million years. Much effort has been expended on searching for their causes because there is no reason why one should not happen again. The most famous extinction event, at the end of the Cretaceous Period, was marked by a meteorite impact and volcanic eruptions. Although many other impacts have been found, none can readily be linked to extinction events. The most common factors in mass extinctions are large-scale eruptions of lava and volcanic gases, and the absence of oxygen in the oceans. Volcanic gases, especially carbon dioxide, are connected with short-term regional climate change, medium-term acid rain, ozone depletion, and long-term global warming. This atmosphere–ocean link is the most likely reason for mass extinctions, including those in which impacts played a role and where glaciation was a factor.

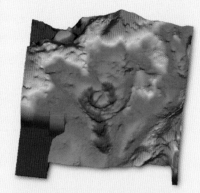

CHICXULUB CRATER, YUCATÁN PENINSULA
When an asteroid plowed into the Gulf of Mexico 65 million years ago, it was around 6 miles (10km) in diameter—large enough to create a crater 150 miles (240km) wide and set off a mass-extinction event.

PERMIAN–TRIASSIC BOUNDARY CLAYS
Scientists take samples from a cliff in Italy. The dark brown notch halfway up is evidence of vegetation destroyed suddenly at the end of the Permian Period.

K–PG BOUNDARY
Also known as the K-T boundary, this thin, gray band is found worldwide. It is rich in iridium, an asteroidal element, and is evidence for the impact at the end of the Cretaceous.

GOSSES BLUFF
The hills of Gosses Bluff in Australia are the remnants of an impact crater that was originally 14 miles (22km) wide, created 142 million years ago. The rocks show a type of fracture associated with catastrophic impacts, yet there is no link to any extinction event. The Chicxulub crater (above right), by comparison, is ten times larger.

Major mass extinctions

Since the end of the nineteenth century, it has been shown that the history of life has included evolutionary "booms" as well as "busts" of catastrophic extinctions alongside a more gradual background of species origination and extinction. The most famous "bust" episode occurred at the end of the Cretaceous Period. It resulted in the extinction of dinosaurs, various marine-reptile groups, flying pterosaurs, and various invertebrates, and has been linked to a major impact event and volcanic activity. An even greater extinction event at the end of the Permian saw the extinction of up to 96 percent of marine species, 70 percent of terrestrial vertebrate species, and 83 percent of insect genera. An earlier mass extinction at the end of the Ordovician coincided with a rapid fall and rise in sea level related to ice-cap growth and melting.

RECORD OF EXTINCTIONS
The state of fossil preservation makes analysis of data associated with extinction very difficult. The most reliable continuous records come from the remains of sea creatures such as ammonites.

LAST OF THE DINOSAURS
Fossil records show that *Triceratops*, a 23ft- (7m-) long herbivore, was one of the last dinosaurs to walk on Earth.

IRON PYRITES
Seabed rocks in Greenland contain iron pyrites formed in the absence of oxygen. They are a record of the oxygen-poor conditions that led to an extinction at the end of Permian times.

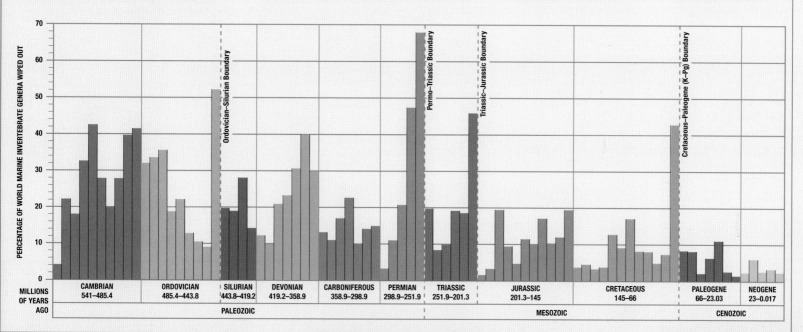

PERCENTAGE OF WORLD MARINE INVERTEBRATE GENERA WIPED OUT

70 — 60 — 50 — 40 — 30 — 20 — 10 — 0

Ordovician–Silurian Boundary
Permo–Triassic Boundary
Triassic–Jurassic Boundary
Cretaceous–Paleogene (K-Pg) Boundary

MILLIONS OF YEARS AGO		
CAMBRIAN 541–485.4	ORDOVICIAN 485.4–443.8	SILURIAN 443.8–419.2
DEVONIAN 419.2–358.9	CARBONIFEROUS 358.9–298.9	PERMIAN 298.9–251.9
TRIASSIC 251.9–201.3	JURASSIC 201.3–145	CRETACEOUS 145–66
PALEOGENE 66–23.03	NEOGENE 23–0.017	

PALEOZOIC — MESOZOIC — CENOZOIC

Recovery from mass extinctions

The fossil record provides evidence of mass extinctions but also of the subsequent recovery of life. For example, coral reefs were particularly affected by the end-Permian extinction. Their collapse had a major impact on oceanic food chains, but new skeleton-forming corals evolved, slowly reestablishing reef structures. Similarly, the collapse of terrestrial ecosystems at the end of the Cretaceous Period is marked by the extinction of the dinosaurs, yet the recovery of vegetation is evident from fern spores: plants adapted to cope with the aftermath of the environmental disaster.

PERMIAN TO TRIASSIC
Land animals were greatly affected by the Permian extinction. Even successful plant-eaters such as these two *Diictodons* (left), found huddled in their burrow, died out, but their evolutionary descendant, *Lystrosaurus* (above), survived.

FOSSIL RANGES OF VERTEBRATE GROUPS IN THE KAROO
The scale of extinction at the end of the Permian Period is reflected in the abundant fossils of land animals preserved in the Karoo region of South Africa. A few groups were completely wiped out, but enough survived for life to continue.

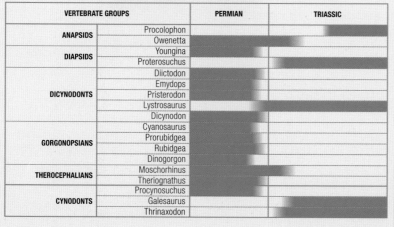

VERTEBRATE GROUPS		PERMIAN	TRIASSIC
ANAPSIDS	Procolophon		
	Owenetta		
DIAPSIDS	Youngina		
	Proterosuchus		
DICYNODONTS	Diictodon		
	Emydops		
	Pristerodon		
	Lystrosaurus		
	Dicynodon		
GORGONOPSIANS	Cyanosaurus		
	Prorubidgea		
	Rubidgea		
	Dinogorgon		
THEROCEPHALIANS	Moschorhinus		
	Theriognathus		
CYNODONTS	Procynosuchus		
	Galesaurus		
	Thrinaxodon		

Recent extinctions

Since the last glacial maximum (about 20,000 years ago), most of Earth's larger land animals have disappeared, except those that remain in Africa. It was thought that a changing climate led to the demise of cold-adapted animals such as the woolly mammoth and giant deer. However, recent dating shows that their extinction coincided with the arrival of modern humans in various regions. Similarly, the New Zealand giant moa did not become extinct until one or two centuries after the arrival of modern humans, around 1250. However, climate change may also have played a role in some areas.

GOLDEN TOADS
Today, amphibians are particularly vulnerable to extinction. Discovered in Costa Rica's mountainous tropical forest in 1966, the golden toad has not been seen since 1989.

WOOLLY RHINO
Coelodonta, the woolly rhinoceros, lived in ice-age North America and Eurasia. It became extinct 10,000 years ago—after humans arrived.

Types of Fossils

Fossil remains are preserved by such a diversity of geological processes that their form and appearance can be very deceptive. In fact, it has taken over 200 years for fossil experts to understand exactly how fossils are formed and how to interpret them accurately.

What is a fossil?

A fossil is the remains of a once-living organism that has been preserved in rock. Fossils were once thought to be the remains of both animate and inanimate things, but by the end of the seventeenth century, it was generally agreed they were of purely organic origin, although the fossilization process has usually resulted in considerable chemical change. There are three fossil types: body, trace, and chemical fossils. Body fossils include the fossilized remains of hard parts such as shells, bones, teeth, and some plant material, while trace fossils are marks preserved in hardened sediment that record evidence of past life, such as footprints. Chemical fossils consist of degraded organic substances from the molecules of living things, such as fossil DNA recovered from bone or teeth. No matter how they form, many fossils can be identified at species level.

OTZI THE ICEMAN
The frozen cadaver of a 5,300-year-old Neolithic hunter, found in glacial ice high in the Tyrolean Alps, was so well preserved that it was initially recovered by police. But, just like the meat in a domestic freezer, this body—despite its great age—is not a true fossil because it has not undergone fossilization processes.

WELL PRESERVED
This Jurassic ichthyosaur, *Stenopterygius megacephalus*, is a true fossil. It might look almost perfectly preserved, but what appears to be skin has been replaced by a microbial film.

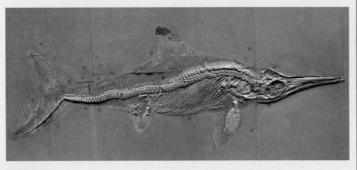

TRACE FOSSILS
Trace fossils, such as these 400-million-year-old tetrapod footprints on Valentia Island, Ireland (top), or fossil feces, known as coprolites, of *Tyrannosaurus rex* (above), provide tantalizing clues about the day-to-day lives of extinct creatures.

BODY FOSSILS
The body fossil of *Homeosaurus* (right), an 8in- (20cm-) long Jurassic reptile, reveals its lizardlike skeleton. A tooth (below) is all that is left of *Carcharodon auriculatus*, an extinct ancestor of the modern great white shark. Sharks' cartilaginous skeletons are not normally fossilized, but one tooth is often enough to identify a shark species.

Body fossils and trace fossils

The most easily visible, familiar fossils are body fossils derived from the shells and skeletons of animals that lived in shallow seas millions of years ago. Microorganisms also leave body fossils. These can only be seen with a microscope but include huge numbers of plant and algae components such as pollen and spores. Trace fossils, such as the burrows of animals and spaces left by plant root systems, are also common. However, these are rarely preserved with the organism that made them, so an entirely different naming system had to be invented to deal with them. Yet trace fossils can be very useful sources of biological and ecological information, especially about extinct animals such as dinosaurs. For example, the study of dinosaur tracks is helping resolve arguments about how fast these creatures could run (see p.38).

How fossils form

Organic remains are fossilized in many ways. There is almost always a loss of information as the remains decay and are buried, compressed, and sometimes heated. At best, soft tissues such as feathers can be fossilized, but at worst, all that is left is a trace, such as a footprint. The chances of any remains being fossilized are actually very slim, especially if organisms have no easily preservable "hard parts" like shell or bone. Organic remains are commonly preserved in places where they will have little or no disturbance, such as in still or slow-moving water. Even delicate tissues can be preserved when decay is slowed or even halted by low oxygen levels, low temperatures, or a natural preservative such as tar or resin.

FOSSILS REVEALED
Wind and rain can expose fossils by removing the surrounding softer rock. Here, the petrified trunks of a Triassic conifer have been brought to light in Arizona.

UNALTERED PRESERVATION

Very occasionally, the unaltered remains of organisms are completely preserved, particularly in cases where the hard parts are made of minerals that do not degrade or recrystallize over time. For example, shells and skeletons made of calcite, found in echinoderms such as starfish, are more stable than those composed of aragonite, which is found in mollusks. Silica is another stable mineral that is found in some sponges and microscopic algae known as diatoms. Organic material is essentially unstable, so shells lose not only soft tissue but also their color, except under special circumstances, such as when they are covered with resin. Even then, however, microbes can still degrade a fossil, although it may look perfect on the outside.

SILICA DIATOM SKELETONS

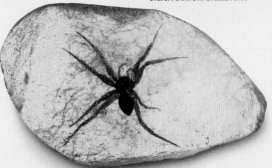

SPIDER PRESERVED IN AMBER

TRITON SHELL   **TRITON FOSSIL**

RECRYSTALLIZATION

Many fossils are preserved as a result of recrystallization (the formation of new crystals), especially when a skeleton or shell is made of an unstable crystalline mineral with a closely related but more stable form, such as aragonite. Aragonite is a very common constituent of mollusk shells and coral skeletons, but it usually recrystallizes into the more stable mineral calcite, which has a slightly different crystal structure. This process does not alter the surface details of the fossil; in fact, even some internal details may be preserved, which allows the fossil to be identified and its structure and growth to be examined.

**THECOSMILIA
CORAL COLONY**

EUOMPHALUS (TOP VIEW)   **EUOMPHALUS (CROSS-SECTION)**

CARBONIZATION

The complex organic compounds that make up plant tissues are prone to a chemical change called carbonization, which results in a loss of their watery components yet leaves a carbon residue that is much more stable over time. When a plant is covered by sediment, the burial compresses and may even heat its remains. Pressure squeezes out any liquid and gas, and flattens the plant tissues; heat may turn these into charcoal, leaving only carbon behind in the form of a compression fossil. Over time, even this carbon residue may be lost, leaving in its place only an impression and outline of the plant tissue. In fine-grained sediment, this is often enough to preserve detailed information so that the species can be identified, and this process is particularly effective in preserving leaves and delicate animal forms.

**NEUROPTERIS
SCHEUCHZERI LEAF**

BASSWOOD STEMS AND BERRIES

REPLACEMENT

Replacement—the complete replacement of an organism's structure with another material—is common during the long, complex process of fossilization or petrification (see pp.36–37). Burial of fossil remains and transformation of the surrounding sediment into rock involves physical and chemical changes, the latter often resulting from the groundwater that saturates the sediment. Groundwater containing dissolved chemicals commonly substitutes them, atom by atom, for original fossil minerals so that fine details are often preserved. Silica often replaces and fills in cell spaces of woody fossil plant stems, while the iron sulfide mineral pyrite replaces the aragonite of shells of mollusks such as ammonites.

PETRIFIED LOG CROSS-SECTION   **AMMOLITE (MOTHER-OF-PEARL PRESERVATION)**   **PYRITIZED AMMONITE**

IMPRESSIONS, MOLDS, AND CASTS

When an organism is pressed into soft sediment, it creates a negative impression, and many of its external details are recorded in negative relief. Because this impression effectively forms an external mold of the fossil, any sediment that later fills it in, or fills in the spaces within the fossil itself, such as the hollow inside a seashell or skull, forms an internal mold with details preserved in positive relief. If fossil material degrades and disappears altogether after the surrounding sediment hardens, this mold still preserves a record of the fossil's internal and external structures. Scientists often use natural molds to make artificial casts of such "missing" fossils in order to study their structure in greater detail.

IMPRESSION OF AN EOCENE POPLAR LEAF

original shell

**EXTERNAL MOLD OF
CYCLOSPHAEROMA**   **INTERNAL MOLD OF
CYCLOSPHAEROMA**   internal mold
of *Myophorella*   external mold
of bivalve shell

SHELLS IN SANDSTONE

SCIENCE
VIRTUAL FOSSILS

Traditionally, paleontologists prepare fossils by laboriously removing sediment stuck to their surfaces. This is difficult and sometimes impossible, especially around delicate structures. Much effort has gone into finding less destructive means of revealing small, fragile fossils. Chemical cleaning can be effective, but even this is not always feasible. New techniques employing nondestructive computed tomography (CT scans) and digital processing generate three-dimensional images from a series of two-dimensional X-ray scans. But where X-rays cannot differentiate between the fossil and its surrounding rock, the physical cutting of slices through the fossil produces a stack of images that can be electronically transformed into a virtual fossil.

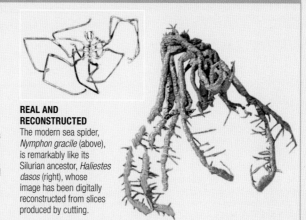

**REAL AND
RECONSTRUCTED**

The modern sea spider, *Nymphon gracile* (above), is remarkably like its Silurian ancestor, *Haliestes dasos* (right), whose image has been digitally reconstructed from slices produced by cutting.

The making of a fossil

In addition to supporting living organisms, bodies of sea water and fresh water can also preserve their remains, buried within sediments at the bottom. Much depends upon conditions at the boundary between the sediment and the water, but wherever fine-grained, muddy sediments are deposited, preservation can be remarkably good. For example, details of the covering of tough, bony scales on the body and fins of a type of marine fish called a coelacanth may be preserved more or less intact, along with the bones of the head, if the carcass comes to rest in water without enough oxygen to support scavengers. The different stages in the fossilization process are shown here. However, in well-oxygenated water, scavengers may consume the body altogether, leaving just some bones and scales preserved in their fossilized feces.

LIVING COELACANTH

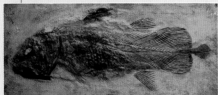

FOSSILIZED COELACANTH

FROM FISH TO FOSSIL
Coelacanths are a primitive kind of bony fish, with fleshy fins and thick, heavy scales. The mineralization of the fish's body after death could take as little as several decades, while the formation of the rock around the fossil may take millions of years.

DEATH

The death of an organism, such as this coelacanth, may be caused by predation, disease, or a natural disaster. Initially, a dead coelacanth will float to the surface, buoyed up by gas released by internal decay. Upon rupture of the body cavity, the body sinks until it lies on top of the sediment. From the moment of the fish's death, scavengers may begin to break down or remove the soft tissues. The scales may fall away, exposing the skeleton. Only the toughest body parts, such as bones, teeth, and scales, are likely to survive long enough to be fossilized, except in exceptional circumstances. In addition to attack by living organisms, the remains may be subjected to physical damage—for example, caused by the action of waves and currents.

BURIAL

Where and how, and how quickly organic remains are buried makes a great difference to their preservation as fossils. If the coelacanth settles in the fine-grained muds of a quiet lagoon, most of its scale-clad body may be preserved. However, more typically, the body will be disturbed by water action and will be broken apart and worn down before it is finally buried. The illustration, right, shows a cross-section of a sediment-covered scale.

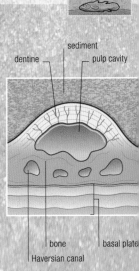

dentine · sediment · pulp cavity · bone · basal plate · Haversian canal

Where and how, and how quickly organic remains are buried makes a great difference to their preservation as fossils.

ECAY

cay is a continuing process from the
oment of death to final fossilization as a
neral residue within the surrounding rock.
pending on the rate of burial, most decay
d loss of soft tissue occurs before burial.
wever, many biodegrading organisms such
bacteria live in sediment, and these continue
process of decay and change after the
dy has been entombed, often removing all
ce of soft tissue. But the sediment's
chemical environment is often different from
t at the surface, so other chemical changes
occur and preserve details of any
naining soft tissues, such as gills, muscles,
d guts. As the body decays, sediment
ntinues to settle, forming separate layers.

REPLACEMENT

The mineral components
of tissues may be replaced
by different minerals, such
as calcium carbonate or
phosphate, supplied by
water in the sediment
(see right). The pressure
of successive layers of
sediment will compress the
coelacanth's body. If the water
enters the body before
compression, its original form
will be preserved. If it enters
after compression, just a
flattened outline of the body will
be preserved.

scale's surface and
cavities become
compressed under
weight of sediment

water percolating through
sediment picks up
minerals in solution and
fills scale's cavities

ROCK FORMATION

Finally, hard, stratified rock
is formed. The sediment
and fossil remains are
cemented together by
minerals, which fill spaces
and replace original
materials. The tough
coelacanth scales fossilize
better than the thin scales
of the many other species
of more advanced bony
fish, such as members
of the salmon family.

sediment turns
into rock

further flattening
due to compression

original bony part of scale has been
replaced with mineral crystallized from
the fluid, atom by atom

less replacement
occurs in the
dentine at the top
of the scale

Information in the Fossil Record

Fossils reveal extraordinary and wonderful details about ancient life. They tell us about what living organisms looked like, how they behaved, and which environments they lived in. They can also be used to estimate the ages of the rocks in which they are found.

Life in the past

While only a small portion of life on Earth has been preserved in the fossil record, fossils still provide most of the data we have about prehistoric life and its development within the planet's changing environments. The fossil record favors creatures with hard, preservable parts, yet it has revealed the outlines of the history and evolution of major groups of organisms, including extinct ones such as the trilobites, ammonites, dinosaurs, seed ferns, and our own earliest ancestors. We now have a record of evolution over at least the last 600 million years, showing how life developed in the seas and emerged onto land, and how ecological relationships were established between the first plants and animals. Despite catastrophic extinction events, life has evolved and adapted to fill Earth's available and habitable niches, from ocean depths to mountaintops.

LARGE EYE
The socket indicates that *Oviraptor* had very large eyes encircled by a bony ridge.

FLEXIBLE NECK
The number and arrangement of vertebrae indicate that this dinosaur had a long and very flexible, curving neck.

FISH-EATING FISH
This predatory fish, *Mioplosus labracoides*, from the Green River Formation in Wyoming, bit off more than it could chew when it tried to swallow its prey 50 million years ago. The scarcity of other *Mioplosus* fossils indicates that this species was a solitary predator.

FOSSIL FOOTPRINTS
Much can be learned about ancient animals' behavior by studying the preserved marks they left in the ground, such as these dinosaur footprints. Found in Middle Jurassic-age strata at Ardley in Oxfordshire, England, they form trackways up to 600ft (180m) long. Trace fossils like these help scientists calculate how fast extinct animals such as dinosaurs walked (see below).

CLAWED FINGER
Curved claws 3¾in (8cm) long were used for ripping and tearing—possibly for defense as well as feeding.

BIRDLIKE FEET
Very large, three-toed feet helped *Oviraptor* to walk on two hind legs, a characteristic of other theropods, such as *T. rex*.

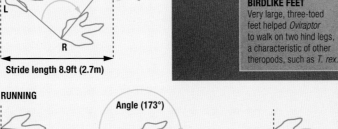

WALKING

Angle (117°–132°)

L
L
R

Stride length 8.9ft (2.7m)

DINOSAUR LOCOMOTION
One of the trackways at Ardley provided the first evidence of the changing gait and speed of a large predatory theropod—the dinosaur group that includes giant carnivores such as *Tyrannosaurus rex*. The footprints record a bipedal theropod (one walking on two legs) breaking into a run from a walk. Calculations based on stride length and foot orientation show that it increased its speed from 4.2 to 18.1mph (6.8 to 29.2kph).

RUNNING

Angle (173°)

L
R
L

Stride length 18.5ft (5.7m)

LONG TAIL
A long tail helped in balance, and evidence from primitive relatives suggests it may have been feathered.

PARENTAL CARE
Fossils tell us not only about the physical form of extinct animals but also how they lived. This adult *Oviraptor*, a small theropod dinosaur, had distinctly birdlike features. It died protecting about 20 eggs in a sandstorm, providing a unique record of parental care.

STRONG BEAK
The parrotlike beak attached to *Oviraptor's* toothless but powerful jaws may have been used to crack nuts or eat hard-shelled prey.

PROTECTIVE POSITION
The forearm spread over eggs is in the same posture used by many ground-nesting birds. If *Oviraptor* had feathered arms, these would have both covered and insulated the eggs.

EGGS IN NEST
Its position, hunched protectively over a clutch of about 20 eggs, suggests that this dinosaur had nesting habits similar to birds.

Dating

Even before Darwin and Wallace proposed their theory of evolution, it was recognized that fossils change over time, with different ones occurring in different rock layers. By the early nineteenth century, fossils were being used to characterize successive intervals and periods of geological time; for example, certain species of sea urchins, ammonites, and bivalves were known to be characteristic of the Cretaceous Period while the Cambrian was noted for particular trilobites and brachiopods. Relative fossil dating has since been highly refined by using "index species" as reference points. Ideally, index species are common, widespread, easily identifiable, and rapidly evolving, such as graptolites and ammonites. Scientists use them to distinguish well-defined intervals of time in rock layers or strata. However, finding an accurate age for these intervals required radioactive dating of suitable rocks in the stratigraphic rock record. This has now been achieved with considerable accuracy for a record of the last 542 million years.

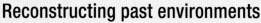

STRATA AT LOCATION A
STRATA AT LOCATION B

INDEX FOSSILS
The succession of fossils in stratigraphic sequences changes with environments and the passage of time. Fossils that characterize particular environments and time intervals are known as index fossils, and they can be used to match strata of the same type and age.

Youngest rocks — TIME — Oldest rocks

C

B

A

Age of bed containing fossils B and C

Age of bed containing fossils A, B, and C

FOSSIL ASSEMBLAGES
In a sequence of rocks without index fossils, groups of other fossils may sometimes indicate age. Such fossils are particularly useful if their age ranges overlap; and the more fossils there are, the more precisely the rock can be dated.

Reconstructing past environments

Interpreting the fossil record is vital for reconstructing environments and their development over time. Such interpretations are based on the knowledge that all organisms have certain inherent constraints, such as the amount of light, heat, or water needed to survive and reproduce. For example, strata containing tropical marine fossils—reef corals, sea lilies, and calcareous algae—were deposited in shallow tropical seas. These show that in Silurian and Early Jurassic times, Britain and Ireland lay at low latitudes with subtropical marine conditions. By contrast, the bones of cold-adapted mammoths and giant deer found in much younger deposits (less than a million years old) are indicative of cold, high-latitude areas. Analysis of pollen and beetle remains from fossil deposits shows evidence of species whose habitats are restricted to high, tundra-type environments, indicating climate change associated with Quaternary ice ages.

MAIDENHAIR TREE
The maidenhair tree, *Ginkgo biloba* (right), is a living fossil whose ancestors (above) served as dinosaur food. Ginkgos prefer warm climates and were abundant in Jurassic Britain. Their subsequent disappearance suggests climate change.

Bias in the fossil record

Fossils do not preserve a fair sample of the life of the past because fossilization favors organisms with preservable parts such as shells, bones, teeth, and certain tough plant tissues. Furthermore, the rock record is full of gaps that reflect past changes in sea level and plate movements. It also shows a bias toward the deposits of shallow-water continental-shelf seas and inland lake basins. Despite the greater size of oceans compared to land and shallow seas, very few ocean-floor rocks or fossils are preserved because they have been subducted by plate movements (see p.18). Overall, the fossil record tends to underrepresent the past life of ocean floors, terrestrial uplands, and any organisms with soft bodies (such as jellyfish), or delicate tissues (such as flowers). However, now that this limitation is understood, paleontologists have redressed the balance by seeking examples of underrepresented environments.

SCAVENGED BONES
Other factors prevent the formation of fossils. In the natural world, almost any food item will be scavenged by some specialist organism. Today, these range from microbes to hyenas (above), whose powerful jaws destroy bones for the marrow.

WALL OF BONES
Granted National Monument status in 1915, Dinosaur National Monument in Utah is one of the world's richest sources of dinosaur bones. Paleontologists have uncovered hundreds of bones from several species of Jurassic dinosaurs in this former river channel, including those of the giant, long-necked, quadrupedal sauropods.

Key Fossil Sites

Centuries of scientific investigation have uncovered a wealth of fossil evidence at sites around the world. Some of these places have provided particularly revealing glimpses into the past, greatly advancing our understanding of evolution. A selection of these sites is featured here.

JOGGINS, CANADA
This UNESCO World Heritage Site is home to a 313-million-year-old Carboniferous coal forest that was once prone to wildfire. Fossils of some of the first egg-laying, lizardlike animals have been found in the hollows of trees preserved here.

BURGESS SHALE, CANADA
This UNESCO World Heritage Site, discovered in 1909 by American paleontologist Charles Walcott, has yielded thousands of Middle Cambrian marine fossils from around 505 million years ago, many with soft parts such as gills preserved.

GUNFLINT CHERT, CANADA
The discovery in the 1950s of 2-billion-year-old marine microfossils was the first convincing evidence of life in Precambrian times.

RHYNIE CHERT, UK
Chert from 408-million-year-old deposits has revealed fossils of the oldest bog community, including primitive plants and early arthropods.

GREEN RIVER, USA
The 54-million-year-old lake deposits of the Green River formation preserve a wide range of Eocene fossils, from turtles, snakes, and fish to plants and insects.

JURASSIC COAST, UK
Fossils of Jurassic marine and flying reptiles were first found here in the early 19th century by the Anning family of professional collectors. A UNESCO World Heritage Site, it preserves Mesozoic strata and fossils in coastal cliffs.

MORRISON FORMATION, USA
Late Jurassic strata (148–155 million years old) have yielded a succession of terrestrial flood plain fossils, including the bones and footprints of giant sauropods such as *Diplodocus* and *Apatosaurus*, as well as the plants they ate.

LA BREA TAR PITS, USA
Pleistocene-age tar trapped thousands of animals, from saber-tooth cats to condors, perfectly preserving their bones, often with entire skeletons.

DOMINICAN AMBER
Paleogene-age plant resin captured a huge number of tropical forest insects, sometimes leaving their original coloring intact. These range from ants and bees to weevils and cockroaches—all alive around 30–40 million years ago.

KEY STUDY
ANTARCTIC FOSSIL SITES

In Permian times, when Antarctica was part of Gondwana, the remains of *Glossopteris* plants became the coal found today in the Transantarctic Mountains. Such deposits continued into the Triassic; the herbivore *Lystrosaurus* has been found with cycad and ginkgo plant fossils. By Cretaceous times, when Gondwana had broken apart, Antarctica was near its present position, but warm climates allowed forests and plant-eating dinosaurs to thrive, even in these high latitudes.

PATAGONIA, ARGENTINA
Recent discoveries have revealed that in the Late Cretaceous, Patagonia was home to some of the largest dinosaurs that ever lived, including *Patagotitan mayorum*. A model based on a 122ft- (37m-) long juvenile's remains can be seen in New York's American Museum of Natural History.

SANTANA, BRAZIL
The Cretaceous Santana and older Crato formations between them have yielded fossils from a diverse group of fish, insects, and plants, including examples of preserved soft tissue. The land is quarried in places for its fossils; many are sold commercially.

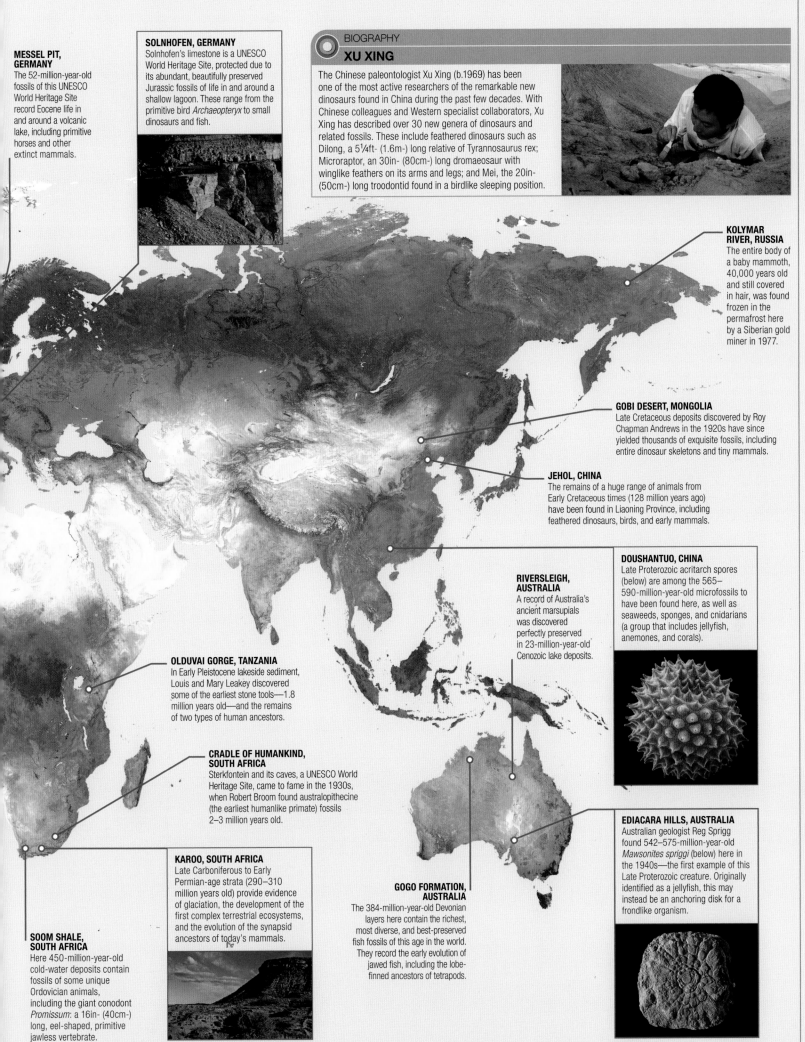

MESSEL PIT, GERMANY

The 52-million-year-old fossils of this UNESCO World Heritage Site record Eocene life in and around a volcanic lake, including primitive horses and other extinct mammals.

SOLNHOFEN, GERMANY

Solnhofen's limestone is a UNESCO World Heritage Site, protected due to its abundant, beautifully preserved Jurassic fossils of life in and around a shallow lagoon. These range from the primitive bird *Archaeopteryx* to small dinosaurs and fish.

BIOGRAPHY

XU XING

The Chinese paleontologist Xu Xing (b.1969) has been one of the most active researchers of the remarkable new dinosaurs found in China during the past few decades. With Chinese colleagues and Western specialist collaborators, Xu Xing has described over 30 new genera of dinosaurs and related fossils. These include feathered dinosaurs such as Dilong, a 5¼ft- (1.6m-) long relative of Tyrannosaurus rex; Microraptor, an 30in- (80cm-) long dromaeosaur with winglike feathers on its arms and legs; and Mei, the 20in- (50cm-) long troodontid found in a birdlike sleeping position.

KOLYMAR RIVER, RUSSIA

The entire body of a baby mammoth, 40,000 years old and still covered in hair, was found frozen in the permafrost here by a Siberian gold miner in 1977.

GOBI DESERT, MONGOLIA

Late Cretaceous deposits discovered by Roy Chapman Andrews in the 1920s have since yielded thousands of exquisite fossils, including entire dinosaur skeletons and tiny mammals.

JEHOL, CHINA

The remains of a huge range of animals from Early Cretaceous times (128 million years ago) have been found in Liaoning Province, including feathered dinosaurs, birds, and early mammals.

DOUSHANTUO, CHINA

Late Proterozoic acritarch spores (below) are among the 565–590-million-year-old microfossils to have been found here, as well as seaweeds, sponges, and cnidarians (a group that includes jellyfish, anemones, and corals).

RIVERSLEIGH, AUSTRALIA

A record of Australia's ancient marsupials was discovered perfectly preserved in 23-million-year-old Cenozoic lake deposits.

OLDUVAI GORGE, TANZANIA

In Early Pleistocene lakeside sediment, Louis and Mary Leakey discovered some of the earliest stone tools—1.8 million years old—and the remains of two types of human ancestors.

CRADLE OF HUMANKIND, SOUTH AFRICA

Sterkfontein and its caves, a UNESCO World Heritage Site, came to fame in the 1930s, when Robert Broom found australopithecine (the earliest humanlike primate) fossils 2–3 million years old.

EDIACARA HILLS, AUSTRALIA

Australian geologist Reg Sprigg found 542–575-million-year-old *Mawsonites spriggi* (below) here in the 1940s—the first example of this Late Proterozoic creature. Originally identified as a jellyfish, this may instead be an anchoring disk for a frondlike organism.

KAROO, SOUTH AFRICA

Late Carboniferous to Early Permian-age strata (290–310 million years old) provide evidence of glaciation, the development of the first complex terrestrial ecosystems, and the evolution of the synapsid ancestors of today's mammals.

GOGO FORMATION, AUSTRALIA

The 384-million-year-old Devonian layers here contain the richest, most diverse, and best-preserved fish fossils of this age in the world. They record the early evolution of jawed fish, including the lobe-finned ancestors of tetrapods.

SOOM SHALE, SOUTH AFRICA

Here 450-million-year-old cold-water deposits contain fossils of some unique Ordovician animals, including the giant conodont *Promissum*: a 16in- (40cm-) long, eel-shaped, primitive jawless vertebrate.

The Geological Timescale

The geological timescale is a scheme for dividing the history of Earth into named units. It is a hierarchical system, in which the largest divisions are called eons, followed in order of size by eras, periods, epochs, and ages. A standard timescale is an essential tool for geologists—working without one would be like studying history without a calendar. The refinement of the timescale using modern dating methods has allowed geologists to construct an accurate chronology of the evolution of life and other events in Earth's history.

The history of the timescale

Early ideas about geological time were driven by curiosity and the practicalities of mining. In the late eighteenth century, German mineralogist Abraham Werner proposed a succession of rocks deposited by a universal flood: first, primitive igneous rocks were laid down, followed by secondary sedimentary layers (strata), then tertiary surface deposits. However, in the nineteenth century, scientists realized that strata could be characterized by their fossils. They used this knowledge to draw the first geological maps, which included vertical sections showing the relative timescale for deposits of successive rock types

in each country—such as coal and chalk, which are widespread in Europe. Although attempts had been made to establish Earth's geological age, there was still no means of dating the rocks themselves. The knowledge that fossil layers could be matched up across national borders led to the development of internationally recognized divisions of geological time. Fossil strata were divided into a series of periods from the Cambrian to the Quaternary, with older rocks classified as Precambrian. Divisions of smaller units (epochs and ages) and larger ones (eras and eons) were added later.

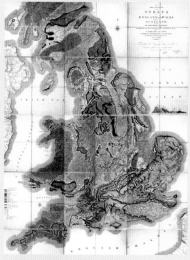

THE FIRST GEOLOGICAL MAP
The first modern geological map of an entire country was compiled by the English geologist and surveyor William Smith in 1815. Smith pioneered the use of fossils to establish the chronology of strata.

The modern timescale

The modern geological timescale provides an internationally recognized chronology of Earth's history for 4.6 billion years. The timescale allows a geologist to go anywhere in the world, examine the strata, identify the fossils within them, and give them an approximate age within the global scheme. It also lets geologists communicate with the confidence that they are referring to the same events, strata, and time periods. The modern timescale combines a number of schemes: lithostratigraphic (based on changing sedimentary rock types and sequences); biostratigraphic (evolving fossils); chronostratigraphic (radiometric dating); magneto-stratigraphic (the changing polarity of Earth's magnetic field); and the history of global changes in ocean and atmospheric chemistry preserved in the sedimentary rock record. The terms "Upper," "Middle," and "Lower" may also be replaced by "Early," "Middle," and "Late."

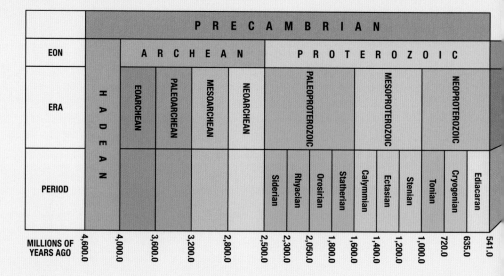

EON		PRECAMBRIAN														
EON		ARCHEAN				PROTEROZOIC									
ERA	HADEAN	EOARCHEAN	PALEOARCHEAN	MESOARCHEAN	NEOARCHEAN	PALEOPROTEROZOIC				MESOPROTEROZOIC			NEOPROTEROZOIC		
PERIOD	HADEAN					Siderian	Rhyacian	Orosirian	Statherian	Calymmian	Ectasian	Stenian	Tonian	Cryogenian	Ediacaran
MILLIONS OF YEARS AGO	4,600.0	4,000.0	3,600.0	3,200.0	2,800.0	2,500.0	2,300.0	2,050.0	1,800.0	1,600.0	1,400.0	1,200.0	1,000.0	720.0	635.0	541.0

EON	PHANEROZOIC																																			
ERA	PALEOZOIC																MESOZOIC																		
PERIOD	Carboniferous						Permian									Triassic							Jurassic								
EPOCH	Mississippian			Pennsylvanian			Cisuralian				Guadalupian			Lopingian		Lower		Middle		Upper			Lower				Middle				Uppe
	Lower	Middle	Upper	Lower	Middle	Upper																										
AGE	Tournaisian	Visean	Serpukhovian	Bashkirian	Moscovian	Kasimovian	Gzhelian	Asselian	Sakmarian	Artinskian	Kungurian	Roadian	Wordian	Capitanian	Wuchiapingian	Changhsingian	Induan	Olenekian	Anisian	Ladinian	Carnian	Norian	Rhaetian	Hettangian	Sinemurian	Pliensbachian	Toarcian	Aalenian	Bajocian	Bathonian	Callovian	Oxfordian
MILLIONS OF YEARS AGO	358.9	346.7	330.9	323.2	315.2	307.0	303.7	298.9	295.0	290.1	283.5	272.95	268.8	265.1	259.1	254.14	251.9	251.2	247.2	242.0	237.0	227.0	208.5	201.3	199.3	190.8	182.7	174.1	170.3	168.3	166.1	163.5

Dating the timescale

By the mid-nineteenth century, geologists realized that the rock layers that form the stratigraphic record of Earth's history must have taken hundreds of millions of years to accumulate, given that they are many tens of miles thick. Yet there was still no reliable means of dating them; for example, British physicist William Thomson's empirical methods suggested that Earth was probably 20–100 million years old. The discovery of radioactivity showed that Thomson's measures were far too short. Radiometric dating made it possible to calculate dates for the crystallization of certain minerals. In 1904, Ernest Rutherford obtained the first mineral-based radiometric age of 500 million years; by 1911 the first radiometric timescale had been compiled. Today's timescale, based on thousands of measurements, has been linked to the stratigraphic rock record to produce an International Stratigraphic Chart, with radiometric-based dates applied to the recognized stratigraphic boundaries.

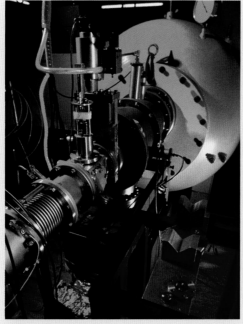

LINEAR ACCELERATOR
This equipment can determine the number of atoms of radiocarbon isotope 14C, also known as carbon-14, which allows carbon-containing materials up to 50,000 years old to be dated.

SCIENCE
MAGNETOSTRATIGRAPHY

In recent decades, a new kind of chronology has been developed, based on periodic reversals in the polarity of Earth's magnetic field. When some igneous and sedimentary rocks form, iron-rich minerals in the rocks align themselves to the prevailing magnetic field. By combining studies of the magnetic alignment of rocks on land with radiometric dating, geologists have produced a timescale of magnetic reversals. This timescale has been used to date rocks from the ocean floor. In particular, the dating of rocks of similar magnetism either side of spreading ridges (see below) has provided key evidence to support the theory of plate tectonics.

symmetrical patterns of iron-rich mineral alignment indicate the changing direction of Earth's magnetic field

transform fault   spreading ridge

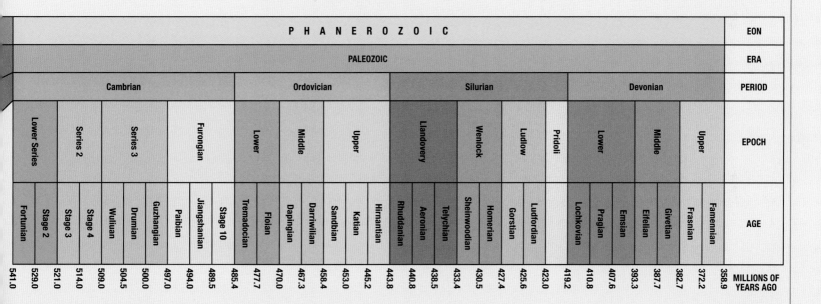

P H A N E R O Z O I C																																		**EON**
PALEOZOIC																																		**ERA**
Cambrian					Ordovician						Silurian								Devonian						**PERIOD**
Lower Series	Series 2		Series 3		Furongian			Lower	Middle	Upper	Llandovery			Wenlock		Ludlow		Pridoli	Lower		Middle		Upper		**EPOCH**
Fortunian	Stage 2	Stage 3	Stage 4	Wuliuan	Drumian	Guzhangian	Paibian	Jiangshanian	Stage 10	Tremadocian	Floian	Dapingian	Darriwilian	Sandbian	Katian	Hirnantian	Rhuddanian	Aeronian	Telychian	Sheinwoodian	Homerian	Gorstian	Ludfordian	Lochkovian	Pragian	Emsian	Eifelian	Givetian	Frasnian	Famennian				**AGE**
541.0	529.0	521.0	514.0	509.0	504.5	500.0	497.0	494.0	489.5	485.4	477.7	470.0	467.3	458.4	453.0	445.2	443.8	440.8	438.5	433.4	430.5	427.4	425.6	423.0	419.2	410.8	407.6	393.3	387.7	382.7	372.2	358.9		**MILLIONS OF YEARS AGO**

P H A N E R O Z O I C																																							
MESOZOIC										**CENOZOIC**																													
Cretaceous												Paleogene										Neogene								Quaternary					
Lower							Upper					Paleocene		Eocene			Oligocene		Miocene						Pliocene		Pleistocene			Holocene	
Berriasian	Valanginian	Hauterivian	Barremian	Aptian	Albian	Cenomanian	Turonian	Coniacian	Santonian	Campanian	Maastrichtian	Danian	Selandian	Thanetian	Ypresian	Lutetian	Bartonian	Priabonian	Rupelian	Chattian	Aquitanian	Burdigalian	Langhian	Serravallian	Tortonian	Messinian	Zanclean	Piacenzian	Gelasian	Calabrian	Middle	Upper		
139.8	132.9	129.4	125.0	113.0	100.5	93.9	89.8	86.3	83.6	72.1	66.0	61.6	59.2	56.0	47.8	41.2	37.8	33.9	27.82	23.03	20.44	15.97	13.82	11.63	7.246	5.333	3.600	2.58	1.80	0.781	0.126	0.0117		

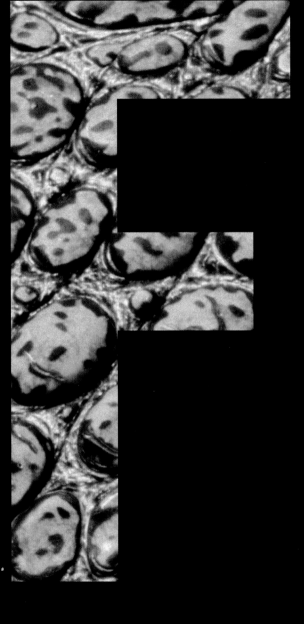
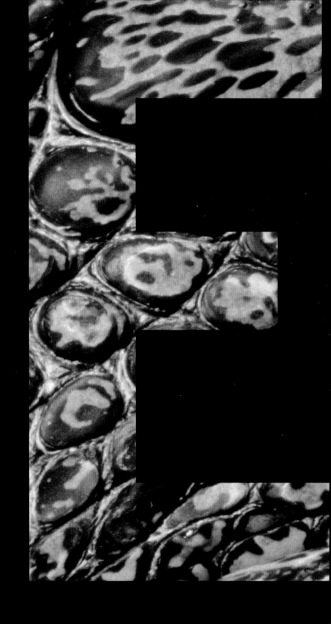

ON EARTH

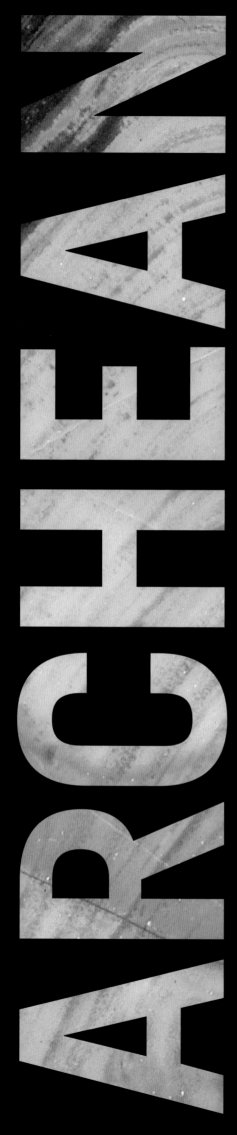

ARCHEAN

Archean

The Archean Eon represents a period in Earth's history from 4 billion to 2.5 billion years ago. While these are arbitrarily chosen figures, the beginning of the Eon coincides with the end of the so-called "Great Cometary Bombardment" of the Earth, while the end coincides with the "Great Oxygenation Event," when Earth's atmosphere became permanently enriched in oxygen.

KOMATI RIVER VALLEY
Rocks from some of Earth's oldest volcanic crust are found here in the Barberton Greenstone Belt of southern Africa, renowned for its gold deposits.

Oceans and continents

The beginning of the Archean was marked by the end of a bombardment of Earth by comets or asteroids. Determining the positions of continents and oceans at this distance in time is an extremely difficult business. Very few rocks of this age remain, and those that do have, in many cases, been profoundly altered. Indirect evidence for the earliest rocks comes from crystals of the mineral zircon, eroded from granites and redeposited in younger sedimentary rocks. These formed during the early Archean, or even in the preceding Hadean period. The oldest of these is a single crystal of zircon from Western Australia that has been dated as more than 4 billion years old. The presence of zircon at this extremely early period in Earth's history suggests that slabs of continental crust were forming and, just as significantly, there was water on the planet. However, it may not have been until about 3 billion years ago that a land mass of any appreciable size was formed. Before this, fledgling continents probably came together and were then rapidly eroded or subducted, leaving only a few traces in the rock record. It has been speculated that the first supercontinent, referred to as "Ur," comprised the ancient cratons, also known as shields, of central-southern Africa, the Pilbara region of western Australia, India, and small parts of Antarctica. About half a billion years later, at the end of the Archean, it has been

PILBARA REGION, AUSTRALIA
Some of the world's oldest sedimentary and volcanic rocks, dating back almost 3.5 billion years, are found in the mineral-rich Pilbara Region in northwest Australia.

suggested that a second continent called "Arctica" formed, consisting of the cratons of present-day Canada, Greenland, and Siberia. Reconstructing the position of these early continents on the globe is very difficult. During the late Archean, Ur may have been situated in high latitudes, whereas Arctica was in low latitudes. The Archean oceans are likely to have been more extensive than they are today because the continents were smaller. They also differed from modern oceans in that they were about 1.5 to 2 times more saline. Moreover, they were devoid of dissolved oxygen. The only life that oceans could support was in the form of single-celled organisms, probably anaerobic bacteria and archaea, which do not require oxygen.

Evidence for their existence comes from fossils of filamentous bacteria and structures called stromatolites found in rocks as old as 3.49 billion years in present-day Australia and South Africa (see opposite page).

FOSSILIZED MUD POOL
Gas escape structures in a fossilized mud pool in the Barberton Greenstone Belt show that geothermal activity was ubiquitous in the early Earth, 3.5 billion years ago, possibly hosting heat-loving bacteria.

Climate

For much of the Archean, the Earth's atmosphere is thought to have been devoid of oxygen. Instead, it was composed primarily of nitrogen, methane, and carbon dioxide. The Sun was much weaker then, emitting less radiation, so had it not been for these greenhouse gases, the world would have been a ball of frozen rock. In fact, research indicates that an atmosphere rich in greenhouse gases helped create extremely high global temperatures, especially early in this period. These could have been within the range 131–176°F (55–80°C). Ocean temperatures would have been correspondingly high, and the only forms of life capable of living in the hot, salty seas were thermophilic (heat-loving) anaerobic bacteria and archaea. Although the climate was very hot overall, about 2.9 billion years ago, temperatures appear to have fallen drastically for a relatively short time and the Earth may even have experienced its first glaciation. Geological evidence suggests that during the last few hundred million years of the Archean, temperatures returned to their previously high levels.

VOLCANIC GASES
Volcanoes, many of them submarine, erupted hot gases that formed the early atmosphere, but there was little free atmospheric oxygen until the end of the Archean.

Life in the Archean

Any life forms that may have lived in the Archean were likely to have been very tough, and very small, like modern eubacteria and archaebacteria. Recent work on the DNA of these modern forms has shown that they are probably the most primitive of life forms found today. They are known to use the energy stored within the rocks and minerals of the Earth as sources from which to create the basic building blocks of life. Fossils from this period are very rare, but of supreme importance for our understanding of how the Earth became a habitable planet. Some of the earliest evidence for once-living cells in rocks of the Archean comes from South Africa and Western Australia. These ancient rocks, dated to 3.49 to 3.43 billion years ago, show that simple bacteria were probably living in settings lacking in oxygen but high in silica and metallic sulfides, especially volcanic environments. All microfossils of this age are rather controversial, but simple filamentous forms have been found in cherts by the Komati River, southern Africa, and around Strelley Pool in the Pilbara region of Australia. Recently, tiny tubular microborings, which may well have been produced by bacteria, have been discovered in basaltic pillow lavas. In Sulphur Springs, Australia, slightly younger clusters of threadlike fossils preserved in iron sulfide (pyrite) have been found. These deposits are

BLACK SMOKER
Columns of black smoke rich in iron sulfides pour out of a vent on the sea floor where temperatures can exceed 752°F (400°C). Life may have emerged and thrived around such vents, which were common in the Archean.

about 3.2 billion years old and comprise sulfide-rich cherts that formed around submarine hot springs, much like those of modern black smokers. The fossils themselves comprise tiny cylindrical threads about one micron wide, trapped between layers of iron sulfide that may have formed part of a black smoker chimney. A second line of evidence for very early life forms is from stromatolites. Microbial stromatolites (see above) often form distinctly crinkly layers, and stromatolites of this kind have only been found in the rock record after about 3 billion years ago. Some of the most ancient examples are known from the Strelley Pool Chert, dated to about 3.43 billion years ago, although these do not show the crinkly layers consistent with a microbial origin. It is only after about 2.6 billion years ago that cyanobacterial fossils appeared, leading to the buildup of atmospheric oxygen by means of photosynthesis. Before this, life may therefore have been widespread, but it seems to have consisted largely of forms living on the enormous supplies of internal energy from the young Earth.

STROMATOLITES
Living stromatolites can be found forming today in Shark Bay, Western Australia, where they are commonly regarded as the work of cyanobacterial communities.

SIMPLE FILAMENT
This microfossil from the Pilbara region of Australia shows a folded filament with cell-like structures made of carbon, comparable to modern iron bacteria.

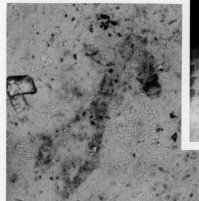

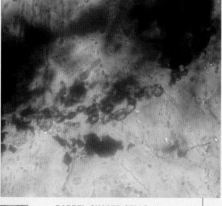

PILLOW LAVA
These rocks on the banks of the Komati River in Barberton Mountain Land in South Africa are about 3.49 billion years old. The pillow lava has intervening black chert that contains tiny structures resembling bacterial microfossils.

BARREL-SHAPED CELLS
These microfossils from the Strelley Pool Chert in the Pilbara region show a chain of cell-like structures similar to modern purple bacteria.

PROTEROZOIC

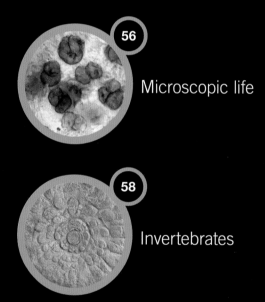

56 Microscopic life

58 Invertebrates

Proterozoic

The Proterozoic Eon covers an unimaginable stretch of time from 2.5 billion to 541 million years ago. The activity of photosynthetic microbes, begun in the Archean, transformed Earth into a planet with an oxygenated atmosphere and oceans. The scene was set for the subsequent evolution of life in the seas, on land, and in the air. Bacteria and archaea were joined by the first simple animals and plants.

GRANITE DOMES
Pink granite outcrops at Enchanted Rock in Texas date back to the Proterozoic. This ancient batholith (body of intrusive igneous rock) is an estimated 1 billion years old.

Oceans and continents

Over the 2 billion years of the Proterozoic, continents evolved and fragmented, but left evidence in the rocks that has enabled geologists to reconstruct how the globe might have looked. A world map can be pieced together by dating early mountain chains, thrown up by continental collisions and long since eroded. This suggests the following history, involving continents on the move. The continent of Ur continued to grow until 1.5 billion years ago, absorbing today's Zimbabwe, northern India, and the Yilgarn block of western Australia. About 1.6 to 1.3 billion years ago, Arctica accreted more continental blocks to form a larger continent called Nena. A third continent, Atlantica, was formed about 2 billion years ago, adding the Tanzanian block about 1.3 billion years ago. Ur, Nena, and Atlantica may have come together to form the continent of Columbia, before separating and reuniting about 1 billion years ago to form the supercontinent of Rodinia. This in turn split 300 million years later, opening the Panthalassic Ocean. By the end of the Proterozoic, these isolated continental blocks had rejoined in the supercontinent of Pannotia. As continents were continually reshaped, the erosion of new mountain chains deposited sediments rich in minerals in the oceans. Microorganisms in the shallower seas gradually changed the composition of the atmosphere, and by the end of the Proterozoic soft-bodied animals began to leave their imprints on the seafloor.

STROMATOLITES
In Shark Bay, Australia, bacterial colonies called stromatolites release oxygen as a by-product when they photosynthesize. Stromatolites flourished in shallow waters surrounding Proterozoic continents.

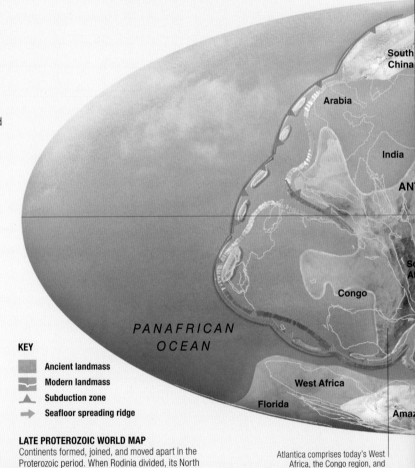

The northern half of Rodinia (Antarctica, Australia, India) rotates and moves north

South China

Arabia

India

ANT

Congo

PANAFRICAN OCEAN

West Africa

Florida

Ama

KEY

- Ancient landmass
- Modern landmass
- Subduction zone
- Seafloor spreading ridge

LATE PROTEROZOIC WORLD MAP
Continents formed, joined, and moved apart in the Proterozoic period. When Rodinia divided, its North American section moved south to the South Pole.

Atlantica comprises today's West Africa, the Congo region, and northeastern South America

---

PALEOPROTEROZOIC

MYA   2500   2000

MICROSCOPIC LIFE

● **2400** First cyanobacteria (photosynthesizing single-celled microorganisms without nuclei—prokaryotes)

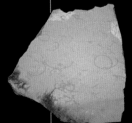

● **1850** First eukaryote (organism with cell or cells containing nucleus and other organelles)

GRYPANIA

— **2700–2300** Oxygenation event — ➡ ●

RED ALGAE

Red algae on the surface of Lake Magadi, Kenya, bear a close resemblance to *Bangiomorpha*, the oldest known multicellular genus in the fossil record, from a locality in arctic Canada.

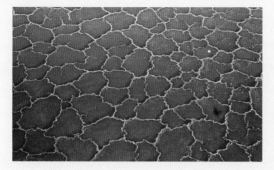

Climate

The early Proterozoic was probably a little cooler than the Archean, but global temperatures may still have been around 104°F (40°C). They appear to have declined rapidly, resulting in a major glaciation event between 2.2 and 2.4 billion years ago. This dramatic cooling may have been triggered by the evolution of photosynthetic cyanobacteria. After a protracted warm phase from 2.2 billion to 950 million years ago, Earth was plunged into another series of major glaciations, many of which may also have been global. Evidence for this has been found on all continents today. Glacial evidence on some continents once located in low latitudes has led to the scenario of a "Snowball Earth," frozen from pole to pole. Yet, during the last few hundred million years of the Proterozoic, conditions were often tropical, especially at low latitudes. Intense glacial periods, perhaps for a million years, were followed by "greenhouse" periods when global temperatures soared. At the end of the Proterozoic, temperatures had returned to preglaciation levels.

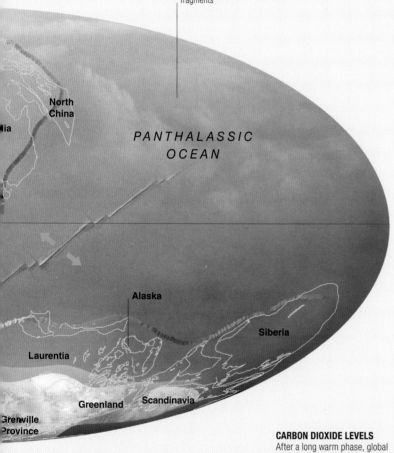

Panthalassic Ocean opens as Rodinia fragments

PANTHALASSIC OCEAN

North China

Alaska

Siberia

Laurentia

Greenland   Scandinavia

Grenville Province

BRACHINA GORGE

A sequence 2.5 miles (4km) thick of tillites (glacial sediments) in the Brachina Gorge, Flinders Ranges, South Australia, provides evidence of extensive glaciation during the late Proterozoic.

CARBON DIOXIDE LEVELS

After a long warm phase, global temperatures declined. During the course of the Proterozoic, carbon dioxide levels increased.

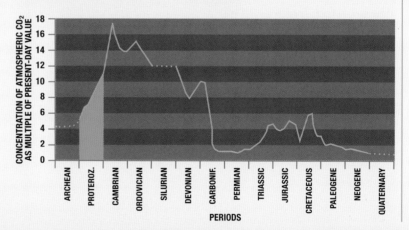

CONCENTRATION OF ATMOSPHERIC CO_2 AS MULTIPLE OF PRESENT-DAY VALUE

PERIODS: ARCHEAN, PROTEROZ., CAMBRIAN, ORDOVICIAN, SILURIAN, DEVONIAN, CARBONIF., PERMIAN, TRIASSIC, JURASSIC, CRETACEOUS, PALEOGENE, NEOGENE, QUATERNARY

MESOPROTEROZOIC

NEOPROTEROZOIC

1000   500   MYA

00 First structurally complex eukaryote, possible fungus

● 1400 Great increase in diversity of stromatolites

● 1200 Microscopic colonization of land; first multicellular red algae (below)

● 1100 Earliest dinoflagellate (single-celled eukaryotic microorganism with flagellae)

● 1000 First vaucherian algae (*Palaeovaucheria*)

● 750 First protozoans (single-celled nonphotosynthesizing eukaryotes) e.g. *Melanocyrillium*

● 750–700 First calcareous algae

● 713–635 First indirect evidence for metazoans (multicelled animals): chemical biomarkers suggest demosponges

● 560 First fungi

STROMATOLITES

BANGIOMORPHA

MICROSCOPIC LIFE

DICKINSONIA

● 565-540 First Ediacaran biota (e.g. *Dickinsonia*)
● 555 First anthozoans
● 550 First evidence for ctenophores (comb jellies), sponges; first anthozoans (group containing corals, sea anemones)

INVERTEBRATES

PROTEROZOIC
MICROSCOPIC LIFE

During the Proterozoic, evolution was almost wholly concerned with making large improvements to the cell. Such changes included the development of a nucleus, oxygenic photosynthesis, and the ability to reproduce sexually. The resulting specialized cells became the ancestors of fungi, plants, and animals.

Early in the Proterozoic, life was entirely microscopic and single-celled. Microfossils of this age are common in rocks called stromatolites—fossilized, cabbagelike growths of bacteria. They can be compared with bacteria found living today in salty lagoons and around hot springs. The cells have no nucleus or evidence for sexual reproduction and are known as prokaryotes (meaning "without a nucleus"). Like many bacteria today, most Proterozoic bacteria did not need oxygen and lived in conditions that would be toxic to animals and plants. Others resembled oxygen-producing "algae" called cyanobacteria. The earliest of these appeared about the same time as a rapid buildup of atmospheric oxygen, which made Earth more habitable for animals and plants.

GUNFLINT CHERT
Early microfossils are preserved in cherts. These rocks contain some of the first recorded fossils. This slice of Gunflint chert shows stromatolite columns that were formed by bacteria.

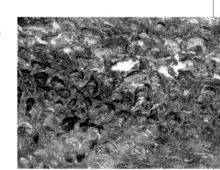

Cells slowly became larger, and more diverse and specialized. These complex cells are called eukaryotes ("with a nucleus"). The transformation of simple cells into more advanced cells was probably the longest and greatest step in evolution. By the end of the Proterozoic, large, soft-bodied organisms were colonizing the seabed.

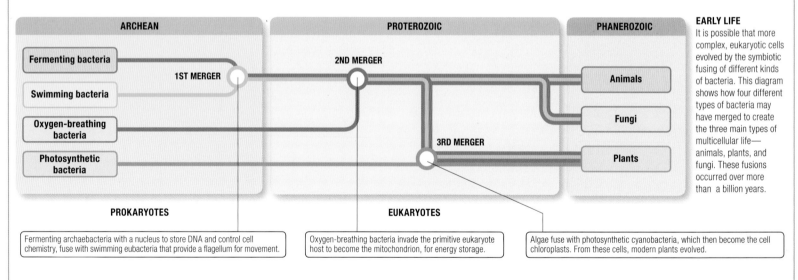

EARLY LIFE
It is possible that more complex, eukaryotic cells evolved by the symbiotic fusing of different kinds of bacteria. This diagram shows how four different types of bacteria may have merged to create the three main types of multicellular life—animals, plants, and fungi. These fusions occurred over more than a billion years.

ARCHEAN	PROTEROZOIC	PHANEROZOIC

Fermenting bacteria
Swimming bacteria
Oxygen-breathing bacteria
Photosynthetic bacteria

1ST MERGER
2ND MERGER
3RD MERGER

Animals
Fungi
Plants

PROKARYOTES
EUKARYOTES

Fermenting archaebacteria with a nucleus to store DNA and control cell chemistry, fuse with swimming eubacteria that provide a flagellum for movement.

Oxygen-breathing bacteria invade the primitive eukaryote host to become the mitochondrion, for energy storage.

Algae fuse with photosynthetic cyanobacteria, which then become the cell chloroplasts. From these cells, modern plants evolved.

GROUPS OVERVIEW

The fossil record for the Proterozoic era reveals a world dominated by microscopic bacteria. Complex eukaryotes only emerged after 1,900 mya, once there was enough oxygen to support them. Later revolutions involved feeding by ingestion at around 750 mya, and the evolution of a through-gut at around 545 mya.

CYANOBACTERIA
Photosynthetic bacterial cells make up the bulk of the early fossil record. Most cells were less than 2 microns across, and they gathered together to form microbial mats that helped build stromatolites. They made it possible for animals to evolve on Earth by ensuring a constant supply of food from sunlight and by producing oxygen.

ALGAE
Larger, more complex photosynthetic algal cells appeared after 1,400 mya. These were over 20 microns across, and progressively evolved to form far larger colonies. They expanded the ecological zone occupied by cyanobacteria by developing big leaves and rootlike systems. By 450 mya, some had moved onto land, eventually evolving into the modern plant groups.

PROTOZOA
Protozoans were single cells that could ingest other creatures. They can be recognized by their microscopic, flasklike shells, and did not appear until after 750 mya. The evolution of ingestion allowed for more rapid recycling of the organic matter grown by late Proterozoic algae. Some protozoans merged into colonies that specialized into early fungi and animals.

SPONGES
Colonies of cells that were differentiated into specific tissues to form simple animals did not appear until the end of the Proterozoic. The earliest recognizable sponge fossils are spicules, which had appeared by about 543 mya. Sponges are filter feeders, and helped "clean-up" the water column and the sea floor at the very end of the Precambrian.

Gunflintia

GROUP Cyanobacteria
DATE Proterozoic to present
SIZE 5 microns wide
LOCATION North America, Australia

Gunflintia is the first fossil to appear abundantly anywhere in the rock record. It was among the first oxygen-producing cyanobacteria. Photosynthetic organisms such as *Gunflintia* helped raise the oxygen in the atmosphere, making the Earth more habitable for later oxygen-users such as protozoans, plants, and animals.

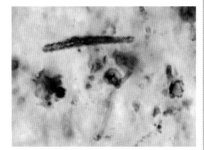

COMMONLY FOUND
Filaments of *Gunflintia* are seen here with rounded spheres of *Huroniospora*. *Gunflintia* is one of the most common microfossils in Gunflint chert.

Eoentophysalis

GROUP Cyanobacteria
DATE Proterozoic to present
SIZE 5 microns wide
LOCATION Worldwide

Fossilized *Eoentophysalis* occurs widely in shallow-water cherty rocks from the Proterozoic. It is very similar to a modern-day coccoid cyanobacterium, *Entophysalis*, which can still be found in shallow, salty lagoons. Although bacteria have simple structures, they are very cosmopolitan in the way they live.

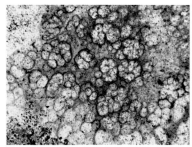

BUNCHES OF GRAPES
Not all cyanobacterial microfossils were filamentous like *Gunflintia*. Some, like *Eoentophysalis*, were arranged in packages, like bunches of grapes.

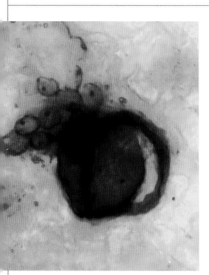

Torridonophycus

GROUP Chlorophyceae
DATE Proterozoic to present
SIZE 20 microns wide
LOCATION Scotland

This algal microfossil is seen here escaping from a round, baglike structure called an acritarch, which helped to protect the enclosed organism from cold, drying out, or lack of oxygen. Their presence suggests the surface of the land was starting to turn green during springtime as far back as a billion years ago.

GREEN ALGA
It is thought the dark blob inside each cell could be the remains of specialized subunits (organelles), suggesting this is a eukaryotic green alga.

Bangiomorpha

GROUP Rhodophyceae
DATE Proterozoic to present
SIZE 50 microns wide
LOCATION Canada

Multicellular filaments, similar to those of modern red algae, first appeared around 1,200 million years ago. *Bangiomorpha* had specialized reproductive structures, and a primitive holdfast that attached it to the sea floor and allowed it to rise upward toward the sunlight. The evolution of sexual reproduction and cellular differentiation can be taken as the start of the long march toward the more complex body plans of plants and animals.

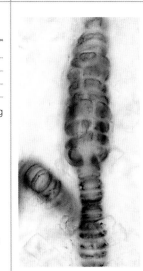

STACKED CELLS
Large cells, similar to those typically found in modern-day red algae, are stacked like a pile of plates in this fossilized *Bangiomorpha*.

Melanocyrillium sp.

GROUP Amoebozoa
DATE Late Proterozoic to present
SIZE 60 microns long
LOCATION America, Northern Europe

Melanocyrillium can be compared to a modern group of single-celled organisms called testate amoebae. Like the amoeba, they had a nucleus, mitochondria, an elastic body wall, and pseudopods for feeding on other organisms. The addition of an outer shell provided protection against dessication and attack.

ORGANIC BAGS
These vase-shaped microfossils typically consist of flasklike bags of organic material. Some have additional mud particles or secreted silica scales.

Megasphaera

GROUP Possible Animalia
DATE Late Proterozoic
SIZE 500 microns wide
LOCATION China

Originally interpreted as green algae, these are now thought to be the eggs and embryos of early animals, or possibly the cells of sulfur-oxidizing bacteria. Despite their exquisite preservation, their affinities are still debated. They may be the embryos of sponges of the Ediacara biota, which appears just above them in the rock record.

ANIMAL EMBRYO?
Inside the sculptured wall of *Megasphaera* can be seen a cluster of cells that looks like the early stages of division in an animal embryo.

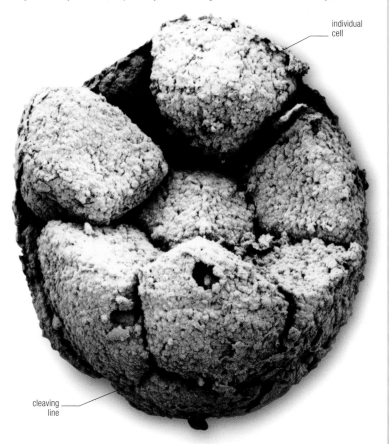

individual cell

cleaving line

Protospongia

GROUP Porifera
DATE Late Proterozoic to Cambrian
SIZE ¹/₁₆in (1mm) long
LOCATION Worldwide

Modern sponges are widely agreed to be descendents of the earliest cell colonies that evolved into modern animals. The earliest fossil evidence for sponges are tiny sponge spicules, which first appeared very close to the end of the Proterozoic era. They probably evolved as protection, as well as support, during the great arms race at the start of the Cambrian Explosion (see p.66).

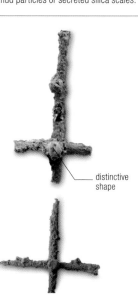

distinctive shape

DISTINCTIVE SHAPES
Some sponge spicules are very distinctive, particularly these cross-shaped forms that first appeared at the very end of the Ediacaran Period.

PROTEROZOIC
INVERTEBRATES

Until 1946, the only known Proterozoic fossils were layered deposits called stromatolites, which were formed by the growth of blue-green algae. However, in 1946, the Ediacaran fauna was discovered in Australia, and an entirely new perspective on the nature of the late Precambrian opened up.

Ediacaran fauna

The late Precambrian was generally a time of simple ecosystems with short food chains, consisting of bacteria, algae, and protists, which were not fossilized. It was also a time of successive glaciations, some of which were so severe that they covered the Earth with ice and gave rise to the "Snowball Earth" hypothesis. But about 580 million years ago, the Ediacaran fauna (see p.61) began to flourish. This consisted of a diverse range of organisms in many shapes and sizes, most of which had a "quilted" surface and lacked a gut. Over 30 genera have been identified, but opinion is still divided as to whether they are remote ancestors of modern animals, such as jellyfish, sea pens, or worms, or

if they are the relics of a failed evolutionary experiment, unrelated to modern organisms—almost as though they were life from another planet. If this is the case, then these Vendozoa, as they are known, may have had photosynthetic algae within their tissues or fed by absorbing nutrients across their surfaces.

PUZZLING DISK
It is not known if this flat, disklike fossil, *Mawsonites*, was a jellyfish or a now-extinct Vendozoan.

EDIACARA HILLS
North of Adelaide, South Australia, are the Ediacara Hills, where a range of late Precambrian fossils, the Ediacaran fauna, were described in 1947.

Fine preservation

Apart from the Ediacara Hills in Australia, these late Precambrian fossils have been found in many other parts of the world, including Canada, Russia, and England. The Ediacara Hills have yielded many thousands of specimens that are preserved in very fine sand—an important factor in the survival of the fossils. They appear at the interface of the thin silt and hard sandstone, which suggests that the living animals were stranded on tidal mudflats or in shallow tidal pools and preserved when the sand covered them. While these Ediacaran organisms lived in shallow, well-lit waters, others lived at depths too great for sunlight to reach.

SEAFLOOR DWELLER
The common Ediacaran fossil *Chamiodiscus* had a "quilted" surface and was anchored to the seafloor, rising vertically.

Charnia

GROUP Unclassified
DATE Late Precambrian
SIZE 6in–6½ft (0.15–2m) tall
LOCATION England, Australia, Canada, Russia

In 1957, a schoolboy, Roger Mason, found a *Charnia* fossil in late Precambrian rocks in Leicestershire, England. It was described a year later as a probable alga, but since then, its similarity to modern sea pens—which are a type of octocoral—has generally been accepted. Another interpretation is that *Charnia* was a type of quilted organism that evolved in the late Precambrian, to which the general name Vendobionta is given. *Charnia* has a feather-shaped frond with a series of side branches that remain in contact with each other along their entire length. These branches are arranged at an angle of about 45 degrees to the vertical axis. They have regular subdivisions that, if the organism were similar to a sea pen, would have housed the polyps. Some specimens have been found with a stem at the base, and in fossils of the related *Charniodiscus*, the stem is connected to a basal disk. This example has led to the theory that the disk-shaped fossils, *Aspidella* (see p.60) and *Medusinites*, are the anchoring disks of *Charnia*-like fronds. Although the fronds typically broke up when the organism died, the disks, which were already buried in the sediment, were fossilized. Recently, the suggestion that *Charnia* was similar to sea pens has been questioned because the mode of budding at the tip of the colony appears to differ from that of modern sea pens.

CHARNIA IN LIFE
It has been suggested that *Charnia*'s fronds were shaped to produce a large surface area over which it could absorb organic carbon or other nutrients from the water column.

CHARNIA SP.
Like most Precambrian fossils, all we know about *Charnia* comes from impressions made by the outside of its body in soft sediments. No internal structures have been preserved.

subdivided
side branch

Haootia

GROUP Possible Cnidarian
DATE Late Precambrian
SIZE 4–6in (10–15cm) tall
LOCATION Canada

Haootia is known from two specimens discovered in Newfoundland in 2008. It has a symmetrical, four-sided, cup-shaped body at the top of a short stalk with an attachment disk at its base. The body walls are made up of numerous long fibers, interpreted as muscle tissue. If this interpretation is correct, *Haootia* shows the earliest fossil evidence for muscle tissue, making it the earliest undisputed Eumetazoan—an animal with true tissues and specialized cells. *Haootia*'s shape and musculature suggest that it is an early jellyfish. Its body plan is very similar to modern stalked jellyfish, which live attached to the seabed, although the different arrangement of the muscle fibers may indicate that *Haootia* had a different feeding method, using a pulsing motion to gulp in water and nutrients.

HAOOTIA QUADRIFORMIS
The prominent muscle fibers extend upward from the four corners of the body as arms, which branch several times.

Aspidella

GROUP Possible Cnidarian
DATE Late Precambrian
SIZE 1/16–2in (0.1–5cm) across
LOCATION Canada, British Isles

When *Aspidella* was discovered in Canada in 1872, it was the first Precambrian fossil to be described. Later, geologists decided it was a pseudofossil—a fossil-like impression produced by sedimentary processes. After the Ediacaran faunas of Australia were described in the 1960s, *Aspidella* was reexamined, and geologists concluded that it was indeed a fossil. Initially classified as a jellyfish, *Aspidella* is now thought to have been the anchoring disk of a frondlike organism, such as *Charnia* (see p.59), although many more disks have been found than fronds. The fossil *Cyclomedusa*, found in 1946 at Ediacara, South Australia, is thought to represent the same organism and is now also called *Aspidella*.

ASPIDELLA SP.
The fossil is a small disk, often with a raised central boss, and radial and concentric ridges.

Yelovichnus

GROUP Unclassified
DATE Late Precambrian
SIZE 3/8–6in (1–15cm) long
LOCATION Canada, Russia, British Isles, China, Australia

YELOVICHNUS SP.
The raised ridges of *Yelovichnus* have to be photographed under low-angle lighting to be visible.

These enigmatic fossils were first thought to be feeding trails. Their meandering nature appeared to indicate a sophisticated method of feeding, whereby an animal, often an annelid worm or a mollusk, grazes methodically over the seabed and removes surface food. However, better-preserved examples of *Yelovichnus* imply a different story. Unlike true feeding trails, there is no evidence of a turning point in the meander. Instead, the fossils appear to be collapsed, segmented tubes. It has more recently been suggested that they represent spirally organized organisms, possibly algae.

Dickinsonia

GROUP Possible Placozoan
DATE Late Precambrian
SIZE 3/8in–3ft 3in (1–100cm) long
LOCATION Australia, Russia

Dickinsonia is one of the most perplexing fossils found in the Ediacaran fauna. At first sight, it appears to be segmented, with a distinct head and tail end. This led to the belief that it was a segmented marine worm. Hundreds of examples showing all growth stages and in various types of preservation have been collected—the largest specimens known are about 3¼ feet (1 meter) in length. However, no convincing gut or other internal structures have ever been found. This has led to the conclusion that *Dickinsonia* fed by absorbing food through its entire undersurface. More recently, it has been suggested that this strange animal may be a placozoan—a group of animals that is represented by only one living species. Placozoans have only four types of cell, which exist in two layers. It is possible that they may be a halfway stage between sponges and Eumetazoa, animals with true tissues and specialized cells.

central groove

mature end

body segment

No one knows which end of *Dickinsonia* **was its head**—or whether it had one at all.

DICKINSONIA COSTATA
Numerous segments radiated from a central groove. The larger segments at one end may have been more mature than those at the opposite end.

Parvancorina

GROUP Unclassified
DATE Late Precambrian
SIZE ³/₈–1in (1–2.5cm) long
LOCATION Australia, Russia

Parvancorina is an Ediacaran fossil with a shield-shaped front end, which appears to be a distinct head. The body has a central axial ridge along its length and shows some evidence of segmentation. Up to 10 pairs of possible appendages have been identified on some *Parvancorina* fossils. Many show a distinct set of growth stages. A large number of specimens were found with their head shields facing in the direction of the water current, implying a type of feeding strategy. *Parvancorina* is similar in shape to the Cambrian *Marrella* (see pp.72–73).

PARVANCORINA MINCHAMI
The shield-shaped head region has led to suggestions that *Parvancorina* was an ancestor of trilobites.

Tribrachidium

GROUP Unclassified
DATE Late Precambrian
SIZE ³/₄–2in (2–5cm)
LOCATION Australia, Russia, Canada

Tribrachidium is a mysterious Ediacaran fossil. It is a disk-shaped organism with three raised "arms" on its surface and a raised border. It also has a suggestion of bristlelike structures around the edge. *Tribrachidium* has triradiate symmetry— it is symmetrical along three lines. No triradiate animals existed throughout the Phanerozoic, but it is possible that an animal with triradiate symmetry may have preceded the echinoderms, which have five-fold radial symmetry. However, *Tribrachidium* lacked the calcareous plates that are a characteristic of echinoderms.

TRIBRACHIDIUM HERALDICUM
One theory is that *Tribrachidium* was a holdfast to a *Charnia*-like specimen (see p.59).

Spriggina

GROUP Unclassified
DATE Late Precambrian
SIZE 1¼in (3cm) long
LOCATION Australia, Russia

Spriggina was a segmented animal with what appears to be a clear head and tail end. The head is horseshoe shaped and the body consists of around 40 segments, with a prominent midline along its back and a small tail. Fossils have been found with various degrees of curvature, which shows that the body was flexible. Debate continues about what *Spriggina* really was. Some compare it to modern marine annelid worms, while others suggest that it could be a prototrilobite, because its head has spines that are similar to the genal (cheek) spines of trilobites. For now, *Spriggina* is probably best placed within the Bilateralia, a group that includes all animals that are bilaterally symmetrical at some point in their life cycle.

⊙ KEY FIND
EDIACARAN FAUNA

In 1946, geologist Reginald Sprigg was examining ore prospects in the Flinders Ranges of South Australia when he discovered a group of fossils in the Ediacara Hills. He described these soft-bodied animals as jellyfish from the Early Cambrian. Their actual Precambrian age was soon established, and this diverse fauna, including *Spriggina*, became known as the Ediacaran fauna.

Is it a **worm**, a **prototrilobite**, or completely **unrelated** to either? The debate over *Spriggina* continues.

SPRIGGINA FLOUNDERSI
One of the big problems with *Spriggina* being a primitive trilobite or other arthropod is that its body segments are not arranged in opposite pairs, but are slightly offset against each other down the midline.

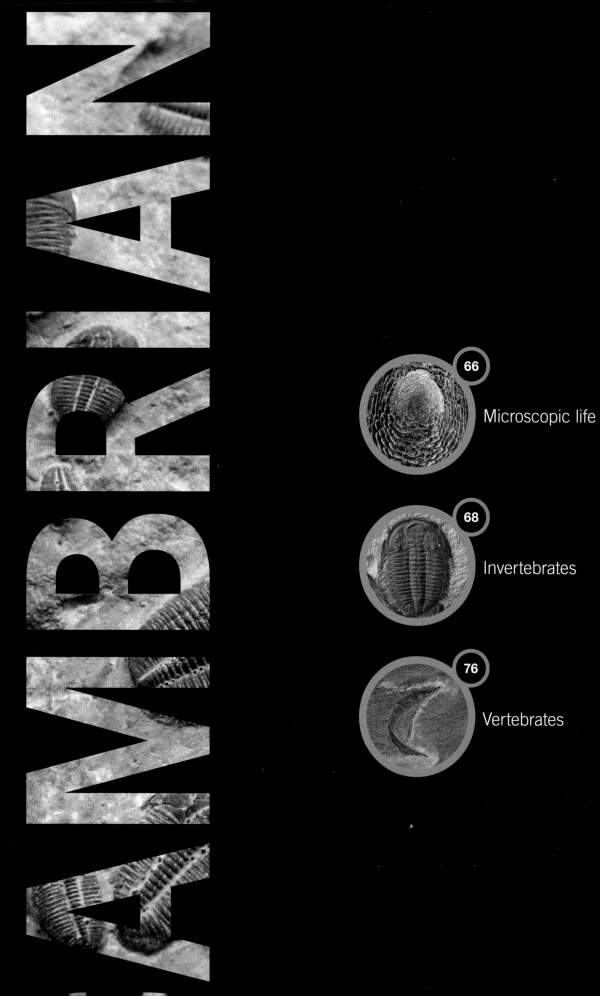

CAMBRIAN

Cambrian

The Cambrian Period, between 541 and 485 million years ago, was one of rapid plate movement. Most importantly, there was an extraordinary explosion in animal diversity. Many of these new forms evolved the ability to biomineralize; this probably occurred in response to the evolution of predators. Soft-bodied animals from this period are preserved in localities such as Canada's Burgess Shale and Chenjiang in China.

KAZAKHSTAN TODAY
Kazakhstan, now a large, dry, landlocked country in Central Asia, was a group of microcontinents in Cambrian times. These arid hills lie east of the shrinking Aral Sea.

Oceans and continents

By the beginning of the Cambrian, Pannotia had begun to break up, while another supercontinent, Gondwana, was being assembled. This comprised much of present-day South America, Africa, Madagascar, India, Australia, and Antarctica. At this time, the continents appear to have been moving across the globe at a relatively fast rate of about 12in (30cm) per year. As Pannotia fragmented, the Iapetus Ocean opened and widened between the continents of Laurentia (present-day North America), Baltica (northern Europe), and Siberia. To the north of Laurentia lay the Panthalassic Ocean. During the Early and Middle Cambrian, Laurentia drifted rapidly from polar to tropical latitudes so that it straddled the Equator. To the east of the other continents, Gondwana—named after the Gondwana region of central northern India—the largest land mass, stretched from the South Pole to the Equator. This rotated through 90 degrees, driven by movement of the Earth's lithospheric plates. Other smaller continental blocks existed, such as Kazakhstan and China, and part of present-day Southeast Asia. Most were of relatively low relief and were surrounded by extensive shallow seas. During periods of sea level rise, continents were inundated, apart from Gondwana, eastern Siberia, and Kazakhstan, which were mountainous. Life proliferated in warm, shallow-water marine environments. Complex animals with mineralized skeletons evolved, such as sponges, brachiopods, corals, mollusks, echinoderms, and arthropods. Changes in ocean chemistry, greater diversity of plankton species at the base of the food chain, and pressure from predators may all have contributed to the Cambrian Explosion (see p.68).

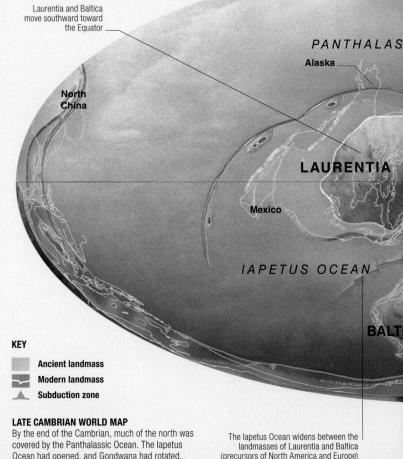

Laurentia and Baltica move southward toward the Equator

PANTHALAS...
Alaska
North China
LAURENTIA
Mexico
IAPETUS OCEAN
BALT...

KEY
- Ancient landmass
- Modern landmass
- Subduction zone

LATE CAMBRIAN WORLD MAP
By the end of the Cambrian, much of the north was covered by the Panthalassic Ocean. The Iapetus Ocean had opened, and Gondwana had rotated.

The Iapetus Ocean widens between the landmasses of Laurentia and Baltica (precursors of North America and Europe)

APPALACHIAN MOUNTAINS
Sediments that consolidated into rocks uplifted in the Appalachians in North America were deposited near the start of the Cambrian Period (542 million years ago) on the shores of the new Iapetus Ocean.

TERRENEUVIAN

SERIES

| MYA | 541 | 530 | 520 |

MICROSCOPIC LIFE

INVERTEBRATES

- **540** First small shelly fossils: tommotiids, halkieriids, wiwaxids
- **535** First trilobites (e.g. below); other arthropods: bradoriids, brachiopods (orthiids), crustaceans: mollusks (monoplacophorans, bivalves), echinoderms, cephalochordates, conularids, lobopodians, priapulids, nematodes, chelicerates, rostroconchs, onychophorans (marine), foraminifers, radiolarians
- **530** First rhombiferans and first edrioasteroids (extinct echinoderms); first arthropod trace fossils on the seafloor
- **525** First graptolites (e.g. *Dictyonema*); Chengjiang biota
- **520** First pentamerid brachiopods (lamp shells)

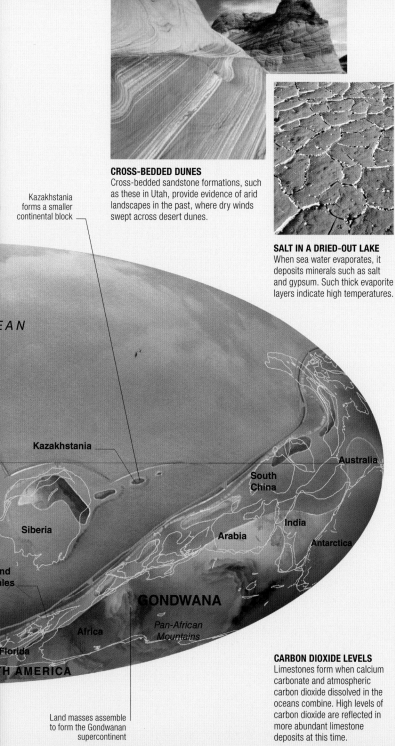

CROSS-BEDDED DUNES
Cross-bedded sandstone formations, such as these in Utah, provide evidence of arid landscapes in the past, where dry winds swept across desert dunes.

SALT IN A DRIED-OUT LAKE
When sea water evaporates, it deposits minerals such as salt and gypsum. Such thick evaporite layers indicate high temperatures.

Kazakhstania forms a smaller continental block

Kazakhstania

Australia

South China

India

Arabia

Antarctica

Siberia

GONDWANA

Pan-African Mountains

Africa

Florida

TH AMERICA

Land masses assemble to form the Gondwanan supercontinent

Climate

Detailed knowledge about Earth's climate during the Cambrian Period is rather limited. Carbon dioxide levels were probably high, up to 15 times present-day levels, and there were huge accumulations of evaporites. Both factors provide evidence for high global temperatures. One estimate suggests that by the Late Cambrian these averaged 122–140°F (50–60°C). Many low- lying continents experienced arid conditions produced by wind blowing from the land and low rainfall. Few indicators for hot, humid climates have been found in the Cambrian geological record, although there are examples in Laurentia of laterite and bauxite, known to form in the tropics in these conditions. Instead, particularly along the western margin of the Asian blocks, appropriate rainfall, topographic, and wind conditions existed for the development of arid climates. Dry and humid regions alike were located in low latitudes near the Equator. There were no continents located directly over the poles, and oceans covered a larger part of the globe (about 85 percent, compared with about 70 percent today). There would have been less variation than today in temperature from the poles to the tropics— conditions were mainly tropical and subtropical, and glaciations did not occur. Evidence for this view is provided by the presence of limestone deposits up to a latitude of 50 degrees.

BURGESS SHALE
The Burgess Shale area, in the Canadian Rocky Mountains, was part of ancient Laurentia. The spectacular Cambrian fauna preserved here would have lived in warm seas just south of the Equator.

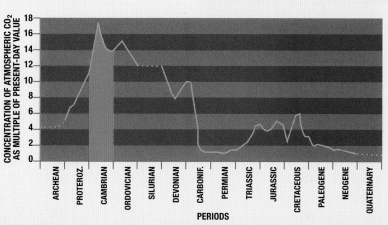

CARBON DIOXIDE LEVELS
Limestones form when calcium carbonate and atmospheric carbon dioxide dissolved in the oceans combine. High levels of carbon dioxide are reflected in more abundant limestone deposits at this time.

CONCENTRATION OF ATMOSPHERIC CO₂ AS MULTIPLE OF PRESENT-DAY VALUE

18 16 14 12 10 8 6 4 2 0

ARCHEAN | PROTEROZ. | CAMBRIAN | ORDOVICIAN | SILURIAN | DEVONIAN | CARBONIF. | PERMIAN | TRIASSIC | JURASSIC | CRETACEOUS | PALEOGENE | NEOGENE | QUATERNARY

PERIODS

SERIES 3

FURONGIAN

510

500

490

MYA

MICROSCOPIC LIFE

● 510 First cephalopods (nautiloids) and chitons (polyplacophorans)

● 505 Fossilization of Burgess Shale (e.g. *Marrella* and *Wiwaxia*)

● 500 First conodonts, archaeogastropods (sea snails); first bellerophontine gastropods, and ostracods (tiny crustaceans); first arthropod trace fossils on land

MARRELLA

WIWAXIA

INVERTEBRATES

VERTEBRATES

CAMBRIAN
MICROSCOPIC LIFE

The beginning of the Cambrian witnessed an incredible event in evolution—in about 20 million years, animal groups, from worms to fishes, appeared. Microscopic remains of these creatures have been remarkably well preserved. These small, shelly fossils are evidence of this avalanche of evolution.

Before the Cambrian, most animals lacked either jaws or a through-gut, which meant they had no anus. The evolution of chewing, and then of predation, started an arms race that rapidly transformed ecosystems around the world.

Jaws and protection

Among the first toothlike elements to appear were those of *Protohertzina*, belonging to a creature similar to a modern arrow worm. External skeletons quickly evolved for protection. The tiny, cap-shaped shells of *Maikhanella*, for example, were made from clusters of spines called sclerites. Later, other animals, such as brachiopods, merged their sclerites to form a single, solid shell. Fossils of soft tissues, including the cyanobacteria and other algae, can be found in Cambrian phosphates, but after the Cambrian, such remarkable preservation became rare, possibly because more scavenging was taking place on the sea floor.

The Cambrian cascade

A multitude of new, small organisms, collectively known as plankton, began to drift freely through the Cambrian seas and oceans, transforming the food chain. The great oceans provided a valuable means of dispersal, as well as a degree of protection from grazers and predators. Radiolarian zooplankton, with their delicate silica skeletons, appeared and fed on other small, drifting organisms. These zooplankton began to form the biological silica deposits called radiolarian cherts. Diverse types of phytoplankton also thrived, using sunlight to make their own food by means of photosynthesis. Many secreted a protective armor called resting cysts and fell to the sea floor, where they were readily preserved. All of these creatures provided an abundant supply of food for the new types of animals that appeared during the Cambrian Explosion.

CAMBRIAN OUTCROP
This red limestone outcrop in Siberia contains a multitude of microfossil remains from the huge number of new species that appeared during the Cambrian Explosion.

GROUPS OVERVIEW

The microfossil record from this period shows a world that was increasingly influenced by multicellular animals with a through-gut system. These organisms are called bilaterians. Their earliest remains can often be found as microfossils when the Cambrian limestones in which they are preserved are digested in weak acids in a laboratory.

PROTOCONODONTS

The tiny jaws of *Protohertzina* are similar to those of some modern, predatory arrow worms. They appeared all over the world at the very start of the Cambrian. It is possible that these fearsome jaws caused other creatures to build protective shells or burrow into the sea floor, starting a kind of "arms race" that turned into the Cambrian Explosion.

SMALL SHELLY FOSSILS

The earliest protective skeletons were often tubular, like tiny drinking straws. It can be difficult to tell what species these remains come from, but it seems likely they were built by creatures that were related to modern coral polyps, marine fan worms, or even deep sea vent worms. They flourished on the sea floor until the trilobites and other scavengers evolved.

MICROMOLLUSKS

The sclerites of enigmatic multielement skeletons and tiny cap-shaped mollusks became abundant soon after the start of the Cambrian period. They created a huge variety of shell shapes, and while they can look like clams or garden snails, evidence suggests that these tiny mollusks lacked many of the features seen in mollusks of the later Paleozoic.

MICROARTHROPODS

Arthropods diversified explosively, leaving a trail of phosphatized remains, sometimes including legs, gills, and soft parts. Scanning electron microscope images have revealed the rapid evolution of a variety of limbs in Cambrian arthropods. Later forms evolved specialized appendages, like those of modern spider crabs and lobsters.

Protohertzina

GROUP Chaetognathans
DATE Early Cambrian
SIZE 1/16in (2mm) long
LOCATION Worldwide

This microfossil appears near the base of Cambrian rocks. Each toothlike "protoconodont" is made of calcium phosphate, and has a hollow basal portion for attachment. It is shield-shaped in cross-section and has a long, curved spine. Occasionally, *Protohertzina* are found in clusters that closely resemble the portcullislike jaws of modern arrowworms. This suggests that this was an early predator.

sharp tip

strengthening rib

EARLY WEAPON
It has been suggested that the earliest predators, such as *Protohertzina*, used clusters of spines, like this one, to tear apart their prey.

Anabarites

GROUP Possible Cnidarian
DATE Early Cambrian
SIZE 3/16in (5mm) long
LOCATION Worldwide

This tubular fossil had a calcareous shell made of aragonite. In cross-section, it is a distinctive trilobed, cloverleaf shape. Like many small, shelly fossils from the early Cambrian, its biological affinities remain uncertain, but it could have been close to the cnidarian groups of jellyfish and corals. It may have lived in colonies, embedded in the sediment, and feeding on organic matter in the water column.

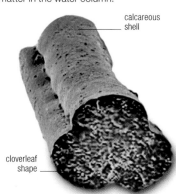

calcareous shell

cloverleaf shape

UNUSUAL SHAPE
Shells of *Anabarites* have a curious trilobed cross-section, which is quite unlike anything alive today. Calcium-phosphate-filled microfossils like this are found at the very start of the Cambrian.

Maikhanella

GROUP Mollusks
DATE Early Cambrian
SIZE 1/16in (2mm) long
LOCATION Worldwide

This cap-shaped fossil had a weakly mineralized shell that was phosphatized after its death. The pineapplelike shell was made from tightly packed spines called sclerites. At first, as seen in this example, these sclerites were not strongly fused together. Sometimes they are found separately, and are called *Siphogonuchites* or *Halkieria*. After several million years of evolution, the spines became more strongly fused into a single, rigid shell that was similar in shape to a modern limpet. In Greenland, examples have been found where the sluglike owner had been carrying two *Maikhanella*-like caps, surrounded by a skirt of *Siphogonuchites*-like sclerites. This complex arrangement is a reminder that much experimentation was still taking place at this time. *Maikhanella* probably grazed on algae that grew on the Cambrian sea floor.

SNAIL ANCESTOR
Maikhanella is one of the earliest mollusks. It is thought that modern snails and cuttlefish evolved from forms like the one shown here, which is just 1/16in (1mm) across.

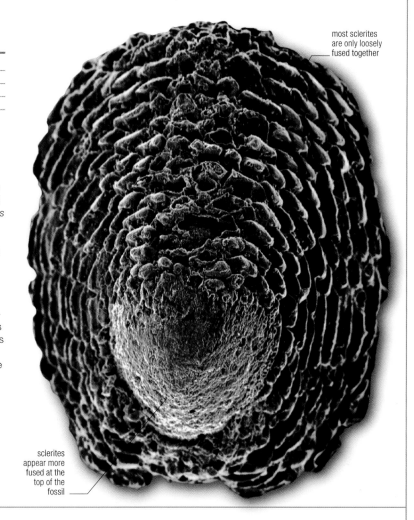

most sclerites are only loosely fused together

sclerites appear more fused at the top of the fossil

Aldanella

GROUP Mollusks
DATE Early Cambrian
SIZE 1/8in (3mm) wide
LOCATION Worldwide

Modern-looking mollusk shells evolved with surprising rapidity during the course of the early Cambrian period. This resulted in coiled forms, such as the one shown here, which appeared across a wide area from Canada to China. It occurred alongside other mollusks that crudely resemble much younger bivalves and even coiled cephalopods. Although little *Aldanella* looks similar to a modern garden snail, there is no evidence that it was closely related to Gastropoda. The owner of this shell probably grazed on detritus near the surface of the Cambrian sea floor.

OXFORD FIND
This *Aldanella* shell was fossilized around 530 million years ago. It was discovered in a borehole in Oxford, England when its discoverers were searching for coal.

Orsten arthropod

GROUP Arthropoda
DATE Late Cambrian
SIZE 1/8in (3mm) long
LOCATION Sweden, China

Although the Cambrian is known as the age of trilobites, many other arthropod groups also existed. Some have been remarkably well preserved within phosphatic nodules. When these nodules are broken up and immersed in weak acids, remarkable fossils, seldom found in younger rocks, are released. Examination under a scanning electron microscope can reveal the details of their tiny legs and featherlike gills. This little ostracodlike fossil probably fed on the souplike layer that had settled on the dark, cold sea floor.

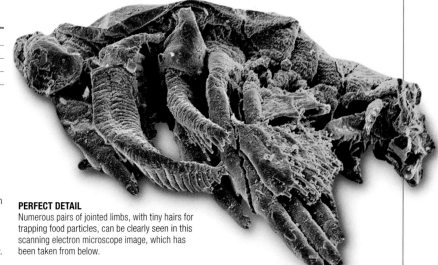

PERFECT DETAIL
Numerous pairs of jointed limbs, with tiny hairs for trapping food particles, can be clearly seen in this scanning electron microscope image, which has been taken from below.

CAMBRIAN
INVERTEBRATES

The Cambrian was a critical time in evolutionary history because diverse marine invertebrates began to appear. Some had eyes and strong jaws that enabled them to live as the first active predators. Other Cambrian evolutionary fauna, such as trilobites, flourished, but declined from the Ordovician onward.

The "explosion" of life

The Cambrian Explosion was the first truly significant evolutionary event in the history of life, heralding the origin and proliferation of hard-shelled marine invertebrates. First, following the extinction of the Ediacaran fauna (see p.61) about 543 million years ago, came the "small, shelly fossils," which are represented by phosphatic horns and coils, plates, and tubes. These persisted until around 525 million years ago, when the first trilobites, inarticulate brachiopods, and other hard-shelled fossils started to appear. But these represent a mere fraction of the vast diversity of Cambrian life. The Middle Cambrian Burgess Shale fauna, discovered in British Columbia, Canada, in 1909 (see p.72), and the Lower Cambrian Chengjiang fauna, found in Yunnan, China, in 1984, testify to a plethora of body plans, especially in arthropods and "lobopods," represented today by the terrestrial velvet worm *Peripatus*, and rarely preserved as fossils.

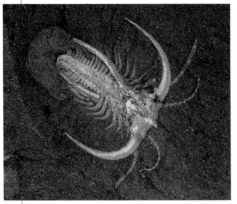

BURGESS SHALE ARTHROPOD
This small arthropod, *Marrella splendens*, is from the Burgess Shale. It is just one of a large number of different arthropods that have been found there.

SHELLY REMAINS
These small, shelly fossils are hyolithids, which formed an integral part of the Cambrian evolutionary fauna, and are often found in great numbers.

Diversity in the Cambrian seas

The Burgess Shale and the Chengjiang faunas were dominated by a diverse range of marine invertebrates with soft bodies or unmineralized shells. They lived alongside the more familiar trilobites, which were common at the time.

The Cambrian seas would have abounded with these soft-bodied animals, but their chances of preservation would normally be slim. By the Late Cambrian crustaceans and other faunas had also started to appear.

MIDDLE CAMBRIAN TRILOBITE
This trilobite, *Ellipsocephalus hoffi*, was found in the Czech Republic. Large numbers of this species are often found grouped together.

GROUPS OVERVIEW

Many marine invertebrate groups first appeared in the Cambrian. The diverse fauna included early forms of sponges that did not survive to the end of the period and brachiopod forms that are, in contrast, still living today. The Cambrian saw the earliest forms of trilobites evolve at the beginning of the period, and echinoderms were also abundant.

ARCHAEOCYATHIDS

These extinct fossil sponges were only in existence during the Early and Middle Cambrian. They were regularly conical in form, with two perforated cones (one inside the other) that were connected by partitions and originally lined with feeding cells. They often formed small reefs.

INARTICULATE BRACHIOPODS

These brachiopods, some forms of which are still living today, have a two-valved shell and a muscular stalk, or pedicle. They are suspension-feeders, filtering out particles of food from the water surrounding them. Some genera, such as the "living fossil" *Lingula*, have hardly changed in 500 million years.

ECHINODERMS

Cambrian echinoderms do not appear to have closely resembled the echinoids, or crinoids, that are still in existence today, yet the structure of their calcitic plates was identical. Echinoderms were an important component of the Cambrian evolutionary fauna.

TRILOBITES

These now extinct creatures first arose in the Early Cambrian and persisted through to the Permian. They were arthropods with a typical three-lobed calcitic shell and compound eyes; the legs and gills were very delicate and were preserved in just a small number of species.

Archaeocyathids

GROUP Archaeocyathids
DATE Early to Middle Cambrian
SIZE 2–12in (5–30cm) long
LOCATION Antarctica, Australia, North America, S. Europe, Russia

Archaeocyathids were the first reef-building organisms. They were confined to tropical areas and were only in existence for a relatively short period, but they diversified rapidly during this time. Some forms were solitary, while others were colonial. They were made up of two calcite cones, arranged one inside the other with a space between them. The cones were held apart by calcitic septa, or partitions, that crossed the space. The outer cone attached to the seabed by a rootlike structure at its base.

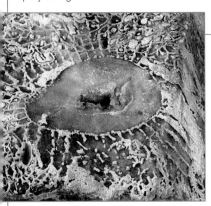

central cavity

septum

ARCHAEOCYATHID
These spongelike organisms were composed of two downward-facing cones that were both perforated. They relied on the flow of water to transport food and oxygen to them, in contrast to some sponges that can generate their own water currents.

Ottoia

GROUP Worms
DATE Middle Cambrian to present
SIZE 1½–3¼in (4–8cm) long
LOCATION Canada

Ottoia is one of the fossils found in Canada's Burgess Shale that can be associated with a living animal group—the marine priapulid worms. *Ottoia* is the most common Burgess Shale priapulid worm, with around 1,500 known specimens. It gathered food with a tubelike organ called a proboscis, which was equipped with spiny hooks and could be turned inside out. Many fossilized specimens have a strongly curved body, which has led to suggestions that, like its modern counterparts, *Ottoia* lived in a U-shaped burrow and extended its proboscis to catch prey. Several specimens have been found with their gut contents intact, which include small, shelly hyolithids and even members of their own species, which may indicate that *Ottoia* was cannibalistic.

MUDDY TOMBS
Ottoia is thought to have lived passively in its burrow. Mudslides may have carried them into deep water, where large numbers were ultimately entombed.

Echmatocrinus

GROUP Possible Echinoderm or Anthozoan
DATE Middle Cambrian
SIZE 1in (2.5cm) wide, below the tentacles
LOCATION Canada

Echmatocrinus is an unusual fossil, found only in the Burgess Shale formation in Canada. The surface of its long, conical body was covered with thin, polygonal plates or scales that were arranged irregularly. There were seven to nine plated arms or tentacles that were attached to the top of the body. Long, thin, nonmineralized branches occurred on alternate sides of these arms. *Echmatocrinus* has been difficult to classify. When it was first described it was thought to be a crinoid (sea lily), but the absence of distinctive echinoderm features, such as five-fold symmetry, has led to other interpretations. Some paleontologists suggest it could actually be an octocoral (a subclass of corals).

Echmatocrinus **has been difficult to classify. When it was first described, it was thought to be a crinoid, but it could actually be an octocoral.**

BRANCHING OUT
Echmatocrinus had a long, conical body with a number of tentacles that were covered with thin branches. Its body tapered toward the bottom, where it would have attached itself to the substrate.

Small Shelly Fossils

GROUP Small Shelly Fossils
DATE Early Cambrian
SIZE Up to ³⁄₈in (1cm) wide
LOCATION Worldwide

Small Shelly Fossils is the name given to a wide variety of different fossils, including closely packed groups of small shells that occur in some of the earliest Cambrian rocks in many parts of the world. Sometimes they form beds about ³⁄₄in (2cm) thick, where the shells had been swept together by seabed currents. While some of these fossils are entire organisms, many are small parts of larger ones, such as plates or spines belonging to mollusks or several enigmatic groups of extinct marine invertebrates.

hyolithids

TINY TUBES
This specimen shows a number of small, flattened, conical tubes, known as hyolithids. The dark coloration toward the top has been caused by an abundance of the mineral phosphate.

Lingulella

GROUP Brachiopods
DATE Early Cambrian to Middle Ordovician
SIZE 3/8–1in (1–2.5cm) long
LOCATION Worldwide

Lingulella was an inarticulate brachiopod, meaning that it used muscles to hold its valves (shells) in place, rather than teeth or sockets. It was attached to the substrate by a fleshy stalk that emerged from an opening in its pedicle (lower) valve. *Lingulella*'s shell was elongated and the beak area was pointed. The surface of the shell had fine growth lines and fossils show delicate radial striation on the inner layers.

fossilized substrate — pointed beak

VERTICAL BURROW
Like its modern-day relatives, *Lingulella* probably lived in vertical burrows. This specimen became fossilized in situ.

LINGULA

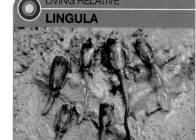

Lingula is a modern brachiopod whose shell shape can be traced back to the Cambrian. Unlike most brachiopods, it lives buried in sediment on the seabed. It is anchored by a long, fleshy pedicle that emerges between its two valves; it has no special opening like *Lingulella*. *Lingula* is adaptable and can reanchor itself if disturbed by current activity.

Bohemiella

GROUP Brachiopods
DATE Middle Cambrian
SIZE 3/8–3/4in (1–2cm) long
LOCATION Czech Republic, Australia, Morocco

Bohemiella was an orthid brachiopod, a group of articulate brachiopods that lived throughout the Paleozoic era. Its valves (shells) were elongated in a crossways direction. The pedicle (lower) valve was less concave than the brachial (upper) valve—or sometimes almost flat. There was a small pair of teeth on the pedicle valve and corresponding sockets in the brachial valve. These had long projections on their inside edges to hold the teeth in position.

INNER VALVE
This inside mold of *Bohemiella*'s brachial valve shows scars where the shell's opening muscles were attached in life.

brachial valve

Wiwaxia

GROUP Mollusks or worms
DATE Middle Cambrian
SIZE 1 1/4–2in (3–5cm) long
LOCATION Canada

Wiwaxia was a striking animal with a body that was symmetrical about its length, elliptical viewed from the top, and squarish in cross-section. Its upper surface was covered with overlapping rows of protective, armorlike plates, called sclerites, and two rows of long spines. Its lower surface lacked any form of protection. *Wiwaxia*'s mouth contained a food-gathering apparatus with two or three rows of rear-facing, conical teeth. These were probably used for scraping algae or bacteria from the seabed, or for collecting food particles from the surrounding water. Because of the similarities between this apparatus and the radula of mollusks, it is thought that *Wiwaxia* may be related to mollusks. An alternative hypothesis suggests that it is related to worms.

STRIKING SPINES
Although only up to 2in (5cm) long, *Wiwaxia* was a striking animal with an armor-plated body and upwardly projecting spines. Its mouth was located on the underside of its body.

MYSTERIOUS FOSSIL
Wiwaxia is a mysterious fossil. Found in the Burgess Shale fossil bed, it dates to about 500 million years ago. Its plates and spines were not mineralized and appear to have been fibrous in texture.

Pojetaia

GROUP Bivalves
DATE Early Cambrian
SIZE 1/16in (1–2mm) long
LOCATION Australia, Canada, China, Greenland, Russia, Turkey, USA

Early Cambrian bivalves are not only very rare—only two genera are known—but also very small. *Pojetaia* is currently the oldest bivalve ever discovered. Its shell was almost circular with well-defined beaks on each valve (shell) and a straight hinge-line. The outside of the shell had growth increments and some faint radial ribs. Internally, the shell had two adductor muscles that opened and closed the valves. The hinge had five or six teeth, and sockets in each valve helped the valves align.

SHALE FOSSIL
Helicoplacus fossils have been found most commonly in shales.

Helicoplacus

GROUP Echinoderms
DATE Lower Cambrian
SIZE 1–1 1/2in (2.5–4cm) long
LOCATION USA

Helicoplacus was a bizarre echinoderm that represents a very early, and ultimately unsuccessful, body plan. Unlike other echinoderms, it lacked radial symmetry of any kind. Tiny plates were arranged spirally around its shell, which in its resting state was pear-shaped. In fossils, the shell plates are usually found separated, suggesting that they were not fused together. Instead, the body had the capacity to expand and contract, and during expansion the plates separated.

EXPANDING BODY

Helicoplacus's plates were not rigidly fused, so they may have allowed its body to expand. Recent ideas suggest that the sharper end of its body was partly buried in the seabed, so expansion of the body may have assisted in feeding or respiration.

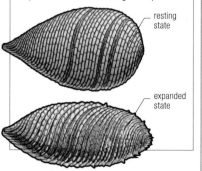

resting state

expanded state

Hallucigenia

GROUP Lobopodians
DATE Middle Cambrian
SIZE Up to 1in (2.5cm) long
LOCATION Canada, China

In the 1970s, the first reconstructions of *Hallucigenia* showed the animal walking on rigid, stiltlike, paired "limbs," with a row of fleshy projections along the top of its back. However, later discoveries showed that earlier reconstructions had turned the organism upside down. *Hallucigenia* had an elongated body, with a rounded "head" at one end, and a long, fleshy "tail." It had seven pairs of stiff, pointed spines along the upper side of its body and a cluster of small projections close to its rear end. Lobopodians are not true arthropods, as their limbs are not jointed, but they are thought to be closely related.

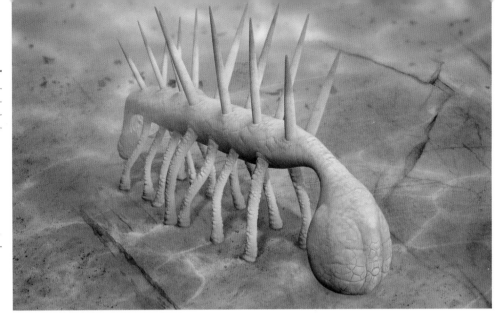

VELVET WORM
Hallucigenia can be recognized as a velvet worm, of which there are living relatives in today's temperate and tropical rain forests.

Opabinia

GROUP Arthropods
DATE Middle Cambrian
SIZE About 2½in (6.5cm) long
LOCATION Canada

Opabinia is one of the strangest animals to have been found in Canada's Burgess Shale fossil bed—unlike any other. Its head had five prominent eyes—two sets of pairs, and one eye that was central. Extending from the front of its head was a long, flexible, trunklike feature, or proboscis, which ended in a pod-shaped organ bearing small spines that were probably used to grasp prey. *Opabinia* would have used its proboscis to pass its prey up to its mouth. Its elongated body was composed of 16 segments, each of which possessed a flaplike lateral lobe with gills on the underside. Its tail had three flaps that projected from either side.

BIZARRE APPEARANCE
Opabinia's form was so bizarre that when a reconstruction was presented to the Paleontological Association at its meeting in Oxford, England, in December 1972, its appearance provoked laughter.

Marrella

GROUP Arthropods

DATE Middle Cambrian

SIZE Up to ¾in (2cm) long

LOCATION Canada

Marrella is the most abundant Burgess Shale fossil (see below), with over 25,000 known specimens. Curiously, this Canadian site is the only place it has been found in the world. *Marrella* was given the informal name "lace crab" by the American paleontologist Charles Walcott (see p.75) due to its "feathery" appearance. It had two distinctive pairs of large, backward-facing spines, one pair running alongside the body, the other above it. Two pairs of antennae arose from the front of its body, one very long, the other shorter and stouter. The body was composed of 20 segments, each of which had pairs of identical limbs, and a gill branch. The identical limbs suggest that it was a primitive arthropod.

MARRELLA SPLENDENS

The headshield and two sets of backward-facing spines are clearly visible on the above specimen. The discovery is particularly significant because it may be a descendent of a common ancestor of the crustaceans, trilobites, and chelicerates.

SEABED SCAVENGER

Marrella is thought to have swum along the seabed or just above it, feeding on tiny particles of organic material. Its jointed legs had feathery gill branches that formed part of its respiratory system. The teethlike projections on the inner pair of spines are evident in this illustration.

 KEY SITE

BURGESS SHALE

The 510 million-year-old Burgess Shale in British Columbia, Canada, is one of the world's most significant sources of fossils, as it contains well-preserved soft-bodied organisms—a very rare occurrence. It provides a unique window into the full diversity of Cambrian life, with a wealth of different types of fossils. The Burgess fauna provides important evidence for the "Cambrian Explosion," which was a rapid diversification of marine life (see p.68).

Olenellus

GROUP Arthropods
DATE Middle Cambrian
SIZE Up to 2¼in (6cm) long
LOCATION North America, Greenland, Mexico

Olenellus was a trilobite with a semicircular headshield and a tapering glabella (central head bulge) with four pairs of backward-pointing furrows. Its eyes were crescent shaped, and their front ends merged with the glabella's frontal lobe. Its thorax had 18 segments, and the third was wider and longer than the others, and ended in long spines.

OLENELLUS THOMPSONI
Although its head was well-developed and its eyes large and crescent shaped, *Olenellus* possessed only a small, poorly formed tail.

Olenus

GROUP Arthropods
DATE Late Cambrian
SIZE Up to 1½in (4cm) long
LOCATION British Isles, Norway, Sweden, Denmark, Newfoundland, Texas, South Korea, Australia, Kazakhstan

Olenus is a trilobite fossil commonly found in dark mudstones, which were deposited on the seafloor in environments with low oxygen levels. It had up to 15 thoracic segments, with very wide pleural or side lobes. These are thought to have supported extended gills, which would have helped the animal absorb the maximum amount of oxygen possible in such an environment. Evidence also suggests that *Olenus* and its relatives may have developed a symbiotic relationship with sulfate-reducing bacteria, either by feeding on them directly or by absorbing nutrients directly from them.

OLENUS GIBBOSUS
Olenus had a rounded headshield with an oblong glabella (central, raised part) and small, crescent-shaped eyes. Its tail was much smaller than its headshield.

Ellipsocephalus

GROUP Arthropods
DATE Middle Cambrian
SIZE Up to 1½in (4cm) long
LOCATION Sweden, Czech Republic, Morocco, Canada, Poland, Norway

This trilobite fossil is commonly found in large clusters in Cambrian rocks in the Prague district of the Czech Republic. Many specimens are complete but lack free cheeks, which suggests that they were molts, and that the animals congregated to shed their old exoskeletons in order to grow in size. *Ellipsocephalus*'s headshield

had a prominent, smooth glabella with slightly concave sides. Like *Olenus* (above), its eyes were small and crescent shaped. The border of the headshield was narrow, defined by a rather narrow, shallow furrow.

ELLIPSOCEPHALUS HOFFI
Ellipsocephalus had a thorax divided into 12 well-defined segments.

Paradoxides

GROUP Arthropods
DATE Middle Cambrian
SIZE Up to 17½in (45cm) long
LOCATION Europe, North America, British Isles, Morocco, Turkey, Siberia

Paradoxides was one of the largest known genera of trilobites, and it was probably a predator high in the Cambrian food chain. Underneath the widest part of the glabella lay the hypostome, a large, platelike structure supporting the stomach whose size and shape suggests predatory activity.

PARADOXIDES PARADOXISSIMUS GRACILIS
Paradoxides's thorax consisted of 18 to 21 segments, all of which ended in long spines, with those at the tail end longer than the others.

Elrathia

GROUP Arthropods
DATE Middle Cambrian
SIZE Up to 1¾in (4.5cm) long
LOCATION W. USA, Canada, Greenland

ELRATHIA KINGII
Elrathia's exoskeleton is usually flattened due to its preservation in shale.

Elrathia is one of the best-known trilobites in North America. Its headshield was semicircular, with a short, conical glabella and two pairs of short, shallow furrows. The crescent-shaped eyes were set some distance from the glabella, near the front. There were up to 14 segments in its thorax, and its tailshield was much smaller than its head. The Pahvant Ute Indians of Utah used *Elrathia kingii* fossils in amulets to ward off evil.

Tomagnostus

GROUP Arthropods
DATE Middle Cambrian
SIZE Up to ¾in (2cm) long
LOCATION Sweden, Denmark, British Isles, Czech Republic, Newfoundland, E. USA, Siberia, Australia, Greenland

Tomagnostus is one of many agnostid trilobites with an almost global distribution, and is thus valuable in correlating Cambrian rocks across wide areas. It probably lived in the open ocean, but it occurs as a fossil in association with more "local" trilobites in different regions. Agnostids were highly specialized and distinctive members of the trilobite group.

TOMAGNOSTUS FISSUS
The head- and tailshields of *Tomagnostus* were almost equal in size and shape.

Anomalocaris

GROUP Arthropods
DATE Middle Cambrian
SIZE Up to 3ft 3in (1m) long
LOCATION Canada, S. China, USA, Australia

Anomalocaris was the largest animal in the Cambrian ecosystem that became fossilized in what is now the Burgess Shale in British Columbia, Canada. The animal's head had two eyes, in front of which was a pair of segmented, downward-curving appendages. Each of the segments of these appendages had a pair of spiny projections on its underside. *Anomalocaris*'s mouth was situated on the underside of the head and consisted of a circle of elongated plates. Behind each eye were three small flaps. The body was divided into eight segments, each of which had side flaps, and the small tail featured an upturned fan of flaps. Fragments of *Anomalocaris* were originally thought to belong to separate animals. The front appendages were interpreted as the segmented abdomen of a shrimplike crustacean, while the circular mouthparts were thought to be a jellyfish.

3ft 3in The **length** of the largest specimens of *Anomalocaris*. Its **size** likely made it a **top predator** of the Cambrian seas.

TOP PREDATOR?

Although it was probably not the swiftest of swimmers, many scientists believe that *Anomalocaris* was at the top of the food chain in the Cambrian seas. However, despite its large size, there is debate about whether its nonmineralized mouthparts were strong enough to break through the tough exoskeletons of animals such as trilobites. Instead, it may have used its circular mouth to suck in soft-bodied prey.

BIOGRAPHY
CHARLES WALCOTT

American invertebrate paleontologist Charles Doolittle Walcott (1850–1927, center) was an enthusiastic fossil collector as a boy in New York. With little formal education, he forged a highly successful career as secretary of the Smithsonian Institution and director of the US Geological Survey. His most famous discovery, in 1909, was of superbly preserved fossils of Cambrian animals, including *Anomalocaris*, in the Burgess Shale in the Canadian Rockies.

ANOMALOCARIS SP.
Viewing this specimen with its segmented body, it is easy to see why scientists thought *Anomalocaris* was a type of shrimp or crustacean when it was discovered in the Canadian Rockies. In fact, its genus name means "anomalous shrimp," so called because its appendages were unsegmented, unlike in true shrimp.

CAMBRIAN VERTEBRATES

The Early Cambrian was a crucial period in the history of life on Earth—witnessing an explosion in animal diversity. About 540 million years ago, the seas were home to the first wave of new animal groups, some of which would eventually die out, but others would lead to the evolution of the first vertebrates.

The start of the Cambrian is marked by the appearance of a vast array of animal forms in marine sedimentary rocks that are unknown in older rocks. This explosion in diversity saw the arrival of most of the major groups of animals or animal body plans. They included the first chordates—animals that at some point in their life cycle possess a notochord, the precursor of a backbone—and the first vertebrates.

The first vertebrates

Vertebrates are a group of animals that possess a backbone (vertebral column), which surrounds the notochord and provides support for the body. They also have a skull (cranium) that encloses the brain, eyes, olfactory organs, and internal ears. Another important characteristic of vertebrates is that parts of the head, gill arches, and nerves form from neural crest cells.

NONVERTEBRATE AND VERTEBRATE
These two schematic diagrams compare the basic body plans of a nonvertebrate chordate and a hypothetical early vertebrate. Note the presence of a brain protected by a cranium, sense organs, and gill arches in the latter.

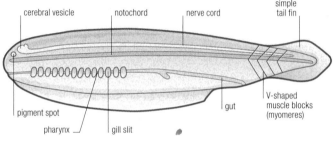

cerebral vesicle — notochord — nerve cord — simple tail fin
pigment spot — pharynx — gill slit — gut — V-shaped muscle blocks (myomeres)

EARLY NONVERTEBRATE CHORDATE

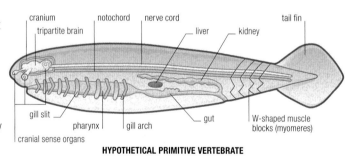

cranium — tripartite brain — notochord — nerve cord — liver — kidney — tail fin
cranial sense organs — gill slit — pharynx — gill arch — gut — W-shaped muscle blocks (myomeres)

HYPOTHETICAL PRIMITIVE VERTEBRATE

These cells migrate from the nerve cord early in the development of the embryo and travel to different parts of the animal to form these structures. Other vertebrate features include the nervous control of the heart, a set of muscles to control eye movement (extrinsic muscles), at least two semicircular canals in the inner ear, and a lateral line system running along the head and body that has sensory organs called neuromasts. All extinct jawless fish are vertebrates. Of the two living groups, lampreys are considered to be vertebrates, but many authorities exclude hagfish because of their poorly developed vertebral column. However, hagfish do have a cranium.

Until quite recently, it was thought that vertebrates arose in the Ordovician. In 1984, a remarkable Early Cambrian fossil site was discovered in Chengjiang, Yunnan Province, China. Most of the 180-plus species found there are invertebrates, but in 1999, two 530-million-year-old vertebrates were announced to the world (see below and opposite).

(see below and opposite).

GROUPS OVERVIEW

The Cambrian Explosion produced the first cephalochordates and vertebrates, both members of the Phylum Chordata. All chordates have a distinct body plan, at least in their embryonic stages. A stiffening rod, called a notochord, extends dorsally from front to back. Above the notochord lies a hollow nerve cord, and below it lies the gut.

CEPHALOCHORDATES

Cephalochordates retain a notochord even as adults, but they lack a vertebral column and neural crest cells. The first cephalochordate, *Pikaia*, swam in the sea about 530 million years ago. It is very similar to the only living cephalochordates, the lancelets (or *Branchiostoma*; see right), in having gill slits and body muscles arranged in a V-shaped pattern.

MYLLOKUNMINGIDS

The first vertebrates, *Myllokunmingia* and *Haikouichthys*, are often grouped in the same taxon. They are known from over 500 specimens found in the Chengjiang shales. They possess small vertebral elements that represent the first vertebral column (backbone). Other vertebratelike features include a clearly defined head and structures that supported the gill slits.

LANCELETS

Resembling long leaves, lancelets usually lie half-buried in sediment with their heads sticking out as they filter feed. The characteristic V-shaped muscle blocks that run along both sides of the body can be seen through their transparent skin.

Pikaia

GROUP Cephalochordates
DATE Early Cambrian
SIZE 2in (5cm) long
LOCATION Canada

A marine animal that lived about 530 million years ago, *Pikaia* is one of the earliest-known chordates and belongs to the subphylum cephalochordata (see opposite). In appearance, *Pikaia* is similar to the modern lancelet, differing only in having a pair of antennalike structures on its head. *Pikaia* was small and delicate, with a dorsal nerve cord running from front to back under which lay the supporting notochord. V-shaped muscles ran along the sides of its flattened body. A narrow fin membrane extended down the rear two-thirds of the body, broadening into a single tail fin that tapered to a point.

CANADIAN FOSSIL

Pikaia was discovered in the Burgess Shale deposits in British Columbia (see p.72), where it lived alongside a variety of invertebrates.

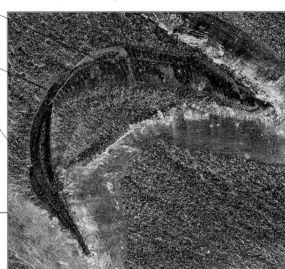

muscle blocks along flattened body

narrow fin membrane

tail tapers to a point

UNLIKELY ANCESTOR

Pikaia may not look very exciting, but it is of great interest to scientists. It clearly shows traces of a notochord, dorsal nerve cord, and muscle blocks— key features in the evolution of all vertebrates.

Myllokunmingia

GROUP Myllokunmingids
DATE Early Cambrian
SIZE ¾–1¼in (2–3cm) long
LOCATION China

The oldest vertebrates by about 30 million years, *Myllokunmingia* and *Haikouichthys* (see below) were tiny, jawless marine fish. *Myllokunmingia* was similar in form to *Haikouichthys* but less slender. It had a distinct head, and its body had backward-facing, V-shaped muscle blocks (myomeres). However, it differed from *Haikouichthys* in having pouchlike structures associated with its five or six gills—*Haikouichthys* had gill bars (and possibly more gills). *Myllokunmingia* fossils suggest that it had a cartilaginous skull and some primitive vertebral elements. Some parts of the digestive system are preserved, but not the mouth, and no tail parts have been found. *Myllokunmingia* had a triangular dorsal fin that inclined gently upward a short distance behind the head. Farther back, on the underside of the body, were fin folds (precursors of fins).

Haikouichthys

GROUP Myllokunmingids
DATE Early Cambrian
SIZE 1in (2.5cm) long
LOCATION China

Haikouichthys is thought to be one of the most primitive fish without jaws (agnathans). Preserved in the 530-million-year-old marine sediments of Chengjiang (see opposite), *Haikouichthys* is unlike any other agnathan. It had a rounded extension to the head which bore sensory organs—the eyes, and possibly nasal sacs and otic capsules associated with hearing. On the sides of its body, it had at least six, and possibly up to nine, gill slits supported by gill bars. V-shaped muscle blocks are also visible in places. *Haikouichthys* had a notochord, but there is also evidence of vertebralike elements similar to those seen in modern lampreys. A number of circular structures on the lower part of the body may have been slime organs because these are also seen in hagfish.

CONTINUOUS FIN

Haikouichthys had a prominent dorsal fin, and there is some evidence that it was supported by radials. It was continuous with the tail fin, which in turn was joined to the narrower ventral fin fold.

530 The age in **millions of years** of *Haikouichthys* and *Myllokunmingia*. They had small **vertebral elements** that represent the **first backbone**.

ORDOVICIAN

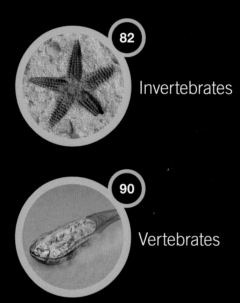

82

Invertebrates

90

Vertebrates

Ordovician

During this period, the nature of marine faunas changed dramatically. Cambrian faunas were replaced by a more diverse assemblage in the mid-Ordovician, and animals, in the form of arthropods, first colonized the land. At the end of the Ordovician, the first of five major mass extinctions of flora and fauna occurred. Trilobites, echinoderms, brachiopods, graptolites, bryozoans, and reef-builders were particularly badly hit.

ARTHROPOD TRACKS
These tracks were made by an early arthropod on wet, rippled sand about 450 million years ago. They are preserved in the Tumblagooda Sandstone, Western Australia.

Oceans and continents

The rapid plate movement and volcanism of the Cambrian Period continued, resulting in a profound reorganization of the positions of the continents and ocean basins. Extensive shallow epicontinental seas with almost flat seabeds surrounded the main landmasses. Into these seas were shed vast amounts of sediment eroded from barren mountain belts still largely devoid of plant life. Rising sea levels and tectonic activity reduced smaller landmasses to a series of archipelagos. Four major continents existed: Gondwana, Laurentia, Baltica, and Siberia. By the Middle Ordovician, Siberia had moved from its Southern Hemisphere position to the Northern Hemisphere, nearer Laurentia (which lay across the Equator). The vast Panthalassic Ocean remained to the north of these landmasses. Gondwana still extended from north of the Equator to the South Pole. During the Ordovician, it rotated counterclockwise, carrying the areas that form today's Australia and part of Antarctica into the Northern Hemisphere.

The small islands of the South China block lay off Gondwana's western margin. In the Paleo-Tethys Ocean, between Gondwana and Laurentia, the smaller landmass of Baltica shifted into lower latitudes. Meanwhile, the Iapetus Ocean continued to widen. The Ordovician landscape was one of mostly barren continents, with frequent volcanic eruptions and earthquakes, and coastlines that were constantly reshaped. The wide, shallow seas were inhabited by extensive coral reefs and a great diversity of marine invertebrates, until many species were wiped out in a mass extinction at the end of the Ordovician.

MOUNT ST. HELENS ERUPTION
Towering columns of ash were thrown up by Mount St. Helens, Washington, when it erupted in 1980. Ordovician volcanism was on an even larger scale, producing the largest ashfalls known.

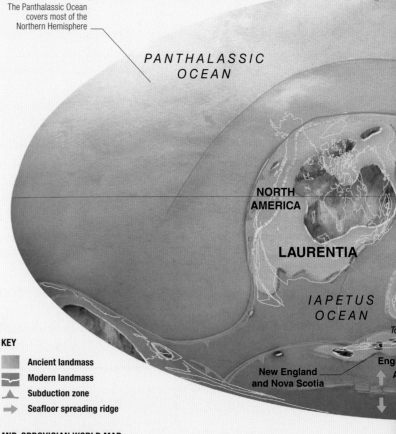

The Panthalassic Ocean covers most of the Northern Hemisphere

PANTHALASSIC OCEAN

NORTH AMERICA

LAURENTIA

IAPETUS OCEAN

New England and Nova Scotia

Engl

KEY

- Ancient landmass
- Modern landmass
- Subduction zone
- Seafloor spreading ridge

MID-ORDOVICIAN WORLD MAP
The movement of continents resulted in intense tectonic activity. Gondwana rotated, Baltica shifted southward, and Siberia moved closer to Laurentia.

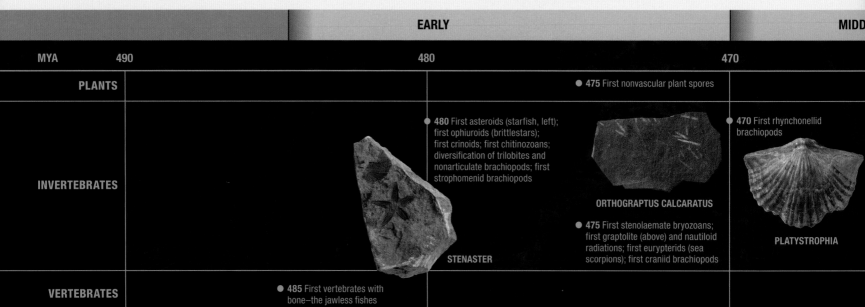

	EARLY		MIDD
MYA	490	480	470
PLANTS			● 475 First nonvascular plant spores
INVERTEBRATES		● 480 First asteroids (starfish, left); first ophiuroids (brittlestars); first crinoids; first chitinozoans; diversification of trilobites and nonarticulate brachiopods; first strophomenid brachiopods **STENASTER**	● 470 First rhynchonellid brachiopods
		ORTHOGRAPTUS CALCARATUS ● 475 First stenolaemate bryozoans; first graptolite (above) and nautiloid radiations; first eurypterids (sea scorpions); first craniid brachiopods	**PLATYSTROPHIA**
VERTEBRATES	● 485 First vertebrates with bone—the jawless fishes		

RED SEA TEMPERATURE
The waters of the Red Sea and the Persian Gulf today can reach a temperature of 108°F (42°C) at their upper limit. Similar temperatures were found in oceans at the beginning of the Ordovician, before they began to cool.

CORAL REEF
Reef-building animals secreted calcium carbonate, building up structures that formed the basis for new ecosystems in Ordovician seas.

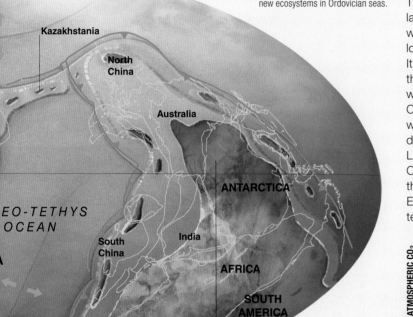

Kazakhstania

North China

Australia

LEO-TETHYS OCEAN

South China

India

ANTARCTICA

AFRICA

SOUTH AMERICA

GONDWANA

Sahara Desert

Gondwana rotates, taking parts of Australia and Antarctica north of the Equator

CARBON DIOXIDE LEVELS
When temperatures were high, carbon dioxide levels would have been considerably higher than today, but they declined when the Ordovician world entered a period of intense cold.

Climate

Like the Cambrian Period that preceded it, the Ordovician was initially very warm, with ocean temperatures of about 108°F (42°C). However, a prolonged period of cooling soon began, with temperatures dropping to 73°F (23°C) by the end of the Early Ordovician. Greenhouse conditions gave way to those of modern equatorial oceans. Temperatures remained reasonably constant for about 25 million years during the Middle to Late Ordovician. This period of relative stability was followed by a period of rapid temperature decline, due to extensive glaciation close to the end of the Ordovician. Evidence for this is widespread, in Gondwanan rocks in Arabia, the Sahara, western Africa, Canada, and South America. Laurentia, situated at a much lower latitudinal position, was not affected by the glaciation. Glacial deposits indicate that continental ice sheets existed in Africa and Brazil, and Alpine-type glaciers formed in the Andes region. The glaciation was at a high latitude, centered in the South Pole, where southern Gondwana was located at the end of the Ordovician. It began and was most intense in the area of the supercontinent that was to become North Africa. Conditions during the glacial period were arid, becoming more humid during the Ordovician's final phase. Life prospered during most of the Ordovician, until Earth reentered this icehouse phase. By the Early Silurian, modern equatorial temperatures had returned.

FOSSILS ON EVEREST
Fossil fragments show that some limestone strata on Mount Everest's summit were laid down in warm Ordovician seas. Uplifted 60 million years ago, rocks on this peak are scoured by glaciers today.

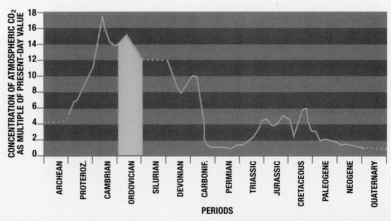

CONCENTRATION OF ATMOSPHERIC CO_2 AS MULTIPLE OF PRESENT-DAY VALUE

18 16 14 12 10 8 6 4 2 0

ARCHEAN | PROTEROZ. | CAMBRIAN | ORDOVICIAN | SILURIAN | DEVONIAN | CARBONIF. | PERMIAN | TRIASSIC | JURASSIC | CRETACEOUS | PALEOGENE | NEOGENE | QUATERNARY

PERIODS

LATE

460 | 450 | 440 | MYA

PLANTS

460 Major echinoderm (below), ostracod, and coral radiations; metazoan reefs develop

455 First spiriferid brachiopods

450 First land arthropod burrows—millipedes; first complete known conodont animal; first echinoids

445 Mass extinction—50% of species go, mainly trilobite (below) and echinoderm families, genera, and species. Also corals, nautiloids, brachiopods, graptolites, conodonts, and acritarchs

AMBONYCHIA RADIATA

65 Gastropod radiation in aurentia; bivalves (above) spread Gondwana; radiation of onodonts, bryozoans, bivalves, critarchs, first articulated rachiopods; first atrypid rachiopods

MALOCYSTITES MURCHISONI

TRIARTHRUS

INVERTEBRATES

VERTEBRATES

ORDOVICIAN
INVERTEBRATES

An important evolutionary phase during the Ordovician records the most significant increase in marine life during Earth's history. However, in contrast with the Cambrian "explosion," relatively few new invertebrate body plans were introduced during this period.

The Great Ordovician Biodiversity Event, during which marine biodiversity tripled, was one of the two most important evolutionary events in the history of life, second only to the Cambrian "explosion" (see p.68).

Great biodiversification

During the Early and Middle Ordovician, there was a phenomenal proliferation and diversification of marine life, phased over some 25 million years. This resulted from both geological and biological processes, and the development of new ecosystems. In

SEA MAT
This "stony" bryozoan had a strong, calcitic skeleton. Bryozoans first arose in the early Ordovician.

particular, there were more continents than at any other time in the Paleozoic, each with surrounding continental shelves. These continents provided a great number of new habitats. Sea levels were high, the continental shelves were flooded, and warm climates prevailed until the end of the Ordovician. Moreover, there were

FILTER FEEDER
This strophomenid brachiopod, typical of the Early Paleozoic, lay flat on the seafloor and had two calcitic valves, which let water currents in and out.

profound evolutionary changes, both in the plankton (possibly fueled by a great input of mineral-rich volcanic material) and in the animals that lived in or near the seabed, especially the suspension feeders. During this time, the first large reefs, constructed by algae, were established.

Mass extinction

Toward the end of the Ordovician, an intense although short-lived glaciation event took place, with ice extending to fairly low latitudes. Sea levels were drastically lowered, much living space was lost on the continental shelves, and tropical faunas suffered particularly through lowered temperatures. The cool-water forms, however, were less affected since they tracked the rims of the ice.

ARTHROPOD ADAPTATION
Trilobites diversified in the Ordovician. This specimen has thin lines crossing the head that enabled the exoskeleton to break easily when the animal molted.

GROUPS OVERVIEW

The Ordovician saw tremendous radiation of species, but the end of this period was a time of mass extinction due to glaciation. The open ocean was never recolonized by trilobites thereafter. The subsequent Silurian invertebrate faunas were, in some ways, impoverished versions of those of the Ordovician.

BRYOZOANS
Bryozoans made their first appearance in the Ordovician. Colonial organisms, they consist of small, soft-bodied animals residing in cavities within a calcite skeleton. While the Ordovician forms are extinct, modern bryozoans are an important component of modern-day ocean fauna.

ORTHID BRACHIOPODS
Orthids are the simplest of the Ordovician articulated brachiopods. They were anchored by a pedicle to the seafloor, with their two calcite valves held upright and open during feeding. Orthids were affected by the extinction that occurred at the end of the Ordovician, but survived until the end of the Permian.

BIVALVES
Ordovician bivalves were very diverse, primarily inhabiting nearshore environments where the otherwise dominant brachiopods were less common. Most of them were filter feeders or fed on bacteria extracted from the sediment, in the same way as the modern *Nucula*.

GRAPTOLITES
These form the main preservable component of the plankton from the Early Ordovician to about the middle of the Devonian. They are colonial animals and are excellent zonal fossils. Their fossils are usually flattened, but rare three-dimensional preservation shows them to be of surprisingly complex form.

Sponge anchorage spines

GROUP Sponges
DATE Cambrian to Devonian
SIZE Up to 1³⁄₄in (4.5cm) long
LOCATION Worldwide

During the Ordovician period, large areas of the seabed were covered with loose mud that was unsuitable for many of the bottom-dwelling animals of the time. However, one group was able to colonize the seafloor—the glass sponges, so called because they possessed skeletons consisting of silica. These simple animals evolved extremely long spicules, or spines, to anchor themselves in the soft mud so that they could exploit the multitude of tiny food particles in the surrounding waters; these spicules also enabled them to maintain their comparatively small sponge bodies above the sediment without being swept away.

long, slender spicules

SPONGE ANCHORAGE SPICULES
Because the spicules of these glass sponges were often very long and fairly thick, they are far more conspicuous fossils than those of the small sponge bodies they supported.

corallite

lacuna

"panpipe" formation

Catenipora

GROUP Anthozoans
DATE Late Ordovician to Late Silurian
SIZE Corallite ¹⁄₁₆in (1–1.5mm) across
LOCATION Worldwide

Catenipora was a colonial tabulate coral (see p.98) composed of elliptical corallites arranged in a single series. A soft-bodied

polyp (a tiny animal with tentacles and a mouth) would have lived in each corallite. The rows of corallites join and divide, forming a network with spaces (lacunae) in between, often filled with sediment. From the side, the corallite structure looks like an array of tiny panpipes. *Catenipora* lived in warm shallow seas, possibly with only the top of the corallites poking above the surface of the sediment.

CATENIPORA SP.
The corallites have thick walls with horizontal partitions, which look like chains when seen in cross-section.

The corallite structure looks like an array of tiny panpipes.

Constellaria

GROUP Bryozoans
DATE Middle Ordovician to Early Silurian
SIZE Branch ³⁄₈–¹⁄₂in (1–1.5cm) across
LOCATION Worldwide

Constellaria was a branching bryozoan colony. The branches were erect and fairly thick but were sometimes compressed in one direction. The surface of the colony was covered in distinctive star-shaped mounds called maculae. These structures have been interpreted as "chimneys" to expel water away from the surface of the colony, after it had been filtered to obtain food for the organism.

CONSTELLARIA SP.
Every branch surface was covered in tiny, star-shaped maculae.

macula

Diplotrypa

GROUP Bryozoans
DATE Ordovician to Silurian
SIZE Colonies up to 4in (10cm) across
LOCATION Worldwide

Diplotrypa was a trepostome bryozoan that formed large dome-shaped colonies. The fossilized colony surface is covered with many circular holes. These were the openings to long tubular structures called autozooecia, which were divided into parts by thin diaphragms. In life, soft-bodied feeding zooids (autozooids) lived within these autozooecia. Each autozooid had a lophophore, which consisted of a ring of tentacles that surrounded the mouth. Food particles were driven toward the mouth by small, hairlike cilia on the tentacles, which created a flow of water by rapidly beating up and down. When the zooids were not feeding, the tentacles were withdrawn into the autozooecia for protection. Between the autozooecia there are smaller tubes that would have housed the nonfeeding zooids.

DIPLOTRYPA SP.
Diplotrypa often featured large holes in the colony, which were produced by another organism boring into it, possibly a polychaete worm.

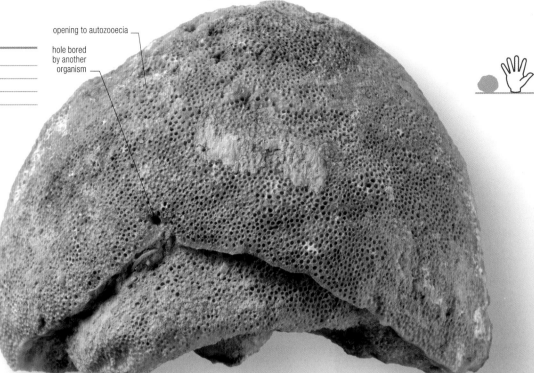

opening to autozooecia

hole bored by another organism

sharp radial rib

beak

Dinorthis

GROUP Brachiopods
DATE Ordovician
SIZE Up to 1½in (3.5cm) long
LOCATION Worldwide

Dinorthis was an orthid brachiopod, a group that ranged throughout the Paleozoic. The most notable feature of *Dinorthis* are the sharp radial ribs, which are equally prominent on both valves. Also notable is that the smaller (brachial) valve is more strongly convex than the larger (pedicle) valve, and in some species, the latter may even be concave. The margin of the shell is rounded, so that in outline, the shell has the shape of a letter D.

DINORTHIS ANTICOSTIENSIS
The ribs are very well defined on both valves of *Dinorthis*. The beaks of both valves project slightly above the straight hinge-line.

Rafinesquina

GROUP Brachiopods
DATE Ordovician to Silurian
SIZE Up to 1½in (4cm) long
LOCATION Worldwide

Rafinesquina was a common brachiopod, ornamented with strong concentric ribs. The larger half of the shell (the pedicle valve), through which its stalk (pedicle) passed, is convex in shape, while the smaller half (the brachial valve) is concave and fits neatly inside it. This indicates that the animal living inside was very thin. The beaks project slightly above the center of the straight hinge-line.

R. ALTERNATA
The hinge-line on this valve is clearly visible in this fossil brachiopod.

hinge-line

Platystrophia

GROUP Brachiopods
DATE Middle Ordovician to Late Silurian
SIZE Up to 1½in (4cm) long
LOCATION Worldwide

Platystrophia was closely related to *Dinorthis*, but unlike *Dinorthis* (see above), both halves of its shell (valves) were strongly convex, which makes it appear quite thick when seen side-on. Both valves had very strong radial ribs, and there was a strong outward fold in the smaller (brachial) valve, which was matched by a corresponding inward fold in the larger (pedicle) valve. Some species developed fine pimples on the surface of both the brachial and pedicle valves, although these are not visible on the specimen below.

PLATYSTROPHIA PONDEROSA
The flexible stalk that anchored *Platystrophia* to the seabed was attached to the shell at the V-shaped point in the center of the hinge-line.

point of pedicle attachment

strong radial rib

Endoceras

GROUP Cephalopods
DATE Middle to Late Ordovician
SIZE Up to 30ft (9m) long
LOCATION North America, N. Europe, Russia, E. Asia

Endoceras belonged to the same group as the modern-day *Nautilus*. It had a long, conical shell that was divided internally into many chambers, which were joined by a long tube (siphuncle) that emptied water from the chambers. The largest chamber at the wider end was undivided and housed the soft, muscular body of the animal. *Endoceras* would have swum horizontally, moving by expelling water rapidly out of a funnel, in the same way as a modern *Nautilus*. It would have been a formidable predator, and at up to 30ft (9m) long, was one of the largest animals of the Ordovician seas.

wide siphuncle space

ENDOCERAS PROETFFORME
This cross-section shows where the siphuncle would have been.

suture line

DIVIDING LINES
This section of shell shows the partitions (septa) that would have divided the internal chambers.

Orthonyboceras

GROUP Cephalopods
DATE Middle to Late Ordovician
SIZE 10in (25cm) long
LOCATION North America, Asia

During the Ordovician, there were many types of nautiloid cephalopods and this group evolved rapidly. *Orthonyboceras* was one of the more modest-sized members of this group, and is similar in some ways to *Endoceras* (see opposite) because both shells are fairly straight and contain many individual chambers. *Orthonyboceras* would have had a problem with excess buoyancy, which it counteracted by laying down calcium carbonate in its internal chambers to make itself heavier.

individual chamber

ORTHONYBOCERAS COVINGTONENSE
The wall (septa) of each chamber had a hole through which a long strand of soft tissue (siphuncle) passed. In this cross-section, the siphuncle looks like a row of beads.

position of siphuncle

Maclurites

GROUP Gastropods
DATE Ordovician
SIZE Up to 2³⁄₄in (7cm) across
LOCATION North America, Europe, N.E. Asia

Maclurites had a large, heavy shell and a distinctive shape. One side of the shell was almost flat, while the opposite side was made concave by a wide umbilicus. Unlike most modern gastropods, it is thought *Maclurites* led a sedentary lifestyle. Rather than grazing, it removed tiny particles of food from the water.

MACLURITES SP.
Unlike most gastropods, *Maclurites'* shallow umbilicus faced upward.

Praenucula

GROUP Bivalves
DATE Middle to Late Ordovician
SIZE Up to 1in (2.5cm) long
LOCATION North America, Europe

Praenucula was a primitive bivalve mollusk, distantly related to the modern genus *Nucula*, which includes small shellfish known as nut clams. Unlike more advanced bivalves, *Praenucula* was a deposit-feeder, using special appendages to sweep food particles from the seabed toward its mouth. Its gills were only involved with respiration, in contrast to more advanced bivalves that have modified gill margins that act like sieves to filter tiny particles of food from the water around them. In *Praenucula* fossils, both halves of the shell (valves) are convex and the beaks are almost centrally placed along the hinge-line.

ambulacrum

Malocystites

GROUP Echinoderms
DATE Middle Ordovician
SIZE 1in (2.5cm) high
LOCATION North America

This globular-shaped echinoderm belonged to a rare, short-lived class called the Paracrinoidea. They differed from crinoids (sea lilies), which still exist today, in that they possessed fixed arms, rather than free-moving ones. They also had fixed, rigid plating over the upper surface of the cup. Unusually, the anal opening (rather than the mouth) was in the central position on the upper surface of the cup. One arm originated from each end of the mouth and they branched irregularly over the entire surface of the cup.

slitlike mouth

MALOCYSTITES MURCHISONI
The cup was made up of around 30 irregular, polygonal plates. The anal opening was in the center, and the slitlike mouth was to one side.

Ambonychia

GROUP Bivalves
DATE Late Ordovician
SIZE Up to 2in (5cm) long
LOCATION USA, Europe

Like modern-day mussels, *Ambonychia* spent its adult life fixed to a rock or similarly hard surface by means of a mass of strong, sticky filaments called a byssus. *Ambonychia* was particularly common in the shallow seas that existed over eastern North America and parts of the Baltic region during the late Ordovician, as the firm shelves provided ideal conditions for byssal attachment. The surface of both of its valves was covered with quite strong radial ribs.

AMBONYCHIA RADIATA
Some of the radial ribs are still visible on this specimen.

radial rib

PRAENUCULA SP.
This internal mold shows the pointed beak, which is positioned almost at the center of the hinge-line.

beak

Isoruphusella

GROUP Echinoderms
DATE Middle Ordovician
SIZE ³⁄₄in (2cm) across
LOCATION Canada

Isoruphusella belonged to an extinct group of echinoderms called Edrioasteroids. Like its modern-day relatives, such as starfish, *Isoruphusella*'s body had five-fold symmetry. On the surface were the five radiating ambulacra that carried the tube feet responsible for both feeding and respiration, though the food grooves and the mouth are hidden under the plates of the ambulacra. Around the periphery was a raised rim of scaly plates.

ISORUPHUSELLA INCONDITA
This species would have been attached by its underside to either the seabed or a larger shell.

flattened, disklike shape

Eodalmanitina

GROUP Arthropods
DATE Middle Ordovician
SIZE Up to 1½in (4cm)
LOCATION France, Portugal, Spain

The headshield of this striking trilobite was dominated by its prominent crescent-shaped eyes, which had large, individual lenses. The central region of the head (glabella) was weakly inflated, with marked lateral indentations. The thorax was composed of 11 segments; after the sixth, they became gradually narrower. On the eighth and ninth segments of the specimen shown, there is some damage that reveals the surface of the joint beneath the preceding segment. The tailshield was triangular and ended in a short, narrow spine.

The eyes of _Eodalmanitina_ were composed in a series of large lenses, arranged in vertical files.

cheek spine

distinct, curved eye

articulating surface under damaged thorax segment

EODALMANITINA MACROPHTALMA
The triangular cheeks of _Eodalmanitina_ terminate in short, backward-facing spines. Part of the right hand one can be seen on this specimen.

Cyclopyge

GROUP Arthropods

DATE Early to Late Ordovician

SIZE Up to 1¼in (3cm) long

LOCATION Wales, England, France, Belgium, Czech Republic, Kazakhstan, S. China

The most characteristic feature of this trilobite was its enormous eyes, which occupied the whole of each side of the headshield. They were composed of hundreds of tiny lenses, and the surfaces were curved in such a way as to give almost 360-degree vision. This is in contrast to most trilobite eyes, where the field of vision was confined to a much narrower band.

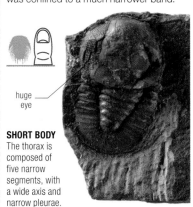

huge eye

SHORT BODY
The thorax is composed of five narrow segments, with a wide axis and narrow pleurae.

limb

TRIARTHRUS EATONI
Unusual conditions of preservation in one fossil bed meant that the antennae and limbs have been preserved.

Triarthrus

GROUP Arthropods

DATE Late Ordovician

SIZE Up to 1¼in (3cm) long

LOCATION NE. USA, Norway, Sweden, SW. China

Triarthrus is one of only several trilobites that have been found with limbs preserved. It had a pair of slender antennae on the headshield, and behind these were three pairs of two-branched (biramous) limbs. Each limb consisted of a walking leg and a gill branch, with which it breathed. Similar limbs were present on each body (thoracic) segment and on the tailshield. The limb segment at the base of each limb (coxa) had an array of spines on its inner side. Acting together, the coxae were probably used to pass food forward to the mouth, which lay underneath the headshield.

Trinucleus

GROUP Arthropods

DATE Middle Ordovician

SIZE Up to 1¼in (3cm) long

LOCATION Wales, England

A trinucleid trilobite was among the first trilobites ever to be illustrated, in 1698. This group was one of the most widespread among trilobites in the Ordovician. The distinctive headshield is made up of three smooth lobes, around which is a fringe. The function of the fringe is a mystery—it may have been a sensory device, or was possibly involved in filtering food particles out of the water. The lack of eyes is not a primitive feature. Many different lineages of trilobites lost their visual organs.

TRINUCLEUS FIMBRIATUS
The style of the fringe varies significantly between different genera of trinucleid trilobites.

pitted fringe

Orthograptus

GROUP Graptolites

DATE Late Ordovician to Early Silurian

SIZE Up to 2¼in (6cm) long

LOCATION Worldwide

Orthograptus is a biserial graptolite, which means that its single branch (stipe) had two rows of cups (thecae), arranged back-to-back on either side. It is commonly found flattened and compressed, but in life, it was oval or rectangular in cross-section. The thecae have either an angular opening or, in some cases, a pair of short, slender spines protruding from the lip. Some species have three short spines at one end. *Orthograptus* specimens are most commonly preserved flattened in shale.

individual *Orthograptus*

ORTHOGRAPTUS CALCARATUS
This graptolite has serrated edges, caused by the flattened openings of the tubular thecae.

sicula

serrated inner edge of the stipe

D. MURCHISONI
The openings of the thecae are only on the inner edges of the stipes.

Didymograptus

GROUP Graptolites

DATE Middle Ordovician

SIZE Up to 2in (5cm) long

LOCATION Worldwide

Didymograptus includes a group of graptolites with two branches (stipes). Examples of this are commonly known as "tuning fork" graptolites because of their shape. The key to identifying graptolites lies in the way in which the first few cups (thecae) in the colony are arranged, and this requires well-preserved specimens. The sicula—the theca that accommodates the first soft-bodied individual (zooid) in the colony to develop from the embryo—lies at the origin of the two branches in *Didymograptus*. The stipes are almost vertical, and up to 40 or more thecae may be present.

Rhabdinopora

GROUP Graptolites

DATE Early Ordovician

SIZE Up to 4½in (12cm) long

LOCATION Worldwide

Rhabdinopora appear fossilized as triangular networks of branches that radiate from an apex. In life, the colony was conical, but it is invariably preserved flattened on its side, except in the case of small, young colonies that can be preserved in plan view, giving them a starlike pattern. *Rhabdinopora* belongs to the dendroid group of graptolites. Most of these lived on the seabed, but *Rhabdinopora* evolved a pelagic lifestyle, inhabiting the upper layers of the open sea. It was widespread in the early Ordovician oceans, and is an important fossil for recognizing Lower Ordovician rocks across the world.

RHABDINOPORA SOCIALIS
At the apex is the sicula, a tiny conical cup that was occupied by the first zooid (soft bodied part) to develop.

sicula

Selenopeltis

GROUP Arthropods
DATE Early to Late Ordovician
SIZE Up to 4¹/₂in (12cm) long
LOCATION British Isles, France, Iberia, Morocco, Czech Republic, Turkey

Selenopeltis was an unusual spiny trilobite. The headshield was almost rectangular and the central region of the head (glabella) tapered slightly toward the front. The eyes were small and situated about halfway between the glabella and the side of the headshield. The cheeks were drawn out into long, slender genal spines that extended back beyond the tip of the tailshield. The thorax was made up of nine segments. The central lobe of each segment had a large, convex lobe at the side. The tip of each segment (pleurae) had distinctive, arched ridges that reached beyond the end of the segments into long, slender spines. Spines at the front extended as far, or even farther, back than the tip of the tailshield. Each segment also had shorter, curved spines underneath the longer spines. The tailshield was short and much smaller than the head. It had a short axis with two or three rings, the first giving rise to a curved, raised rib that extended to form a backward-pointing spine.

SPINY TRILOBITES
This slab of rock contains examples of three different trilobite genera: *Selenopeltis*, which has long spines on both flanks of the body; *Dalmanitina*, which is smaller with one long spine extending back from the tip of the tail; and *Calymenella*, which is large and elongated with an even, round outline (bottom, right).

Selenopeltis

Dalmanitina

Selenopeltis may have favored **cooler waters** along the margins of Gondwana, vanishing from the more northerly latitudes after the Mid Ordovician, when **plate tectonic movements** took them farther south.

EARLY		MIDDLE		LATE		
Tremadocian	Floian	Dapingian	Darriwilian	Sandbian	Katan	Hirnantian

ORDOVICIAN
VERTEBRATES

Many organisms increased in size, strength, and speed during the Ordovician, although most marine life remained quite small. The seas saw the arrival of jawless fish that exhibited a major development in vertebrate evolution: bony plates and scales.

Microvertebrate faunas contain animals that are known from bits of bone, scales, or teeth, generally obtained by breaking down large quantities of rock (often by acid preparation), then patiently sorting through what remains. Ordovician rocks from around the world have proved to be a rich source of microvertebrate fauna, which have revealed new lines of vertebrate evolution.

The origin of bones and scales

Bony plates, scales, and possibly teeth appeared for the first time in vertebrates in the Ordovician, but the reasons for the evolution of these characteristic vertebrate features are still poorly known. Bone is composed of calcium phosphate and some scientists have suggested that it evolved as a way for the animal to remove excess phosphorus from its body, and perhaps to then use the bony skeleton as a phosphorous reservoir, available to be reabsorbed when needed (phosphorus is an important

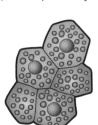
ASTRASPIS

PORASPIS

TOLYPELEPIS

DEVELOPMENTS IN FISH DERMAL PLATE ORNAMENTATION
The headshield of *Astraspis* had polygonal plates with a central tubercle around which successive zones of tubercles were produced as the fish grew. In the Silurian fish *Poraspis*, the plates had longitudinal dentine ridges and only appeared when it was fully grown. *Tolypelepis* (Late Silurian) had both types of plate.

component in several physiological processes). Bony plates and scales could also have evolved as a means of providing protection against predators or parasites, or as insulation for electroreceptors (sensory organs sensitive to electrical signals).

The conodont question

Conodont elements are microfossils that are widespread in marine sedimentary rocks from the Late Cambrian to the Late Triassic. They were first described in 1856 as tiny toothlike elements, and became a useful tool in the dating of rocks. However, their origin remained a mystery until 1982, when a fortuitous discovery in a collection finally associated conodont elements with an actual animal. This "conodont animal," from Carboniferous rocks in Scotland, possessed V-shaped muscle blocks and what appeared to be eyes at the front of the body. Another conodont animal, from Ordovician rocks in South Africa, appears to have had muscles associated with the eyes. Held within the body, in the approximate region of the gill arches, were the conodont elements. Their assembled arrangement and the wear patterns on these elements were suggestive of a feeding mechanism. Conodont elements are said to be made of bone and enamel, and all of these features are characteristic of vertebrates. Despite this, the identification of conodonts as true vertebrates remains controversial.

GROUPS OVERVIEW

The jawless fish comprise eight groups of fossil fish and two living groups, the hagfish and lampreys (see p.76). They arose in the Ordovician with the arandaspids and astraspids, followed by the anaspids, thelodonts, galeaspids, heterostracans, osteostracans, and pituriaspids. None of these fossil agnathans survived beyond the Devonian (see p.33).

ARANDASPIDS

Arandaspids are the earliest-known jawless fish, appearing about 470 million years ago. Until the discovery of *Sacabambaspis*, *Arandaspis* was the only known Ordovician vertebrate from the Southern Hemisphere. Their headshields consisted of bony dorsal and ventral plates separated by a row of smaller plates. Like the astraspids, they died out in the Late Ordovician.

ASTRASPIDS

Astraspids were first described by the renowned American paleontologist Charles Walcott in 1892, after he found fragments of bony plates in the Harding Sandstone in Colorado. *Astraspis*, whose name means "star shield," is distinguished by the ornamentation on the plate's surface (see above). Astraspids appeared about 450 million years ago.

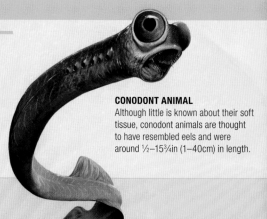

CONODONT ANIMAL
Although little is known about their soft tissue, conodont animals are thought to have resembled eels and were around $\frac{1}{2}$–$15\frac{3}{4}$in (1–40cm) in length.

Arandaspis

GROUP Agnathans
DATE Early Ordovician
SIZE 8in (20cm) long
LOCATION Australia

Arandaspis is named after the Aranda Aboriginal people, who live in the Alice Springs area of Australia where it was discovered. It was preserved as impressions in 470-million-year-old shallow-water sandstone deposits, making *Arandaspis* one of the earliest known jawless fish. Shaped like a flattened oval, its bony headshield was separated into upper and lower halves by a series of about 14 small, square, horizontally running plates that covered and protected its gill openings. The eyes were small and faced forward, and some paleontologists think that nostrils opened between them. *Arandaspis* had no teeth in its ventrally situated mouth and it probably fed by filtering organic debris and microorganisms from the sea floor. Its body was covered with long scales. To date, no tail has been found.

Astraspis

GROUP Agnathans
DATE Late Ordovician
SIZE 5–6in (13–15cm) long
LOCATION Central North America

Among the oldest-known American vertebrate, *Astraspis* had a distinctive headshield and tapering body. The upper half of the shield had five conspicuous longitudinal ridges and numerous polygonal plates (see opposite). The plate ornamentation and closely overlapping scales may have enhanced the fish's streamlined shape, which suggests that *Astraspis* lived in a marine habitat with strong currents or a tidal influence.

UNPROTECTED GILLS
Unlike *Sacabambaspis* and *Arandaspis*, the gill openings of *Astraspis* were not covered and the eyes were situated on the sides rather than at the front of the head.

Sacabambaspis

GROUP Agnathans
DATE Early Ordovician
SIZE 12in (30cm) long
LOCATION Bolivia

Discovered in 1986, *Sacabambaspis* was a jawless fish that inhabited coastal areas of a shallow sea that once extended across parts of South America. The head region was broad, with front-facing, close-set eyes. The upper and lower parts of the headshield were clearly demarcated by about 20 small plates, between which the gill openings were hidden. The tapering body ended in a unique tail fin: in addition to a dorsal and ventral fin web, a long, rodlike extension of the notochord had a small fin web at its tip.

SALTWATER FISH
Sacabambaspis was a bottom-feeder that lived in salty coastal waters. The positioning and accumulation of the fossils found suggest the fish were killed by a sudden influx of freshwater that lowered the salinity beyond their tolerance level.

Because it lacked **paired fins,** *Sacabambaspis* **was probably a poor swimmer.**

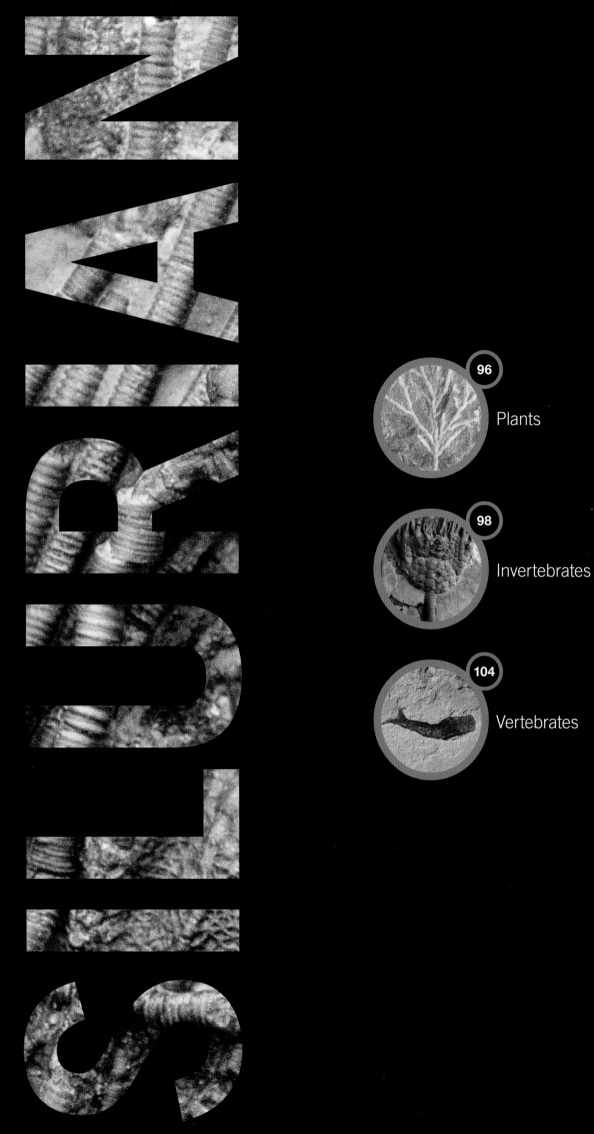

SILURIAN

Silurian

The Siluran period was a time of slower development for the many forms of marine invertebrates that had evolved, and of recovery from the Ordovician extinction event. Huge coral reefs, comparable in extent to Australia's Great Barrier Reef, flourished in sunlit tropical seas. New genera of fishes appeared in the oceans and freshwater, and the first small vascular plants began to colonize coastal areas of the continents.

NIAGARA DOLOMITE
Water rushes over hard dolomites at Niagara Falls, part of a long escarpment around the Michigan Basin. These rocks formed in tropical Silurian seas.

Oceans and continents

During the Silurian, the Iapetus Ocean began to close, as Baltica and another smaller land mass, Avalonia, moved north to collide obliquely on the southern and eastern margins of Laurentia. Avalonia essentially comprised present-day southern Britain and Ireland. Many preexisting island arcs were lost during this collision, dragged down beneath plate margins by the powerful forces of subduction. The Rheic Ocean opened and widened through the Silurian to the south of this new, enlarged landmass and to the north of Gondwana. One of the results of the northerly movement of Baltica into Laurentia was a continued northward movement of Siberia into higher latitudes in the shrinking Panthalassic Ocean. The North and South China blocks started to shift away from the northern margin of Gondwana during the Silurian, and moved north across the Paleo-Tethys Ocean. Gondwana rotated into an even more southerly, poleward orientation, carrying present-day Australia southward to lie across the Equator and moving all of Antarctica into the Southern Hemisphere. The closing of ocean basins and the rapid melting of ice sheets brought about a significant rise in sea levels, helping expand shallow sea environments for corals and fishes. Graptolites reestablished themselves after the Ordovician extinction, and many species of these colonial animals were pelagic, drifting on global ocean currents. Their rapid evolution, short species lifespan, and widespread distribution make them invaluable index fossils for correlating rocks and past marine environments around the world.

DINGLE PENINSULA
Silurian clastic rocks in these spectacular cliffs on Ireland's Dingle Peninsula were formed when sediments were eroded, accumulated, and consolidated in ancient Avalonia.

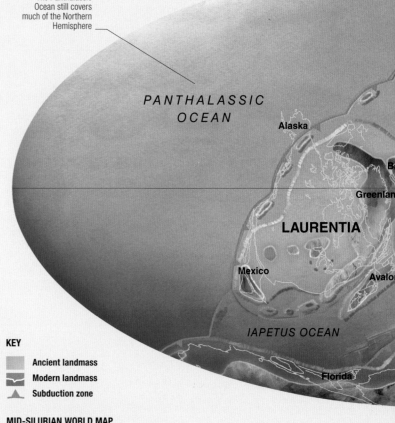

The Panthalassic Ocean still covers much of the Northern Hemisphere

PANTHALASSIC OCEAN

Alaska

Bar

Greenland

LAURENTIA

Mexico

Avaloni

IAPETUS OCEAN

Florida

KEY
	Ancient landmass
	Modern landmass
	Subduction zone

MID-SILURIAN WORLD MAP
Gondwana occupied the south, while Baltica and Avalonia continued northward to collide with Laurentia, closing the north of the Iapetus Ocean.

LLANDOVERY　　　　　　　　　　　　　　　**WENLOCK**

MYA	440		430

PLANTS ● 440 Earliest spores of vascular (higher) plants

● 430 First lucinid bivalves, e.g. *Cardiola*

CARDIOLA

INVERTEBRATES ● ◀ — 440–435 Progressive decrease in graptolite diversity — ▶

● ◀ — 430–425 Major decrease in graptolite diversity

GLACIAL MELT
Glaciers advanced and retreated several times during the early Silurian. As the climate became warmer, the large-scale melting of glacial ice and ice caps brought about a rise in sea levels.

Climate

Estimating paleotemperatures is extremely difficult. One study of fossil brachiopods puts global temperatures during the Silurian at between 93 and 147°F (34–64°C), while other studies suggest 70–113°F (21–45°C). However, the general consensus is that conditions were very warm. Recent research has also shown that the climate may have been more variable than was once thought. Glaciation that began in the Late Ordovician continued, with four major episodes of advancing ice during the first 15 million years of the Silurian. When the ice was at its maximum extent, it was cold at high latitudes and cool and humid nearer the Equator, whereas in the interglacial periods high latitudes were cool, and lower latitudes were warm and arid. During the later Silurian, limestones extended from their earlier limit of up to 35 degrees latitude to reach 50 degrees by the Carboniferous. Their wide distribution, particularly when deposited in low reef environments, indicates that warm oceans, and tropical and subtropical conditions expanded. Reefs were also widespread, extending northward as far as 50 degrees.

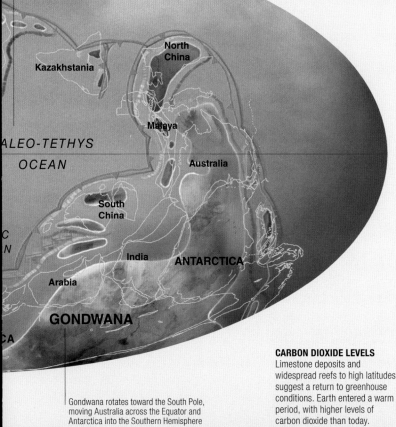

The Paleo-Tethys Ocean begins to open

Kazakhstania

North China

Malaya

PALEO-TETHYS OCEAN

Australia

South China

India

ANTARCTICA

Arabia

GONDWANA

Gondwana rotates toward the South Pole, moving Australia across the Equator and Antarctica into the Southern Hemisphere

LIMESTONE FORMATIONS
Fossil-rich limestones (above) on the Swedish island of Gotland date back to the late Silurian. Monumental eroded sea stacks (left), 26–33ft (8–10m) high, stand in ranks on the beach of the tiny island of Faro, off Gotland's northern tip.

CARBON DIOXIDE LEVELS
Limestone deposits and widespread reefs to high latitudes suggest a return to greenhouse conditions. Earth entered a warm period, with higher levels of carbon dioxide than today.

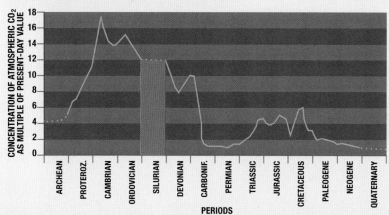

CONCENTRATION OF ATMOSPHERIC CO₂ AS MULTIPLE OF PRESENT-DAY VALUE

18 16 14 12 10 8 6 4 2 0

ARCHEAN | PROTEROZ. | CAMBRIAN | ORDOVICIAN | SILURIAN | DEVONIAN | CARBONIF. | PERMIAN | TRIASSIC | JURASSIC | CRETACEOUS | PALEOGENE | NEOGENE | QUATERNARY

PERIODS

420

410   MYA

Oldest land plant
COOKSONIA

PLANTS

420 First trigonotarbid arachnid; first land scorpion

DIMEROCRINITES

INVERTEBRATES

420 First ray-finned fish

VERTEBRATES

SILURIAN
PLANTS

The first evidence of plants on land is from the Ordovician, but fossils of land plants become steadily more common through the Silurian. By the end of the Silurian, the diversification of plants on land and the creation of new kinds of land ecosystems was well underway.

There is clear evidence for a massive explosion of animal life in the sea around the beginning of the Cambrian, but the first evidence of well-developed ecosystems on land, with complex animals and plants, came much later, in the Silurian.

First steps onto land

The earliest indication that plants were beginning to colonize the land is provided by microscopic spores isolated from Ordovician rocks. These spores have a tough and resistant wall, which would have helped the living cell survive desiccation and dispersal through the air. The resistant wall also enhances the preservation of spores in the fossil record. From the Ordovician and Silurian onward, spores, and later pollen grains—which are their equivalent in seed plants—are a very important part of the plant fossil record.

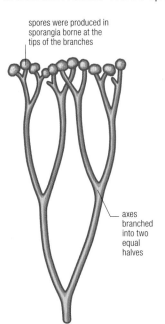

spores were produced in sporangia borne at the tips of the branches

axes branched into two equal halves

COOKSONIA
Several different kinds of early land plants have been included in the genus *Cooksonia*, but they all have small, slender, equally branched stems that bear sporangia at their tips.

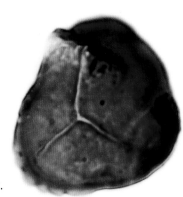

LAND PLANT SPORE
The spores of living and ancient land plants are produced in groups of four (tetrads). Each spore has a three-armed mark on one surface that is formed by contact with the other three spores in the tetrad.

Spores of early land plants become more common during the Silurian and are joined eventually by plant fossils assigned to the genus *Cooksonia*. Although these fossils are usually tiny, their branching axes and terminal sporangia (see left) are visible to the naked eye. Some *Cooksonia* fossils show the first evidence of specialized, water-conducting cells in the center of their axes. These cells have special internal thickenings of the cell wall that prevented them from collapsing as water was carried through them to the aerial parts of the plant.

Freshwater origins

Several lines of evidence suggest that the land surface was colonized from freshwater rather than directly from the oceans. The earliest land plants may have been a special kind of green alga that was adapted to living in temporary pools. The resistant-walled spores may have been an early adaptation that allowed these plants to disperse from pool to pool.

GROUPS OVERVIEW

Fossil plants from the Silurian are relatively rare and usually rather small and poorly understood, but there is no doubt that the Silurian was a critical time in the early diversification of land plants. Some of the pioneering land plants of the Silurian may have been closely related to living mosses, liverworts, and hornworts, but few of the early fossils from this time are known well enough to assess this in detail.

RHYNIOPHYTES
Rhyniophytes are a very varied group of extinct primitive land plants from the Silurian and Devonian. Different rhyniophytes may be related to different modern groups and include the early genus *Cooksonia*, as well as better-preserved fossil plants from the Devonian, such as *Rhynia*.

LYCOPHYTES
Lycophytes were the first group of land plants to diverge from rhyniophytelike ancestors. They include living clubmosses, as well as an important group of extinct plants known as the zosterophylls. The early lycophyte *Baragwanathia* is known from the Late Silurian of Australia.

The **colonization** of the land by **freshwater plants** appears to have only happened **once in the history** of plant life.

Psilophyton

GROUP Invertebrates
DATE Silurian
SIZE 8in (20cm) long
LOCATION Sweden

Fossils of early plants are sometimes difficult to distinguish from other objects, such as mineral growths or animal remains. *Psilophyton hedei* is a classic example of the confusion this can create. When it was found, it was identified as the world's earliest plant, but it is now thought to be a colony of animals related to graptolites. This is only one of several fossils classified as *Psilophyton*. Others, such as *P. burnotense* (see p.116), are known to be plants.

PSILOPHYTON HEDEI
Found on the Swedish island of Gotland, this plantlike fossil is probably a colony of marine invertebrates, borne on a branching "stalk."

Macroalgae

GROUP Algae
DATE Precambrian to present
SIZE Up to 260ft (80m) across
LOCATION Worldwide

Algae make up a large, heterogenous group of organisms that include the closest relatives of green plants. They evolved in water, where most still live today. The earliest forms were microscopic and single-celled and ancestral to larger algae, including seaweeds. Red algae often form hard deposits of calcium carbonate that fossilize well, as do some green algae. Brown algae leave fewer fossils because they are uncalcified and less robust.

FOSSILIZED MACROALGA
Calcified green seaweeds are the most common fossilized macroalgae. They grow in shallow seas, where sunlight penetrates, enabling them to photosynthesize.

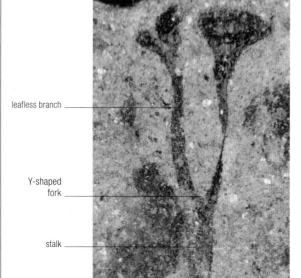

leafless branch

Y-shaped fork

stalk

Cooksonia

GROUP Rhyniophytes
DATE Late Silurian to Devonian
SIZE 3/8–2in (1–5cm) high
LOCATION Worldwide

Growing not much taller than a pin, *Cooksonia* is one of the world's earliest-known land plants. It had slender, leafless branches with Y-shaped forks, topped by capsules that released microscopic spores. Some fossils have a dark stripe in their stems. This may be the remains of vascular tissue—the internal plumbing that most living plants have for moving water. *Cooksonia* grew on estuarine mud and other damp, low-lying habitats, forming dense mats. The first fossil specimen was found in Wales in 1934. Since then, a number of species have been discovered around the world.

COOKSONIA PERTONI
Spores developed in the trumpet-shaped ends of the leafless branches of this early land plant.

Baragwanathia

GROUP Lycophytes
DATE Late Silurian to Early Devonian
SIZE 10in (25cm) high
LOCATION Worldwide

simple leaves

For the Silurian, *Baragwanathia* was an unusually large and complex plant, with upright branches that grew from ground-hugging stems. Its surface was covered with simple, spirally arranged leaves—a feature that has persisted in lycophytes through the present day. Like all early land plants, *Baragwanathia* reproduced by growing spores. These formed in capsules tucked into its axils—the points where the leaves meet the stems.

BARAGWANATHIA LONGIFOLIA
Baragwanathia's simple leaves were up to 1½in (4cm) long. Its stems were several inches across, making this plant much larger than most others in Silurian times.

stem

The entire surface of the plant was covered with simple, spirally arranged leaves.

SILURIAN
INVERTEBRATES

The Silurian was a short period of time lasting only about 28 million years, but during this time the land was invaded by both plants and animals. It was a "greenhouse" period, with warm tropical seas bearing rich and diverse faunas, but it was also punctuated by a number of short-lived glaciations.

Plants colonized the land during the Silurian, and the earliest, such as *Cooksonia*, were just a couple of inches high, with photosynthetic stems and bearing spore capsules. These were confined to damp places. Where there were plants, animals followed, and Silurian millipedes are known both from fossils and the tracks they made when walking. Marine Silurian scorpions also eventually made the transition onto land.

Extensive reef development

During the Silurian, extensive reef systems developed, following those of the Ordovician (see p.82). Silurian reefs are particularly well known from North America, the Swedish island of Gotland, and the Welsh borderland. They grew as elongated barrier reefs or as smaller patch reefs—the latter can often be seen in vertical quarry faces, one stacked above the other, and consisting of pale, largely unbedded, fine,

TABULATE CORAL
In this type of coral, the polyps were set in calcareous corallites; each corallite comprised a tube with a flat platform (tabula) below the polyp. Living tissue ran through the pores, connecting the individual polyps.

corallite — septum
tabula
inter-connecting pore between polyps

calcareous sediment, within a normal bedded sequence. These are not strictly coral reefs, because like most Paleozoic buildups—as they are properly known—they were constructed by algae. Corals and stromatoporids (calcareous layered structures—see *Stromatopora*, opposite) grew on the surface of these buildups; likewise crinoids, brachiopods, trilobites, snails, and bivalves, which also inhabited the seafloor between the buildups. The ecosystem was greatly diverse, with patrolling cephalopods as the top predators. These Silurian environments have been the source of some of the world's best-preserved fossils.

REEF FORMATION
Algal reefs, such as the one above, are formed as one layer of carbonate after another is deposited. Paleozoic buildups probably only rose a few feet above the level of the surrounding seafloor.

RETIOLITES GEINITZIANUS
This was a planktonic graptolite, whose skeleton consisted of a network of thin collagen strings, presumably surrounding the soft-bodied zooids. This may have aided buoyancy.

see p.82

GROUPS OVERVIEW

Some invertebrate groups began to make the transition to land during the Silurian, but it was also a time of great adaptation for marine invertebrates. Colonies of tabulate corals emerged, creating important habitats for many other invertebrate species. Nautiloid, graptolite, and some arthropod groups adapted successfully to the changing environments during this period.

TABULATE CORALS
One of the two major Paleozoic coral groups (the other being rugose corals), tabulate corals became extinct in the Permian. They are always colonial, never solitary, and were an important component of Silurian shallow-water fauna. Septa, the vertical partitions which divide the corallites, are usually small or absent.

NAUTILOIDS
Most of the Silurian nautiloid cephalopods had straight or curving shells; a few were fully coiled. These were the top predators of the time. They are distinguished by their straight, simple sutures (lines where the internal partitions join the inside of the shell), which are visible on internal molds.

GRAPTOLITES
These continued to flourish throughout the Silurian and into the Devonian. Some Ordovician types—with thecae (which housed the zooids) on both sides of the stem—lasted to the Early Silurian; most Silurian graptolites, however, had thecae on one side only (monograptids). Some became large and elaborate.

ARTHROPODS
Trilobites continued through the Silurian—often quite modified in form—and adapted to various environments. There were also eurypterids (water scorpions) living in shallow and often brackish or freshwater, and pod-shrimps, crustaceans like the living *Nebalia*.

Stromatopora

GROUP Sponges
DATE Silurian to Devonian
SIZE Colonies 2in–6½ft (5cm–2m) wide
LOCATION Worldwide

Stromatoporoids were spongelike marine creatures that ranged from small to massive, and examples are commonly found in both Silurian and Devonian limestones. Made up of vertical tubes intersected by transverse structures, *Stromatopora* had closely spaced calcitic plates with strong vertical pillars and star-shaped elements called astrorhizae. (These cannot be seen in this example.) Each astrorhiza had a central, circular opening from which furrows radiated in an irregular series of markings. For years, paleontologists were unsure of how to group stromatoporoids. A modern group of sponges, the sclerosponges, have exhalant canals similar to astrorhizae, but they also have an aragonite skeleton, while that of stromatoporoids is calcitic. Therefore, stromatoporoids are best regarded as a separate class of sponge.

calcitic plate

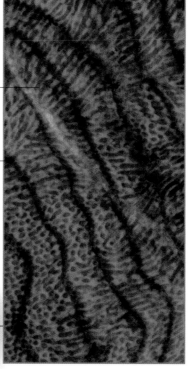

vertical tubes

growth band

STROMATOPORA CONCENTRICA
The porous and closely packed calcitic tubes that made up the skeleton of *Stromatopora concentrica* show spongelike qualities.

Favosites

GROUP Anthozoans
DATE Late Ordovician to Middle Devonian
SIZE Corallite 1/16in (1–2mm) in diameter
LOCATION Worldwide

Favosites was a tabulate coral that lived in warm, shallow seas and formed colonies in a variety of shapes; this example is a flattened hemispherical form. The individual skeletons or corallites that made up the colony were many-sided in cross-section. Their walls were thin and closely packed, giving the colony a honeycomb appearance. The corallite walls also had up to four longitudinal rows of pores. The septa, the vertical partitions that divided the inside of the corallites, were short and commonly in the form of rows of spines. Horizontal partitions (tabulae) were also present. It is unusual to find fossilized remains of the soft body parts, although calcified polyps have been found in examples of Silurian *Favosites*.

corallite

FAVOSITES SP.
The corallites or skeletons gave the colony a honeycomb appearance.

Heliolites

GROUP Anthozoans
DATE Middle Ordovician to Middle Devonian
SIZE Corallite 1/16in (1–2mm) in diameter
LOCATION Worldwide

Heliolites was a tabulate coral that lived in warm, shallow seas. It existed in different forms, which could be branching or in massive colonies. The soft-bodied, coral polyps would have rested on top of cylindrical skeletons called corallites. In cross-section, corallites were smooth and circular or had a scalloped edge made up of 12 segments. Twelve small, longitudinal partitions (septa) may have been present, extending a short distance from the corallite wall. There were horizontal partitions (tabulae) in the central regions of the corallites. Between the corallites, the skeletal tissue was made up of small, polygonal tubes divided by transverse diaphragms.

corallite

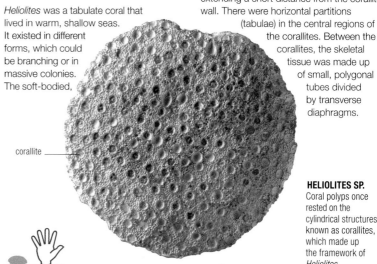

HELIOLITES SP.
Coral polyps once rested on the cylindrical structures known as corallites, which made up the framework of *Heliolites*.

Goniophyllum

GROUP Anthozoans
DATE Early to Middle Silurian
SIZE Calice ½in (1.5cm) in diameter
LOCATION Europe, North America

Goniophyllum was a solitary rugose coral, with a distinctive four-sided top section. The soft-bodied coral polyp lived in a concave cavity, called a calice. A lidlike structure or operculum, made of four thick, triangular plates, covered the calice, but this is not usually preserved in fossils. Contracting muscles shut the lid so that the polyp was protected when not feeding. The calice was deep, with short, thick partitions (septa), and its horizontal partitions (tabulae) and small, curved plates were thickened in zones.

calice

GONIOPHYLLUM PYRAMIDALE
Four plates, not shown here, on the top of the coral could close so that the polyp was protected.

Ptilodictya

GROUP Bryozoans
DATE Late Ordovician to Early Devonian
SIZE Branch ¹⁄₁₆–½in (2–15mm) wide
LOCATION Worldwide

Ptilodictya was an upright bryozoan colony, made up of one straight or slightly curved branch. At the base was a conical socket that allowed the colony to articulate. Rectangular holes on the surface opened to the autozooecia, where the individual soft-bodied animals known as autozooids lived. These feeding autozooids had a lophophore—a ring of tentacles surrounding their mouths, which could be withdrawn into the autozooecia when the zooid was not feeding.

PTILODICTYA LANCEOLATA
Small openings on the colony surface led to chambers where the autozooids lived.

Leptaena

GROUP Brachiopods
DATE Middle Ordovician to Devonian
SIZE ½–1½in (1.5–4cm) long
LOCATION Worldwide

Leptaena was a genus of brachiopods that had an outward-curving pedicle valve or shell with an inward-curving brachial valve just inside it. The hinge-line was straight, and the beaks of both shells projected above it. The pedicle opening, where the fleshy stalk attaching the animal to a surface protruded, was located just below the pedicle valve beak. The shell's surface had strong concentric ribs crossed with fine radial ribs. Toward the front edge of the shell was a sharp bend in both valves.

LEPTAENA RHOMBOIDALIS
Leptaena's smaller brachial valve sat closely inside the pedicle valve, leaving room for only a very compressed body.

concentric ribs

Pentamerus

GROUP Brachiopods
DATE Silurian
SIZE 1–2¼in (2.5–6cm) long
LOCATION North America, British Isles, northern Europe, Russia, China

Pentamerus was large, with two outward-curving valves or shells; its length was usually greater than its width. The pedicle valve beak was prominent; the opening where the animal's fleshy stalk protruded lay just beneath it. Its shell surface was almost smooth, but featured fine radial ribs and fine growth lines. Inside, a thin dividing wall (septum) was located nearest the pedicle beak, and supported a large, spoonlike structure that acted as a muscle-attachment area. This feature is characteristic of pentamerid brachiopods. *Pentamerus* often lived in large groups.

Atrypa

GROUP Brachiopods
DATE Early Silurian to Late Devonian
SIZE ¾–1¼in (2–3cm) long
LOCATION Worldwide

Atrypa was a brachiopod or lamp shell with a flat or slightly outwardly curved pedicle or larger shell (valve) and a strongly outwardly curved smaller or brachial valve. The beaks of both shells projected slightly above the hinge-line. The shell surface was characterized by strong, concentric ribs crossed by radial ribs. There was a slight outward fold in the brachial valve and a corresponding fold or sinus in the pedicle valve. The supports for the animal's feeding organ, the lophophore, were located inside the brachial shell (see panel, right), and were arranged in spirals directed toward the center of the shell.

radial ribs

concentric ribs

ATRYPA SP.
Layers of concentric ribs characterize *Atrypa*, and these are crossed by radiating ribs—a trait found in modern bivalves.

ANATOMY
LOPHOPHORE

lophophore supports (spiralia)

Like all brachiopods, *Atrypa* fed by means of an organ called the lophophore, which carried large numbers of minute, tentaclelike cilia that trapped food particles and swept them into the mouth. The longer the lophophore, the better the animal's chance of collecting more food. Arranging the lophophore in a spiral provided an efficient way of getting a long lophophore into a small area, but it needed some support to keep turns of that spiral separated. Thus, in *Atrypa*, the lophophore supports took the form of two delicate, calcite spirals, the spiralia, located in the interior of the brachial valve.

Gomphoceras

GROUP Cephalopods
DATE Middle Silurian
SIZE 3–6in (7.5–15cm)
LOCATION Europe

Gomphoceras was a marine mollusk, which belonged to the same group as the modern-day *Nautilus*. The thin walls inside the shell were closely spaced, and it had a large body chamber that narrowed toward the shell opening at maturity. The aperture is missing from this example, but in other fossils final opening was complete but small, with little space for tentacles. This suggests that, at maturity, the animal could no longer feed and died after mating, as do many modern cephalopods.

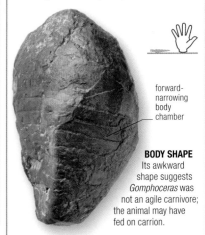

forward-narrowing body chamber

BODY SHAPE
Its awkward shape suggests *Gomphoceras* was not an agile carnivore; the animal may have fed on carrion.

Cardiola

GROUP Bivalves
DATE Silurian to Middle Devonian
SIZE ³⁄₈–1¼in (1.2–3cm) long
LOCATION Africa, Europe, North America

Cardiola was a strongly ribbed bivalve with well-marked growth rings. The shells (valves) were of equal size, with prominent beaks. A smooth, triangular area lay beneath each beak, with fine growth-parallel ribs across it. Internally, two adductor muscles controlled closure of the valves, but there were no hinge teeth. A ligament would have opened the valves when the adductor muscles relaxed. *Cardiola* was probably attached to the seabed by threads known as byssus, which it retained from its initial settling as a larva.

growth-parallel ribs

ribbed shells

CARDIOLA INTERRUPTA
Cardiola fossils such as these are rarely found with other fossils, suggesting a specialized environment.

Gissocrinus

GROUP Echinoderms
DATE Middle Silurian to Early Devonian
SIZE Cup with complete arms 2³⁄₄in (7cm) high
LOCATION Europe, USA

Gissocrinus was a crinoid, a marine animal related to sea urchins and starfish. It had a small cup made of three circles of plates. In the upper circle, the five radial plates had wide, crescent-shaped upper surfaces; these served as articulation points for the animal's arms, which extended upward from this point. A sixth plate in this circle, at the back of the cup, marked the base of a tubular structure, which ended with the anal opening at its top. The arms branched several times, and the branches were always equal in length. In this example, however, only the lower branches have been preserved.

branching arm

GISSOCRINUS INVOLUTUS
Although only its lower parts are shown here, *Gissocrinus* had up to 32 branches.

Pseudocrinites

GROUP Echinoderms
DATE Late Silurian to Early Devonian
SIZE Cup ½–1¼in (1.5–3cm) across
LOCATION Europe, North America

Pseudocrinites was shaped like a round or oval inflated disk, with a flat rim formed by two, thin-plated zones called ambulacra. It was anchored to the seabed by a jointed stem. The mouth was set opposite the stem attachment in the center of the oral surface. The ambulacra began at each side of the mouth, and in some specimens reached down as far as the stem. In well-preserved examples, a number of short, articulated appendages called brachioles can be seen. The sockets where they were attached show that they were arranged alternately on each side of the ambulacra. The anus, an opening into the water vascular system, and a gonopore (for release of eggs or sperm) were also located on the oral surface. The hard, external shell features rhomboid-shaped areas that are situated across plate junctions, and pairs of slitlike openings that penetrate the shell; the function of these was probably respiratory. Respiratory structures scattered over the cup surface are characteristic features of the class Cystoidea, to which *Pseudocrinites* belongs.

respiratory structures

ambulacrum

PSEUDOCRINITES BIFASCIATUS
This fossil shows rhomboid-shaped respiratory structures that characterize *Pseudocrinites*, but some cystoids had many pairs of pores over the cup surface.

Pseudocrinites **used its jointed stem to anchor** itself **to the seabed, where it lived as a suspension-feeder.**

Lapworthura

GROUP Echinoderms
DATE Late Ordovician to Middle Silurian
SIZE 4–4¹/₂in (10–12cm) across
LOCATION Europe

Lapworthura was a brittlestar with a large, central disk area. All of its body openings were located on the animal's lower (oral) surface, and its mouth was situated centrally. *Lapworthura's* arms were relatively stout for a brittlestar, and its internal skeleton of calcareous plates, known as ossicles, were arranged in pairs opposite each other. Fossilization of brittlestars is usually dependent on the animal being buried alive by the sudden influx of sediment: for example, during a storm. If this does not happen, the soft, outer skin of the organism decays, and the ossicles separate and disperse. Brittlestars, starfish, and sea urchins are all echinoderms; like modern brittlestars, *Lapworthura* was probably a carnivore.

stout arm

LAPWORTHURA MILTONI
Unlike modern brittlestar species such as *Ophiothrix fragilis*, *Lapworthura miltoni*'s arms were relatively short and stout.

central disk

CENTRAL STAR
Lapworthura miltoni's large, starlike disk was where the animal's mouth was located.

GREEN BRITTLESTAR

Brittlestars date back to the Ordovician and are represented today by 2,000 species found from polar to equatorial regions. Their tube feet are primarily sensory. They occur mainly below 1,640ft (500m), ranging to abyssal depths. There are, however, some shallow-water species, one of which, the green brittlestar (*Ophiarachna incrassata*), is sometimes kept in aquariums—although, as owners have discovered to their cost, it has a penchant for small fish and crustaceans.

Dalmanites

GROUP Arthropods
DATE Early to Late Silurian
SIZE Up to 4in (10cm) long
LOCATION Worldwide

The trilobite *Dalmanites* had a large headshield with a prominent, forward-widening glabella in the center that featured two pairs of narrow furrows as well as deep, oblique ones. Its eyes were large and set at the back of the head, and had prominent lenses. Its cheeks were gently convex, and extended into robust, broad-based spines. Under the headshield, and attached to it beneath the glabella, was a large hypostome: a calcified structure thought to have enclosed the stomach. The thorax was segmented, and the broadly triangular tailshield ended in a spine.

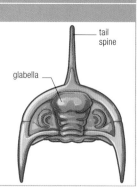

segmented thorax

DALMANITES CAUDATUS
The thorax of *Dalmanites* was made up of 11 separate segments, clearly visible here.

ANATOMY
ENROLLMENT

Like woodlice, most trilobites had the ability to roll themselves into a ball in order to protect their soft undersides. This image depicts *Dalmanites* rolled up and seen from above, with its tail spine pointing forward from underneath the head. This spine may have aided molting: all trilobites periodically shed their hard exoskeleton to accommodate growth. In fact, most trilobite fossils are molts rather than dead carcasses.

tail spine

glabella

Encrinurus

GROUP Arthropods
DATE Silurian
SIZE Up to 2in (5cm) long
LOCATION Worldwide

Encrinurus was a small trilobite that was widespread in the world's seas throughout the Silurian. As a fossil, it is immediately recognizable from the many prominent pustules on its headshield, which have given it a nickname—the "strawberry-headed trilobite." The central part of the headshield, the glabella, was inflated and pear-shaped, widening toward the front. It had several furrows, but these are not easy to see on this specimen. The eyes (only the bases are preserved here) sat on the ends of short stalks, suggesting this trilobite may have spent much of its time half-buried in soft sediment on the seabed, with only its eyes protruding above the surface.

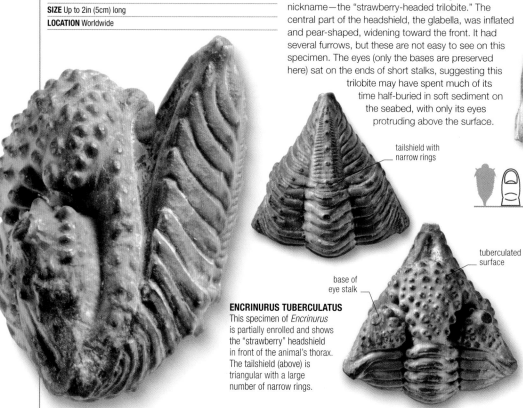

tailshield with narrow rings

base of eye stalk

ENCRINURUS TUBERCULATUS
This specimen of *Encrinurus* is partially enrolled and shows the "strawberry" headshield in front of the animal's thorax. The tailshield (above) is triangular with a large number of narrow rings.

EXALLASPIS SP.
Although only the head and a small part of the thorax are visible here, the backward-curved spines extending from the ends of the thorax can clearly be seen.

backward-curved spine

tuberculated surface

Exallaspis

GROUP Arthropods
DATE Middle to Late Silurian
SIZE Up to 1in (2.5cm)
LOCATION Worldwide

This trilobite's headshield had a prominent central raised area (glabella) that narrowed toward the front, with three distinct lobes; the hindmost was the largest. A curved, raised ridge ran from the front of the glabella to the rather small eyes, which were located on the back part of the headshield, about halfway between the glabella and the lateral edge. The thorax was composed of 10 segments, which laterally tapered into long, curving spines on both sides. The tailshield was small, and had two large, backward-pointing spines with a row of four small ones between them. There were also two short spines to the front of the large ones. On the specimen shown above, however, most of the thorax and the tailshield are missing.

Calymene

GROUP Arthropods
DATE Silurian
SIZE Up to about 2¼in (6cm) long
LOCATION Worldwide

One of the most familiar trilobites, *Calymene* had a roughly semicircular headshield dominated by a prominent glabella (the raised, central part) that narrowed toward the front and was broadly bell-shaped. The glabella had three lobes running along each of its two outer edges, and the hindmost of these was the largest. Its eyes were small and located roughly opposite the midpoint of the glabella. This was such an abundant fossil in the limestone quarries of Dudley, England, that it became a local emblem.

Calymene is often referred to as the "Dudley Locust" or "Dudley Bug."

glabella

CALYMENE BLUMENBACHII
Calymene's thorax was made up of 13 segments, and its tailshield was smaller than its rounded head.

PROTECTIVE ROLL
Many *Calymene* fossils, like the one above, have been found rolled up. Like modern woodlice, most trilobites did this for protection.

Monograptus

GROUP Graptolites
DATE Early Silurian
SIZE Up to 2in (5cm) long
LOCATION Worldwide

Monograptus is characterized by having armlike structures (thecae) on only one side of its stipe (branch). The thecae housed the individual soft-bodied zooids of the colony. This genus appeared in the geological record at the beginning of the Silurian period, about 443 million years ago, and a large number of species of *Monograptus* and related genera evolved rapidly over the following 30 million years. Many of these are of worldwide distribution. Graptolites were colonial hemichordates, a small group that is allied to vertebrates. Living hemichordates are comparatively rare, but the abundance of graptolite fossils in many rocks shows that they were once widespread in the planet's oceans.

SCIENCE
ZONE FOSSILS

Most graptolites were planktonic, and were therefore widely distributed in Ordovician and Silurian oceans. They evolved rapidly and individual species existed for a relatively short time, usually about 2 million years. These attributes make graptolites ideal zone fossils. Individual species or groups of species can be used to calibrate rock successions, and correlate them across the world. In Silurian rocks, around 40 graptolite zones have been recognized, each with an average length of about 0.7 million years.

long, tubular thecae

MONOGRAPTUS TRIANGULATUS
In this species, the thecae are long, tubular, and widely spaced on the colony's outer curve.

thecae with triangular base

MONOGRAPTUS CONVOLUTUS
Here, the colony forms a spiral, and its long thecae, on the outside of the curve, give it the appearance of a watch spring.

Eusarcana

GROUP Arthropods
DATE Late Silurian
SIZE Most a few inches long, but giants of over 3ft 3in (1m) long are known
LOCATION Europe

Eurypterids like *Eusarcana* were scorpionlike animals, and the majority inhabited freshwater or brackish water. In some, the first pair of limbs ended in pincers armed with sharp, pointed teeth on their inner edges. The animal's small headshield is not well preserved on this specimen, but was roughly rectangular, and its small eyes were located close to the front. The mouth lay underneath, and the first pair of appendages, the small chelicerae, lay in front of it. *Eusarcana* had four pairs of spiny walking legs, and the sixth, hindmost pair were swimming legs with paddlelike ends. Its abdomen was divided into two parts: a broad, oval preabdomen, made up of seven segments; and a narrower, nearly cylindrical postabdomen, with five segments. Behind the abdomen was a pointed terminal segment called a telson.

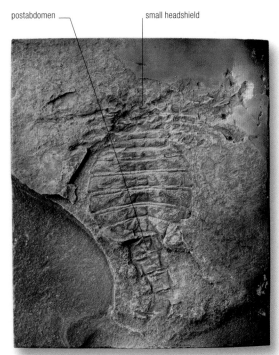

postabdomen

small headshield

EUSARCANA OBESA
Eusarcana fossils are sometimes referred to as "water scorpions"—and the similarity of shape is apparent here.

SILURIAN
VERTEBRATES

Jawless fish (agnathans) were the most common vertebrates of the Silurian. However, new vertebrate groups appeared and diversified during the period. The oceans were home to a key development—jaws—and by the end of the Silurian, all the major groups of vertebrates had evolved.

The period of glaciation that ended the Ordovician extended into the Silurian, but gradually the ice melted, sea levels rose, and the oceans became warmer. While plants and invertebrates sought to establish themselves on land, vertebrates continued to take advantage of the hospitable environment offered by the oceans. And it was the Silurian seas that saw what is probably the most important advance in vertebrate history: the evolution of jaws.

The advantages of jaws

The possession of jaws enabled the development of new types of behavior. With jaws, an object can be grasped firmly, and if those jaws have teeth, food can be sliced into pieces small enough to swallow, while hard items can be ground down. Eating plants was now an option, and many jawed

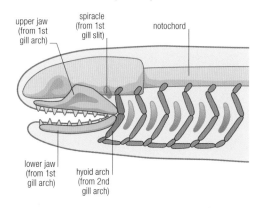

AGNATHOUS (JAWLESS) VERTEBRATE

braincase · 1st gill slit · notochord · mouth · 1st gill arch · 2nd gill arch

upper jaw (from 1st gill arch) · spiracle (from 1st gill slit) · notochord · lower jaw (from 1st gill arch) · hyoid arch (from 2nd gill arch)

GNATHOSTOME (JAWED) VERTEBRATE

EVOLUTION OF THE VERTEBRATE JAW
In agnathans, the first and second gill arches support the first gill slit. In jawed vertebrates, the first gill arch has become a pair of jaws and the first gill slit a spiracle, through which water can be pumped over gills. The jaws can also bear teeth.

vertebrates became larger than their jawless contemporaries. With or without teeth, a strong pair of jaws can be a powerful weapon—for killing or defense—but jaws can also be used to hold on tight to a partner during courtship or mating and to carry young. Objects can be manipulated with jaws—holes can be dug, stones can be moved, and pieces of plants can be used to build nests. The possibilities for jawed vertebrates were many and varied.

Fossil finds in China

The sarcopterygians, or lobe-finned fish, are important jawed vertebrates because they are the group from which tetrapods (four-legged animals) evolved (see p.122). They were thought to have first appeared in the Early Devonian, but in 1997, parts of the skull and lower jaw of a new fish, *Psarolepis*, were found in China's Yunnan Province, proving that sarcopterygians first evolved even earlier, in the Late Silurian. However, the main radiation of sarcopterygians, including several other new and primitive forms found more recently in China, occurred in the Early Devonian.

GROUPS OVERVIEW

New jawless fish groups in the Silurian included the anaspids, thelodonts, osteostracans, and galeaspids. Osteostracans had pectoral fins supported by a shoulder girdle, a feature shared with jawed vertebrates but absent in all other jawless vertebrates. This suggests that osteostracans are the sister taxon to the jawed vertebrates. Of the jawed fish, acanthodians, placoderms, and chondrichthyans were present in the Early Silurian, while the actinopterygians (ray-finned fish) arose in the Late Silurian.

ANASPIDS
These jawless fish are found primarily in Silurian deposits in the Northern Hemisphere. Some anaspids, such as *Birkenia*, possess large dorsal scales but lack paired fins. Others, such as *Pharyngolepis*, have paired finlike structures extending along the ventral surface of the body. *Jamoytius* and *Euphanerops* are unusual in having more than 30 gill arches.

OSTEOSTRACANS
Osteostracans first appeared in the Middle Silurian and, like the anaspids, were restricted to the Northern Hemisphere. They are characterized by a large, bony headshield, with dorsal eyes. The headshield was drawn out laterally into spines, forming a horseshoe, although these spines were absent in primitive members of the group, such as *Ateleaspis*.

THELODONTS
Thelodonts arose in the Early Silurian and survived until the Late Devonian, and these jawless fish had a worldwide distribution. They are well known from isolated scales (microvertebrates), but more complete specimens have also been found, such as *Loganelia*, showing the head, tail, gill arches, and evidence of paired flaps that may have been fins.

Birkenia

GROUP Agnathans
DATE Late Silurian to Early Devonian
SIZE 6in (15cm) long
LOCATION Europe

A small jawless fish, *Birkenia* had a laterally compressed, "fusiform" body (wider in the middle than at both ends). It was covered in elongated scales arranged in distinct rows. Those on the hind dorsal flanks sloped down and back instead of down and forward. Large scales ran along the top of the body; some pointed forward, others backward, and a double-headed scale in the middle pointed in both directions. *Birkenia* had a well-developed anal fin, and small eyes with a single nasal opening between them. It lived in freshwater and was an active swimmer that fed on detritus (the remains of dead plants and animals).

UNIQUE SCALES
Birkenia's distinctive arrangement of dorsal scales is unique among anaspids, the group to which it belongs.

double-headed scale

head covered in small scales

laterally compressed body

UNUSUAL TAIL
Birkenia's hypocercal tail, in which the larger lower lobe was supported by the downwardly directed notochord, initially confused an early paleontologist, who reconstructed the fish upside down.

Ateleaspis

GROUP Agnathans
DATE Early Silurian to Early Devonian
SIZE 6–8in (15–20cm) long
LOCATION Scotland, Norway, Russia

Ateleaspis was a primitive osteostracan and is thought to have lived in sheltered seas or river mouths. It is the earliest-known fish with paired appendages—its pectoral fins. It had two dorsal fins: the front one was covered in scales; the hind fin was larger, with webbed spines. *Ateleaspis*'s head was protected by a bony shield with sensory fields. Bony scales covered the body. The mouth was on the underside of the head, which suggests that *Ateleaspis* was a bottom-feeder.

10 The number of **gill pairs** that were positioned **beneath** the body of *Ateleaspis*.

FLATTENED BOTTOM-FEEDER
Ateleaspis's body was flattened, and the forward portion was flatter than the hind section, suggesting it was a bottom-feeder.

wide, flattened head

pectoral fin

tail fin

Loganellia

GROUP Agnathans
DATE Late Silurian
SIZE 4–18in (10–44cm) long
LOCATION Europe

Loganellia was a thelodont, a flattish fish, with a body entirely covered in scales and a hypocercal tail. Its eyes were small and far apart. The position of its mouth on the underside of its head suggests that it fed on the sea bed by sifting through the mud. Paired flaps on the side of the head may have functioned as fins.

FORKED TAIL
Loganellia had a wide, flat head and a long, forked tail, in which the lower lobe was much longer than the upper one (hypocercal).

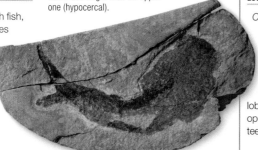

Climatius

GROUP Acanthodians
DATE Late Silurian to Early Devonian
SIZE 4¹⁄₂in (12cm) long
LOCATION British Isles

Climatius was a short fish with many short spines. Both pectoral fins had a spine, as did the front and rear dorsal fins and the anal fin, and there were five pairs of spines underneath the body. Bony armor covered its "shoulders." Its tail fin was sharklike, with a larger upper lobe; its eyes were large, and its nasal openings small. *Climatius* had small grinding teeth, which suggest it fed on tiny creatures.

Andreolepis

GROUP Actinopterygians
DATE Late Silurian
SIZE 3¹⁄₂in (9cm) long
LOCATION Europe, Asia

One of the earliest known ray-finned fish, *Andreolepis* had sharp, pointed teeth that were interspersed with smaller toothlike projections and arranged in several rows. Old teeth were not shed; instead, new ones were added along the jaw's inner edge. Its scales were diamond shaped and ridged and carried a thin layer of a substance called ganoine, which is similar to the enamel found on a tooth.

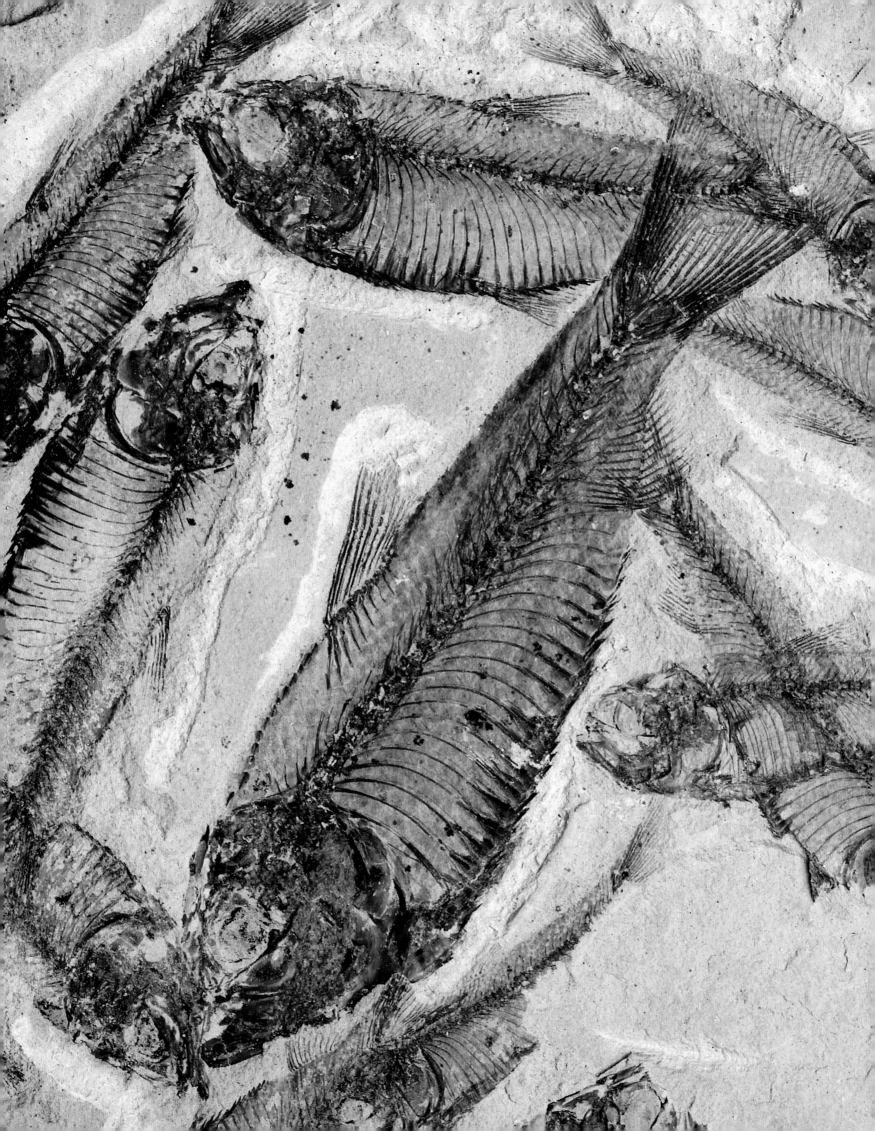

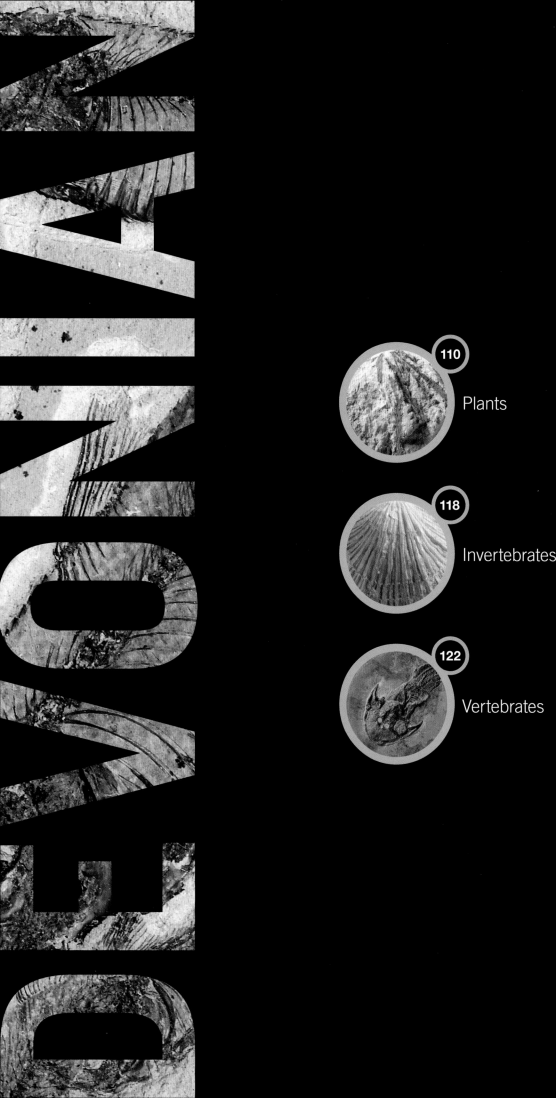

DEVONIAN

Devonian

The Devonian Period, often called the "Age of Fish" in Earth's history, is characterized by an evolutionary radiation in some fish groups, including strange jawless fish and armored placoderms. Equally significant was the great increase in the diversity of plants, which spread from the coastlines of the continents across the land and created new terrestrial ecosystems: the world's first forests.

CALEDONIDES MOUNTAINS
The once massive Caledonides are heavily eroded, but still extend from Scandinavia to the Appalachians in North America.

Oceans and continents

By the middle of the Devonian, about 400 million years ago, the Iapetus Ocean had completely disappeared, as Baltica and Laurentia collided. This collision unleashed massive tectonic forces that continued to reshape the continental landscape, forcing up the Caledonide mountains in today's Scandinavia, northern Britain, and Greenland, and the beginning of the Northern Appalachian chain in eastern North America. Gondwana rotated in a clockwise direction around an axis centered on Australia, bringing the western margin of the continent closer to the Equator and to Laurentia. Land masses gradually became greener as ferns and treelike plants formed forests and swamps, adding organic material to soils and creating new habitats for invertebrates. The Rhynie Chert in Scotland provides an exceptional insight into the flora of a Mid-Devonian peat bog. During the Late Devonian, there were major changes in oceanic geochemistry, with periods when the oceans became depleted of oxygen. Rapid plant diversification may have enriched rivers with high levels of nutrients that were then carried to the seas. Because reefs prefer low-nutrient conditions, this may have contributed to a decline in their diversity during the Devonian mass extinction.

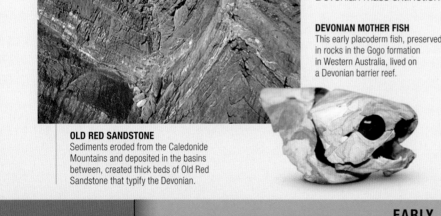

OLD RED SANDSTONE
Sediments eroded from the Caledonide Mountains and deposited in the basins between, created thick beds of Old Red Sandstone that typify the Devonian.

DEVONIAN MOTHER FISH
This early placoderm fish, preserved in rocks in the Gogo formation in Western Australia, lived on a Devonian barrier reef.

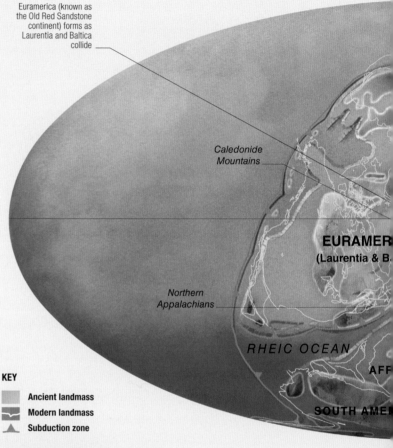

Euramerica (known as the Old Red Sandstone continent) forms as Laurentia and Baltica collide

Caledonide Mountains

Northern Appalachians

EURAMER
(Laurentia & B

RHEIC OCEAN

AFF

SOUTH AME

KEY
- Ancient landmass
- Modern landmass
- Subduction zone

MID-DEVONIAN WORLD MAP
Baltica and Laurentia combined to form Euramerica and close the Iapetus Ocean. This led to the creation of the Caledonide mountain chain.

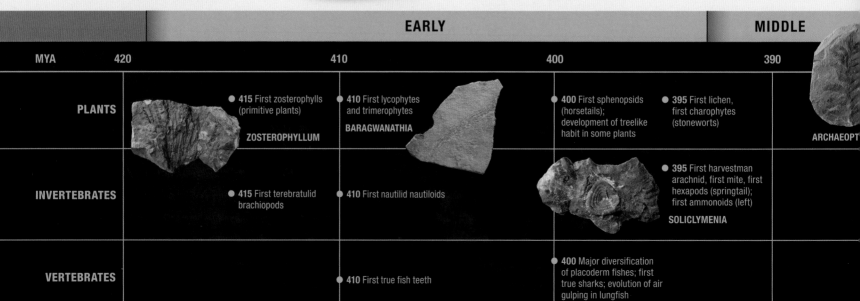

	EARLY			MIDDLE
MYA 420	410	400	390	
PLANTS	• 415 First zosterophylls (primitive plants) **ZOSTEROPHYLLUM**	410 First lycophytes and trimerophytes **BARAGWANATHIA**	• 400 First sphenopsids (horsetails); development of treelike habit in some plants	• 395 First lichen, first charophytes (stoneworts)	ARCHAEOPT
INVERTEBRATES	• 415 First terebratulid brachiopods	• 410 First nautilid nautiloids		• 395 First harvestman arachnid, first mite, first hexapods (springtail); first ammonoids (left) **SOLICLYMENIA**	
VERTEBRATES		• 410 First true fish teeth	• 400 Major diversification of placoderm fishes; first true sharks; evolution of air gulping in lungfish		

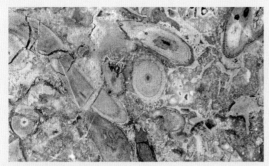

RHYNIE CHERT FOSSILS
Silicified fossils of Early Devonian age are found beautifully preserved in Scotland's Rhynie Chert. This thin section shows stems of one of the first land plants, *Rhynia*.

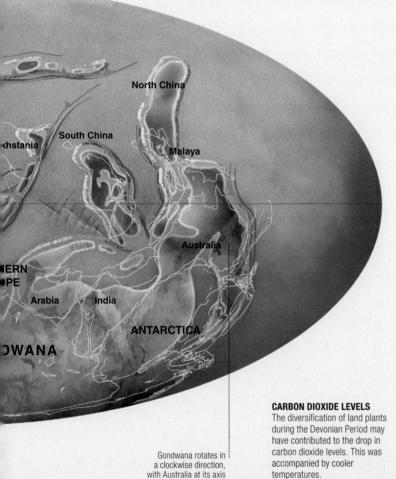

North China

South China

Malaya

Khstania

Australia

ERN
PE

Arabia   India

ANTARCTICA

WANA

Gondwana rotates in a clockwise direction, with Australia at its axis

Climate

The global climate was generally relatively warm, arid, and dry, and the overall Equator to pole temperature gradient was less than today. In the Early Devonian, much of Laurentia and the northeastern part of Gondwana were hot and arid. Large parts of what are now Australia, North America, and Siberia were covered by warm, shallow seas. Analysis of fossil brachiopods using oxygen isotopes (see p.23) suggests that in the Early Devonian ocean temperatures were about 79°F (26°C), comparable with modern-day oceans. Later in the Devonian they could have risen to about 86°F (30°C). In higher southern latitudes across Gondwana the climate was warm temperate. The polar regions were cool, but without any ice cover. As Laurentia and Gondwana moved closer together, a wider tropical belt became established north of the Equator. Here, the first thick coal deposits reveal that rain forests stretched across what is now the Canadian Arctic and southern China. The Siberian continent to the north, and the northern parts of the converging continents of Laurentia and Gondwana remained arid, but the more southerly parts of Gondwana in higher latitudes became more temperate and more humid, with some of the higher mountainous regions close to the Devonian South Pole, in what is now the Amazon Basin, becoming cold enough for glaciers to develop.

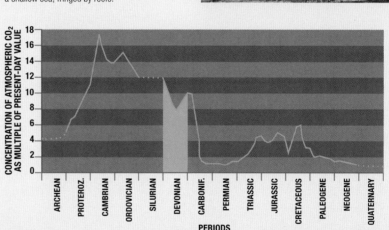

WINDJANA GORGE, THE KIMBERLEY
Spectacular Windjana Gorge in the Kimberley region, in Australia's northwest corner, exposes the remains of a large barrier reef from Devonian times, when the area was a shallow sea, fringed by reefs.

CARBON DIOXIDE LEVELS
The diversification of land plants during the Devonian Period may have contributed to the drop in carbon dioxide levels. This was accompanied by cooler temperatures.

CONCENTRATION OF ATMOSPHERIC CO₂ AS MULTIPLE OF PRESENT-DAY VALUE

(Y-axis: 0, 2, 4, 6, 8, 10, 12, 14, 16, 18)

PERIODS: ARCHEAN, PROTEROZ., CAMBRIAN, ORDOVICIAN, SILURIAN, DEVONIAN, CARBONIF., PERMIAN, TRIASSIC, JURASSIC, CRETACEOUS, PALEOGENE, NEOGENE, QUATERNARY

LATE

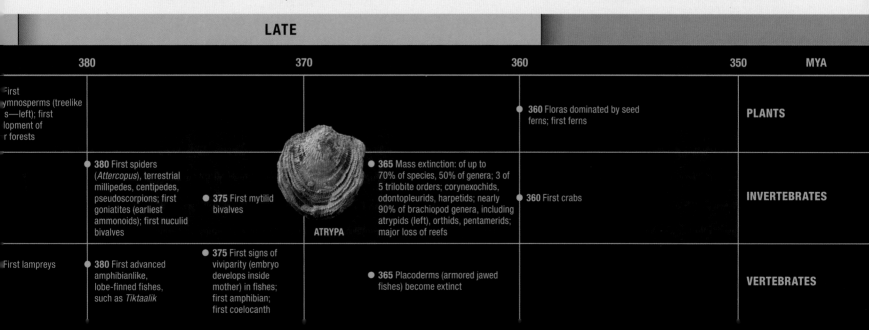

	380	370	360	350	MYA
PLANTS	First gymnosperms (treelike ...s—left); first ...lopment of ...r forests		**360** Floras dominated by seed ferns; first ferns		
INVERTEBRATES	**380** First spiders (*Attercopus*), terrestrial millipedes, centipedes, pseudoscorpions; first goniatites (earliest ammonoids); first nuculid bivalves	**375** First mytilid bivalves	**365** Mass extinction: of up to 70% of species, 50% of genera; 3 of 5 trilobite orders; corynexochids, odontopleurids, harpetids; nearly 90% of brachiopod genera, including atrypids (left), orthids, pentamerids; major loss of reefs	**360** First crabs	
VERTEBRATES	First lampreys; **380** First advanced amphibianlike, lobe-finned fishes, such as *Tiktaalik*	**375** First signs of viviparity (embryo develops inside mother) in fishes; first amphibian; first coelocanth	**365** Placoderms (armored jawed fishes) become extinct		

ATRYPA

DEVONIAN PLANTS

The development of land plants initiated during the Ordovician and Silurian continued rapidly during the Devonian, with an explosion of diversity and the appearance of many new and different kinds of plants. By the end of the Devonian, the forerunners of many groups of living plants had already evolved.

Land plants evolved from ancestors that lived in freshwater and key adaptations that help control water loss are present in various forms in early land plants. Early animals faced the same difficulty as they followed plants into the new kinds of ecosystems that were forming on land. Several different groups of animals moved onto land independently, but the colonization of land appears to have occurred only once in the evolutionary history of plants.

Adapting to life on land

The most important fossil occurrence for understanding the structure of early land plants, as well as for studying many aspects of early land ecosystems, was discovered in 1914 near the village of Rhynie, Scotland. The Rhynie Chert is an ancient geothermal wetland that is preserved in silica. Nearby volcanic activity appears to have been the cause of the remarkable preservation. In the Rhynie Chert, early land plants, along with fungi, algae, and diverse arthropod fauna, are preserved in their position of growth, exactly where they once lived over 400 million years ago. The fossil plants in the Rhynie Chert are exquisitely preserved. They include several different kinds of rhyniophytes, such as *Aglaophyton* and *Rhynia*, as well as the early lycopod *Asteroxylon*. Rhyniophytes have simple branching axes, similar to those of *Cooksonia*, but with more

variation in the form of the branching. Spores are produced in elongated sporangia at the branch tips. Axes of *Asteroxylon* are covered with small, flattened, bristlelike leaves, occasionally with small, kidney-shaped sporangia scattered among them. *Asteroxylon* is very similar to the modern clubmoss *Huperzia*.

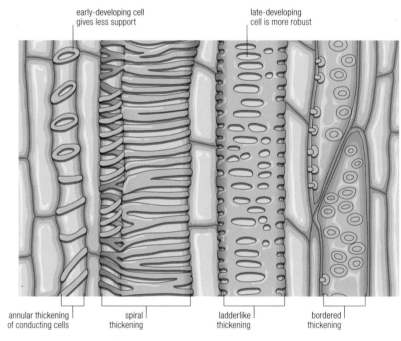

early-developing cell gives less support
late-developing cell is more robust

annular thickening of conducting cells | spiral thickening | ladderlike thickening | bordered thickening

WATER-CONDUCTING CELLS
The specialized cells that originally evolved for water conduction later became coopted to provide support, resulting in several different kinds of cells specialized for different functions.

GROUPS OVERVIEW

The Devonian saw the rapid appearance of many new and different kinds of plants. All were probably derived from a single common ancestor, exploiting the new ecological opportunities made possible by the ability to maintain a sufficient supply of water. The Devonian diversification of land plants is the botanical equivalent of the Cambrian Explosion (see p.68).

GREEN ALGAE
Many different lines of evidence indicate that various groups of freshwater green algae, especially the stonecrops and similar charophytes, are the closest living relatives of land plants. Beautifully preserved fossil algae very similar to living stonecrops are known from the Early Devonian Rhynie Chert.

RHYNIOPHYTES
Several different kinds of simple, early land plants are grouped together as rhyniophytes. The earliest forms were intermediate between mosses and more complex land plants. Branching of the sporophyte or spore-producing phase, which is seen in all rhyniophytes, resulted in the production of many sporangia rather than just one, which increased spore output.

ZOSTEROPHYLLS
Zosterophylls are a very prominent group of early land plants during the Early and Middle Devonian. Their kidney-shaped sporangia and the development of their water-conducting tissues indicate that they are closely related to living clubmosses. They differ from clubmosses in their flattened stems and the absence of leaves.

CLUBMOSSES
The earliest clubmosses are from the Late Silurian but the group diversifies rapidly during the Devonian, quickly differentiating into the three main subgroups that are recognized today. *Asteroxylon* is one of the best-known early clubmosses. Clubmosses have a continuous fossil record from the Devonian up to the present day.

On land, **water is scarce. Key problems faced by all land plants** are how to get water, how to keep it, and how to **use it efficiently.**

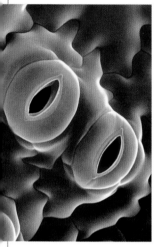

STOMATA
Land plants need to take in carbon dioxide for photosynthesis, but must also not lose too much water. Stomata are adjustable valves that open and close to control the exchange of gases between air spaces inside the plant and the atmosphere.

None of these early land plants are very large, but all have elongated cells in the center of their axes that helped conduct water from their lower parts, which were rooted in the soil, to the upper photosynthetic parts growing in the air. In some rhyniophytes, such as *Aglaophyton*, these water-conducting cells are similar to those of living mosses, which lack internal thickenings of their cell walls. In other early land plants, such as *Rhynia* and *Asteroxylon*, these internal thickenings are present, making these cells more similar to the specialized water-conducting cells of living ferns and lycopods. The outer surfaces of early land plants from the Rhynie Chert and other fossil localities are covered by a waxy cuticle, which helped reduce water loss from the plant into the air.

The struggle for light

During the Middle and Late Devonian, rhyniophytes were eventually displaced by larger plants with more complex branching, such as *Psilophyton*. Larger plants may have been more successful as the struggle for light intensified in early land ecosystems. *Psilophyton* and similar plants also show a new kind of growth form, with a single main axis and smaller lateral branches. These lateral branches subsequently became modified to form the different kinds of leaves seen in living ferns, horsetails, and seed plants. As the size of land plants continued to increase during the Devonian, more conducting cells were needed to keep the upper parts of the plants supplied with water. Increasingly, these cells also provided the internal support needed to keep the plants upright. In the Middle and Late Devonian, a new innovation was the ability to produce large quantities of water-conducting cells throughout the life of the plant. This gave some plants the capacity to produce significant amounts of woody tissue. The earliest true trees with woody trunks appear for the first time in the Late Devonian.

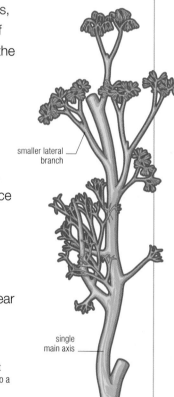

smaller lateral branch

single main axis

BRANCHING SYSTEMS
Psilophyton from the Early and Middle Devonian was larger than most rhyniophytes, and had more complex branching. This developed into a new growth pattern in which one branch dominated while others became lateral branch systems with limited growth capacity.

PRIMITIVE EUPHYLLOPHYTES

During the Devonian, the relatively simple rhyniophytes were replaced by larger forms—primitive euphyllophytes—with much more complex patterns of branching. It was these more complex forms that ultimately evolved into living horsetails, ferns, and seed plants.

HORSETAILS

Toward the end of the Devonian, there were several different kinds of primitive euphyllophytes with complex branching that were the forerunners of modern-day horsetails. By the end of the Devonian, *Archaeocalamites* showed many clear similarities to members of the living genus *Equisetum*.

PROGYMNOSPERMS

A key innovation seen among several groups of fossil plants from the Late Devonian was the capacity to produce additional water-conducting tissues throughout the life of the plant. These tissues comprise the wood of modern trees. One of these groups of Late Devonian plants was the progymnosperms.

PRIMITIVE SEED PLANTS

The oldest seed plants are known from the Late Devonian and probably evolved from progymnosperm ancestors. All the early seed plants are similar, making it likely that seeds evolved only once in the history of land plants. The origin of seeds was an important step toward new kinds of reproduction that were less dependent on water.

Prototaxites

GROUP Fungi
DATE Late Silurian to Devonian
SIZE Up to 26ft (8m) high
LOCATION Worldwide

Prototaxites is one of the world's most enigmatic macrofossils. It was first described in the late 1850s by Canadian geologist John William Dawson, who thought it was fossilized rotting wood. He named it *Prototaxites*, "first yew." *Prototaxites* fossils are sometimes found intact, or broken into short, cylindrical sections that typically have concentric markings resembling tree rings. However, under a microscope, their structure is clearly unlike wood; instead of plant cells with clearly defined walls, they have microscopic tubes that run vertically inside the "trunk." Individually, these tubes are thinner than a human hair, but they form a dense mass that can be over 3ft 3in (1m) across. Dawson interpreted these as fungal threads feeding on dead wood. However, later researchers dismissed his suggestion that *Prototaxites* was a tree, or even a plant. At various times, it has been interpreted as a kelplike alga, a lichen, and most recently, as the giant fruiting body of an ascomycete fungus.

barklike
"trunk"

concentric markings

PROTOTAXITES LOGANI CROSS-SECTION
It is easy to see why *Prototaxites* was taken for fossilized wood. Its surface texture strongly resembles tree bark, just as its internal structure mimics a tree's concentric growth rings. Only under the microscope do the differences become clear.

PROTOTAXITES LOGANI
At up to 26ft (8m) high, *Prototaxites* must have been quite imposing. Whatever its true classification, there can be no doubt that it dominated the Devonian landscape.

MYSTERIOUS IDENTITY

In 1872, botanist William Carruthers identified Prototaxites as the stalk of a giant, kelplike alga. In the 20th century, this theory was largely accepted, but a 2001 study by Francis Hueber of the Smithsonian Institution showed that a fungal identity was more likely. Recent findings of reproductive structures lend support to this idea, pointing to a relationship with sac fungi or ascomycetes.

Parka

GROUP Algae
DATE Late Silurian to Early Devonian
SIZE Up to 2¾in (7cm) across
LOCATION Worldwide

With a rounded outline and netlike surface, *Parka* is an intriguing fossil. In the past, it was mistaken for many objects, including the eggs of prehistoric arthropods or fish. However, its microscopic structure reveals that it was an alga, with a hard outer coat covering a flat body only a few dozen cells thick. The fossil's surface disks are tiny compartments, each filled with thousands of spores. Anatomical and chemical features link *Parka* to a group of algae called the coleochaetes. These are among the closest living relatives of green plants.

rounded, disklike
clusters

Parka was made up of many **disklike structures**, each containing **thousands of microscopic spores**.

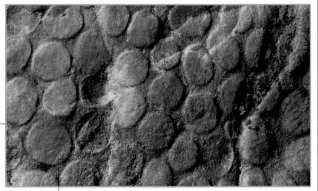

covered
disk

PARKA DECIPIENS
Each individual disk contained thousands of microscopic spores. These would have been dispersed and then each would have germinated to produce a new plant.

Rhynia

GROUP Rhyniophytes
DATE Early Devonian
SIZE 7in (18cm) high
LOCATION Scotland

Rhynia is one of the best-known early plants, with horizontal and upright branches but no true roots or leaves. The horizontal branches, called rhizomes, spread out as they colonized the ground and its upright branches divided repeatedly, creating a low-growing tangle that intercepted a maximum amount of light. It was one of the first vascular plants, with specialized tissues for carrying water and dissolved substances. It also had a waterproof outer covering and stomata: microscopic pores that could be opened or closed to regulate water loss and gas exchange. The tiny nodules scattered over its stems are less easily explained: these have been identified as signs of damage, dormant branches, and secretory structures.

RHYNIE FLORA
Rhynia and other primitive plants grew close to silica-rich hot springs. The springwater would periodically engulf them, preserving them as it cooled and its silica crystallized.

THE RHYNIE CHERT

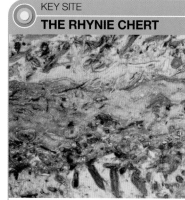

In 1914, Dr. William Mackie noticed some fossils in a garden wall in the Scottish village of Rhynie. The fossils were in fine-grained chert; when examined, they revealed an ecosystem containing some of the world's earliest-known terrestrial plants, as well as fossilized fungi and many extinct arthropods.

silicified stem

layered sediment

Aglaophyton

GROUP Rhyniophytes
DATE Early Devonian
SIZE 7in (18cm) high
LOCATION Scotland

One of the earliest terrestrial plants, *Aglaophyton* flourished near hot springs about 407 million years ago. Anchored by microscopic hairs, its creeping rhizomes produced upright branches that repeatedly divided into two. *Aglaophyton* had a waterproof outer covering to prevent dehydration, and microscopic pores (stomata) that could be closed by their special guard cells. Spore-producing, egg-shaped capsules grew on its tips.

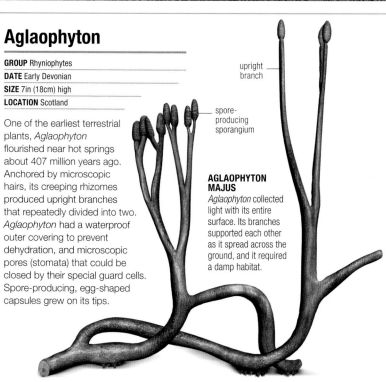

upright branch

spore-producing sporangium

AGLAOPHYTON MAJUS
Aglaophyton collected light with its entire surface. Its branches supported each other as it spread across the ground, and it required a damp habitat.

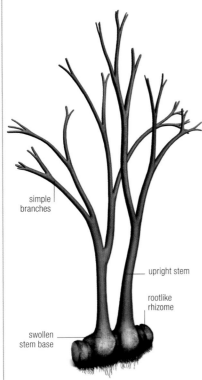

simple branches

upright stem

rootlike rhizome

swollen stem base

Horneophyton

GROUP Rhyniophytes
DATE Early Devonian
SIZE 8in (20cm) high
LOCATION Scotland

With its slender branches, *Horneophyton* resembled several other early plants found in Scotland's Rhynie Chert. However, its fossils show two unique features: swollen stem bases and lobed spore-producing organs (sporangia). Each of the lobes was cylindrical with a central internal column. This structure is found in living mosses, but the spore-bearing part of the *Horneophyton* life cycle is free-living—an evolutionary breakthrough absent in mosses. This set of features has made it difficult to classify. However, there is no doubt that it was successful in the Early Devonian, when it grew in miniature thickets on damp ground.

HORNEOPHYTON LIGNIERI
Growing at or just beneath ground level, *Horneophyton*'s rootlike rhizome may have helped keep the plant in place. They probably also facilitated growth in dense thickets.

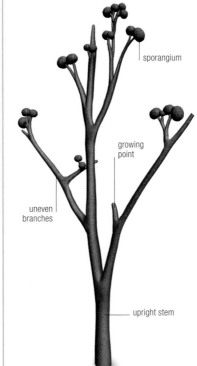

sporangium

growing point

uneven branches

upright stem

Renalia

GROUP Rhyniophytes
DATE Early Devonian
SIZE 12in (30cm) tall
LOCATION Canada

Many early land plants had dichotomous growth, meaning that their stems divided into equal halves. However, *Renalia* exhibited a different kind of growth, with unequal branching creating more complex shapes. Each of its stems had a leading growing point, with small branches splitting off on either side. These side branches bore the plant's spore-forming organs (sporangia). Fossils of *Renalia* come from the Gaspé Peninsula in Quebec, where they were first found in the 1970s. Exactly where it fits in plant evolution is uncertain, although its spore capsules may be a link with simple plants called zosterophylls.

RENALIA HUEBERI
Renalia's kidney-shaped sporangia were several millimeters across—a substantial size for an Early Devonian plant. When the spores inside were ripe, the capsules split open to release them into the air.

Zosterophyllum

GROUP Zosterophylls
DATE Late Silurian to Middle Devonian
SIZE 10in (25cm) high
LOCATION Worldwide

Preserved in fine-grained rock, fossils of *Zosterophyllum* have been found in many parts of the world. They first appeared in the Silurian, and by the Devonian, many different species had evolved. Known collectively as zosterophylls, they covered large areas of damp, low-lying ground.

Fossil zosterophylls were first thought to be related to *Zostera*, or eel-grass, which is one of the few flowering plants that grows in the sea. However, we now know they are not closely related. Zosterophylls may have been ancestors of lycopods, the plant group that included the giant clubmosses.

flattened branches

cluster of sporangia

ZOSTEROPHYLLUM RHENANUM
Zosterophylls reproduced by releasing spores. The spore-forming organs, or sporangia, were located along the sides of stems, not at the tips. Some were grouped together in clusters or primitive cones.

branching stem

ZOSTEROPHYLLUM RHENANUM
Zosterophyllum had a smooth stem that branched repeatedly from near the base. It also had an internal vascular system like the ones used by most of today's plants to carry water.

Discalis

GROUP Zosterophylls
DATE Early Devonian
SIZE 12in (30cm) high
LOCATION China

Pressed flat by ancient silt, fossils of *Discalis* show a plant that flourished nearly 400 million years ago. This zosterophyll was discovered in China in the late 1980s. Like other zosterophylls, *Discalis* did not have true roots or leaves and was made up of heavily branched stems. It grew by uncurling across the ground. The stems often formed H-shaped or K-shaped branches, as well as simply dividing in two. The branches may have helped bind the plant into a sturdy tuft. *Discalis* had spore capsules (sporangia) arranged in open spirals along the sides of its stems. Each sporangium was the size of a pea.

DISCALIS LONGISTIPA
The stems of *Discalis* were covered with minute spines, known as enations. These enations had button-shaped tips.

> *Discalis's* **spiky stems** are a **mystery.** Perhaps they were an early example of a plant **protecting itself** from animals.

siltstone

spine on stem

Sawdonia

GROUP Zosterophylls
DATE Devonian
SIZE 12in (30cm) high
LOCATION Northern Hemisphere

During the nineteenth century, *Sawdonia* was the subject of a case of mistaken identity, when two very different fossil plants were combined. Today, it is recognized as a typical zosterophyll, with creeping, rootlike rhizomes and upright stems. These grew by uncoiling, and formed paired branches. *Sawdonia's* water-conducting system may have given it support—a double role that is a key feature of modern vascular plants. *Sawdonia's* kidney-shaped spore capsules split into two halves to release their spores.

spiny stem

SAWDONIA ORNATA
Like many zosterophylls, *Sawdonia's* stems are covered with tiny spines. Even with the naked eye, the spines often show up well on fossils, giving the stems a distinctive saw-toothed edge.

Sciadophyton

GROUP Rhyniophytes
DATE Early Devonian
SIZE 2in (5cm) high
LOCATION Worldwide

The spore-producing form, or sporophyte, was just one phase in the life cycle of a primitive plant. The sporophyte dispersed spores, which grew into gametophytes, a second phase concerned with sexual reproduction. Gametophytes are seldom preserved as fossils, but *Sciadophyton* is one rare example. It had stalks ending in cups where the male and female sex cells (gametes) were produced. After rain, mature male cells swam through the water to fertilize the female ones—and sprout into new sporophytes. Gametophytes and sporophytes generally look very different. Several different types of primitive land plants seem to have produced *Sciadophyton*-like gametophytes, possibly including some zosterophylls.

SCIADOPHYTON
At least a dozen stalks grew from *Sciadophyton's* central point, and the upper surface of each stalk's cup held the gametangia: the organs that produced the eggs or the sperm.

Asteroxylon

GROUP Lycophytes

DATE Early to Middle Devonian

SIZE 20in (50cm) high

LOCATION Europe

With its covering of small, scalelike leaves, *Asteroxylon mackiei* is the most complex plant that has been found in the Rhynie Chert in Scotland; a second species found elsewhere in Europe is probably a misattributed zosterophyll. Its name, meaning "star wood," is a reference to the shape of its water-carrying vascular system, which was starlike in cross-section. The rootlike rhizomes under the ground were thinner than the stems above it—an arrangement that echoes the proportions of living plants. *Asteroxylon*'s stems were covered in leaflike scales. Like leaves, their purpose was to collect light, but the scales were constructed differently from true leaves. Instead, they were probably an early form of microphyll—the leaflike flaps that are a feature of modern lycophytes.

LIVING RELATIVES
LYCOPHYTES

Living lycophytes follow the pattern pioneered by *Asteroxylon* and have simple leaves crowded around their stems. The 1,290 living lycophyte species include clubmosses, as shown here; spikemosses (which are not mosses at all); and a group of pond-dwelling plants called quillworts.

ASTEROXYLON MACKIEI
Asteroxylon was one of the tallest plants of the Early Devonian. Unlike most plants from this time, its stems had an obvious main shoot with smaller branches at intervals along its sides.

Psilophyton

GROUP Psilophytes

DATE Early to Late Devonian

SIZE 24in (60cm) high

LOCATION Worldwide

First identified in 1859, *Psilophyton* was one of the earliest of the euphyllophytes, a group that includes the great majority of living plants. Many different species have been found across the world. Some were smooth-stemmed, and others were covered in spines. However, one species, *P. hedei*, described from Silurian rocks, turned out to be a colony of marine invertebrates rather than a plant (see p.97).

PSILOPHYTON BURNOTENSE
Like most early plants, *Psilophyton* was leafless. It grew in a twisted mesh of stems. Spores were produced from spore capsules borne on the fine tips of the side branches.

Eospermatopteris

GROUP Cladoxylopsids

DATE Middle Devonian

SIZE 26ft (8m) high

LOCATION North America

Eospermatopteris is the name given to the trunks that were previously known as *Wattieza*. It is a relative of *Calamophyton* (see below) and often described as the world's first tree. *Eospermatopteris* had a similar growth form to *Calamophyton* but was slightly larger, reaching a height of 26ft (8m). A grove of *Eospermatopteris* trunks was unearthed in New York in the 1870s, but the true nature of these trees remained unknown until the discovery of whole plants with attached leaves and reproductive organs in 2007.

EOSPERMATOPTERIS
Large natural sandstone cast of base of tree trunk. The longitudinal furrows reflect the root mantle.

Calamophyton

GROUP Cladoxylopsids

DATE Middle Devonian

SIZE 13ft (4m) high

LOCATION Worldwide

These large stems bearing numerous, tiny, finely divided branches belong to *Calamophyton*, which was among the earliest trees. *Calamophyton* trees were small, with slender, usually unbranched trunks growing to 13ft (4m) in height.

These were crowned with tufts of large leaflike branches, giving them a tree-fern-like growth habit. The spore capsules were at the branch tips. When ripe, the spores escaped through a lengthwise split in the capsule. *Calamophyton* grew in dense stands along with an understory of lycopods and aneurophytaleans (see opposite). Leaves were shed continuously, carpeting the floor in leaf litter that became home to numerous arthropods, including millipedes, centipedes, and the ancestors of modern spiders.

CALAMOPHYTON PRIMAEVUM
At first glance, the flattened structures that surround the stem of *Calamophyton* appear to be leaves or fronds. However, they are actually fine branches, which are subdivided many times from the plant's central stem.

rhizome

central stem

flattened branches

Archaeopteris

GROUP Progymnosperms
DATE Late Devonian
SIZE 26ft (8m) high
LOCATION Worldwide

With its tall trunk and spreading branches, *Archaeopteris* is the first tree to have formed forests on a truly global scale. It is also one of the first plants known to have had dense wood and true leaves. Study of *Archaeopteris* began in the second half of the 19th century, when its fossilized leaves were thought to belong to a low-growing fern. They were named *Archaeopteris*, or "ancient wing," because of their feathery appearance—a name similar to the fossil bird *Archaeopteryx* (see p.252). Nearly a hundred years passed before it was linked with fossilized trunks that showed it to be a tree. *Archaeopteris* had a similar outline and type of wood to many modern conifers. However, it belonged to an earlier group known as the progymnosperms.

FROND STRUCTURE
Archaeopteris's sweeping fronds bore thousands of flat leaflets. Its spores developed in cone-shaped clusters of capsules, usually at the base of each frond.

Xinicaulis

GROUP Cladoxylopsids
DATE Late Devonian
SIZE 20ft (6m) high
LOCATION Worldwide

The earliest trees had trunks that grew in a different way to any modern plant. Each trunk had a ring of hundreds of individual strands of xylem (water-conducting cells) that were interconnected in many places and an outer mantle of roots. Recent discoveries of trunks with anatomical preservation show how these trees could grow to a large size by the production of large amounts of soft tissues and new wood around the individual xylem strands. This general proliferation of tissues caused the trunks to expand in girth, forcing the wood strands farther apart. This form of growth is unique but bears some similarity to living palms.

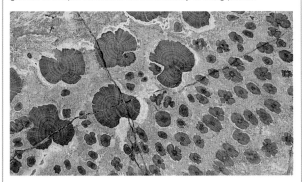

XINICAULIS LIGNESCENS
A cross-section of a partial trunk reveals large individual cylinders of wood surrounded by an outer zone of roots around 4in (10cm) thick.

Aneurophyton

GROUP Progymnosperms
DATE Middle to Late Devonian
SIZE 9¾ft (3m) high
LOCATION Northern Hemisphere

This bushy plant was related to *Archaeopteris* (see left), but it represents an earlier stage in progymnosperm evolution before the development of true leaves. One sign of this is that its stems branched in many different planes; this feature helped leafless plants like *Aneurophyton* trap more light. Unlike many other progymnosperms, *Aneurophyton* had only a small amount of wood, which suggests that it was a low-growing or straggling plant rather than a substantial tree. Elongated, complex organs (sporangia) that were borne in clusters produced spores. Divided into two groups, these organs, which were attached to shared stalks, looked like stubby fingers in a pair of hands.

branched stem

spore-producing organ

branching in many different planes

woody stem

ANEUROPHYTON SP.
Like other progymnosperms, *Aneurophyton* had a growth layer, or cambium, beneath the surface of its stems. This made the stems thicken as they lengthened—a characteristic shown by nearly all living trees.

Elkinsia

GROUP Primitive seed plant
DATE Late Devonian
SIZE 3ft 3in (1m) high
LOCATION USA

Elkinsia marks a great turning point in evolution because it is one of the earliest plants that is known to have produced seeds. It had straggling stems with two different kinds of fronds: some spread out to catch the light, while others divided into fine branches that ended in structures called cupules. The cupules contained seeds arranged in groups of four; each of these seeds were about ¼in (7mm) long. Unlike simple spores, seeds are complex packages of living cells that contain food reserves and an embryo plant.

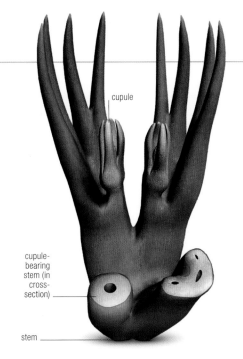

cupule

cupule-bearing stem (in cross-section)

stem

DEVONIAN
INVERTEBRATES

The Devonian was the time when the land turned green. From the small land plants of the Early Devonian came the tall forests of later periods, and with them came animal inhabitants, including invertebrates. But this period was a time of change in the marine realm, culminating in the Late Devonian mass extinction.

The plants of the Early Devonian were small, vascular, and generally leafless, although some had "scale leaves" like those of clubmosses living today. At Rhynie, in Scotland, an area of preserved hot springs (see p.113), many early plants have been fossilized in chert, a type of sedimentary rock.

Spread of terrestrial organisms

The simple plants of the Devonian were confined to damp environments, although by the Late Devonian a greater range of habitats had evolved and a number of animals appeared, including spiderlike arachnid forms (trigonotarbids), flightless insects, myriapods, and mites. Many invertebrates colonized the new, tall-tree environments that were the precursors of the great Carboniferous forests.

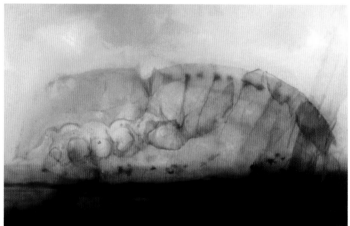

TRIGONOTARBID
This spiderlike *Palaeocharinus* was preserved within the Rhynie chert. The large fangs and stout legs suggest that it was a ground predator. Some specimens have been found within the fossilized stems of plants, which would have provided excellent cover for ambushing prey.

Late Devonian mass extinction

Throughout the Devonian, the globe cooled slowly. Representatives of several invertebrate groups, such as trilobites, gradually disappeared in stages through time due to a combination of environmental changes. Before the end of the Devonian, there was a dramatic catastrophe, the second of the great five Phanerozoic mass extinctions. The causes of this are still much debated. Sea levels had risen, but there is also some evidence for changes in ocean currents that may have disturbed stagnant bottom water and poisoned the upper waters of the sea. There is also increasing evidence for multiple meteorite impacts around the time of the extinction event.

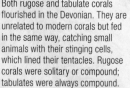

DEVONIAN SPIRIFERID
The two valves in this brachiopod opened along a straight hinge so that seawater could be sucked in for feeding and respiration.

hinge-line

brachial valve, seen from below

pedicle valve, seen from above

GROUPS OVERVIEW

Some invertebrate groups, such as sponges, became much more abundant in the Devonian. Corals continued to evolve throughout the period, with tabulate and rugose forms becoming widespread. Spiriferid brachiopods flourished, and some species grew very large during this time of global cooling. However, other groups, such as the trilobites, did not fare so well.

SPONGES
These originated in the Cambrian and were widespread by the Devonian. Tiny, spikelike structures called spicules supported the body, and these are often found in sediment. Sponges suck water in through pores in their baglike body and strain off food particles using specialized cells.

ANTHOZOANS
Both rugose and tabulate corals flourished in the Devonian. They are unrelated to modern corals but fed in the same way, catching small animals with their stinging cells, which lined their tentacles. Rugose corals were solitary or compound; tabulates were always compound.

SPIRIFERID BRACHIOPODS
These were typical of the Devonian and Carboniferous and superficially resembled bivalves. Inside the valve were two conical spires, pointing opposite ways, through which seawater was filtered and food particles strained off. A stout pedicle attached the brachiopod to the seafloor.

EURYPTERIDS
These large "water scorpions," often measuring at least 6½ft (2m) in length, inhabited fresh and brackish waters, as well as marginal marine environments. Most of them were ferocious predators. Some may have been able to crawl on land for short distances.

Heliophyllum

GROUP Anthozoans

DATE Early to Middle Devonian

SIZE Average diameter of cup 1¼in (3cm)

LOCATION Worldwide

Heliophyllum was a rugose coral that lived in the oceans more than 385 million years ago. As a fossil, it is normally found singly. Its skeleton, the corallum, is conical, although slightly irregular in shape. There are fine growth ridges on the outer walls. The shape of *Heliophyllum's* corallum indicates its growth history. Decreases in its diameter are the result of environmental stress, and bends reflect changes in growth orientation. The soft-bodied coral polyp would have lived in a shallow concave cavity at the top, known as a calice.

HELIOPHYLLUM SP.
The shallow cup (calice) at the top of the fossil is where the coral animal sat in life.

shallow cup

hard, ridged outer layer

ANATOMY
RUGOSE CORAL

Rugose corals existed in solitary or colonial forms. Solitary ones were often horn-shaped and are informally known as horn corals. The soft-bodied polyp rested in the concave calice at the top of the corallum. The inside of the coral was subdivided with radial plates (septa). The septa were typically of two kinds, major and minor, which alternated with each other. As the coral grew, horizontal partitions (tabulae) and small arched plates (dissepiments) were secreted.

soft-bodied polyp

horizontal tabulae

outer wall of corallum

Hydnoceras

GROUP Sponges

DATE Late Devonian

SIZE Up to 10in (25cm) long

LOCATION Europe, USA

Hydnoceras was a vase-shaped glass sponge that lived in the oceans between 385 and 360 million years ago. Seen in cross-section, it had twin walls, one inside the other. The inner wall surrounded a large central chamber. Like living glass sponges (see panel, right), the body of *Hydnoceras* had a skeleton made up of numerous similarly shaped structures known as spicules, made from silica. *Hydnoceras* spicules had six rays, each ray forming an angle of 90 degrees with the next, so that every spicule looked like a cross with a vertical bar running through its center. Some large outer spicules are visible on *Hydnoceras*, forming a surface pattern of squares. Inside each square are smaller squares.

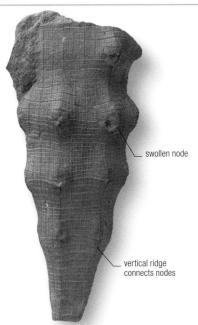

swollen node

vertical ridge connects nodes

HYDNOCERAS TUBEROSUM
Hydnoceras fossils, such as this, have a distinctive shape, with several waistlike constrictions around the body. Between these "waists" are expanded areas, each carrying about eight swollen nodes.

LIVING RELATIVE
GLASS SPONGES

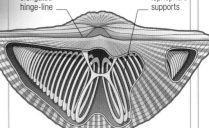

Hexactinellid sponges have a skeleton composed of typically six-rayed spicules of clear opaline silica fused together. The clear spicules give these animals their name of glass sponges. Perhaps the most famous of the living species is Venus's flower basket, *Euplectella aspergillum*. In some parts of the world, glass sponges form large reefs, such as those off the coast of British Columbia. They live at a range of depths but are mainly found in waters of about 660ft (200m).

Mucrospirifer

GROUP Brachiopods

DATE Middle Devonian

SIZE Up to 3¼in (8cm) across

LOCATION Worldwide

Mucrospirifer was a spiriferid brachiopod with an unusually long hinge-line. The widest point of the fossil shell, the hinge-line, runs between the two valves, coming to a point at each end.

In the center of the smaller (brachial) valve is a clear, V-shaped fold; there is a corresponding infold on the shell's larger (pedicle) valve. On either side of the brachial valve's central fold, there are strong radial ribs. The ribs themselves are crossed with growth lines. The growth lines are far stronger near the lower edge of the specimen shown here, suggesting that there was a pause in growth later on in this brachiopod's life.

ANATOMY
COILED LOPHOPHORES

Spiriferid brachiopods such as *Mucrospirifer* were elongated along their hinge-line. Shell models have shown that if open shells are situated facing the current, then internal spiral eddies (whirlpools) are produced. This meant that *Mucrospirifer's* coiled, calcite lophophore supports were able to maximize the food-gathering capacity of the animal's feeding structure, called the lophophore. This structure filtered food particles, such as organic matter or minute organisms, from the surrounding seawater.

elongated hinge-line

lophophore supports

upper valve's beak

growth lines

MUCROSPIRIFER MUCRONATA
Its distinctive winged shape gave *Mucrospirifer* its modern nickname of "butterfly shell."

fold

Cheiloceras

GROUP Cephalopods
DATE Late Devonian
SIZE Up to 1¾in (4.5cm) long
LOCATION Europe, Africa, Australia

Cheiloceras was a small goniatite. Its shell was tightly coiled, with a small, navel-like hollow, known as the umbilicus, at its center. The coiling of its shell was involute, meaning that the inner coils were hidden by the outermost one. In cross-section, *Cheiloceras* was somewhat flattened. Around the whorl were irregularly spaced constrictions, almost as if someone had pulled a string tight around the shell and left an impression. The significance of these constrictions is unknown, although it has been suggested that they may represent short periods of restricted growth. The shell also had faint ribs running around it, much less visible in *Cheiloceras* fossils than in those of many other ammonites. The suture line was gently curved.

constriction of whorl

CHEILOCERAS VERNEUILI
Cheiloceras was so tightly coiled that it seemed to be made up of only one large outer coil, as seen here.

Murchisonia

GROUP Gastropods
DATE Ordovician to Triassic
SIZE Up to 2in (5cm) long
LOCATION Europe, North America, Asia, Australia

Murchisonia existed on Earth longer than almost any other gastropod genus: a period of more than 200 million years. It was a sea creature that fed on algae and similar marine growth. *Murchisonia*'s shell was the only body part that fossilized and is notable for the spiral ridge, known as the slit band, that runs around the center of the whorl. This ridge can be traced forward to a slit at the edge of the mouth of the shell, a feature shown by many primitive gastropods, including the Jurassic *Pleurotomaria* (see p.228). In the specimen shown here, the whorls are ornamented with long nodes, although not all *Murchisonia* fossils exhibit these. Indeed, *Murchisonia* generally showed very little in the way of shell ornamentation. In the Middle Devonian, however, there was a sudden burst of evolution that produced many more ornamented forms.

MURCHISONIA BELINEATA
The slit band, which runs around the middle of each whorl, is clearly visible on this specimen. A slit in the shell's mouth would have enabled the animal to direct the exhalant current from the body cavity away from the head region.

node ornamentation

slit band

Eldregeops

GROUP Arthropods
DATE Middle to Late Devonian
SIZE Up to about 2¼in (6cm) long
LOCATION USA and Morocco

Eldregeops, along with its similar close relative *Phacops*, was one of the most characteristic trilobites of the Devonian period. Its headshield was dominated by a large, bun-shaped, central raised area known as a glabella, which in many species had a bumpy surface. The eyes, which were located close to the back of the glabella, were large and crescent shaped. They carried large, individual lenses arranged in a hexagonal pattern across the surface, and can easily be seen with the naked eye. Outside the eye, the cheek fell away at a steep angle. The hypostome, a broadly oval structure, lay beneath the front part of the glabella and was rigidly attached to the rest of the headshield at its front end. The thorax had 11 segments, and the tailshield was smaller than the head, with eight or nine well-defined segments on its axis and five to six ribs on the flanks, defined by deep furrows.

crescent-shaped eyes

tuberculated glabella

LARGE GLABELLA
One of *Eldregeops*'s most prominent features was its bun-shaped glabella, the raised area between its eyes, which in this specimen had a tuberculated surface.

ELDREGEOPS AFRICANUS
Like modern woodlice, most trilobites could roll up to protect their softer undersides, and *Eldregeops* was no exception.

glabella

tailshield

SEGMENTED BODY
The thorax was divided into 11 segments. The pleurae (the side parts of each segment) had rounded ends. The tailshield was smoothly curved.

⊙ ANATOMY

SOPHISTICATED VISION

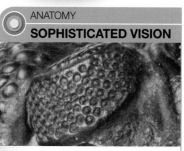

Eldregeops was characterized by eyes that had individual lenses arranged in a hexagonal pattern, each separated by thin walls of cuticle. Research in the early 1970s revealed that, in life, these lenses were composed of calcite crystals, which behaved like glass. Each lens had a bowl-like lower unit and an upper unit that was convex on both sides. Experimental models have demonstrated that the slight difference in refraction of the two units produced a sharp focus. Thus, these trilobites had sophisticated vision, and it is possible that it was used in a predatory feeding mode.

wide, shallow umbilicus

SOLICLYMENIA PARADOXA
This specimen clearly shows the distinctive triangular coiling. Not all *Soliclymenia* species exhibit this—some have the more usual circular coiling.

Soliclymenia

GROUP	Cephalopods
DATE	Late Devonian
SIZE	Up to 2in (5cm) across
LOCATION	Europe, N. Africa

The strange-looking *Soliclymenia* was one of very few ammonoids with triangular coiling. The whorls in its concave shell barely overlap, resulting in a wide and shallow umbilicus (navel-like indent) at the center of the coil. The whorls are ornamented by densely spaced ribs. *Soliclymenia* was a member of a group of early ammonoids that evolved in Europe or North Africa. Fossils from this group are distributed only in these areas, and only in the Late Devonian period. Unlike other ammonoids, their siphuncle—the internal tube that connected the living shell to the earlier chambers—migrated from the ventral (underside) of the shell to the dorsal (upper side) during early growth.

MINIATURE SQUID
With its soft, muscular body, the living *Soliclymenia* resembled a tiny squid. The animal used its long tentacles for capturing prey.

DEVONIAN VERTEBRATES

Often called the "Age of Fish," the Devonian saw a major radiation of fish groups both in marine and freshwater environments. By the Late Devonian, the stage was set for an invasion of the land, with the arrival of the first four-legged vertebrates, the tetrapods.

Devonian seas teemed with life, and a wide variety of fish evolved and spread around the globe, including the first true sharks. Jawless fish, such as the osteostracans, were the first to explore estuaries and lagoons and were followed by jawed predators, including placoderms and the sarcopterygians (bony fish with lobed fins). It was the sarcopterygians that gave rise to the first vertebrates to venture onto land.

The origin of tetrapods

Tetrapods are four-legged animals and include amphibians, reptiles, birds, and mammals. The earliest-known tetrapods appeared in the Late Devonian about 365 million years ago, in the form of *Acanthostega* and *Ichthyostega*. They are assigned to the Class Sarcopterygii and evolved from the lobe-finned fish. Recent research has established that the fish most closely related to the tetrapods is *Tiktaalik roseae*, whose fossils were found in 375-million-year-old rocks on Ellesmere Island in the Canadian Arctic in 2004. *Tiktaalik* and another closely related sarcopterygian fish, *Panderichthys rhombolepis*, which lived about 380

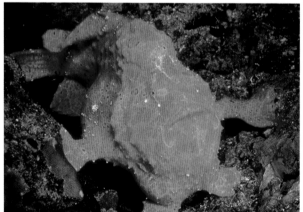

million years ago, have flattened skulls and dorsally placed eyes, similar to *Acanthostega*. One feature that links *Tiktaalik* more closely to *Acanthostega* is that its skull had become separated from the bony shoulder girdle so that the head could be moved more freely; *Panderichthys* retained the fishlike connection to the shoulder girdle.

The discovery of the fossils of *Acanthostega* and *Ichthyostega* overturned some entrenched misconceptions about the evolution of the first tetrapods. Originally, it was thought that tetrapods had five fingers and toes ("digits") when hands and feet evolved from fish fins. However, *Acanthostega* has eight digits, while *Ichthyostega* has seven (see right), indicating that the number of digits was variable and only became fixed at five later in tetrapod evolution. In addition, the idea that tetrapods first evolved when they lumbered onto the land also proved to be incorrect. The fossil specimens of *Acanthostega* and *Ichthyostega* were found in rocks in Greenland that were clearly deposited

BRACED FOR WALKING
Frogfish are actinopterygians that live in coral reefs. Instead of swimming, they use their pelvic and pectoral fins to "walk." As they move around, their left and right pelvic fins touch the coral independently, in a similar way to the walk of a tetrapod.

GROUPS OVERVIEW

The Age of Fish saw the beginning—and the end—of many groups of fish. Placoderms were a hugely successful and diverse group, but they survived for only 50 million years, whereas sharks became important fish and are still represented by about 400 species today. But the main event in terms of vertebrate evolution was the arrival of the tetrapods.

OSTEOSTRACANS

By the Early Devonian, osteostracans had diversified and were widely distributed in North America, Europe, and Asia. These jawless fish typically had bony, horseshoe-shaped headshields. They lived in marine or estuarine environments and were mostly bottom-dwellers. Osteostracans died out in the Late Devonian, about 370 million years ago.

HETEROSTRACANS

These heavily armored jawless fish arose in the Early Silurian. The most common subgroup in the Devonian was the pteraspids, characterized by large dorsal and ventral bony shields, but they were found only in the Northern Hemisphere. The psammosteids, which arose in the Early Devonian, were the only surviving heterostracans in the Devonian and form the largest known members of the group.

ARTHRODIRES

Arthrodires are one of the largest groups of placoderms. These early jawed fish had articulated armored plates on their head and neck and were found worldwide in the Devonian. One of the most famous is *Dunkleosteus*, which at 20ft (6m) long was one of the largest Devonian vertebrates. This fearsome predator had one of the most powerful bites ever known.

ANTIARCHS

Another group of placoderms, antiarchs had unusual front fins that were covered in bony plates; their exact function is still a matter of speculation. Antiarch jaws were small and weak compared to those of arthrodires, and they are thought to have scooped up mud containing tiny invertebrates. Antiarchs, along with all the rest of the placoderms, became extinct at the end of the Devonian.

The **earliest tetrapods** evolved in **estuaries** in the Late Devonian. They retained **tail fins** and **internal gills**—basically, they were **fish with legs**.

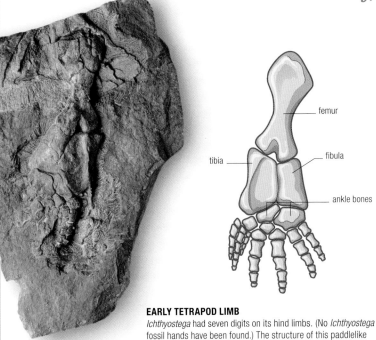

EARLY TETRAPOD LIMB
Ichthyostega had seven digits on its hind limbs. (No *Ichthyostega* fossil hands have been found.) The structure of this paddlelike limb indicates that it was probably of more use for swimming than walking, somewhat like a seal's flipper.

femur
fibula
tibia
ankle bones

ICHTHYOSTEGA FOSSIL LIMB

The origins of internal fertilization

The Gogo Formation in Western Australia is the remains of an extensive Devonian reef system (see p.43). Many placoderms have been preserved there in limestone nodules. These nodules can be dissolved with acetic acid, revealing the fossilized fish within. The preservation is so fine that placoderm muscle tissue has recently been discovered, as well as evidence of embryos within the body of the mother. In *Materpiscis attenboroughi*, the embryo is connected to its mother by an umbilical cord. These embryos provide some of the earliest evidence in the fossil record of internal fertilization and gestation. In external fertilization, sperm from the male mixes with eggs in the water and the embryos develop outside of the female.

in a river environment, indicating that these early tetrapods lived in streams or small rivers. The fossils are well preserved and nearly complete rather than being disarticulated, indicating that these animals died and were buried very close to where they lived, rather than being transported any great distance. Also, their paddlelike limbs were more suited to swimming than walking, and *Acanthostega* had a fishlike tail. Although they both had lungs and could breathe air, they retained internal gills. It seems that tetrapods in the Devonian were gradually adapting to the final transition from water to land; their conquest of the land would not be completed until the Carboniferous.

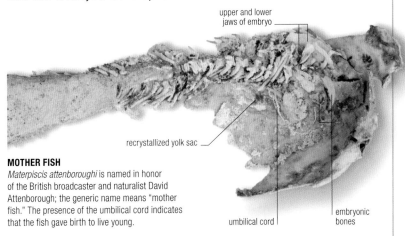

upper and lower jaws of embryo
recrystallized yolk sac
umbilical cord
embryonic bones

MOTHER FISH
Materpiscis attenboroughi is named in honor of the British broadcaster and naturalist David Attenborough; the generic name means "mother fish." The presence of the umbilical cord indicates that the fish gave birth to live young.

ACANTHODIANS

Acanthodians are characterized by the large, bony spines in front of their fins. (They are also known as "spiny sharks," but they are not true sharks.) Early acanthodians were exclusively marine, but by the Middle Devonian, these jawed fish had colonized freshwater systems. Their distinctive diamond-shaped scales are common in the fossil record, but skull and jaw bones are rare.

ACTINOPTERYGIANS

Actinopterygians, or ray-finned fish, first appeared in the Late Silurian; today, this highly diverse group has more than 28,000 species. As their name suggests, these bony fish have ray fins built from a fan of narrow bone or cartilage rods. *Cheirolepis*, from the Middle Devonian, is a primitive actinopterygian because its teeth lack the mineralized cap found in other actinopterygians.

ACTINISTIANS

These sarcopterygians arose in the Devonian and were highly successful. Coelacanths were thought to have died out with the last-known fossil genus, *Macropoma*, from the Cretaceous, until the spectacular discovery of a living coelacanth, *Latimera chalumnae*, in a fisherman's catch off eastern South Africa in 1938. In 1997, a second species was found in Indonesia.

DIPNOANS

This group of sarcopterygians are known as lungfish. Their "lung" is a modified swimbladder, which as well as aiding buoyancy (its usual function in fish) also absorbs oxygen and removes wastes. Dipnoans had a worldwide distribution in the Devonian. Today, there are three genera of living lungfish found in Australia (*Neoceratodus*), Africa (*Protopterus*), and South America (*Lepidosiren*).

TETRAPODS

Tetrapods include all four-legged vertebrates. When they evolved in the Late Devonian, they were still reliant on water for survival. Current research shows that today's lungfish are more closely related to tetrapods than any of the coelacanths. However, numerous fossil sarcopterygians are even closer relatives, the most recent discovery being *Tiktaalik roseae*.

Stethacanthus

GROUP Chondricthyans
DATE Late Devonian to Early Carboniferous
SIZE 5ft (1.5m) long
LOCATION North America, Scotland

Although the outline of *Stethacanthus* was similar to that of modern sharks, there are striking differences. The tail fin is almost symmetrical—in modern sharks, the upper lobe is usually much larger than the lower lobe. The pectoral fins had a long, distinctive rod, called the "whip," which trailed behind the fin. A third distinctive feature is the large structure projecting from its back. Sometimes called the "ironing board" because of its shape, its upper surface was studded with a dense patch of denticles—tiny, toothlike granules embedded in the skin. The only other place denticles appeared was in a patch on the head. These features, probably present only in males, may have been important for gripping females during mating.

Weigeltaspis

GROUP Agnathans
DATE Devonian
SIZE Approximately 4in (10cm) long
LOCATION Europe, North America

A primitive heterostracan (see p.122), *Weigeltaspis* lacked both jaws and paired appendages. The body scales, which were largest behind the single pair of branchial openings (gills) and smallest on the tail, bore a complex surface ornament. The substantial tail fin suggests that *Weigeltaspis* was likely to have been a relatively strong swimmer—the fin web was interleaved with at least five horizontal, scaly, fingerlike projections, which would have aided propulsion. *Weigeltaspis* was a relatively small fish that lacked the spines seen in later genera. It was probably a plankton feeder.

EFFICIENT SWIMMER
This *Weigeltaspis* fossil from the Ukraine clearly shows the streamlined shape and large tail of an active swimmer.

five scaly projections in fin web

Protopteraspis

GROUP Agnathans
DATE Late Silurian to Early Devonian
SIZE Unknown
LOCATION North America, Europe, Australia

The jawless pteraspids can be divided into five families, including Protopteraspididae, to which *Protopteraspis* belongs. This somewhat unspecialized fish had a narrow, rounded snout that was less elongated than those of many related genera. It also lacked the large, rear-facing projections, called cornua, that extended back from the sides of the bony headshields of other pteraspids. Instead, it had small points on each side, just behind the single gill opening. Like its relatives, however, *Protopteraspis* did possess a medium-sized dorsal spine. The plates covering the headshield had concentric, serrated ridges of dentine (calcified tissue). The headshield formed at an early age and grew by addition to the plate edges. In adults, these plates fused together. *Protopteraspis*'s body scales were small and diamond shaped.

Protopteraspis gosseleti

BOTTOM-DWELLING FISH
The fish on the right is *Protopteraspis gosseleti*, but the identity of its companion in death is not known. Because it was slightly flattened from top to bottom, *Protopteraspis* is thought to have been a bottom-dweller. It lived in freshwater.

Drepanaspis

GROUP Agnathans
DATE Early Devonian
SIZE 14in (35cm) long
LOCATION Europe

One of the earliest known psammosteid fish, *Drepanaspis* was small compared to its Late Devonian descendents, some of which reached 6$\frac{1}{2}$ft (2m) in length. Its flattened headshield featured a mosaic of small, scalelike plates that separated the midline plates from the paired ones on either side. These small plates appeared as the fish matured and were entirely absent in juveniles. Its flattened shape suggests it was a bottom-dwelling fish that foraged on the ocean floor. However, it is unclear how it fed because its jawless mouth pointed upward. *Drepanaspis* had no fins other than the ones on its tail.

PADDLE-SHAPED FISH
Distinctly paddle shaped when viewed from above, *Drepanaspis* was heavily armored and had a flattened headshield. This fossil of *Drepanaspis gemuendenaspis* is from Germany.

central crest

close-set eyes

long, narrow spine (cornua)

Zenaspis

GROUP Agnathans

DATE Early Devonian

SIZE 10in (25cm) long

LOCATION Europe

Zenaspis was a large osteostracan (see p.122) with a horseshoe-shaped head, which is also seen in the more primitive *Ateleaspis* (see p.105). *Zenaspis* differed from *Ateleaspis* in having long, narrow cornua (backward-pointing projections) on the head, and the lateral sensory fields were also somewhat narrower. The rear edge of the bony headshield was drawn up into a central crest that continued as a row of dorsal-ridge scales, then formed a single dorsal fin toward the back of the body. The pectoral fins, missing in this specimen, were smaller than those of *Ateleaspis* and had narrow, rather than broad, bases. The body scales were also much larger. The mouth (on the underside of the fish) lacked teeth and instead was lined with elongated oral plates. These merged into the scales covering the chamber that housed paired gill openings. Like *Ateleaspis*, *Zenaspis* was a bottom-dweller that lived in shallow seas or river mouths.

DISTINCTIVE HEADSHIELD
The headshield of *Zenaspis* can be clearly distinguished from its body, which is covered in bony scales. The eyes were close together and situated on top of the head where, as a bottom-dweller, *Zenaspis* could keep a lookout for predators.

Osteostracans **first appeared** in the Middle Silurian. They had **diversified** into **many forms** by the Early Devonian, but became extinct during the Late Devonian.

Lunaspis

GROUP Placoderms
DATE Early Devonian
SIZE 4–12in (10–30cm) long
LOCATION Germany

The heavily armored *Lunaspis* lived in the shallows of a sea that covered parts of Europe 400 million years ago. This distinctive flattened fish had plates that were heavily ornamented with concentric rings, and separate head- and trunk shields linked at a rudimentary hinge joint. On the headshield, the centrally located nuchal plate was very long and extended forward as far as the backs of the eye sockets, which were just on the top of its head. Because the jaws have not been preserved, little can be said of its diet, but its flattened shape suggests that *Lunaspis* lived on the seabed. Rigid pectoral spines projected at about 45 degrees from the body and extended to the back edge of the trunk shield. Instead of a dorsal fin, *Lunaspis* had three large ridged scales behind the median dorsal plate. The rest of the body was covered with bony scales that decreased in size toward the fish's tail, which appears to have been divided into two equal parts.

Rhamphodopsis

GROUP Placoderms
DATE Middle Devonian
SIZE 4½in (12cm) long
LOCATION Scotland

Unlike many placoderms, *Rhamphodopsis* was not heavily armored. Its body tapered into a whiplike tail. This small freshwater fish has the distinction of being one of the earliest examples of a fish in which the sexes can be distinguished by the form of their pelvic fins. Like sharks, the males had claspers—rodlike structures that aided internal fertilization. In females, the pelvic fins were covered in large scales. *Rhamphodopsis* had strong, crushing tooth plates, which initially led scientists to think that ptyctodontids, such as *Rhamphodopsis*, were closely related to sharks. However, because anatomically they share many more features with placoderms, ptyctodontids are assigned to that class.

Dicksonosteus

GROUP Placoderms
DATE Early Devonian
SIZE 4in (10cm) long
LOCATION Norway

This primitive arthrodire (see p.122) had a long trunk shield and long, curving, spinal plates that formed the front of an opening (the pectoral fenestra) for the pectoral fin. In *Dicksonosteus*, the median dorsal plate of the trunk shield was long and narrow and ended in a rounded lobe, rather than the spinelike tapering seen in *Coccosteus* (see opposite). The headshield had a small nuchal plate centered at the back and a pair of larger central plates. The armor as a whole was flattened dorsoventrally and adorned with concentric rows of tubercles. The two pairs of upper tooth plates were covered in robust, toothlike denticles, corresponding to those at the front of the lower jaw. The body tapered, but the tail is unknown.

Gemuendina

GROUP Placoderms
DATE Early Devonian
SIZE 10–12in (25–30cm) long
LOCATION Germany

A flattened fish with large, winglike pectoral fins, *Gemuendina* was a bottom-dweller with a shape similar to modern rays. Its eyes were on the top of its head, with nostril openings between them, and it had a wide mouth. Two large plates, the submarginals, covered the gill region on its sides. The only other identifiable plates on the head were the suborbitals on the front and outsides of the eye sockets, and the paranuchals on the headshield, which carried sensory canals. The rest of the cranium did not have a complete covering of bony plates, but was instead surrounded by a mosaic of small plates called tesserae. The very short trunk shield did not extend beyond the opening for the narrow-based pectoral fin. The pelvic fins comprised small, semicircular flaps, while the dorsal fin was reduced to a single dorsal spine. The anal fin was absent, and the tail, like the body, was long, tapering, and covered in scales. Larger toothlike scales were also present, randomly scattered over the fins.

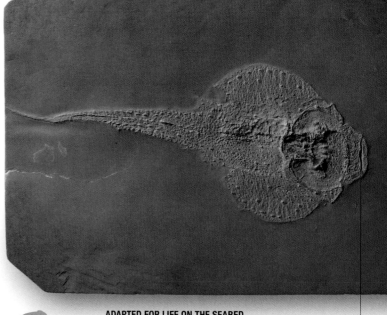

ADAPTED FOR LIFE ON THE SEABED
On this well-preserved specimen of *Gemuendina stuertzi*, the dorsally situated sense organs (eyes and nostrils) and mouth are clearly evident. Their positions are adaptations for living on the seafloor.

mouth

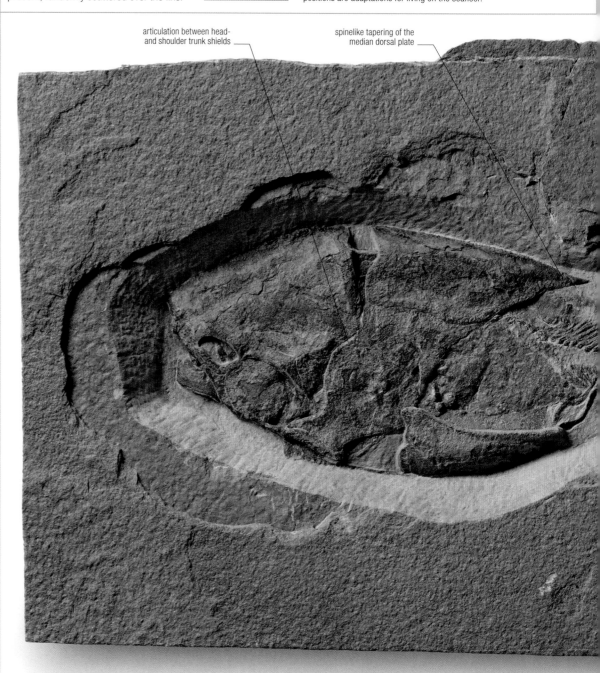

articulation between head- and shoulder trunk shields

spinelike tapering of the median dorsal plate

Rolfosteus

GROUP Placoderms
DATE Late Devonian
SIZE 12in (30cm) long
LOCATION Australia

Known from a single complete specimen and a few bone fragments, *Rolfosteus* was a bizarre, long-snouted fish that lived among the reefs of Western Australia. The rostral plate was greatly elongated, forming a tapering tube that comprised about half the length of the headshield. The function of this tubular snout is unknown. It may have been used to expose buried prey on the sandy seabed or to provide better streamlining so the fish could hunt shrimp living near the surface. *Rolfosteus* had flattened, crushing tooth plates at the back of the mouth, suggesting that small crustaceans and shellfish formed at least part of its diet. The snout could also have been a sexual characteristic of males that was used for display to attract females during the breeding season.

Coccosteus

GROUP Placoderms
DATE Middle to Late Devonian
SIZE 15½in (40cm) long
LOCATION North America, Europe

Probably the best known of all arthrodires, *Coccosteus* was first described in 1841, after which 49 species were identified, although many were later reclassified. This armored fish had large pectoral, dorsal, and tail fins and a powerful tail, suggesting that it was an able swimmer. The two pairs of tooth plates in the upper jaw initially had teeth, but these were gradually worn down by contact with the lower jaw to form sharp, bladelike cutting edges. Stomach contents reveal that *Coccosteus* was a predator of both acanthodians (see p.123) and juvenile arthrodires. It may have lain in wait on the seabed and ambushed prey, using its tail to power the attack. *Coccosteus* had a long, tapering median dorsal plate.

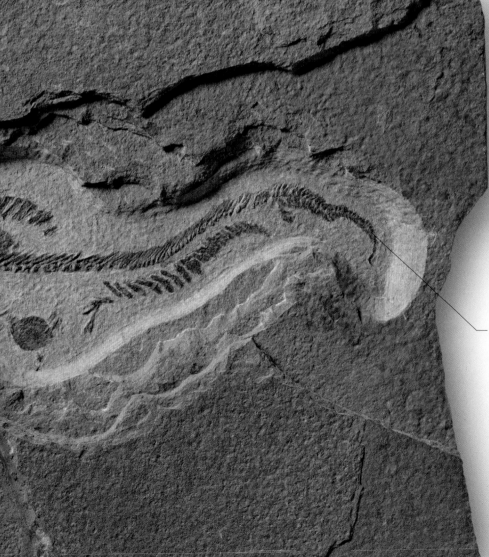

SKULL PLATES
The different plates on the roof of the skull are clearly evident in this specimen, as are some of the sensory lines that may be seen as grooves running across the surface of some plates.

eye socket

long, tapering tail

plate suture

DOUBLE JOINT
The head- and trunk shields of arthrodires are linked by a simple articulation on each side that allows the head to move only in a vertical plane (up and down). The "socket" is on the headshield and the "hinge" is on the trunk shield.

Dunkleosteus

GROUP Placoderms
DATE Late Devonian
SIZE 20ft (6m) long
LOCATION USA, Europe, Morocco

Dunkleosteus was a heavily armored fish, but its armor did not enclose its pectoral fins at the rear. This development allowed the fin base to increase in length and aided mobility, suggesting that *Dunkleosteus* was an active hunter in the shallow seas in which it lived. The upper and lower parts of its trunk shield were connected only by small, nonprojecting spinal plates. Its lower jaw and front upper tooth plate eventually wore down, leaving fanglike points. *Dunkleosteus* lacked ornament in most species, but its plates often bear bite marks and puncture wounds, indicating that, despite its size, it was itself hunted.

DISTINCTIVE PLATES
Dunkleosteus was covered in bony plates that were up to 2in (5cm) thick. The median dorsal plate (top) contrasts with *Coccosteus* (see p.127) in that it has a rounded posterior margin and lacks a posteriorly directed spine. The headshield (below) is equally distinctive, with its inwardly curving rear.

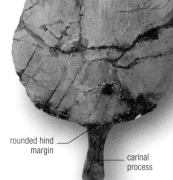

rounded hind margin

carinal process

eye socket

neck joint

FORMIDABLE PREDATOR
This huge fish had a massive bladelike lower jaw and upper tooth plates with sharp shearing edges. *Dunkleosteus* could open widely when taking hold of prey, its gape aided by the unarmored space between the head- and trunk shields.

Bothriolepis

GROUP Placoderms
DATE Middle Devonian to Early Carboniferous
SIZE 12in (30cm) long; a few reached 3ft 3in (1m)
LOCATION Australia, North America, Europe, China, Greenland, Antarctica

One of the best-known antiarchs (see p.122) with over 60 species, *Bothriolepis* had a somewhat clumsy appearance. It had a boxlike trunk shield with almost vertical sides and a flat bottom. The "roof" may have varied among species. The pectoral appendages were not finlike, but were gently curving spinelike structures, covered with small bony plates.

opening for eyes and nostrils

pectoral appendage

LONG APPENDAGES
The long pectoral appendages of *Bothriolepis* often extended beyond the rear of the trunk.

Pterichthyodes

GROUP Placoderms
DATE Middle Devonian
SIZE 8–12in (20–30cm) long
LOCATION Scotland

Pterichthyodes was a small, heavily armored fish with distinctive pectoral appendages, not unlike *Bothriolepis* (left) in appearance. However, the head-to-trunk proportions were different in that *Pterichthyodes* had a longer body and relatively smaller head, and the pectoral appendages were shorter in length. Having its eyes on the top of the head (and its ventrally flattened trunk shield) suggests that *Pterichthyodes* was a bottom-dweller. The body behind the armored parts tapered and was covered with overlapping cycloid scales. The single dorsal fin was triangular and was supported anteriorly by large arrow-shaped scales.

LAKE DWELLER
It has been suggested that *Pterichthyodes* may have used its pectoral appendages to crawl along the bottom of the ancient Scottish lakes in which it lived.

long body armor

short pectoral appendage

heterocercal tail

Cheiracanthus

GROUP Acanthodians
DATE Middle Devonian
SIZE 12in (30cm) long
LOCATION Scotland, Antarctica

Cheiracanthus was a freshwater fish. An active swimmer, it patrolled the middle depths of lakes and rivers, feeding on small prey items caught in its gaping jaws. It had no teeth, so presumably it filtered water through its long gill rakers to extract food. *Cheiracanthus* had large eyes, a large mouth, a deep body, and an upturned (heterocercal) tail. One of its defining characteristics is the lack of intermediate spines on the belly. Each fin was protected by a spine on the anterior edge. Its scales, which were small, nonoverlapping, and ribbed, have been found at many sites. Fin spines, which were loosely attached to the body, have also been discovered. Complete specimens are preserved only in the Old Red Sandstone deposits of Scotland.

dorsal fin spine

anal fin spine

pelvic fin spine

pectoral fin spine

Ischnacanthus

GROUP Acanthodians
DATE Late Silurian to Early Devonian
SIZE 1½–6½in (4–16cm)
LOCATION Scotland, Canada

Ischnacanthus had stout teeth that were larger toward the front. There was a whorl of smaller teeth where the two jawbones met. This freshwater-lake fish was slender and streamlined and had well-developed fins. The spines, which were long and narrow, are found only in association with fins; there were no intermediate spines. The armoring of the head and shoulder area is reduced compared to some earlier acanthodians, making *Ischnacanthus* capable of quick maneuvering while hunting or fleeing capture. The body was covered with small, polygonal scales, and the tail was sharklike.

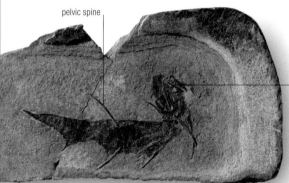

pelvic spine

teeth in upper and lower jaws

TRIANGULAR TEETH
Ischnacanthus had triangular-shaped teeth in its upper and lower jaws. Teeth were added at the front of the jaws, so they were larger, as they had not been worn down.

Cheirolepis

GROUP Actinopterygians
DATE Middle to Late Devonian
SIZE Up to 20in (50cm) long
LOCATION Scotland, Canada

A ray-finned fish, *Cheirolepis* is the most primitive member of the actinopterigian group known from complete specimens. It lived in shallow lakes that existed about 380 million years ago and had a long body covered in tiny interlocking diamond-shaped scales containing ganoine, a substance similar to tooth enamel. Its pectoral fins were fleshy and more like those of a modern lungfish. *Cheirolepis* was probably capable of swimming at high speeds to catch its prey, which it trapped in the many sharp teeth that lined its long jaws. The length of the jaws suggests that it could open its mouth wide and take in sizable food items, possibly up to two-thirds of its own size. Stomach contents have revealed a diet of fish—including its own kind.

dorsal fin

heterocercal tail

RAY-FINNED FISH
Cheirolepis had only one dorsal fin and a sharklike tail with a notochord (skeletal rod) extending into the longer upper lobe. Most of the fin web hung downward.

Dipterus

GROUP Sarcopterygians
DATE Middle Devonian
SIZE 14in (35cm) long
LOCATION Scotland, North America

Lungfish were widespread and diverse in Devonian times. Early forms were found in coastal saltwater sediments, but by the Late Devonian, they had moved almost exclusively to freshwater. *Dipterus* had a well-armored head, the top of which was covered with a mosaic of small bones. Some bones carried lines of pits that may have had a sensory or nutritional function. There was a large pair of tooth plates in the upper and lower jaws, across which ran rows of robust teeth that were probably used for crushing shellfish. A large plate covered the gill chamber, suggesting that gill breathing in lungfish was more important in Devonian times than breathing air.

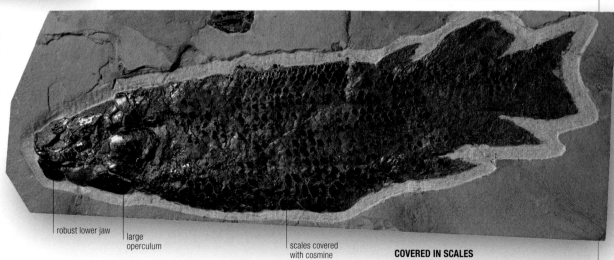

robust lower jaw

large operculum

scales covered with cosmine

COVERED IN SCALES
Rounded scales, coated with a thin, shiny layer of cosmine (a type of dentine), covered *Dipterus*. It had two dorsal fins far down its back; one anal fin; and long, thin pelvic and pectoral fins. The tail was sharklike in structure.

Fleurantia

GROUP Sarcopterygians
DATE Late Devonian
SIZE 10in (25cm) long
LOCATION Canada

Named after Point Fleurant in Miguasha National Park, Canada (see p.133), *Fleurantia* was an early lungfish. Its snout was longer and more slender than those of modern forms, and it also lacked the robust tooth plates that are typical of many lungfish species. Instead, it had rows of conical teeth and small denticles (toothlike projections) lining its palate. The body scales were large and round and lacked cosmine (see *Osteolepis*, p.133). *Fleurantia*'s largest fins were on its back (see below). Like *Dipterus* (see p.129), *Fleurantia* had an anal fin; long, thin pectoral and pelvic fins; and a heterocercal tail (the upper lobe is larger then the lower one). The body was deep and laterally compressed, which suggests that *Fleurantia* was a fast swimmer that hunted prey rather than foraging along the bottom for food. *Fleurantia* lived in fresh or brackish water along with another lungfish, *Scaumenacia*, with which it was initially confused.

DORSAL FINS
Fleurantia had two dorsal fins. The front fin was small and lobe-shaped, while the second was considerably elongated, running along about 25 percent of the fish's total body length.

large round scales

long rear dorsal fin

Holoptychius

GROUP Sarcopterygians
DATE Late Devonian
SIZE 6½ft (2m) long
LOCATION North America, Greenland, Latvia, Lithuania, Estonia, Russia

Holoptychius was a large fish that was a dominant and clearly very successful predator. Its remains have been found throughout the Late Devonian in several parts of the world. The main type of fossils that are found are sheets of large, bony scales, preserved as the body of the fish disintegrated—a single *Holoptychius* scale can be as big as a saucer. *Holoptychius* had long, paired fins rimmed by a fringe of fin rays, and its skull had a joint across the top of its head that allowed the snout to be raised even further when the mouth was opened. *Holoptychius* lived alongside fish such as placoderms, on which it may have fed, and it also ventured into some of the same habitats as the tetrapods, the earliest creatures with four legs (see pp.122–123).

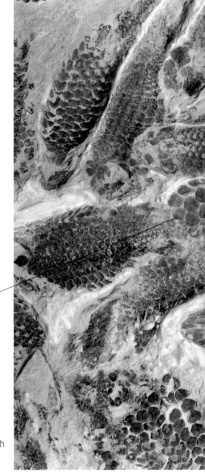

bony scales

MASS GRAVE
Here, many *Holoptychius* have died together along with other fish. The cause of death is unknown. *Holoptychius* belongs to the porolepiforms, whose nearest relatives are the modern lungfishes, although there is little more than a superficial resemblance between *Holoptychius* and any living species.

Tristichopterus

GROUP Sarcopterygians
DATE Middle Devonian
SIZE 12in (30cm) long
LOCATION Scotland, Latvia

Tristichopterus was a tetrapodomorph fish and the most primitive member of the Tristichopteridae, a family of about 10 genera that included *Eusthenopteron* (see p.132) and a huge predator called *Hyneria*. At 12in (30cm) long, *Tristichopterus* was a relatively small member of the family. Its primitive features included a very slightly asymmetrical tail, although the tail did have the three-lobed (trifid) pattern characteristic of the tristichopterids. Other features—aside from its age and location—that differentiate *Tristichopterus* from the better-known *Eusthenopteron* include the proportions of its head: the front part of its skull is relatively short compared to that of *Eusthenopteron*.

> Although they shared some features with **primitive tetrapods, tristichopterids** are not thought to be their direct ancestors.

paddle-shaped pectoral fin

Panderichthys

GROUP Sarcopterygians
DATE Middle to Late Devonian
SIZE 5ft (1.5m) long
LOCATION Latvia, Lithuania, Estonia, Russia

Panderichthys is one of three known Devonian fish that show many transitional, tetrapodlike features. Compared to other tetrapodomorph fish of the time, such as *Eusthenopteron* (see p.132), *Panderichthys* was more flattened top to bottom. And compared with most other fish, it had a longer snout and large eyes placed close together on the top of its head. Similarly, the rear part of the skull that covered the ear region and brain was comparatively short relative to the snout. The hinged joint across the skull that was present in many primitive tetrapodomorph fish was absent in *Panderichthys*. It is likely that it breathed air using a lunglike air bladder to a great extent, supplementing the use of its gills for respiration. *Panderichthys* retained fin rays on its paired fins, and scales all over its body, but fins on the back and tail were absent or reduced in size.

rear dorsal fin

three-lobed tail

LOBED FINS

As a lobe-finned fish, *Tristichopterus* had fins that contained robust bony supports rather than flimsy rays of bone or cartilage as in the fins of most modern fish. The pectoral fins—toward the front of the body—appear to be the most rounded, while the fins farther back are more pointed and streamlined.

Eusthenopteron

GROUP Sarcopterygians
DATE Late Devonian
SIZE 5ft (1.5m) long
LOCATION North America, Greenland, Scotland, Latvia, Lithuania, Estonia

The skeleton of *Eusthenopteron* exhibits clear evidence of the link between fish and land vertebrates. It was a late member of the Tristichopteridae, an extinct family of lobe-finned fishes. The small bones at the base of the pectoral and pelvic fins of *Eusthenopteron* can be traced to the bones—by now greatly enlarged—of the front- and hind limbs of early tetrapods,

such as *Ichthyostega* (see opposite). The structure of its vertebrae and its skull—especially the nostrils—also link this fish to the tetrapods. Long before modern scanning techniques and computer analysis were available, the details of *Eusthenopteron*'s skull were obtained using a more painstaking method. A fossil skull was ground down, millimeter by millimeter, and several photographs were taken at each stage. The sequence of images was used to make replicas of every thin section of skull out of slices of wax. The fossil was destroyed in the process, but the complete wax model produced allowed scientists to study the fish's skull, inside and out.

FOSSIL IMPRESSION
The soft internal parts of *Eusthenopteron* have not been preserved along with the hard skeleton. However, this specimen, which is an impression of the outside of the body, reveals that the fish was covered in large scales.

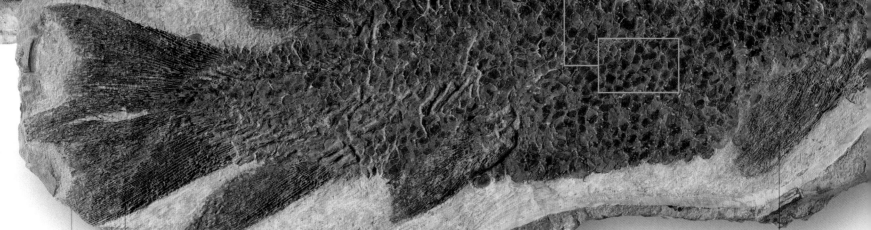

trilobed tail

strong pectoral fin

Tiktaalik

GROUP Sarcopterygians
DATE Late Devonian
SIZE 9¾ft (3m) long
LOCATION Canada

In many respects, *Tiktaalik* is a fish that shows even more tetrapodlike features than *Panderichthys* (see p.131). For example, *Panderichthys* has a series of bones that joins the skull to the shoulder girdle. There is also a series of bones that covers the gill region. These bones

are absent in tetrapods, and most of them are also absent in *Tiktaalik*—at least, they have never been found. And many of the tetrapod features that *Tiktaalik* does share with *Panderichthys* appear to be at a more advanced stage in the transition to land vertebrates. For example, it has an even longer snout and shorter rear portion of the skull than *Panderichthys* and a wider round notch (the spiracular notch) at the back of the skull.

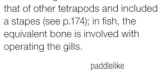

Acanthostega

GROUP Early tetrapods
DATE Late Devonian
SIZE 3ft 3in (1m) long
LOCATION Greenland

Acanthostega is a famous Devonian tetrapod. It had four limbs that ended in eight fingers or toes, but it also had some primitive features, such as fin rays along both the top and bottom of its tail. *Acanthostega* had a well-developed gill skeleton, so it probably used its gills for respiration, as well as having had an air bladder for breathing air. The ear was similar to that of other tetrapods and included a stapes (see p.174); in fish, the equivalent bone is involved with operating the gills.

paddlelike limb

BEST AS PADDLES
The limbs do not look like they were capable of bearing *Acanthostega*'s weight and were probably used as paddles. *Acanthostega* was probably an ambush predator that lived in shallow rivers.

location of hinge in skull

The name *Eusthenopteron* means "good strong fin."

AMBUSH PREDATOR
Eusthenopteron is believed to have been an ambush predator, hiding in aquatic vegetation to attack passing prey, as pike do today. Like *Osteolepis* and *Holoptychius*, it had a hinge across the top of its skull to aid in prey capture.

CASE STUDY

MIGUASHA NATIONAL PARK

Miguasha in Quebec, Canada, is one of the richest sites for Late Devonian fish fossils. The fish lived in coastal to brackish marginal marine environments, and representatives of all the major fish groups of the time, except chondrichthyans, are preserved in the rocks. Tetrapodomorphs such as *Eusthenopteron* have been discovered there, including a particularly important one called *Elpistostege*, a fish that seems to have been very like *Tiktaalik*. In addition to fish, many invertebrates, plants, algae, and other microorganisms combine to give a very detailed picture of Devonian life of the time.

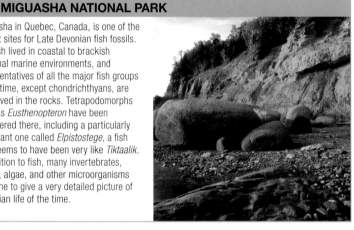

Osteolepis

GROUP	Sarcopterygians
DATE	Middle Devonian
SIZE	20in (50cm) long
LOCATION	Scotland, Latvia, Lithuania, Estonia

Osteolepis is one of the best-known early members of the tetrapodomorphs, the group that gave rise to the four-legged tetrapods. It was one of the first Devonian fish to be discovered and as a result lends its name to the Osteolepididae, an extinct family of lobe-finned fish that are a sister group to the tristichopterids, such as *Eusthenopteron* (opposite). Its fossils are common in the Middle Devonian rocks of Scotland, where it lived in a large, shallow lake known as the Orcadie Basin. Its bony scales and skull bones were covered with a kind of glossy, enamel-like substance called cosmine. Cosmine was full of pores, which are thought to have been the openings to a sensory system for detecting water currents around the fish. *Osteolepis*'s skull had a mobile joint across the top of the head, like that of *Holoptychius* (see p.130), which allowed the fish to open its mouth very wide. *Osteolepis* lived alongside many other kinds of fish, including the lungfish *Dipterus* and several placoderms.

HETEROCERCAL TAIL
The tail of *Osteolepis* had a distinctive heterocercal form, in which the upper lobe was longer than the lower one.

mouth

row of scales

SQUARE SCALES
The name *Osteolepis* means "bone scale," and the fish had roughly square scales that were arranged in rows running from top to bottom.

Ichthyostega

GROUP	Early tetrapods
DATE	Late Devonian
SIZE	5ft (1.5m) long
LOCATION	Greenland

Ichthyostega was the first Devonian tetrapod to be discovered. Its fossils come from rocks in East Greenland, and it lived at the same time as *Acanthostega*. *Ichthyostega* has many unique features that have puzzled paleontologists for decades, but new research is clarifying some aspects of its structure. Its ear region, for example, appears to have been adapted for hearing in water rather than in air. *Ichthyostega* certainly spent some of its time in water, but its vertebral column seems to have been adapted to a form of terrestrial locomotion. Its body was encased in broad, overlapping ribs, which together with its robust shoulders and forearms suggest that strong muscles were used to pull the animal along on land. Its paddlelike hind limbs were better-adapted for swimming.

SEVEN-TOED FOOT
Ichthyostega had three small toes at the leading edge of its foot, followed by four stout ones. The foot was held so that the leading edge would have cut into the sand or mud on which it walked to give it some purchase.

Ichthyostega's **large teeth** point to it being a **predator**, but did it **hunt** on land, in water— **or both**?

stout middle toe

leading edge of foot

CARBONIFEROUS

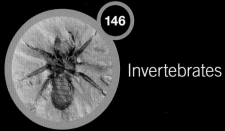
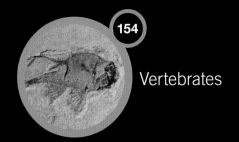

Carboniferous

Carboniferous, meaning "carbon-bearing," is an appropriate name for a period when lush swamps laid down huge coal deposits across the globe. Yet this was largely an icehouse world, with extensive ice sheets lasting tens of millions of years. Earth's crust continued to shift as two massive continents, Euramerica and Gondwana, slowly came together to form an even larger continental land mass, Pangea (Greek for "whole Earth").

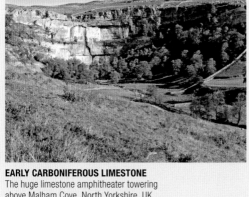

EARLY CARBONIFEROUS LIMESTONE
The huge limestone amphitheater towering above Malham Cove, North Yorkshire, UK, consists of fossilized creatures deposited on the beds of Carboniferous seas.

Oceans and continents

During the Early Carboniferous the continent of Euramerica moved even closer toward Gondwana, reducing the Rheic Ocean to little more than a narrow seaway between the western end of Gondwana and the southwestern tip of Euramerica. The mountain-building that resulted culminated in the formation of the Appalachian and Variscan chains. The Paleo-Tethys Ocean was encircled by Euramerica and Gondwana to its west, and the smaller islands of North China and South China to the east. Gondwana extended from low latitudes in the Southern Hemisphere to the South Pole, where a major ice cap began to develop in the Early Carboniferous. By the Late Carboniferous, Euramerica and Gondwana had merged completely to form the supercontinent of Pangea, extending from high in the Northern Hemisphere across the Equator to the South Pole. The ice caps around the South Pole spread to cover much of the former continent of Gondwana. At this time the Panthalassic Ocean lay to the west of Pangea, and the Paleo-Tethys Ocean to the east. On the continents, forests continued to expand and, with them, land animals diversified, including invertebrates, amphibians, and the first vertebrates capable of laying eggs on land. Large-scale glaciation affected sea levels, which rose and fell with the advances and retreats of ice, affecting coastal and offshore habitats. There were periods when seawater flooded into coastal swamps, and others when shallow bays, deltas, and inlets dried out. These fluctuations are reflected in Carboniferous sedimentary deposits, including alternating bands of shale and coal.

URAL MOUNTAINS
The Ural Mountains, now extending 1,500 miles (2,500km) across west–central Russia were created by the collision between Kazakhstania, Siberia, and Euramerica 250 to 350 million years ago.

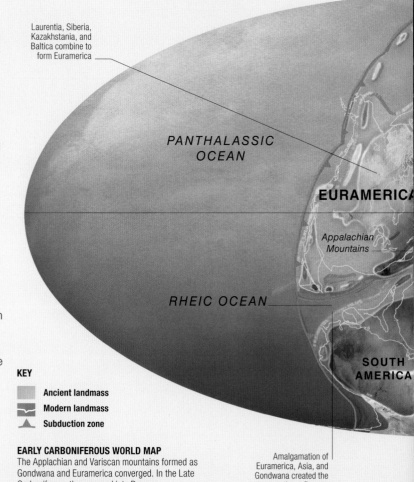

Laurentia, Siberia, Kazakhstania, and Baltica combine to form Euramerica

PANTHALASSIC OCEAN

EURAMERICA

Appalachian Mountains

RHEIC OCEAN

SOUTH AMERICA

Amalgamation of Euramerica, Asia, and Gondwana created the supercontinent Pangea

KEY

- Ancient landmass
- Modern landmass
- Subduction zone

EARLY CARBONIFEROUS WORLD MAP
The Applachian and Variscan mountains formed as Gondwana and Euramerica converged. In the Late Carboniferous they merged into Pangea.

MISSISSIPPIAN

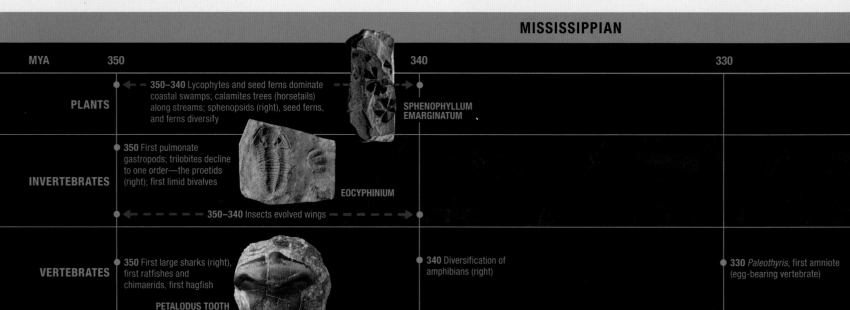

MYA	350	340	330
PLANTS	350–340 Lycophytes and seed ferns dominate coastal swamps; calamites trees (horsetails) along streams; sphenopsids (right), seed ferns, and ferns diversify	SPHENOPHYLLUM EMARGINATUM	
INVERTEBRATES	350 First pulmonate gastropods; trilobites decline to one order—the proetids (right); first limid bivalves EOCYPHINIUM		
	350–340 Insects evolved wings		
VERTEBRATES	350 First large sharks (right), first ratfishes and chimaerids; first hagfish PETALODUS TOOTH	340 Diversification of amphibians (right)	330 *Paleothyris*, first amniote (egg-bearing vertebrate)

ICE SHEET
Massive ice sheets advanced and melted at intervals during the Carboniferous. The maximum extent of glacial ice in this icehouse world was between 305 and 315 million years ago.

Climate

During the Early Carboniferous overall global temperatures continued to increase. A broad tropical belt extended from Euramerica across the Paleo-Tethys to eastern Gondwana. This was separated by a narrow arid belt from the cooler and more temperate conditions that extended into higher latitudes. Although there is some evidence for localized glaciations in southern Gondwana beginning around 350 million years ago, these were of limited extent. The formation of Pangea had a major impact on air circulation patterns and therefore on climate, because such a vast land mass would have experienced climatic extremes. A major glaciation began about 330 million years ago, centered on the southern Gondwanan part of the supercontinent of Pangea. Through the Carboniferous the size of the glaciated regions grew, stretching from South America and Australia into other regions of the Gondwanan part of Pangea, namely present-day India, Africa, and Antarctica, and reaching as far as 35 degrees latitude.

As the ice sheets expanded, the tropical belt in low latitudes reduced in width. An arid belt separated this from the cool temperate region, the warm temperate belt having disappeared. The end of this great ice age is not known with any certainty, but it was at some time between 255 and 270 million years ago, in the Permian period. Despite the glaciation, in lower latitudes close to the Equator and during the many interglacial episodes, conditions were warmer and more humid, resulting in the formation of substantial coal beds. Widely deposited evaporites in some areas of lower latitude also demonstrate the existence of these warmer regions.

WESTPHALIAN COAL BED
As cold and warm periods alternated, coastal swamps were flooded, forming peat bogs that became coal seams, like this one in Germany.

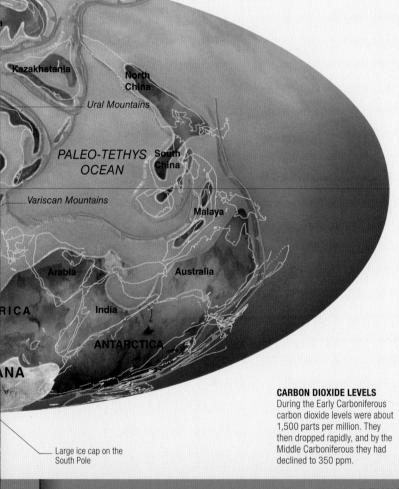

Kazakhstania

North China

Ural Mountains

PALEO-TETHYS OCEAN

South China

Variscan Mountains

Malaya

Arabia

Australia

...RICA

India

ANTARCTICA

...ANA

Large ice cap on the South Pole

CARBON DIOXIDE LEVELS
During the Early Carboniferous carbon dioxide levels were about 1,500 parts per million. They then dropped rapidly, and by the Middle Carboniferous they had declined to 350 ppm.

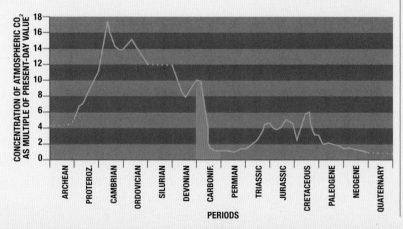

CONCENTRATION OF ATMOSPHERIC CO_2 AS MULTIPLE OF PRESENT-DAY VALUE

18 16 14 12 10 8 6 4 2 0

ARCHEAN | PROTEROZ. | CAMBRIAN | ORDOVICIAN | SILURIAN | DEVONIAN | CARBONIF. | PERMIAN | TRIASSIC | JURASSIC | CRETACEOUS | PALEOGENE | NEOGENE | QUATERNARY

PERIODS

PENNSYLVANIAN

320   310   300   MYA

● **320** First conifers and glossopterids; lianas and epiphytes develop in forests dominated by lycopsids, cordaites, and seed plants; small ferns (right) dominate groundcover

ZEILLERIA

PLANTS

325 First dragonflies; diversification of winged insects and of plant-eating by insects

● **320** First giant arthropleurids (large millipedelike arthropods); millipedes, scorpions, spiders diversify; blattoids (roaches, e.g. *Archimylacris*) dominant insects

● **310** Winged insects become more diverse; giant dragonflies (with 25in/63cm wingspan); last edrioasteroids

ARCHIMYLACRIS

INVERTEBRATES

● **310** First herbivorous tetrapods (diadectid cotylosaurs); dominant amphibians are temnospondyls; first synapsids

● **305** First diapsids (a type of reptile, e.g. *Petrolacosaurus*)

VERTEBRATES

CARBONIFEROUS
PLANTS

By the Carboniferous, many different kinds of large and complex plants had evolved. Giant clubmoss trees dominated the coal swamps. Tree ferns and giant horsetails, as well as several groups of primitive seed plants, grew on riverbanks and in drier areas.

Carboniferous plants reached levels of structural diversity and biological sophistication well beyond that of their Devonian precursors. Treelike forms developed in several groups. There were also herbaceous plants, as well as scramblers and climbers. Carboniferous plants formed new and more complex ecosystems.

COAL BALLS
Much of our knowledge of Carboniferous plants comes from the study of coal balls—calcified pieces of peat that contain the preserved remains of coal swamp plants. This specimen shows a cross-section through the fossil stem *Medullosa*.

Seed plants

Several different groups of seed plants are recognized for the first time during the Carboniferous. Lyginopterids had fernlike leaves and generally produced relatively small seeds, similar to those of the earliest seed plants from the Devonian. As in these very early seed plants, the pollen grains of lyginopterids were indistinguishable from the spores of primitive euphyllophytes (see p.111). Medullosans were generally larger plants, often with large, fernlike leaves. They had unusually large pollen grains, and some also produced very large seeds. Cordaites, which may be closely related to living conifers, had strap-shaped leaves with parallel veins. They produced small seeds and, like conifers, their pollen grains had either one or two attached air sacs, which would have given them additional buoyancy both in air and in fluids. They may have had a pollination system similar to that of living conifers. Some cordaites were trees, while others were scrambling plants. None of these Carboniferous seed plants appears to have produced seeds that could remain dormant for long periods.

Coal swamps

Giant lycopods, such as *Lepidodendron* and *Lepidophloios*, dominated the tropical coal swamps of eastern North America and Europe during the Carboniferous. These remarkable plants grew in a very strange way, quite unlike that of any living tree. They also reproduced in several different ways. The coal swamps, and the strange plants that inhabited them, declined in importance toward the end of the Carboniferous as the wet tropical conditions in which they flourished gradually became increasingly restricted.

LEPIDODENDRON
Impressions of the bark of *Lepidodendron* and other clubmoss trees are among the most common and characteristic Carboniferous fossils. The diamond-shaped marks were formed by the cushions on which the leaves were borne.

GROUPS OVERVIEW

In the Carboniferous, further diversification of land plants resulted in several groups that can be clearly related to living plants. However, all the Carboniferous seed plants appear to represent a relatively primitive level of organization that is less specialized than what is seen in any seed plant living today.

CLUBMOSSES
The diversification of clubmosses continued through the Carboniferous. In particular, ancient relatives of modern quillworts (*Isoetes*) evolved into very large and treelike forms. Unlike most modern trees, these lycopods had relatively little wood and were instead supported by barklike tissue on the outside of the stem.

HORSETAILS
By the Carboniferous, horsetails had evolved into treelike forms, although these were much smaller than the trees of the giant clubmosses. The stems of these plants are usually assigned to the genus *Calamites*, while the leaves are assigned to *Annularia*. These ancient horsetails grew in similar kinds of habitats to their modern relative *Equisetum*.

FERNS
Ferns probably arose from among the euphyllophytes of the Middle and Late Devonian, but by the Carboniferous, they had evolved into many different forms. Especially important were a group of tree ferns with stems assigned to the genus *Psaronius* and leaves assigned to the genus *Pecopteris*.

MEDULLOSANS
Medullosans were one of several groups of primitive seed plants that flourished during the Carboniferous. They were shrubs or small trees with large, fernlike leaves. The name of the group comes from the name given to their stems: *Medullosa*. The leaves are usually assigned to the genera *Neuropteris* and *Alethopteris*.

Thallites

GROUP Bryophytes
DATE Devonian to present
SIZE Spread up to 6in (15cm)
LOCATION Worldwide

Only a few cells thick when alive, *Thallites* is often little more than a film of carbon when fossilized. Precise identification is often uncertain: some *Thallites* specimens are probably algae, while others are liverworts. Liverworts are among the simplest of land plants and do not have roots, stems, or leaves. Their plant body—often a flat green ribbon known as a thallus—clings to solid surfaces by tiny hairs or rhizoids and spreads by branching into two. Plants this simple do not have an internal system for moving water, so they need to live in damp habitats.

THALLITES HALLEI
The delicate pattern seen in this Permian species shows how the plant repeatedly divided and spread across the surface of the rock to which it clings.

flat ribbon

Sigillaria

GROUP Lycophytes
DATE Carboniferous to Permian
SIZE 82ft (25m) high
LOCATION Worldwide

Like *Lepidodendron* (see below left), *Sigillaria* is known first and foremost for its fossilized bark. In one of the most common forms, this looks like an embossed replica of a honeycomb, with six-sided cells arranged with almost mathematical precision. Each cell—or scar—marks the position of a leaf cushion, each of which bore a single grasslike leaf. As the tree grew, the oldest leaves were shed, to be replaced by new ones growing higher up the trunk. As a young plant, *Sigillaria* echoed *Lepidodendron*'s growth pattern. But as they reached canopy height, the two trees matured in very different ways. *Lepidodendron* formed a high crown, with many leafy branches, but *Sigillaria* forked only once, or occasionally twice, producing limbs that looked like huge bottlebrushes. Like all giant clubmosses, it reproduced by forming spores inside cones.

narrow stem with parallel ribs

SIGILLARIA ALVEOLARIS
In this species, the leaf bases sit in furrows, separated by vertical ribs.

SIGILLARIA MAMMILLARIS
The fossilized bark of *Sigillaria* can often be found among coal-bearing rocks. In this fossil, the regular pattern formed by the scars where the tree's leaves were once attached is clearly visible.

six-sided cushion with a leaf scar

Lepidodendron

GROUP Lycophytes
DATE Carboniferous
SIZE 130ft (40m) tall
LOCATION Worldwide

With its scaly bark and towering trunk, *Lepidodendron* is one of the most famous of all prehistoric plants. A giant clubmoss, its fossils show many features linking it with its smaller relatives from Devonian times. Its narrow leaves were spirally arranged, and it reproduced by forming spores inside cones. It spent half of its life as a branchless pole that emerged from the forest floor. As it reached maturity, its upward growth slowed, and branches appeared. Finally, cones formed at the branch tips, and upward growth stopped, as the plant put all its energy into making and shedding spores. In some species, the tree died once this task was complete.

TOWERING TRUNKS
In the lush conditions of the Carboniferous, *Lepidodendron aculeatum* could reach about 130ft (40m) in just 20 years.

LEPIDODENDRON ACULEATUM
Lepidodendron is, strictly, the name given to the bark, which has a pattern of embossed diamond-shaped scars.

cigar-shaped cone

LEPIDOSTROBO-PHYLLUM
This scale, with a sporangium at its base, is one of the smallest visible parts of a *Lepidodendron*.

STIGMARIA FICOIDES
Stigmaria are fossilized stumps and rooting systems of giant clubmoss trees, buried under seams of coal.

LEPIDO-STROBUS OLYRI
Lepidostrobus is the generic name used for the cones of *Lepidodendron*, which grew at the branch tips.

ULODENDRON MAJUS
The name *Ulodendron* is used for stems that show scars. Unlike most living trees, giant clubmosses did not heal scars by growing new bark.

FOREST GIANT
Giant clubmosses such as *Sigillaria* dominated the swampy forests of the Carboniferous, but had dwindled dramatically by the end of the period.

SELAGINELLITES SP.
Easily mistaken for a branch tip, *Selaginellites* had a flat, creeping form with small leaves.

small leaves

branching stem

Selaginellites

GROUP	Lycophytes
DATE	Carboniferous to present
SIZE	6in (15cm) tall
LOCATION	Worldwide

Tree-sized clubmosses were not the only successful lycophytes during the Carboniferous period. There were creeping forms like *Selaginellites*, which is a relative of modern *Selaginella* (see p.219). With its small, spirally arranged leaves of different sizes and dichotomous, branching growth, *Selaginellites* had a strong resemblance to the low-growing clubmosses with the same name that have survived to the present day. It is a relatively rare fossil, and the earliest specimens have been dated to the Early Carboniferous. In Coal Age rocks, shoots of *Selaginellites* should not be confused with the tips of branches from giant clubmosses.

Lepidophloios

GROUP	Lycophytes
DATE	Carboniferous
SIZE	82ft (25m) tall
LOCATION	Worldwide

Bearing a striking resemblance to a snake's skin, this fossilized bark comes from a giant clubmoss that lived at the same time as *Lepidodendron* (see p.139). Although not as tall as its more famous relative, *Lepidophloios* was just as widespread. It grew in peat-rich forests rather than muddy estuarine swamps. *Lepidophloios* produced the largest spores of any primitive vascular plant, and these were probably fertilized aquatically following dispersal.

leaf-scar

LEPIDOPHLOIOS SCOTICUS
The shape of the scars shows that *Lepidophloios*'s leaves were relatively flat and grew in a tight spiral.

Sphenophyllum

GROUP	Sphenophytes
DATE	Carboniferous to Permian
SIZE	3ft 3in (1m) tall
LOCATION	Worldwide

During the Carboniferous period, not all horsetails evolved into treelike forms. Some—including *Sphenophyllum*—were low-growing or climbing plants that lived in damp places and on forest floors. Despite being soft-stemmed, they were ideally placed to form fossils, and their preserved remains are common in Coal Age rocks all over the world. *Sphenophyllum*'s leaves were arranged in whorls, but they had a wide variety of shapes. Some resemble the "pressed flower" appearance of *Annularia* leaves (see opposite), while others end in small hooks that would have helped them climb.

wedge-shaped leaf

SPHENOPHYLLUM EMARGINATUM
Sphenophyllum literally means "wedge-shaped leaves"—an indication that this plant looked quite different to the horsetails alive today.

Bowmanites

GROUP	Sphenophytes
DATE	Carboniferous to Permian
SIZE	Cones up to 3¼in (8cm) long
LOCATION	Worldwide

This name is used for the fossilized cones of *Sphenophyllum* (see right). These soft cones formed on short, fertile shoots, with up to 20 closely set whorls of bracts, or sporangium-bearing leaves. The spores were microscopic, and a single cone could produce many thousands. Those few that were able to germinate produced the gamete-forming body, which, in turn, produced male and female cells. Then, once fertilization occurred, a new spore-forming plant started to grow.

spore-bearing bract

BOWMANITES SP.
In most *Bowmanites* cones, the bracts are sharply upturned at their tips. This fossil shows a mature cone, which has opened out to release its spores.

Calamites

GROUP Sphenophytes
DATE Carboniferous to Permian
SIZE Up to 66ft (20m) tall
LOCATION Worldwide

Once the name given to its stem fossils, but now commonly used for the whole plant, *Calamites* was a conspicuous, bushy tree of Carboniferous swamps. *Calamites* had the typical horsetail outline, with a ridged stem divided into jointed segments, carrying branches and leaves arranged in circular whorls. Mature specimens could reach heights of 66ft (20m), with stems over 24in (60cm) across.

CALAMITES CARINATUS
Molds of *Calamites*'s stems, such as the one shown here, formed when the tree died and sediment infiltrated its core.

ridged stem

ASTEROPHYLLITES EQUISETIFORMIS
Literally meaning "star leaves," *Asterophyllites* is fossilized foliage left by *Calamites* and its relatives. The needle-shaped, upswept leaves are arranged in whorls of 4 to 40.

whorl of leaves

ANNULARIA SINENSIS
Another species of fossilized foliage from *Calamites*, this well-known fossil has elegant, evenly spaced whorls of leaves that can look like printed designs.

Psaronius

GROUP Ferns
DATE Late Carboniferous to Early Permian
SIZE Up to 33ft (10m) tall
LOCATION Worldwide

With its 9¾ft- (3m-) long fronds and elegantly flaring trunk, *Psaronius* would have been one of the most graceful ferns of Carboniferous times. Like today's tree ferns, it had a single, unbranched trunk, topped by an arching crown of fronds that uncoiled as they grew. The trunk grew differently from those of most other trees. Its core was made of a mass of leaf vascular segments, surrounded by thick, fibrous roots that ran down from higher on the trunk. This jacket of roots worked like wood, thickening as the tree gained height. The name *Psaronius* was originally applied to these fossilized stems, but has since been used for the entire plant. *Psaronius* seems to have grown in a wide variety of habitats, from wet, low-lying ground to much drier environments.

husk formed by matted roots

vascular system to leaves

PSARONIUS INFARCTUS
The unusual structure of *Psaronius*'s trunk shows a central mass of woody segments surrounded by a thick, outer covering of roots.

PECOPTERIS SP.
Named after the Greek word for a comb, *Pecopteris* are the fossilized fronds of *Psaronius*. Each frond had rows of symmetrical leaflets, with parallel edges and a central vein. The fronds unfurled like a shepherd's crook from a central growing point.

central vein

symmetrical leaflet with parallel edges

PECOPTERIS MAZONIANUM
This form of foliage is named after Mazon Creek, an exceptionally rich fossil site near Morris, Illinois. During the Carboniferous, this region was part of a tropical flood plain that built up deep layers of fine sediment. Its shale and ironstone have yielded over 100 genera of fossil plants.

Zeilleria

GROUP Ferns
DATE Late Carboniferous
SIZE 20in (50cm) tall
LOCATION Worldwide

During the Carboniferous period, many plants evolved finely divided fronds, and *Zeilleria*'s were some of the most delicate. *Zeilleria* grew in shady, swampy ground. It produced spores on the underside of its leaves, but it may also have spread by shedding specialized leaflets, which would then have taken root. This way of growing is used today by some ferns with highly divided fronds, particularly in wet habitats.

finely
divided
frond

ZEILLERIA SP.
As is clearly shown in this beautiful fossil above, *Zeilleria*'s numerous fronds created dense patterns that look like crystalline growths between layers of mudstone or shale.

Sphenopteris

GROUP Ferns / Seed plants
DATE Late Devonian to Cretaceous
SIZE Frond up to 20in (50cm) long
LOCATION Worldwide

Sphenopteris foliage grew on two distinct kinds of plants. Some grew on ferns, where the fronds produced spores, but most belonged to seed plants with fernlike leaves. One of the best-known of these primitive seed plants is *Lyginopteris oldhamia*, which comes from Carboniferous rocks in Britain. It was covered with tiny club-shaped glands, distinguishing it from all other prehistoric plants. *Lyginopteris* has an important place in botanical history. Its discovery, made in the early 1900s, proved the existence of previously unrecognized groups of extinct seed plants.

alternately
arranged branch

SPHENOPTERIS ADIANTOIDES
Whether it was produced by a fern or a seed plant, *Sphenopteris* foliage has a feathery outline, with alternately arranged leaf segments that are winged at their base. It can be found in rocks throughout the Carboniferous period.

winged leaflet

The **foliage** of *Sphenopteris* grew on two **distinct kinds** of plants: **ferns** and **seed plants**.

Medullosa

GROUP Medullosans
DATE Carboniferous to Permian
SIZE 10–17ft (3–5m) tall
LOCATION Europe, North America

With seeds as big as eggs, *Medullosa* is one of the most intriguing plants of the Carboniferous. Unlike most other seed plants, it combined fernlike foliage with a seed-forming lifestyle, although it did not bear flowers. It grew by producing new fronds at the tip of its stem while the oldest fell away. To reproduce, it formed large seeds and huge numbers of unusually large pollen grains. The grains would have been too heavy to drift far in the wind. Instead, it has been suggested that insects may have carried them between plants, in an early example of the partnership that was to become so important to flowering plants.

TRIGONOCARPUS SP.
Medullosa produced large seeds with three prominent ribs, which hung beneath its fronds.

MEDULLOSA LEUCKARTII
Medullosa grew by producing new fronds at the tip of its stem. Old fronds fell away, leaving a spiral of leaf bases filled with resin, as seen in this stem cross-section, which together formed the trunk.

IMPOSING FERN
Medullosa was one of the tallest seed plants with fernlike leaves. It had a strong, stable trunk and finely divided fronds that could grow to more than 3ft 3in (1m) long.

Potoniea

GROUP Medullosans
DATE Late Carboniferous
SIZE Up to 16½ft (5m) tall
LOCATION Worldwide

Medullosans produced their pollen in specialized organs that hung from their leaves. Many of these, including *Potoniea*, are known from fossils. They grew in a wide variety of shapes, from simple branching structures to objects that look like figs or showerheads several inches wide. All had the same function: to disperse pollen efficiently so that seeds could be formed. The pollen-producing organs are often found in isolation, which makes it hard to link them to particular parent plants.

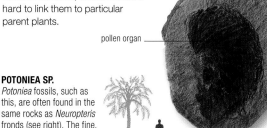

pollen organ

POTONIEA SP.
Potoniea fossils, such as this, are often found in the same rocks as *Neuropteris* fronds (see right). The fine, fingerlike, pollen-producing structures are just visible.

Macroneuropteris

GROUP Medullosans
DATE Carboniferous
SIZE Frond up to 8ft (2.5m) long
LOCATION Worldwide

Medullosans were not the largest plants in the Carboniferous period, but their leaves are some of the most common fossils from that time. *Macroneuropteris* is the name given for one particular form of their foliage, which includes some of the earliest known leaves. In *Macroneuropteris* foliage, the leaflets have heart-shaped bases and are only attached by their stalks, rather than by their blades. They have a fine tracery of spreading veins.

TEXTURED SURFACE
The finely veined surface typical of *Macroneuropteris* foliage and the prominent midrib are clearly visible on this fossilized leaflet.

damage to leaflet edge

MACRONEUROPTERIS SCHEUCHZERI
This fossilized leaflet has tiny bite marks along its edges, perhaps caused by a small, leaf-eating arthropod.

Cyclopteris

GROUP Medullosans
DATE Carboniferous
SIZE Leaflet up to 4in (10cm) across
LOCATION Worldwide

With their rounded outlines, these isolated leaflets look like tiny shells, but they have veins that radiate out like the ribs of a fan. Several kinds of medullosans seem to have produced *Cyclopteris* leaflets. One theory is that these leaflets grew on young plants, in the dim light of the forest floor. Once the plant had grown taller, it produced more typical, fernlike fronds. Another theory is that the rounded leaflets protected the rest of the frond as it uncoiled.

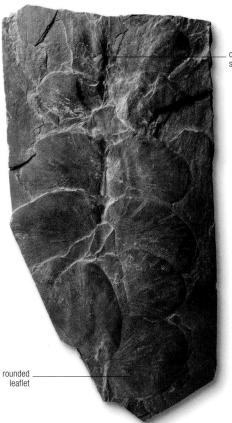

common stem

rounded leaflet

CYCLOPTERIS ORBICULARIS
Cyclopteris fossils are very different from the leaflets of typical seed ferns. This fossil shows a small group fixed to a common stem, but they are also found as isolated leaflets in Carboniferous rocks.

Alethopteris

GROUP Medullosans
DATE Carboniferous to Early Permian
SIZE Frond up to 25ft (7.5m) long
LOCATION Worldwide

Compared to ferns, medullosans were robust plants, with strong stems and a crown of large leaves. *Alethopteris* leaves were the largest of all. One specimen, found in a coal mine in northern France, was nearly 25ft (7.5m) long and 6½ft (2m) wide, making it one of the largest leaves ever discovered from Paleozoic times. *Alethopteris* leaflets could be up to 2in (5cm) long, and they typically flared out near the stem. Leaves of this type were produced by *Medullosa* and its relatives. The tough leaf bases merged to form strong, woody stems. At certain times of the year, the leaves would have been weighed down by a crop of developing seeds.

leaflets joined at base

ALETHOPTERIS SULLIVANTII
The leaflets in this fossil are not separate, but joined together at their bases, creating a continuous flap of tissue around the entire leaf segment.

ALETHOPTERIS SERLII
The thick, strongly veined leaflets typical of *Alethopteris* are clearly seen in this fossilized leaf fragment. The individual leaflets are all connected at their base.

strong middle vein

thick leaflet

Callipteridium

GROUP Medullosans
DATE Late Carboniferous to Early Permian
SIZE Fronds up to 9¾ft (3m) long
LOCATION Worldwide

Meaning "beautiful fern," *Callipteridium* foliage was strikingly convex when seen from above. Individually, *Callipteridium* fossils are often less than 6in (15cm) long, but in life, the leaves could be 20 times that length, subdividing three or four times. Their leaflets were attached by broad bases and had a deeply sunken midvein. Like most fossil leaves, *Callipteridium* are almost always found detached from their parent plant, and unlike the fronds of true ferns, none of them shows structures for making spores. Instead, at least two species have been found with ovules, showing that they were grown by seed ferns or pteridosperms. Compared to fern fronds, the leaves of seed plant fronds had a tough outer cuticle or "skin." This made them more resistant to dry conditions and increased their chances of forming fossils.

Mariopteris

GROUP Lyginopterids
DATE Late Carboniferous
SIZE Fronds up to 20in (50cm) long
LOCATION Worldwide

Usually found as isolated fragments, *Mariopteris* foliage belonged to a collection of climbing forms of seed plants, which grew up other plants in their quest for light. These plants had narrow, pliable stems, and many of them also had grappling hooks on the ends of their leaflets, in some cases up to 1½in (4cm) long. *Mariopteris* fronds also had long, bare stalks between the leaflets—the ideal shape for locking the plant in place once the frond had fully formed. Fossils show that the remainder of the leaf was divided into four roughly equal parts—an unusual feature that helps identify this group of plants.

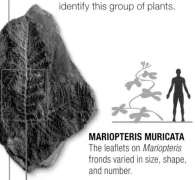

MARIOPTERIS MURICATA
The leaflets on *Mariopteris* fronds varied in size, shape, and number.

Odontopteris

GROUP Medullosans
DATE Late Carboniferous to Early Permian
SIZE Fronds up to 3ft 3in (1m) long
LOCATION Europe, North America, China

Over a thousand different seed plant leaves have been identified from fossil foliage, the majority coming from Coal Age rocks. *Odontopteris* is one of the most common types. Most of these fronds were less than 3 feet long and were attached to plants with *Medullosa* stems (see p.142). However, these plants were much smaller in size than the *Medullosa* trees and would have scrambled through low-growing vegetation or used the larger trees for support. They were common in both swamps and on drier ground, especially toward the end of the Carboniferous.

ODONTOPTERIS SUBCRENULATA
Often found in shale, this specimen is from Shanxi Province in China.

spreading veins

insect damage to rounded leaflet

Plagiozamites

GROUP Noeggerathialeans
DATE Late Carboniferous to Permian
SIZE 3ft 3in (1m) high
LOCATION Worldwide

With their two matching ranks of leaflets, *Plagiozamites* could easily be mistaken for the leaf of a cycad. However, this kind of foliage was also produced by noeggerathialeans—a group of extinct plants that have proved difficult to classify. Noeggerathialeans have been linked to horsetails and to ferns, but recent research suggests that they were more closely allied to progymnosperms.

symmetrical leaflet

PLAGIOZAMITES OBLONGIFOLIUS
Plagiozamites leaves have been found close to fossilized cones, making it likely that both were grown by the same plant. So far, no stems have been found.

Noeggerathialeans **reproduced** by forming **spores**, which developed on the scales of long, **cylindrical cones**.

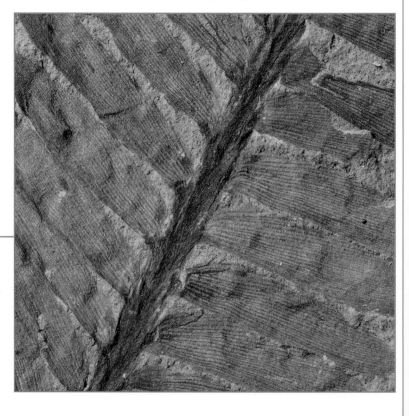

Utrechtia

GROUP Conifers
DATE Late Carboniferous to Early Permian
SIZE 33–82ft (10–25m) high
LOCATION Worldwide

More than a dozen species of *Utrechtia* have been identified from fossilized twigs. A tree with a slender trunk, its branches were arranged in widely spaced whorls and were covered with a spiral of small needles. These were flanked by short shoots that created a spreading, feathery outline. Like other early conifers, *Utrechtia* lived in dry ground rather than the low-lying swamps of Carboniferous times.

female cone

short shoot

small needles

UTRECHTIA PINIFORMIS
Utrechtia's side branches, such as the one shown here, housed cones on their tips. The male cones were pendulous, and the female ones were borne upright.

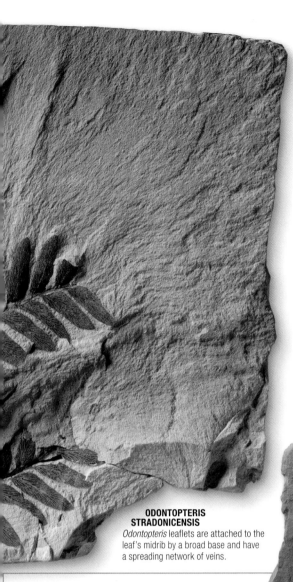

ODONTOPTERIS STRADONICENSIS
Odontopteris leaflets are attached to the leaf's midrib by a broad base and have a spreading network of veins.

Walchia

dwarf shoot

GROUP Conifers
DATE Late Carboniferous to Early Permian
SIZE 33–82ft (10–25m) high
LOCATION Worldwide

The name *Walchia* is used for foliage of this type in which preservation of detail is insufficient to enable accurate assignment to an extinct family. Walchian conifers were common forest trees in the late Carboniferous period. They had a straight, upright trunk, with leafy branches growing out in horizontal tiers. Each branch was covered with needlelike leaves, just a fraction of an inch long. Conifers became more widespread toward the end of the Coal Age, as the global climate dried.

WALCHIA ANHARDTII
On its recent growth, *Walchia* had opposite ranks of dwarf shoots, giving the tips of the branches a feathery outline.

needlelike leaves

Noeggerathia

GROUP Noeggerathialeans
DATE Late Carboniferous to Early Permian
SIZE 3ft 3in (1m) high
LOCATION Worldwide

Named after the German geologist Johann Jacob Noeggerath (1788–1877), this plant was first identified in the 1820s from its distinctive compound leaves. Each one has two rows of leaflets, arranged alternately or in opposite pairs, creating a frond up to 12in (30cm) long. The leaves are thought to have grown on a short trunk and may have been tough and leathery despite their fragile appearance.

rounded leaflets arranged alternately

NOEGGERATHIA FOLIOSA
Unlike other species of the genus, *Noeggerathia foliosa* has distinctive, rounded leaflets.

Cordaites

GROUP Gymnosperms
DATE Carboniferous to Permian
SIZE Up to 148ft (45m) high
LOCATION Europe, North America, China

The cordaites were a major group of plants that shared many features with conifers. They had woody stems and produced seeds in cones, but instead of scales or needles, they grew strap-shaped leaves up to 3ft 3in (1m) long. Their male and female cones grew on separate branches, and they often produced heart-shaped (cordate) seeds. As a group, cordaites grew in a wide range of habitats. Many were low-growing shrubs, but they also included towering, tropical trees that could reach heights of 148ft (45m).

scale

female cone

CORDAITANTHUS
This fossil genus name is used for the cone-bearing structures of cordaites.

female cone

scale

CORDAITANTHUS CONES
Seen close to life size, *Cordaitanthus* cones reveal a spiral of short, overlapping scales. In these female cones, the farthest scales would each have formed a single seed.

CARBONIFEROUS
INVERTEBRATES

The Carboniferous was a time of great change. Global cooling resulted in a series of glaciations across the immense Gondwanan continent in the Southern Hemisphere, a cycle that continued into the Permian. However, marine life continued to evolve vigorously, and massive forests emerged.

In the Carboniferous, great forests spread over huge areas, with giant clubmosses, horsetails, conifer relatives, ferns, and seed ferns flourishing. They culminated in the coal-swamp forests of the Late Carboniferous, which provided havens for amphibians and early reptiles, and also for invertebrates, including eurypterids, dragonflies, and millipedes. Some arthropods grew very large, since the oxygen levels in the atmosphere reached 28 percent or more.

FERNY FOREST
Invertebrates, especially arthropods, flourished in new Carboniferous ecosystems, among which were coal-swamp forests that were dominated by tree ferns.

Forest hazards

The hazards of forest fires increased greatly, and plants adapted to rapid regeneration once the flames had died down. In the Lower Carboniferous of central Scotland, a complete ecosystem was preserved, including scorpions up to 35in (90cm) long, mites, and harvestmen, along with intact amphibian and early reptile skeletons. These, together with many plants, were deposited in a stagnant lake fed by hot springs, into which the animals fled during a forest fire.

RUGOSE CORAL
This Carboniferous coral shows many calcite-walled corallites in which the polyps originally resided.

Marine ecosystems

The seas of the Carboniferous were dominated by brachiopods, rugose and tabulate corals, crinoids, bryozoans, and fish. Such ecosystems began in the Devonian and continued into the Permian. Goniatite fossils are common in algal reefs and in dark shale (a type of sedimentary rock), where they are often found with bivalves. Successive glaciations of the Gondwanan continent, then centered on the South Pole, are recorded in the Northern Hemisphere as rhythmic sedimentary cycles, a result of fluctuating sea levels as ice sheets expanded and shrunk.

COCKROACH
This fossilized early cockroach, *Archimylacris eggintoni*, flourished in moist forest habitats.

GROUPS OVERVIEW

Some invertebrate groups flourished in the changeable conditions of the Carboniferous. On land, lush conditions allowed some insect species to reach vast sizes. Of the marine groups, rugose corals and blastoids were abundant and spread over wide areas. Goniatites thrived in extensive reef habitats, and new forms of echinoderms appeared.

RUGOSE CORALS
Rugose corals were very abundant in the Carboniferous, often forming extensive beds and thickets. Unlike modern corals, the main septa (radial partitions) were laid down on four lines only, so that the coral was biradially symmetrical.

GONIATITES
Goniatites were cephalopods whose spiral shells had gas-filled internal chambers and a body chamber in which the main part of the animal resided. The suture lines where the internal partitions met the inside of the shell had a zigzag form.

BLASTOIDS
These were relatives of crinoids, but generally smaller. They had short stalks and a bud-shaped, tentacled crown within which was a remarkably complex respiratory system. Blastoid fossils are often found abundantly in thin beds, traceable over wide areas.

CRINOIDS
These "sea-lilies," some forms of which still exist today, may look like plants, but they are actually calcite-plated echinoderms. Crinoids have long stalks of stacked calcite disks, ending in a flowerlike crown with extended arms for feeding. Some limestones consist almost entirely of crinoid debris.

Essexella

GROUP Scyphozoans

DATE Late Carboniferous

SIZE 3¼–4½in (8–12cm) across

LOCATION USA

Essexella is one of the rarest types of fossils—a jellyfish. Many other supposed jellyfish in the fossil record are of dubious affinity, but *Essexella* is beyond doubt. It is one of the most common fossils in the Mazon Creek fauna of Illinois, and some specimens have been obtained with a preserved rim of tentacles around the disk. Like other fossils in this fauna, they owe their exceptional preservation to rapid burial and then replacement with siderite (ferrous carbonate) following initial bacterial action. This mineralization left the fossils within siderite concretions and protected them from further decomposition and decay. *Essexella* would have hunted and fed much like modern jellyfish—by using the stinging cells along the length of its tentacles to gather prey before transferring them to the mouth.

ESSEXELLA ASHERAE
Jellyfish rarely fossilize, but in Mazon Creek, they are commonly found in concretions and called "blobs" by local collectors.

Zaphrentoides

GROUP Anthozoans

DATE Early Carboniferous

SIZE Calice ⅜–½in (1–1.5cm) across

LOCATION Europe

Zaphrentoides was a small, solitary coral with a conical skeleton (corallum) that was slightly curved in shape. The soft-bodied coral polyp rested in the concave dip (calice) at the top of the corallum in the living animal. There was a distinctive large, deep space (cardinal fossula) on the convex side of the corallum. This was located between the vertical plates, or septa, which were arranged in a radial pattern. The major septa were elongated and joined together around the cavity of the cardinal fossula.

cardinal fossula

eroded coral wall

radial septa

ZAPHRENTOIDES SP.
The walls of these fossilized small, horn-shaped, solitary corals have been worn down.

Siphonophyllia

GROUP Anthozoans

DATE Carboniferous

SIZE Up to 3ft 3in (1m) long

LOCATION Europe, North Africa, Asia

Siphonophyllia was a very large, solitary rugose coral. It was curved and conical in early stages of growth, often becoming straighter and cylindrical later in life. The soft-bodied coral polyp was situated in the concave calice at the top of the skeleton (corallum). Within the corallum, there were many vertical plates (septa) arranged in a radial pattern. Steeply inclined, curved plates (dissepiments) were found at the edge of the corallum. On the Sligo coast, in Ireland, *Siphonophyllia*'s abundance and its snakelike appearance has earned one rock the nickname "Serpent Rock."

The **snakelike appearance** of *Siphonophyllia* has led to one rock being named **"Serpent Rock."**

curved plates (dissepiments) at edge of corallum

numerous thin septa

SIPHONOPHYLLIA SP.
This fossilized coral is found in shallow water or in calcareous shales and limestones. The long, slender septa can be seen running through the corallum toward the outer edge in this specimen.

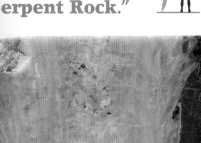

tabulae in center of coral

PARTITIONS
This fossil shows the closely spaced horizontal partitions (tabulae) in the center of the corallum.

Syringopora

GROUP Anthozoans

DATE Late Ordovician to Carboniferous

SIZE Corallites ¹⁄₁₆in (1–2mm) across

LOCATION Worldwide

Syringopora was a tabulate coral (see p.98), consisting of cylindrical tubes (corallites) that were roughly parallel to each other. The tubes were not in direct contact with each other but were connected by smaller horizontal tubes (tubuli). The corallites, which housed the soft-bodied, tentacled polyps at the top, contained spiny partitions (septa) arranged in longitudinal rows. Along the length of the corallite were funnel-shaped plates (tabulae). As the colony matured, it would have formed new corallites and expanded sideways.

corallite

horizontal tube
that connects
corallites

SYRINGOPORA RETICULATA
This coral was made up of long, cylindrical corallites that were connected to each other by smaller horizontal tubes.

As the colony **matured**, it would have **formed new corallites** and expanded sideways.

Siphonodendron

GROUP Anthozoans

DATE Devonian to Carboniferous

SIZE Corallites up to ³⁄₈in (1cm) across

LOCATION Worldwide

Siphonodendron is a colonial rugose coral that lived in shallow seas. The colony, which could grow to more than 5ft (1.5m) in size, was comprised of long cylindrical tubes called corallites, which were roughly parallel and sometimes in contact with one another. These calcite-walled corallites housed the soft-bodied, tentacled polyps. Within the corallites, many vertical plates, or septa, were arranged in a radial pattern with curved plates (dissepiments) at the edge. There were additional horizontal partitions (tabulae) in the central region of the corallites.

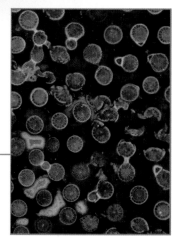

SIPHONODENDRON SP.
This is a very common fossil coral found in Carboniferous rocks. The colony is made up of long corallites in which the soft-bodied polyps lived.

Archimedes

GROUP Bryozoans
DATE Early Carboniferous to Early Permian
SIZE Colony height over 8in (20cm)
LOCATION Worldwide

Archimedes was a fenestrate bryozoan colony (see p.171) that would have stood upright on the sea floor. It is thought to have lived in shallow waters on carbonate muds. Its structure is immediately recognizable due to the distinctive screw-shaped axis. Thin, netlike structures called fronds were attached to the axis of *Archimedes*, forming a continuous spiral. The fronds were made up of branches and on one side, the autozooecia—chambers that housed the soft-bodied individuals (zooids) of the colony—were found in two rows. The branches were connected by partitions, which formed rectangular open spaces between them called fenestrules, hence the name fenestrate bryozoan.

fronds, consisting of thin branches

distinctive screw-shaped axis

ARCHIMEDES SP.
The fronds are very delicate structures, which means that usually only the screw-shaped axis of the fossil is preserved.

WHAT'S IN A NAME?
Archimedes is named after the Greek mathematician because the screw-shaped axis resembles the water pump he invented.

Productus

GROUP Brachiopods
DATE Carboniferous
SIZE Up to 3in (7.5cm) long
LOCATION Europe, Asia

Productus was a slightly unusual brachiopod. As an adult, it had no anchoring pedicle, but instead relied on the weight of its pedicle valve (the half of its shell from which a pedicle would emerge in most brachiopods) to keep it stable on the seabed. Some species of *Productus* had spines on this valve that helped to hold them firmly in position.

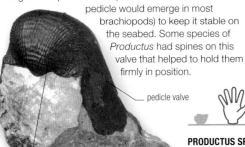

pedicle valve

ribs

PRODUCTUS SP.
On the outside of the pedicle valve are obvious ribs, which are crossed by shallow folds.

Pugnax

GROUP Brachiopods
DATE Devonian to Carboniferous
SIZE Up to 1½in (4cm) long
LOCATION Europe

Like most brachiopods, *Pugnax* was attached to the seabed by a flexible stem or pedicle emerging through the pedicle valve. The outside of most of the shell is essentially smooth and devoid of all ornamentation except for some very fine growth lines and short ribs around the margin of each of its valves.

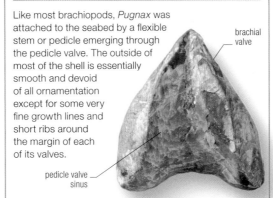

brachial valve

pedicle valve sinus

PUGNAX ACUMINATUS
This view from beneath shows the sinus (infold) of the pedicle valve and the outward fold of the brachial valve.

Fossundecima

GROUP Worms
DATE Late Carboniferous
SIZE Up to 2¼in (6cm) long
LOCATION North America

Fossundecima was a bristle worm (or polychaete) that lived around 300 million years ago in the seas over what is now North America. One distinctive feature was a flexible feeding organ (proboscis) that it could push out from its mouth. This was equipped with powerful jaws, which strongly suggests that it was a predatory carnivore. Behind the head were between 15 and 20 abdominal segments, each carrying bristles, and a final, smaller tail segment. *Fossundecima* moved by use of the paired lobes (parapodia) on each of its abdominal segments; the parapodia were also thought to be used in respiration.

FOSSUNDECIMA KONECNYI
This worm was found preserved in an ironstone nodule.

Vestinautilus

GROUP Cephalopods
DATE Early Carboniferous
SIZE Up to 4¾in (12.5cm) across
LOCATION Europe, North America

Vestinautilus was a coiled, nautiluslike cephalopod. The whorls of its shell overlapped only slightly, resulting in a wide umbilicus (navel-like structure); this had a central perforation because the initial coil was not tight. *Vestinautilus* fossils show a prominent spiral ridge around the outside of each whorl and distinctive, V-shaped suture lines. The ventral region (outside margin) of the whorl is extremely broad, and in mature specimens, the whorls sometimes lose contact with each other as the spiral opens out.

Vestinautilus fossils show a **prominent spiral ridge** around the outside of **each whorl** and distinctive, **V-shaped** suture lines.

VESTINAUTILUS CARINIFEROUS
This internal mold has been preserved in limestone. It clearly shows the suture lines and the ridge, but the body chamber has broken off.

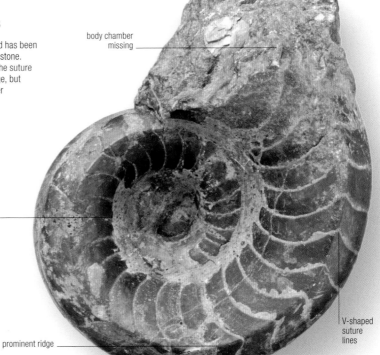

body chamber missing

little overlap between whorls

prominent ridge

V-shaped suture lines

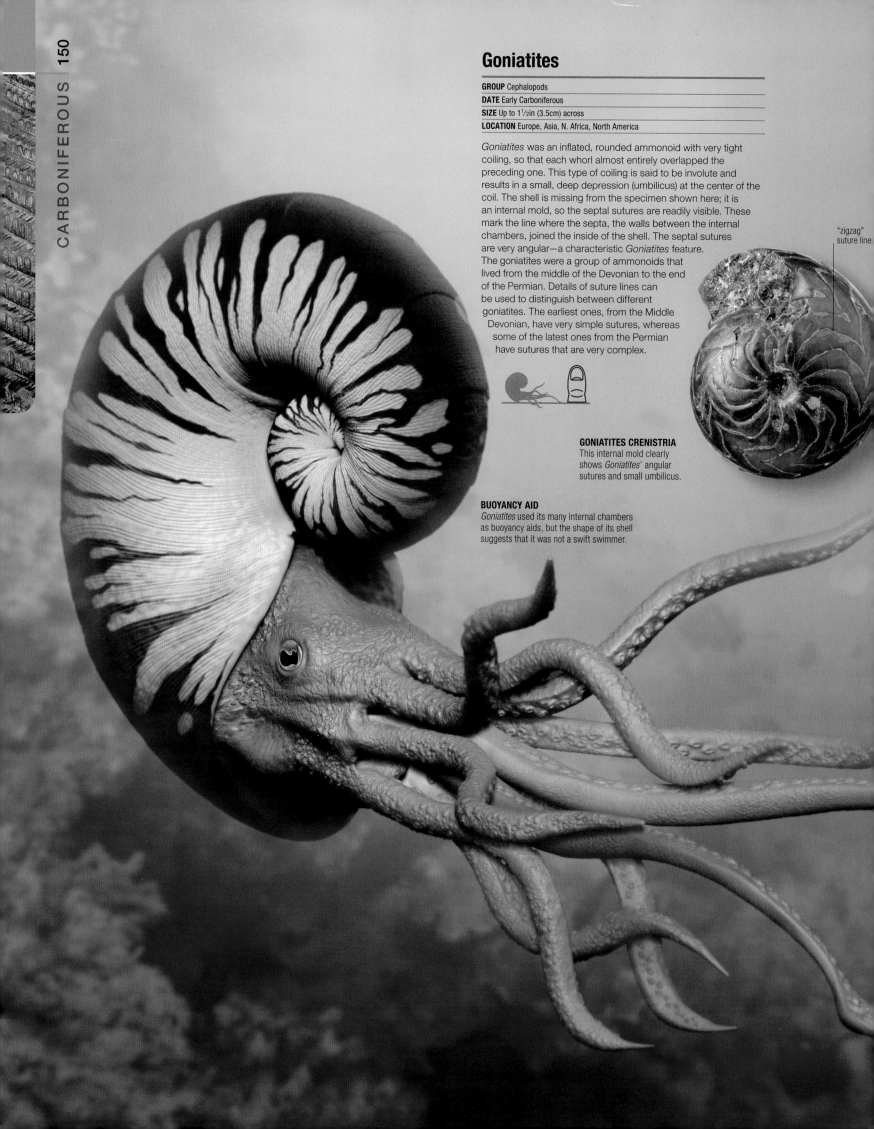

Goniatites

GROUP Cephalopods

DATE Early Carboniferous

SIZE Up to 1½in (3.5cm) across

LOCATION Europe, Asia, N. Africa, North America

Goniatites was an inflated, rounded ammonoid with very tight coiling, so that each whorl almost entirely overlapped the preceding one. This type of coiling is said to be involute and results in a small, deep depression (umbilicus) at the center of the coil. The shell is missing from the specimen shown here; it is an internal mold, so the septal sutures are readily visible. These mark the line where the septa, the walls between the internal chambers, joined the inside of the shell. The septal sutures are very angular—a characteristic *Goniatites* feature. The goniatites were a group of ammonoids that lived from the middle of the Devonian to the end of the Permian. Details of suture lines can be used to distinguish between different goniatites. The earliest ones, from the Middle Devonian, have very simple sutures, whereas some of the latest ones from the Permian have sutures that are very complex.

"zigzag" suture line

GONIATITES CRENISTRIA
This internal mold clearly shows *Goniatites*' angular sutures and small umbilicus.

BUOYANCY AID
Goniatites used its many internal chambers as buoyancy aids, but the shape of its shell suggests that it was not a swift swimmer.

The **pattern of the suture lines** identifies different goniatites.

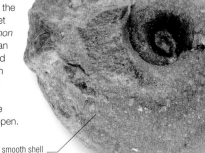

STRAPAROLLUS SP.
Like some modern snails, the whorls of *Straparollus* are relatively loose with an open spiral and completely lack ornamentation.

rounded whorls with an open spiral

Straparollus

GROUP Gastropods
DATE Carboniferous
SIZE Up to 2¼in (6cm) across
LOCATION North America, Europe, Australia

Straparollus was a conical gastropod mollusk with whorls that are rounded in cross-section, but have a slight "shoulder." The shell surface lacks ornamentation, apart from fine growth lines. In some Carboniferous limestones, mollusk fossils are rare because the calcium carbonate shells of most are made of the mineral aragonite, which later dissolved. *Straparollus*, however, had an outer layer of calcite, the other mineral form of calcium carbonate.

Bellerophon

GROUP Gastropods
DATE Silurian to Early Triassic
SIZE Up to 2in (5cm) across
LOCATION Worldwide

Bellerophon is unusual for a gastropod in that its shell was coiled in a plane spiral. The whorls were very rounded, and later whorls enveloped the preceding ones. The shell was essentially smooth. Scars in some fossils show paired muscles (again, a departure from the gastropod norm), and this has led to suggestions that *Bellerophon* belongs to a more primitive mollusk group, the tergomyans. Yet some *Bellerophon* species show an aperture slit and slit band, which suggest it may have been a gastropod. The debate is still open.

smooth shell

BELLEROPHON SP.
Bellerophon's smooth, unadorned shell shows a plane spiral symmetry that is rare in other species of gastropods.

Conocardium

GROUP Rostroconchs
DATE Devonian to Permian
SIZE Up to 6in (15cm) long
LOCATION Worldwide

Conocardium looks like a bivalve but is in fact a rostroconch. Rostroconchs differ from bivalves in that they developed from a single-valved protoconch (initial shell), whereas bivalves developed from a bivalved protoconch. Also, rostroconch shells were fused along the front midline—the hinge region of bivalves—so the two valves gaped open permanently. Like many bivalves, rostroconchs could use a muscular foot to burrow or move. They probably fed on plankton and other particles of organic matter from the surrounding seawater, but exactly how they did this is unknown.

finely ribbed ornamentation

CONOCARDIUM SP.
The shell surface of *Conocardium* was ornamented by fine ribs, and the shell itself was thick.

fine ribbing

DUNBARELLA SP.
With its rounded outline and fine, radiating ribs, *Dunbarella* is similar in some ways to scallops of the present day.

Dunbarella

GROUP Bivalves
DATE Carboniferous
SIZE Up to 1½in (4cm) long
LOCATION Europe, North America

Dunbarella's shell was rounded but had a straight dorsal line produced by earlike shell extensions. Many fine ribs radiated from the center of the line, crossed by growth lines parallel to the outer valve edges. All *Dunbarella* were attached to the seabed with a byssus (like a mussel's "beard") when young, but some broke free at maturity, resting on the seafloor with their right valve lowermost. The ribs of the right valve split in two as the shell grew, while new ribs arose on the left valve in between previously formed ones.

Carbonicola

GROUP Bivalves
DATE Late Carboniferous
SIZE Up to 1½in (4cm) long
LOCATION W. Europe, Russia

A freshwater bivalve that lived in lakes and swamps, *Carbonicola's* shell usually had beaks of differing heights. The shell tapered toward one end, and inside each valve were two adductor muscle scars: one circular and deep, the other slightly larger but shallower. Each valve usually had one tooth and one socket, but some forms had two and others none at all. The ligament that held the shell open lay behind the beaks.

CARBONICOLA PSEUDOROBUSTA
Carbonicola's shell tapered at one end. Its exterior was ornamented only with fine growth lines, which are just visible on this specimen.

Woodocrinus

GROUP Echinoderms
DATE Carboniferous
SIZE Cup and arms 2¼–4in (6–10cm) high
LOCATION Europe

Woodocrinus was a crinoid with a small cup consisting of three circlets of plates. Each arm normally branched twice, giving 20 arms in total. The ossicles or segments that made up the arms were stout, but short, so that each arm was composed of numerous short, circular plates stacked on top of each other. Different arms branched at various heights above the cup, and as a whole, they appeared blunt and thick. A plated anal tube rose from the base of the cup. The column that supported the cup was circular in cross-section and featured regular annual thickening of its component ossicles, which gave the stem a ribbed appearance.

Woodocrinus grew in **extensive "crinoid forests"** in shallow seas.

branching arm

arm ossicle

stem

WOODOCRINUS SP.
The currents that supplied crinoids with food were also responsible for breaking up their bodies after death. This fossil shows the arm and stem ossicles of *Woodocrinus* beginning to separate and drift apart.

Actinocrinites

GROUP Echinoderms
DATE Carboniferous
SIZE Cup up to 1¼in (3cm) across
LOCATION Europe

This picture shows the upper surface of the crinoid *Actinocrinites*. The junctions between the plates have been inked in to show them clearly. In the center is the slightly domed upper surface of the cup; in most crinoids, this is leathery with some calcite plates, but in *Actinocrinites*, it was made up entirely of fused calcite plates. The side of the cup is enlarged by the lower arm (brachial) plates and extra plates (interbrachial) between them, and the bases of four arms can be seen. *Actinocrinites* would have lived on reefs, anchored by a stalklike structure.

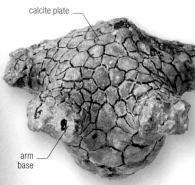

calcite plate

arm base

ACTINOCRINITES PARKINSONI
From the five arm bases (one is hidden here) of *Actinocrinites*, about 30 arms branched out in total.

Pentremites

GROUP Echinoderms
DATE Early Carboniferous
SIZE Cup up to 1in (2.5cm) high
LOCATION North America, South America, Europe

Pentremites was in the blastoid group of echinoderms, which lived fixed to the seabed by a root at the base of a jointed stem. Its cup, shown here, comprised three circlets of plates and five petal-shaped ambulacra. On the edges of the ambulacra were numerous long, slender tube feet that conveyed food particles to a central food groove. From here, food was moved to the star-shaped mouth, shown at the top of this fossil. The tube feet in *Pentremites* were used entirely for feeding.

PENTREMITES PYRIFORMIS
The five openings around *Pentremites*'s mouth are known as spiracles. These served as the exits from the animal's respiratory chambers.

Xyloiulus

GROUP Arthropods
DATE Late Carboniferous
SIZE Up to 2¼in (6cm) long
LOCATION North America, Europe

This millipede is one of several kinds known from Carboniferous rocks. Its head is not well preserved in this specimen, but the long, tubular, segmented body is clearly visible. Each segment is similar and has two pairs of slender appendages; parts of many of these can be seen. Fossil millipedes from rocks of mid-Silurian age in Scotland are the oldest land animals known. Some, from the Carboniferous coal swamps, reached a huge size, sometimes more than 6½ft (2m) long.

XYLOIULUS SP.
This fossil *Xyloiulus* is preserved in a clay-ironstone nodule.

Archegonus

GROUP Arthropods
DATE Early to Middle Carboniferous
SIZE Up to 1½in (4cm) long
LOCATION England, C. Europe, Spain, Portugal, N.W. Africa, Asia, S. China

Archegonus had a rectangular or conical glabella (the central part of the headshield) with shallow, narrow furrows. Its eyes were crescent-shaped, rather small, and located close to the glabella. Its cheeks had short spines on their posterior corners. *Archegonus* had nine thoracic segments, and its tailshield was similar in size to its head. The broad sides had at least six pairs of rather weak ribs. This trilobite is found in sediments that were deposited in fairly deep water; its small eyes also suggest that it probably inhabited this type of environment, with low light levels.

ARCHEGONUS NEHDENENSIS
Archegonus species have been used to zone some Carboniferous rocks.

Hesslerides

GROUP Arthropods
DATE Middle Carboniferous
SIZE Up to 1¼–1½in (3–4cm)
LOCATION North America

Hesslerides was a trilobite with a large, granulated glabella (central raised part) and crescent-shaped eyes, set close to the rear of its headshield. On the inner part of its cheeks, a narrow ridge ran around the lower margin of the eye. The cheeks sloped down toward a flattish border. Its thorax consisted of nine segments, and its tailshield was about as long as its head; the middle lobe had 12 narrow segments, and the side lobes had 10 pairs of ribs, each with a deep furrow. "Three lobes"— two side or pleural ones, and a central or axial lobe—are the definition of the word "trilobite."

HESSLERIDES BUFO
The eyes of this species were set to the back of its headshield, close to the large glabella.

Archimylacris

GROUP Arthropods
DATE Late Carboniferous
SIZE ¾–2in (2–5cm) long
LOCATION Europe, North America

This fossil of *Archimylacris*, a cockroachlike insect, shows two pairs of wings and parts of its body. The front part of the rather crushed thorax can also be seen between the wings, although the head itself is missing. The forewings are the smooth areas on either side of the fossil, and these formed a protective, toughened covering for the body when the animal was not flying. Between them, the thinner, more delicate hind wings show a distinctive vein pattern. This is important in identification.

ARCHIMYLACRIS EGGINTONI
The wings of this specimen of *Archimylacris* are fully developed, indicating that this was an adult; immature stages did not have full wings.

hind wing with delicate veins

Pleophrynus

GROUP Arthropods
DATE Early Devonian to Late Carboniferous
SIZE ½–1½in (1.5–4cm) long
LOCATION North America

Pleophrynus was an extinct, spiderlike arachnid called a trigonotarbid. Unlike true spiders, it lacked poison and silk glands. Its hard outer shell (or carapace) was pear shaped and tapered to a point at the front, close to which

two small eye clusters were situated. Remains of some of its eight limbs can be seen here originating underneath the head (prosoma). Its abdomen (opisthosoma) was covered in rigid plates, which provided protection for the animal's underlying soft tissues, and each of its six segments was ornamented with rows of bumps or tubercles. The back margin of the carapace had four short, spurlike projections.

abdominal plates

PLEOPHRYNUS SP.
Pleophrynus protected its soft body parts with rigid plates that covered its abdomen, as well as with four short, spiky projections.

Tealliocaris

GROUP Arthropods
DATE Early Carboniferous
SIZE Up to 2in (5cm) long
LOCATION Scotland

Tealliocaris was a shrimplike crustacean that inhabited very shallow, probably brackish, water on tidal flats. The hard upper shell (or carapace) occupied the left half of this specimen, which is flattened from the side. The pair of first antennae had two short branches and lay just above the very long, slender, and delicate second antennae (one is preserved here). Some of the thoracic legs can be seen projecting from underneath the carapace; each had two branches. The curved, flexible abdomen was about the same length as the carapace and was composed of five segments, under which were five pairs of swimming limbs or "swimmerets."

lobed tail fan

TEALLIOCARIS WOODWARDI
Although flattened, this fossil is well preserved, showing outlines of some of the five lobes of the distinctive tail fan.

Euproops

GROUP Arthropods
DATE Carboniferous
SIZE Up to 2in (5cm) long
LOCATION Europe, USA

Euproops was an early relative of the horseshoe crab. Its headshield was crescent shaped, with broad, pointed spines at the cusps. Two side ridges on its head bowed outward at the front ends, where its tiny eyes were situated. These ridges tapered into long, pointed spines.

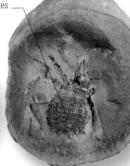

EUPROOPS ROTUNDATUS
The oval abdomen had a narrow, segmented central ridge, and these segments ran onto the animal's flanks. Each rib ended in a short spine.

CARBONIFEROUS
VERTEBRATES

From their origins in the Devonian, tetrapods diversified within the coal swamps of the Carboniferous, adapting to different lifestyles, including those of aquatic opportunists and terrestrial predators. In the oceans, highly armored fish were replaced by more active swimmers with much thinner scales.

The Carboniferous period could be called the "Age of Coal," for the thick swamps of coal-forming plants that grew on land. The end of the Devonian saw the extinction of placoderms, heterostracans, and osteostracans (see pp.122–123). In the Carboniferous, advanced jawed fish—including chondrichthyans, spiny sharks (acanthodians), and ray-finned fish in particular—dominated the seas, lakes, and rivers. On land, tetrapods flourished, perhaps aided by the arthropod colonization. The Carboniferous opens with a 20-million-year period of poor fossilization known as Romer's Gap, after the American paleontologist Alfred Romer. The fossil record then reveals a wide diversity of fauna, so there must have been a major radiation of species during those Gap years.

The first terrestrial fauna

The earliest tetrapods evolved in estuaries during the Late Devonian, and their anatomy suggests they had to live in water. They retained tail fins and internal gills—basically, they were fish with legs. During and shortly after Romer's Gap, the tetrapods changed; whereas they were still adapted to live in the water, many, including the embolomeres and *Crassigyrinus* (see p.157), retained their tail fins but seem to have lost their internal gills,

AMPHIBAMUS
This little temnospondyl had advanced features seen today in frogs and salamanders, including a very large tympanic ear, large eyes, and specialized, hinged teeth. And, just like frogs and salamanders, *Amphibamus* had only four fingers.

which suggests they were more capable of coming onto land. The famous fossil site of East Kirkton in Scotland (see p.160) provides ample evidence of the world's oldest-known vertebrate terrestrial community. The fossil fauna recovered includes a highly diverse collection of tetrapods, with many animals more specialized to live on land. Temnospondyls first appear here, including a species that lacks many of the hallmarks of life in the water. Lepospondyls are also present, represented by the snakelike aïstopod *Ophiderpeton*, which would have been as comfortable on land as in the water. But most striking are the reptilelike animals, including *Westlothiana* (see p.160), which was hailed as the first amniote (see opposite), although this interpretation is disputed. If correct, it would have represented the first time reproduction was freed from needing water—a necessary step toward fully occupying the land. However, in the Late Carboniferous, undisputed amniotes became prominent members of a truly terrestrial fauna.

GROUPS OVERVIEW

The first major radiation of tetrapods took place in the Early Carboniferous, and another followed 40 million years later in the Late Carboniferous. This two-step diversification permitted the full occupation of the land and would lead to the emergence of the modern herbivore-dominated fauna in the Permian.

BAPHETIDS

Appearing in the Early Carboniferous, baphetids (also known as loxommatids) are immediately recognizable by their keyhole-shaped eye sockets. *Megalocephalus* and other baphetids retain the sensory canals on the skulls that indicate an aquatic way of life. Skeletal fossils currently being studied will provide more information since this group is mostly known from skulls.

WHATCHEERIIDS

Another of the Early Carboniferous tetrapods, Whatcheeriids were perhaps the first accomplished terrestrial tetrapods. Animals such as *Pederpes* had tall eye sockets and lacked the sensory structures on the skull that are common in aquatic animals. These tetrapods have only been discovered recently, and more remains to be learned about them.

CRASSYGYRINIDS

A large, Early Carboniferous aquatic predator, this is one of the more bizarre tetrapods. With a gigantic head, an elongate trunk, a tail fin, and tiny forelimbs, *Crassigyrinus* probably lurked in the murky shallows waiting for fish to swim by. However, it lacks the structures associated with internal gills, so it is more advanced than the first tetrapods.

TEMNOSPONDYLS

One of the most successful groups of early tetrapods, temnospondyls lived in most Carboniferous environments. Some were specialized aquatic bottom-dwellers, others lurked at the water's edge (like crocodiles), and still others were terrestrial. They ranged in length from 5in (12cm) to 5ft (1.5m). One type, represented by *Amphibamus*, gave rise to frogs and salamanders.

The **development** of the **amniotic egg** was a **critical step** in the **expansion** of **tetrapods** across **all parts** of the **land**.

Evolution of the shelled egg

The first tetrapods inherited simple eggs from their fish ancestors that were fertilized outside the body. Water was critical for reproduction because the thin membrane surrounding the egg would not protect it against drying out on land. Water was also necessary to carry sperm to the eggs. This mode of reproduction is called "amphibious" and is still seen in frogs and salamanders. In order to become fully terrestrial, it was necessary to protect the egg from drying. This was accomplished by the amniotes (reptiles, birds, and mammals) with the addition of three new membranes to the egg: the

OPHIACODON
One of the better-known early amniotes, this crocodilelike predator is related to the sail-backed *Dimetrodon*, which lived during the Permian. *Ophiacodon* is a synapsid, which is the evolutionary lineage that leads to mammals.

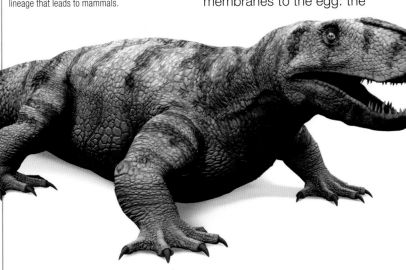

NO WATER REQUIRED
The hard shells and extra internal membranes allow this tortoise's eggs to be laid in the ground without drying out. This was a critical step in the tetrapod radiation onto all parts of the land.

chorion, amnion, and allantois. These membranes, together with the yolk sac, also help support the growing embryo by providing nutrients and collecting waste. Around the outermost membrane lies the shell, which may be leathery and flexible (as in many turtles and lizards) or mineralized and hard (best-known in birds, but also found in crocodiles and many tortoises). Amniotes further became independent from water by their mode of reproduction. Instead of releasing sperm into water over eggs, the male places sperm within the reproductive system of the female, where it travels to and fertilizes the egg. After fertilization, the extra membranes and shell are formed, and the egg is laid on dry land. Among amniotes, mammals have gone one step further, doing away with the shell and retaining the egg inside the mother (monotremes are the exception, since they still lay eggs; see p.337). The extra membranes are modified to interact with tissues of the mother in order to directly nourish the developing fetus. This modification is called the placenta.

LEPOSPONDYLS	EMBOLOMERES	AMNIOTES	DIAPSIDS	SYNAPSIDS
Another diverse group of early tetrapods, lepospondyls were usually small, but they assumed many body forms. One of the earliest-known tetrapods was, ironically, a limbless aïstopod, and the first microsaurs were mistaken for early amniotes (thus their name). The aquatic nectrideans were newtlike, while the elongate lysorophians would curl into burrows and estivate during dry spells.	These large, aquatic tetrapods, such as *Proterogyrinus*, were ubiquitous during the Carboniferous. Some early forms retained tail fins, and their moderately elongate skulls would help for catching fish. Features in their limb bones show more adaptations toward terrestrial life, and many scientists have considered these tetrapods to be early members of the lineage leading to amniotes.	Highly adapted for life on land, the first amniotes appeared in the Late Carboniferous, making their homes in hollow tree stumps. Amniotes are comprised of three groups: reptiles, synapsids (see right), and a collection of early species from before the reptile–synapsid split. Some of the first amniotes were also the first tetrapod plant-eaters (herbivores).	The largest group of reptiles, diapsids are named for two openings in the back of their skulls. Diapsids alive today include lizards, snakes, crocodilians, and birds. Turtles, which arose in the Middle Triassic and lack the skull openings, are also probably diapsids. The first diapsids are found in Late Carboniferous rocks from the central United States.	The lineage leading to mammals, synapsids are named for a single opening in the back of their skull. The first fossils are isolated fragments found in the tree stumps with the first amniotes (see left), meaning that amniotes rapidly diversified soon after they evolved. Carboniferous synapsids include active predators and some of the first herbivores.

symmetrical
tail fins

Falcatus

GROUP Chondrichthyans
DATE Early Carboniferous
SIZE 12in (30cm) long
LOCATION USA

This genus, which is closely related to *Stethacanthus* (see p.124), is known from numerous small specimens preserved in the same limestone seabeds as *Discoserra* (see opposite). *Falcatus* is named for the large, forward-pointing, sickle-shaped spine that is armed with pointed denticles located above the head. This spine is only known in specimens above a certain size, indicating that it was a sexual feature typically found in males and was most likely used for display. The almost symmetrical tail was used for cruising efficiently through the water.

GROUP SWIMMERS
Many specimens of *Falcatus* have been found together, which suggests these sharks traveled in schools.

ANATOMY

SENSE ORGANS

Falcatus had well-adapted senses for life in the murky depths of the sea. It had large eyes ringed by rectangular calcified plates, which probably helped support the eye and acted as areas for muscle attachment. *Falcatus* also possessed a soft snout, or rostrum, which probably housed special sense organs that detected prey by picking up electrical impulses caused by movement of the prey's muscles.

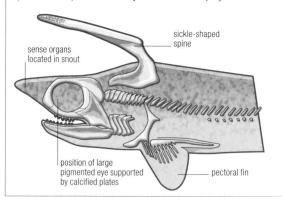

sense organs
located in snout

sickle-shaped
spine

position of large
pigmented eye supported
by calcified plates

pectoral fin

Echinochimaera

GROUP Chondrichthyans
DATE Early Carboniferous
SIZE 12in (30cm) long
LOCATION USA

Echinochimaera is known from two species that can be distinguished by differences in their upper dentition. Both have been found in marine shallow-water sediments. It was a small fish built for maneuverability rather than speed, with a tapered body, relatively large pectoral and pelvic fins, and a paddle-shaped tail. It had two dorsal fins, the front one with a mobile spine on its leading edge. The body was completely covered with scales, and pairs of enlarged scales, called scutes, extended from the rear dorsal fin to the base of the tail. There were also pairs of backward-projecting scutes above the eyes. Males were much larger than females and had different features, including antlerlike eye scutes.

Scute translates as "shield" and explains the protective quality of these special scales.

large dorsal
fin spine

antlerlike
scute over
the eye

tapering body

clasper used to
transfer sperm to
female when mating

SHOWING OFF
The larger males had some elaborate features that could have been used for recognition or displaying during the breeding season.

body covered
in sharp
scales

tall, narrow
body

SLIM FISH
The unusual shape of *Belantsea* made it well adapted for moving slowly through the water while grazing.

large
muscular
fins

Belantsea

GROUP Chondrichthyans
DATE Early Carboniferous
SIZE 28in (70cm) long
LOCATION USA

Belantsea belongs to a rather unusual group of fish with cartilaginous skeletons—the petalodonts. These relatives of sharks had a distinctive set of teeth. Rather than having lots of teeth arranged in numerous rows, *Belantsea* had only seven teeth in the top and bottom jaws. Furthermore, the teeth were not shed while feeding but were kept throughout life. Each tooth had specially reinforced, serrated cutting edges, ideal for grazing on hard materials such as sponges. The leaflike body shape of *Belantsea* is very distinctive. Its small tail suggests slow motion, but the large fins and the deep body indicate good maneuverability.

Discoserra

GROUP Actinopterygians
DATE Early Carboniferous
SIZE 24in (60cm) long
LOCATION USA

Discoserra was a tropical fish that inhabited a shallow bay now known as Bear Gulch in Montana. It had a deep, disk-shaped body that was laterally compressed—it was not built for speed or endurance, but for precise maneuvering. The head was small, with large eyes and a small mouth. *Discoserra* had many large scales (scutes) on the front half of its body, and each dorsal scute had a forward-facing hook. The scales were held together by peg-and-socket joints.

dorsal fin

forward-facing hook on dorsal scute

rounded tail

anal fin

short, pointed snout

Discoserra was not **built** for speed or endurance, but for **precise maneuvering.**

CORAL FEEDER
Discoserra had many features that indicated that it was adapted for feeding among sponges and corals, including a short snout with long, comblike teeth for sucking or grazing.

Megalocephalus

GROUP Early tetrapods
DATE Late Carboniferous
SIZE 5ft (1.5m) long
LOCATION British Isles

Megalocephalus, like most baphetids, is known only from its skull. The baphetids were a family of presumably aquatic predators from the Carboniferous that combined some widespread primitive features with several unique ones.

The skull of *Megalocephalus* (the name translates as "big head") is about 12in (30cm) long and has large, pointed teeth. One unique feature of baphetids, including *Megalocephalus*, is the strange elongated extension on their eye sockets (orbits), the reason for which is still unknown.

keyhole-shaped eye socket

UNUSUAL SKULL
Suggested functions for the extended eye socket include housing for a salt gland or electric organ, or to allow for extra movement of jaw-closing muscles.

Whatcheeria

GROUP Early tetrapods
DATE Early Carboniferous
SIZE 3ft 3in (1m) long
LOCATION USA

Whatcheeria is a close relative of *Pederpes*, but it lived at a later time. It is known from a large amount of fossil material, including several almost complete skeletons and many isolated bones. Many other tetrapods and fish have been found at the same site in Iowa. *Whatcheeria* seems to have been more aquatic than *Pederpes*, with a powerful tail that provided propulsion. Much of its anatomy still remains to be described, including its limbs.

Pederpes

GROUP Early tetrapods
DATE Early Carboniferous
SIZE 3ft 3in (1m) long
LOCATION Scotland

Pederpes is known from a single specimen discovered in the west of Scotland near Dumbarton and was once the only tetrapod known from an articulated skeleton dating from the period known as Romer's Gap (see p.154). It shows a mix of primitive and advanced features. It probably had the usual five toes on its hind limb and a foot that appears adapted for terrestrial walking. By contrast, its hands may have had more than five fingers, although only two tiny ones have been found. Its ear region was like that of *Acanthostega* (see p.132).

FEET FORWARD
The joints in *Pederpes'* feet pointed forward, which suggests that it could walk on land.

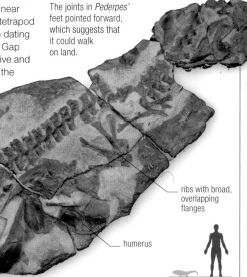

eye socket

ribs with broad, overlapping flanges

humerus

Crassigyrinus

GROUP Early tetrapods
DATE Middle Carboniferous
SIZE 9¾–13ft (3–4m) long
LOCATION Scotland, possibly USA

Crassigyrinus was one of the largest predatory animals in the swamps of the Middle Carboniferous. Its enormous gaping mouth housed rows of very large teeth, positioned mostly near the front so that prey would have been firmly impaled when it closed its jaws in a fast, snap-trap movement. Its large head, elongated body, and deep tail suggest that *Crassigyrinus* was fully aquatic, as do its tiny forelimbs and relatively small hind limbs. In some respects, the skull of *Crassigyrinus* is quite primitive, and its evolutionary relationships are still debated.

LYING IN WAIT
Crassigyrinus may have been an ambush predator, hiding in vegetation or among rocks while waiting for its prey to swim past.

Balanerpeton

GROUP	Temnospondyls
DATE	Early Carboniferous
SIZE	20in (50cm) long
LOCATION	Scotland

Balanerpeton was the earliest member of a large group of fossil amphibians called the temnospondyls. It would have looked like a large salamander, but its middle ear and tympanic membrane more closely resembled those of a frog. *Balanerpeton* was the first known land animal to have had an ear capable of hearing airborne sound waves, so it could have heard its insect prey moving nearby or its mate calling. Its name means "the creeping animal from the hot springs" because its remains were found in what is believed to have been a volcanic crater lake surrounded by hot springs.

FLATTENED SKULL
The indentations for the tympanic membranes can be seen on either side of the back of the skull.

Cochleosaurus

GROUP	Temnospondyls
DATE	Late Carboniferous
SIZE	5ft (1.5m) long
LOCATION	Czech Republic, Canada

Many *Cochleosaurus* fossils have been found at the Nýrany coal mine in the Czech Republic, which contains the remains of a Carboniferous swamp. *Cochleosaurus* was one of the first crocodilelike temnospondyls.

It lurked at the edge of the swamp and ambushed prey such as fish and smaller amphibians. It was the most common large animal living in the swamp, making it the top predator in this type of habitat. Similar temnospondyls filled this niche until the end of the Triassic, when they were replaced by true crocodylomorphs (see p.232). The name *Cochleosaurus* means "spoon-reptile" and refers to two spoonlike bony extensions to the back of the skull, which were hidden under the skin.

Branchiosaurus

GROUP	Temnospondyls
DATE	Late Carboniferous
SIZE	6in (15cm) long
LOCATION	Czech Republic, France

The *Branchiosaurus* specimens found at the Nýrany coal mine had gills—the name means "gilled reptile"—and it is not clear whether these individuals would have remained aquatic throughout their lives or would have become terrestrial as adults.

In some populations of *Apateon* (a later relative of *Branchiosaurus*), the animals remained aquatic as they grew, while in other populations they transformed into terrestrial forms. These alternative life cycles can be seen in some species of newt today, in which metamorphosis into an air-breathing terrestrial form occurs only when conditions are right. For many years, branchiosaurs were all thought to be the larvae of much larger temnospondyls, such as *Eryops* (see p.176), but they are now known to be a distinct group.

6 inches The average length of *Branchiosaurus*, making it one of the **smallest** of the temnospondyls.

Phlegethontia

GROUP	Aïstopods
DATE	Late Carboniferous
SIZE	28in (70cm)
LOCATION	USA, Czech Republic

Phlegethontia is the best-known aïstopod, a group of legless amphibians that resembled snakes. It had rows of small spiked teeth, similar to those found in some nonvenomous snakes today. Its backbone had more than 200 vertebrae, two-thirds of which were in the tail.

SNAKELIKE AMPHIBIAN
Phlegethontia was probably terrestrial, living at the edges of ponds and lakes and venturing out onto aquatic plants in search of prey.

Amphibamus

GROUP Temnospondyls
DATE Late Carboniferous
SIZE 4½in (12cm) long
LOCATION USA

Amphibamus was one of the first Carboniferous amphibians discovered. It was the first temnospondyl to have the type of teeth that are seen in frogs and salamanders today and is thought to be closely related to the ancestor of these two groups. Its small teeth each had two tiny spikes at the tip and were hinged partway down their length. *Amphibamus* means "equal legs," owing to its long front and hind limbs being the same size, like those of salamanders. But it was small by temnospondyl standards.

At sites where *Amphibamus* has been discovered, paleontologists have also found several tiny amphibian fossils that look like *Branchiosaurus* (see opposite) but share features with *Amphibamus*. These may be *Amphibamus* larvae, which suggests some degree of metamorphosis was already happening in this early amphibian. For example, branchiosaur fossils have a long tail with a fin, like a tadpole, while the adult *Amphibamus* is thought to have had a stumpy tail that was more in common with the missing tail in modern frogs and toads.

This little temnospondyl had **advanced features** seen today in **frogs** and **salamanders**.

DELTA-DWELLER
Amphibamus is from a site called Mazon Creek in Illinois, which represents a Carboniferous river delta. It is likely that *Amphibamus* lived in the creeks and along the banks of the delta.

Microbrachis

GROUP Microsaurs
DATE Late Carboniferous
SIZE 12in (30cm) long
LOCATION Czech Republic

More than 100 specimens of *Microbrachis*, of all ages and sizes, have been found at Nýřany. All of them had internal gills and a lateral-line sensory system. Clearly this microsaur was aquatic and spent its entire life in the waters of the swamp, unlike most microsaurs, which were terrestrial.

AT HOME IN THE WATER
Microbrachis, meaning "small arms," had tiny limbs, a flattened tail to aid swimming, fishlike gills, and lateral-line canals along its flanks. All of these are adaptations for a fully aquatic lifestyle.

Proterogyrinus

GROUP Stem amniotes
DATE Early Carboniferous
SIZE 5ft (1.5m) long
LOCATION USA, Scotland

Proterogyrinus was one of the earliest anthracosaurs ("coal lizards"), a group of amphibians that were top predators for much of the Carboniferous. Although it ate fish, *Proterogyrinus* was not fully aquatic. It was one of the first vertebrates to have the long curved ribs and muscles that are used to pump air in and out of the lungs. This evolutionary adaptation developed after vertebrates had moved onto the land.

EARLY WRIGGLER
Proterogyrinus had a long, undulating body and tail—hence its name, which means "early wriggler." Its robust limbs indicate that *Proterogyrinus* could walk on land, so it was not restricted to a wholly aquatic lifestyle.

badly crushed skull

curved rib

long tail

five-toed foot

LIZZIE THE LIZARD
At first glance, *Westlothiana* resembles a lizard, hence its nickname of "Lizzie the lizard," but its ankles are characteristic of an early tetrapod.

Westlothiana

GROUP Early tetrapods
DATE Early Carboniferous
SIZE 10in (25cm) long
LOCATION Scotland

Westlothiana was a small terrestrial animal whose remains have been found at East Kirkton in the West Lothian area of Scotland. It had an elongated body with very short but slender limbs, strongly built vertebrae, and curving ribs that would have enclosed most of the body. Only about five specimens are known, and the skull is badly preserved in all of them. When first discovered, *Westlothiana* was hailed as "the earliest reptile," but key parts of the skull, such as the palate and ear bones, are needed to confirm its evolutionary position. However, *Westlothiana* does appear to be one of the earliest relatives of the modern amniotes (see p.155).

KEY SITES
EAST KIRKTON

The Early Carboniferous rocks from East Kirkton have given us a window into an assemblage of early terrestrial tetrapods. Volcanic activity at the time led to ash falls, fires, warm mineral-rich springs, and poisonous gases, killing and preserving many plants, arthropods, and tetrapods. Among the tetrapods were the earliest members of lineages leading to modern amphibians and reptiles, including *Westlothiana* and *Balanerpeton*.

Ophiacodon

GROUP Synapsids
DATE Latest Carboniferous to Early Permian
SIZE 9¾ft (3m) long
LOCATION USA

Ophiacodon was one of the largest early synapsids (see p.155). It resembled a crocodile, with long jaws and numerous pointed teeth that were ideal for catching fish. Flattened finger and toe bones suggest that it was partly aquatic and

scales protect against predators

strong tail

flattened finger

Paleothyris

GROUP Eureptiles
DATE Late Carboniferous
SIZE 10in (25cm) long
LOCATION Canada

A primitive reptile, *Paleothyris* was a small, lizardlike insectivore that lived in forested regions. Its skeletons have been found in the hollow trunks of lycopsids (giant early relatives of clubmosses), buried in sediment from a flash flood. *Paleothyris* is similar to another early reptile, *Hylonomus*, in being anapsid (having no holes in the back of the skull), but it is 5 million years younger.

AGILE HUNTER
The limbs, body, and teeth of *Paleothyris* indicate that it would have been an agile hunter, easily able to track down and capture its insect prey.

long, curved rib

small, sharp teeth

hind limb

Spinoaequalis

GROUP Diapsids
DATE Late Carboniferous
SIZE 10in (25cm) long
LOCATION USA

Spinoaequalis is known from a single skeleton found in a quarry at Hamilton, Kansas. It was an early diapsid (see p.155) and had long legs and toes and a distinct neck. More unusual, instead of narrowing to a point, its tail remained deep along its length and was vertically flattened. Long, bony spines above and below the vertebrae provided extra muscle-attachment points, making the tail a powerful tool for swimming.

BACK IN THE WATER
Spinoaequalis was the first reptile to return to the water and live alongside amphibians, although it was not fully aquatic and would have returned to dry land to breed.

could use its limbs as paddles. Unlike later pelycosaurs such as *Dimetrodon* (see pp.180–181), *Ophiacodon* was not well adapted for killing large animals, but it probably lived at the water's edge, catching fish. Despite the similarities, being a synapsid, *Ophiacodon* was much more closely related to mammals than to modern crocodiles. This is an example of convergent evolution, where adaptations to a particular way of life evolve separately in different groups of animals.

BIGGER THAN ITS BITE
With its large size and crocodilelike jaws, *Ophiacodon* looks like a fearsome predator, but it seems to have caught mostly fish and small vertebrates rather than large prey.

KEY SITES
JOGGINS FOSSIL CLIFFS

These famous cliffs in Nova Scotia preserve the oldest fossil amniotes, namely *Protoclepsydrops* (a close relative of *Ophiacodon*) and *Hylonomus*, one of the earliest true reptiles. On this site, 300 million years ago, small animals used hollowed-out tree stumps as nesting sites. Today, their remains give us our first clear insight into the record of small terrestrial vertebrates.

coloring adapted
for camouflage

powerful legs aid
swimming

sharp teeth

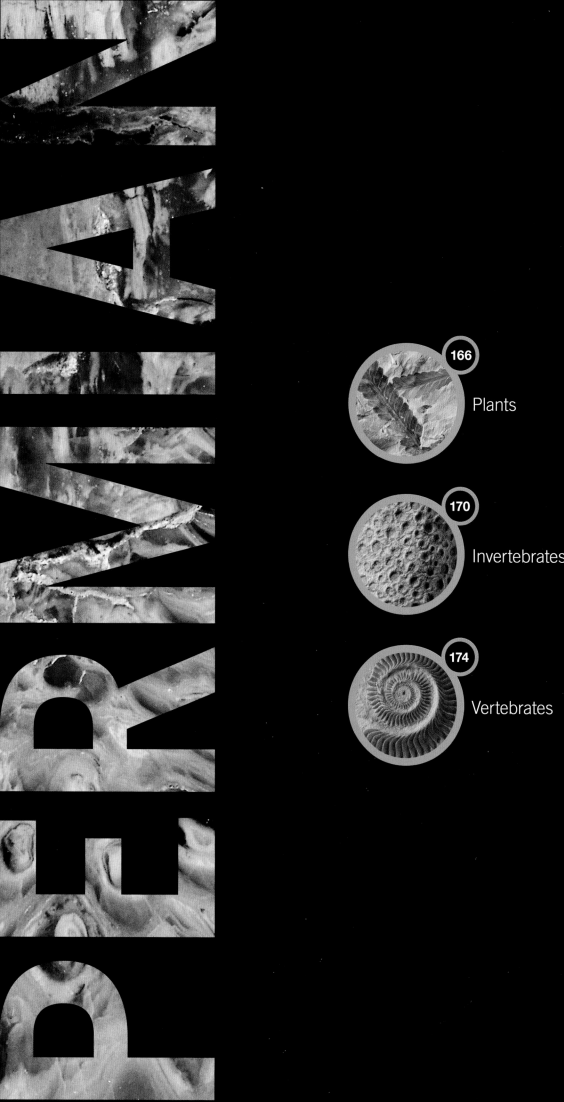

PERMIAN

Permian

The last period of the Paleozoic era was dominated by the giant supercontinent of Pangea, into which Laurasia (Euramerica and Asia) and Gondwana finally merged. The Permian was named after the Russian region of Perm, because its limits were defined by strata found in the Ural Mountains. It ended with the largest mass extinction of the last half a billion years, when at least 90 percent of land and oceanic organisms disappeared.

URAL MOUNTAINS
The forest-covered western slopes of the Urals, which stretch from north to south in west–central Asia, are especially noted for their extensive Permian strata.

Oceans and continents

The sheer size of the Pangean supercontinent had significant implications for the world's climate and the spread of fauna and flora. At low latitudes, the Central Pangean Mountains extended in an east–west orientation, which greatly influenced atmospheric currents and hence weather patterns. During the Late Carboniferous, this mountain belt had been located across the tropics and was the source of the coal deposits formed from the lush rain forests that grew in the wet, tropical climate. By Permian times, it had moved farther north into more arid regions. The mountains blocked the humid, equatorial winds, with the result that deserts formed in the northern part of Pangea (now the central part of North America and northern Europe). Sandstone strata reveal the existence of huge expanses of dunes. As the climate became drier, large mammal-like reptiles appeared, better adapted to coping with the new conditions. In less arid regions, plants continued to diversify. Upland forests of conifers, ferns, and seed ferns established themselves and large expanses of the plant *Glossopteris* fringed cool polar regions. Most of Pangea extended south of the Equator, and included the former continent of Gondwana. Its vast land mass gave some faunas the opportunity to spread over a wide area. On the northern section of Pangea, in Siberia, one of Earth's largest eruptions of magma covered a huge area with flood basalts and was accompanied by outpourings of ash and gases, especially sulfur dioxide and water vapor. This coincided with the Permian mass extinction, when the oceans became starved of oxygen and most marine species died out.

TROPICAL RAIN FOREST
Lush tropical forests, similar to those that flourished during the Carboniferous, still existed around the Equator during the Permian, in areas that are now part of China.

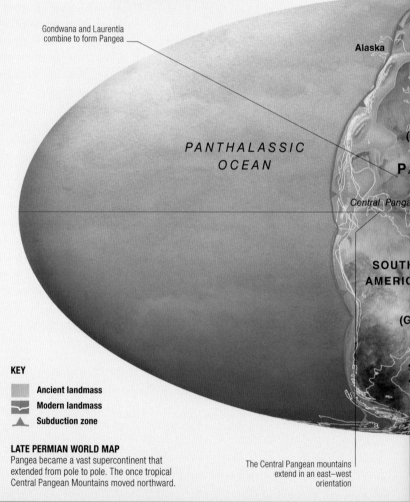

Gondwana and Laurentia combine to form Pangea

Alaska

PANTHALASSIC OCEAN

P

Central Panga

SOUTH AMERIC

(G

S

KEY

▦ Ancient landmass

〰 Modern landmass

▲ Subduction zone

LATE PERMIAN WORLD MAP
Pangea became a vast supercontinent that extended from pole to pole. The once tropical Central Pangean Mountains moved northward.

The Central Pangean mountains extend in an east–west orientation

CISURALIAN

MYA	300	290	280
PLANTS		● 290 Conifers and seed plants diversify	● 280 Seed plants and conifers increase in diversity; lepidodendrids and sphenopsids decrease in diversity
			● 280 First beetles
INVERTEBRATES			

DWYKA TILLITE
Tillites are conglomerates composed of boulders, cobbles, and sand scraped up by glaciers. The Dwyka Tillite in the Karoo Basin in South Africa dates from the Permian period.

Flood basalts erupt from the Siberian section of Pangea

Siberia

Kazakhstania

North China

EA

PALEO-TETHYS OCEAN

South China

tains

RICA

Indochina

Turkey

Iran

Malaya

Tibet

TETHYS OCEAN

India

Australia

RCTICA

a)

a)

The fledgling Tethys Ocean begins to develop east of Pangea

Climate

The glaciation that so influenced the Late Carboniferous world continued, coming to an end in the early part of the Late Permian. Ice sheets covered all the continents that once formed part of Gondwana. Many glacial centers were involved, with repeated ice advances and retreats. Evidence for this comes from glacial deposits in South Africa, India, and Tasmania. The major Permian glacial episode spanned about 2 million years, coming to an end in a period of just a few thousand. From the Late Carboniferous to the Permian, the major centers of ice dispersal migrated with the movement of Pangea over the South Pole. The main centers of glaciation in the Late Carboniferous were in South America, India, the southern part of Africa, and western Antarctica, but they now shifted toward Australia. While in the Early Permian much of the Southern Hemisphere was covered by ice, the climate in mid-latitudes in southern and northern Pangea was relatively cool temperate to arid, with only a narrow tropical belt in equatorial regions. During warmer interglacial periods, coal-forming rain forests were quite widespread in lower latitudes. By the Late Permian, the ice sheets that had covered the Southern Hemisphere landmass had all but disappeared, although small ice caps had developed at the North Pole. Much of Pangea was very dry, with vast deserts in areas that are now South America and Africa in the south, and North America and northern Europe in the north.

SIBERIAN TRAPS
Vast outpourings of basaltic magma continued for thousands of years to form what is now the Siberian Traps. The environmental stress this caused may have contributed to the mass extinction at the end of the Permian.

CARBON DIOXIDE LEVELS
Atmospheric carbon dioxide rose from levels similar to those today in the Early Permian to fairly high. Extensive flood basalt eruptions in Siberia may have contributed to an increase in greenhouse gases.

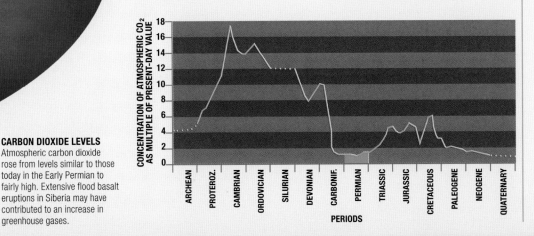

CONCENTRATION OF ATMOSPHERIC CO_2 AS MULTIPLE OF PRESENT-DAY VALUE

18 16 14 12 10 8 6 4 2 0

ARCHEAN | PROTEROZ. | CAMBRIAN | ORDOVICIAN | SILURIAN | DEVONIAN | CARBONIF. | PERMIAN | TRIASSIC | JURASSIC | CRETACEOUS | PALEOGENE | NEOGENE | QUATERNARY

PERIODS

GUADALUPIAN | LOPINGIAN

270 | 260 | 250 | MYA

260 Decrease in *Glossopteris* flora

GLOSSOPTERIS

250 Mass extinction: 50% decline in plant species diversity, mostly affecting woodlands

PLANTS

275 Ammonoid diversity begins to reduce; first ceratitid ammonoids

XENODISCUS

255 First pectinid and carditid bivalves

250 Mass extinction: 96% of all species; all fusulinid foraminifera, rugose and tabulate corals, trilobites, pareiasaurs, eurypterids, receptaculitids; goniatites; strophomenid brachiopods; 5 orders of insects. Almost extinct: echinoids, crinoids, stenolaemate bryozoans, articulated brachiopods

INVERTEBRATES

275 First pareiasaur reptiles below); first therapsid (mammal-ike) reptiles; extinction of pelycosaur reptiles

260 Decrease in diversity of terrestrial and aquatic amphibians; increase in therapsid diversity, especially herbivores like dicynodonts

250 Two-thirds of amphibians, reptiles, and therapsids become extinct

VERTEBRATES

PERMIAN
PLANTS

The drying climates of the Permian finally destroyed the ancient coal swamps that were characteristic of the Carboniferous. This led to the extinction of many characteristic Carboniferous plants. However, new kinds of habitats provided fresh opportunities for the expansion of other plant lineages.

During the Permian, the giant tree lycopods and the habitats in which they thrived declined and eventually disappeared. Medullosans, cordaites, and many other groups of archaic plants also became extinct during the Permian. However, conifers and other new kinds of plants diversified and set the stage for further evolutionary innovation through the Mesozoic.

New kinds of seed plants

Several new groups of seed plants appeared during the Permian. Most are pretty poorly understood, but glossopterids are one of the better known of these groups. They are characteristic of areas that were once at high latitudes in the Southern Hemisphere. Recognition of their distinctive leaves in Australia, South Africa, South America, and Antarctica was important early evidence for continental drift (see p.21). Cycads also made their first appearance in the Permian and

SEED-BEARING LEAF OF CYCAS
Living *Cycas*, one of about 10 genera of living cycads, has seed-bearing leaves that are very similar to those of fossil cycads from the Permian.

subsequently became important in the Mesozoic. Several groups of poorly understood Permian seed plants may be early relatives of these true cycads. In China, an enigmatic group called gigantopterids are especially interesting. Gigantopterids had large leaves with specialized net venation very similar to that seen in angiosperms. Some of these gigantopterids were probably climbers in Permian equatorial forests.

FOSSIL CONIFER
The shoots and leaves of the earliest conifers were very similar to those of some living conifers in the monkey-puzzle family, such as the Norfolk Island Pine.

The rise of conifers

The earliest conifers are from the Late Carboniferous. They had normal, coniferlike leafy shoots, but their cones were much less compact than those of any modern conifer. The cones of early conifers have been compared to the cones of cordaites (see p.145). The early conifers of the Carboniferous occurred in drier environments outside the lowland coal swamps, and as the climate dried in the Permian, conifers and other groups from these areas became more important. Conifers may also have been the first seed plants in which the seed could remain dormant prior to germination.

GROUPS OVERVIEW

The Permian was a time of transition in the evolution of land plants. The ancient plant lineages of earlier in the Paleozoic declined and were replaced by other groups that were apparently better adapted to the drier conditions of the Permian. Among seed plants, the lyginopterids, medullosans, and cordaites had all disappeared by the end of this period.

FERNS
The marattialean tree ferns of the Carboniferous persisted into the Permian. They eventually became much less prominent, but nevertheless persisted through the Mesozoic and Cenozoic to the present. In the drier climates of the Permian, new kinds of ferns appeared for the first time, including forms that are similar to the modern royal ferns.

CONIFERS
The earliest conifers are from the Carboniferous, although they were rare at this time. Conifers flourished and diversified in the drier climates of the Permian. The earliest conifers had normal-looking conifer foliage, but their cones, and perhaps other aspects of their reproduction, were very different from any living conifer.

CYCADS
Fossil cycads that are clearly related to the living group appeared for the first time during the Permian. There is much about these earliest cycads that is still not well understood, but their seed-bearing leaves were very similar to those of the living genus *Cycas*.

GLOSSOPTERIDS
This was one of several new seed plant groups that appeared in the Permian. Glossopterids' characteristic leaves are common in fossil floras from high latitudes in the Southern Hemisphere. This group was especially important in the vegetation of cooler climates during the major Permian glaciation of the Southern Hemisphere.

Oligocarpia

GROUP Ferns
DATE Carboniferous to Permian
20in (50cm) tall
LOCATION Worldwide

This ground-hugging fern is similar to a family of living ferns that are mainly found in the tropics. Known as forking ferns, they have fronds that grow in pairs and creeping rhizomes or underground stems. Like all ferns, *Oligocarpia* spread from place to place by spores, which developed on the undersides of its fronds. However, recent fossil finds show that its rhizomes may have been just as important, allowing it to invade new areas of ground in the way that bracken does today. Unusual for a prehistoric species, these fossils also show the same plant at many different stages of development. These include adult ferns and tiny plantlets with minute rosettes of leaves.

Oligocarpia spread from place to place by **spores dispersed** through the **air**.

midrib of leaf

blade of leaf segment

stems and roots

OLIGOCARPIA GOTHANII
This beautifully preserved specimen was collected in China in the 1920s. Fossils in the same strata suggest that it lived in a wetland habitat, such as a low-lying river plain. Here, periodic flooding may have created the ideal conditions for fossilization—almost all of the plant has been preserved.

Protoblechnum

GROUP Peltasperms
DATE Permian
SIZE Frond 6½–16½ft (2–5m) long
LOCATION Worldwide

Protoblechnum fronds have broad, webbed leaflets—a feature found in some living hard ferns or deer ferns of the genus *Blechnum*, which gives this type of foliage its scientific name. However, instead of belonging to ferns, *Protoblechnum* fronds belonged to a group of plants called peltasperms. Most of these plants were shrubs or small trees. They grew in a variety of habitats, but their foliage would have made them particularly well adapted to areas with an annual dry season, such as subtropical river plains. Their seeds were usually ¼in (6mm) long.

PROTOBLECHNUM SP.
This fossilized *Protoblechnum* species shows the typical webbed leaves. Peltasperms bore their seeds in small, umbrella-shaped organs that were often arranged in open columns or spikes.

Tubicaulis

GROUP Ferns
DATE Late Carboniferous to Permian
6½ft (2m) high
LOCATION Worldwide

Tubicaulis fossils are the preserved trunks of an ancient group of ferns dating back over 300 million years. In cross-section, the fossils show a dense mass of stems embedded in a matrix of roots. Each stem is enclosed by an oval sheath and contains a C-shaped stalk. *Tubicaulis* ferns were upright or scrambling, anchoring themselves with roots that grew from the stems near the base of their leaves. Most lived on shady swamp floors, but some species may have grown high up on trees to collect light.

TUBICAULIS SOLENITES
This cross-section reveals that, unlike in living ferns, *Tubicaulis* stems were constructed "back to front"—the "C" faced outward rather than inward.

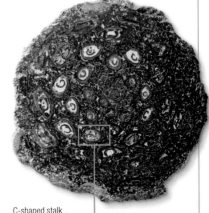

C-shaped stalk

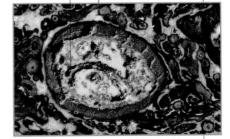

Glossopteris

GROUP Glossopterids
DATE Permian
SIZE 13–26ft (4–8m)
LOCATION Southern Hemisphere

Since its discovery in 1828, by French paleobotanist Adolphe Brongniart, *Glossopteris* has become one of the most studied prehistoric plants. The first *Glossopteris* fossils were found in India, but others have been unearthed in all the land masses of the Southern Hemisphere, providing striking biological evidence for continental drift. *Glossopteris* means "tongue fern," which describes its leaf shape. Dozens of species have been identified, most being shrubs or small trees. Despite its name, *Glossopteris* reproduced by seed, not spores, which developed in stalked organs attached to the leaves.

LOW-LYING FORESTS
With its height and foliage, *Glossopteris* was the dominant plant across much of the Southern Hemisphere, forming immense forests in low-lying ground. Some species were subtropical, but many grew at high latitudes, where winters were long and cold.

defined growth rings

AUSTRALOXYLON SP.
Australoxylon is fossilized wood of glossopterids. It has been found in many parts of the Southern Hemisphere. In structure, it was similar to the wood found in modern conifers.

jointed roots

VERTEBRARIA INDICA
With their cylindrical shape and conspicuous joints, *Vertebraria* fossils resemble the vertebrae of many animals instead of the fossilized roots of *Glossopteris* trees. They are found from India and Africa to Australia, South America, and Antarctica.

leaf fossil

TONGUE-SHAPED LEAVES
Glossopteris is identified mainly by its leaves, which were up to 12in (30cm) long, with a strong midrib and a network of diverging veins. They were arranged in spirals around the ends of the branches.

Callipteris

GROUP Peltasperms
DATE Late Carboniferous to Permian
SIZE Leaf 30in (80cm) long
LOCATION Worldwide

Belonging to the same group of seed ferns as *Protoblechnum* (see p.167), callipterids are well known from fossils of their leaves and seed-producing organs. Their leaves were up to 3ft 3in (1m) long. In addition to having leaf segments arranged in opposite pairs, they often had small leaflets attached to the central midrib. *Callipteris* bore its seeds on the underside of fan-shaped scales, which were arranged in a spiral around a vertical stalk.

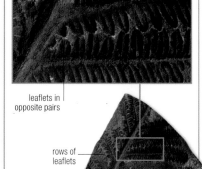

leaflets in opposite pairs

rows of leaflets

central midrib

CALLIPTERIS CONFERTA
Callipteris conferta was one of the most common seed plants of the Permian. It is easy to identify because of the regular sequences of paired leaf segments attached to a large, central rib.

Gigantopteris

GROUP Gigantopterids
DATE Permian
SIZE Leaf up to 20in (50cm) long
LOCATION Southeast Asia, North America

The gigantopterids are an enigmatic group of plants, known mainly from fossils found in Asia and North America. They had woody stems, sometimes with spines, and many had leaves like those of flowering plants. They died out in the Early Triassic, and little is known about how they reproduced, which makes it hard to judge exactly where they fit in plant evolution.

symmetrical leaves

GIGANTOPTERIS NICOTIANAEFOLIA
The fossil above is a species called *Gigantopteris nicotianaefolia*, which was so named because its leaf segments resemble those of tobacco plants.

Tingia

GROUP Noeggerathialeans
DATE Early Permian
SIZE Cone up to 6in (15cm) long
LOCATION China

This foliage belonged to fernlike plants that reproduced by growing spores. In *Tingia elegans*, the leaves bore leaflets in four rows—the two rows on the upper surface were largest, masking the smaller ones underneath. The plant's cones were attached at the base of the leaves. As fossils, noeggerathialeans are relatively rare, although cones and leaves have been found together, showing that they came from the same plant. They appeared in the Late Carboniferous and became extinct during the Triassic. Although they show links with progymnosperms, their place in plant evolution is not clear.

TINGIA ELEGANS
Tingia is known only from the fossil remains of its leaves, like those shown here, and of its cones. *Tingia* reproduced by means of spores, rather than through the production of seeds.

Ullmannia

GROUP Conifers
DATE Late Permian
SIZE 33ft (10m) high
LOCATION Worldwide

Conifers first appeared in the Carboniferous period, but they became more widespread during the Permian, with its cooler, drier global climate. *Ullmannia* was one of these Permian conifers, with narrow, widely spreading leaves. Its male cones produced pollen grains that had a microscopic, winglike sac to help them drift on the wind. Like most conifers, its seed-bearing cones were woody; measuring up to 2¼in (6cm) long. The seeds developed on the upper surface of rounded scales.

narrow leaflet on narrow leaf

leaf stem

ULLMANNIA BRONNI
Ullmannia had narrow, widely spreading leaves, not dissimilar to those of the modern monkey puzzle tree. The leaves were arranged in a spiral around a thick central stem.

Psygmophyllum

GROUP Ginkgos
DATE Permian
SIZE Leaf up to 4in (10cm) long
LOCATION Worldwide

From the Triassic onward, many fossils show how ginkgos evolved variations of fan-shaped leaves. Earlier fossils are far rarer. *Psygmophyllum* may have been an ancient forerunner of living ginkgo, since its leaves superficially have the classic ginkgo outline. However, this resemblance is probably misleading: no fossils have been found of its fruits or seeds, which makes it likely that the leaves come from a more primitive plant.

fan-shaped leaves

PSYGMOPHYLLUM MULTIPARTITUM
Although the leaves of *P. multipartitum* have the usual fan-shaped outline of living ginkgo, it is unclear whether this species and ginkgo were closely related.

PERMIAN
INVERTEBRATES

Most Permian sediments encountered are red beds of desert or river origin, with marine sediments less common. This is because of a drastic lowering of sea level, especially during later Permian times, and the emergence of the continental shelves. The consequences for invertebrate marine life were disastrous.

Sea-level lowering

The Carboniferous–Permian glaciation, which covered so much of the land in the Southern Hemisphere, was largely over by mid-Permian times, and marine faunas proliferated. Yet by the late Permian, the Earth witnessed the most acute biological crisis of all time. The mainly tropical crinoid-, brachiopod-, and bryozoan-dominated ecosystems of the later Permian were destroyed completely—a process that took place over some 10 million years. All the continents were assembled into the giant supercontinent of Pangea, which must have affected climate and oceanic circulation, and it was no doubt an important factor in generating extinction. But most importantly, sinking, inactive mid-oceanic ridges caused the global sea level to fall by up to 920ft (280m), with great loss of habitat, and little sedimentation: warm, shallow seas became reduced to certain

FENESTELLA PLEBEIA
This common Carboniferous–Permian bryozoan fed by filtering food from currents flowing through colonies that grew upward on the near vertical reef-front.

parts of the world only. Massive volcanic eruptions contributed to environmental instability and thus to ecosystem collapse. It was several million years before normal marine sedimentation was restored.

Giant algal reefs

In northern England, Texas, and elsewhere, vast algal reefs—rich in unusual brachiopods and bryozoans—bordered inland seas, connected to the main ocean by narrow channels. As the sea level fell, the reefs died and were stranded high and dry. The inland seas dried out, with great quantities of salt deposited as evaporites, sometimes to be flooded again, at least for a time. This caused a biological catastrophe on an unparalleled scale, and the worst crisis through which life on Earth ever passed.

PERMIAN REEF
This gigantic reef in Texas was built up by calcareous algae but had its own fauna of marine invertebrates and fish.

SCHIZODUS SP.
Permian bivalves, such as this mold of *Schizodus*, are quite common, and were not greatly affected by the Permian mass extinction event.

GROUPS OVERVIEW

Throughout this period, most mollusk groups suffered comparatively little; bivalves, gastropods, and cephalopods continued to do well. Bryozoans, however, did not fare as well during the Permian extinction. Brachiopods were abundant in the early Permian, and ammonoids continued to evolve, becoming increasingly complex and diverse.

BRYOZOANS
Permian bryozoans were normally marine, colonial invertebrates, but there were some freshwater genera. They all usually have calcite skeletons, which house large numbers of tiny zooids; most of these feed with a ring of tentacles, some zooids are specialized for other functions. Several groups were badly affected by the Permian extinction.

BRACHIOPODS
These marine invertebrates were seafloor suspension feeders, and although they also had two valves, they had quite a different body plan from the bivalves (see right). They were a dominant component of the seafloor ecosystem until the mass extinction at the end of the Permian; some species are still in existence to this day.

AMMONOIDS
Several groups of these highly successful cephalopods flourished in the Permian, with "goniatitic" (strongly angled) suture lines (see p.194), and they gave rise to the ceratitid ammonoids that dominated the Triassic.

BIVALVES
Seafloor dwelling or burrowing bivalve clams, which fed by sucking in currents of water laden with food particles, proliferated during this period. With the decline of the brachiopods at the end of the Permian, these became the dominant seafloor suspension feeders.

Rectifenestella

GROUP Bryozoans

DATE Devonian to Permian

SIZE 2in (5cm) average height

LOCATION Worldwide

Rectifenestella was an upright, fenestrate bryozoan colony. It was conical or fan-shaped, with a netlike appearance, and was composed of thin branches, each of which divided into two. These branches were linked by small, usually horizontal partitions, which produced rectangular holes in the colony. Circular openings on one side of the colony were arranged in two rows.

DIRECTION OF FLOW
Filtered water flowed through *Rectifenestella*'s fenestrules and out of the back of the colony.

fenestrule

RECTIFENESTELLA RETIFORMIS
This netlike colony can be seen next to *Acanthocladia anceps*. Both species were wiped out in the end-Permian extinction.

These were the openings to the autozooecia, where the colony's individual soft-bodied animals (autozooids) lived. Each autozooid had a ring of tentacles (lophophore) around its mouth. Small, hairlike extensions (cilia) on these tentacles would beat to create a flow of water that drove food particles toward the autozooid's mouth. The filtered water then flowed out of the rectangular holes to the reverse side of the colony. *Rectifenestella* was supported by rootlike spines, which occurred at the base of the colony on the opposite side to the feeding autozooids, but these cannot be seen on the specimen below.

netlike appearance

ANATOMY
FENESTRATE COLONIES

Bryozoan colonies form a variety of different shapes. Some grow as massive lumps that encrust the seafloor, while others are branching or globe shaped. The extinct fenestrate bryozoans were named according to the way they grew—in distinctive netlike patterns. Their thin branches were connected by partitions called dissepiments, which formed open spaces called fenestrules, from the Latin word for "window." The upright branches often divided in two to create a fan. Small openings, in two or more rows, were located on the front surface of these branches. These were the entrances to the chambers (autozooecia) that housed the colony's tiny, soft-bodied animals.

fenestrule   dissepiment

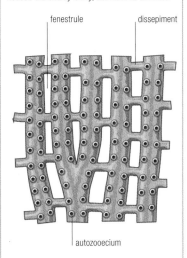

autozooecium

Acanthocladia

GROUP Bryozoans

DATE Carboniferous to Permian

SIZE ³⁄₄in (2cm) average height

LOCATION Europe, North America, Asia

Acanthocladia was an upright, often bushlike, fenestrate bryozoan colony. The branches grew from a central axis and split in two. Sometimes, smaller branches fused together and irregular-shaped holes (fenestrules) would develop between them. There were rows of circular openings on one side of the colony that housed soft-bodied autozooids. When the autozooids were not feeding, their tentacles could be withdrawn. The reverse of the colony was smooth, or had fine striations.

ACANTHOCLADIA ANCEPS
The colony had a bushlike appearance with branches growing from a central axis.

corallite

Thamnopora

GROUP Anthozoans

DATE Devonian to Permian

SIZE Corallite ¹⁄₁₆in (1–2mm) wide

LOCATION Worldwide

Thamnopora was a branching tabulate coral. The corallites, which in life housed the soft-bodied polyps, were circular in cross-section and curved outward from the center of the branch. The polyp would have rested on the thin horizontal partitions (tabulae). The walls of the corallites were perforated with pores and increased in thickness toward the edge of the colony.

THAMNOPORA WILKINSONI
The walls between the corallites of this tabulate coral were very thin.

---

auricle

pedicle valve

sharp infold

HORRIDONIA HORRIDUS
This brachiopod was widest between the auricles at each end of the hinge-line.

Horridonia

GROUP Brachiopods

DATE Permian

SIZE 1–3¹⁄₄in (2.5–8cm) long

LOCATION Europe, Asia, Arctic, Australia

Horridonia was a strophomenid brachiopod, related to *Productus* from the Carboniferous (see p.149). The slightly concave brachial valve, sharply curved at its anterior margin, sat inside the larger, convex pedicle valve. The immature *Horridonia* was attached to the seabed with a stalklike appendage (pedicle). This withered as the animal matured, so the adult relied largely on the weight of its pedicle valve to stay put on the seabed.

Coledium

GROUP Brachiopods

DATE Middle Devonian to Middle Permian

SIZE ⁵⁄₁₆–¹⁄₂in (0.8–1.5cm) wide

LOCATION USA, Indonesia

Coledium was a small rhynchonellid brachiopod with an opening beneath the beak of its larger (pedicle) valve. There was an outward fold on the forward edge of the smaller (brachial) valve, which corresponded with an infold (sinus) on the forward edge of the pedicle valve. The inside of the shell had a thin wall that rose above the floor of the pedicle valve. In life, this would have supported a small, spoon-shaped muscle-attachment site in the middle of the valve.

pedicle opening

faint radial ribs

COLEDIUM HUMBLETONENSIS
This species had weak radial ribs on each valve. Here, the brachial valve is in the foreground, with the beak of the pedicle valve visible at the top.

Xenodiscus

GROUP Cephalopods
DATE Late Permian to Triassic
SIZE 2¼–4in (6–10cm) across
LOCATION Pakistan, Indonesia

Xenodiscus was an ammonoid with a compressed, disk-shaped (discoidal) shell and flattened sides. The whorls on the shell overlapped slightly, and it had a wide, shallow center (umbilicus). Its weak radial ribs were developed on the early whorls but were more faint on the mature body chamber. The point where the internal partitions (septa) were joined to the shell is called the suture line, and it is the form of this line that separated *Xenodiscus* from the goniatites

(a group of ammonoids existing at the same time). The suture line shown below is "ceratitic," and *Xenodiscus* is a ceratitid ammonoid—a group that evolved in the Middle Permian. Unlike the goniatites, it survived the end-Permian extinctions, but the ceratitids eventually died out at the end of the Triassic.

XENODISCUS SP.
The configuration of the whorls and the broad umbilicus mean that this genus is referred to as having an evolute shell. The body chamber has natural cracks infilled by minerals.

broad umbilicus

SUTURE LINES
The suture-line folds that face the aperture are rounded, while those facing backward are finely serrated.

tightly spaced suture lines

body chamber

Juresanites

GROUP Cephalopods
DATE Early Permian
SIZE 2¾–4in (7–10cm) across
LOCATION Russia, Australia

Juresanites was a stout, round goniatite. Each whorl partly overlapped the one before it, and it had a broad and fairly deep center. This type of shell is known as moderately involute. The shell surface was smooth, and there were no visible patterns. Although many Permian goniatites had sutures similar to *Juresanites*, others had much more complex sutures, closer to some of the Mesozoic ammonoids (see pp.279–281).

suture line

rounded venter

JURESANITES JACKSONI
The suture lines on this specimen are visible and are folded as in many of the goniatites. The outside edge of the whorls (venter) is rounded.

Permophorus

GROUP Bivalves
DATE Early Carboniferous to Permian
SIZE ¾–1½in (2–4cm) long
LOCATION Worldwide

impression left by anterior adductor muscle

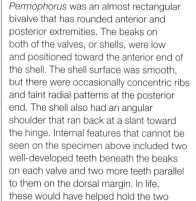

PERMOPHORUS ALBEQUUS
This internal mold shows the impression where the anterior of the two muscles that close the shell was anchored.

Permophorus was an almost rectangular bivalve that has rounded anterior and posterior extremities. The beaks on both of the valves, or shells, were low and positioned toward the anterior end of the shell. The shell surface was smooth, but there were occasionally concentric ribs and faint radial patterns at the posterior end. The shell also had an angular shoulder that ran back at a slant toward the hinge. Internal features that cannot be seen on the specimen above included two well-developed teeth beneath the beaks on each valve and two more teeth parallel to them on the dorsal margin. In life, these would have helped hold the two valves together.

Deltoblastus

GROUP Echinoderms
DATE Permian
SIZE ½–1in (1.5–2.5cm) long
LOCATION Indonesia, Sicily, Oman

Deltoblastus was one of the last blastoid echinoderms to have existed before the group became extinct. In life, *Deltoblastus* would have been attached to the ocean floor by a thin stalk. Compared with the similar-looking *Pentremites* (see p.152), which existed in the Carboniferous, there are some notable differences, such as the more developed upper circlet of plates of *Deltoblastus*. Although it occurred much later than *Pentremites*, *Deltoblastus* is a survivor of a more primitive group of blastoids.

ambulacrum

V-shaped region

DELTOBLASTUS PERMICUS
The five sunken ambulacra carried numerous long, slender extensions used in feeding.

Ditomopyge

GROUP Arthropods

DATE Late Carboniferous to Late Permian

SIZE 1–1¼in (2.5–3cm) long

LOCATION North America, Europe, Asia, W. Australia

Ditomopyge had a distinctively prominent central region of the head or glabella that widened toward the front. Deep furrows in the rear of the glabella divided it into three lobes, a feature that distinguished it from the Carboniferous trilobite *Hesslerides* (see p.153), which was otherwise quite similar. The thorax had nine segments, with a raised axis and side parts (pleurae) that turned downward. The tailshield was about as long as the head and had 14 well-marked rings. The last trilobites, including *Ditomopyge*, became extinct at the end of the Permian following a long decline. They formed only a minor part of marine Permian faunas, although they have been found locally (for example, in the Crimea) in fairly large numbers.

REINFORCING THE HEAD
The long, broad-based projection running from the side of the head is the genal spine, which bolstered the structural strength of the head. The eyes are the kidney-shaped mounds located toward the side of the head.

PERMIAN
VERTEBRATES

Opening with a murmur and ending with a bang, the Permian saw the rise of advanced synapsids. In the oceans, ray-finned fish continued to diversify, while terrestrial vertebrates evolved a new type of predator, the cynodont. Meanwhile, herbivores were perfecting sophisticated chewing mechanisms.

The rise of advanced synapsids (mammal-like reptiles) was the most important evolutionary event of the Permian, because it was critical to the emergence of mammals and, ultimately, modern humans. As the Permian progressed, the terrestrial fauna took on a more familiar setup, with relatively few predators feeding on the more numerous herbivores.

Cynodonts and mammalness

The top predators of the Late Carboniferous were sprawling in posture, demonstrating a less active mode of life, similar to many modern reptiles. By the Late Permian, synapsid predators had evolved features that enabled active hunting: their limbs had elongated and were placed more directly under the body; their teeth had differentiated into incisors, canines, and cheek teeth; and their eyes had moved farther forward, improving stereoscopic vision. These early cynodonts had nearly all the characteristics of mammals. Nowhere is the evolution of "mammalness" more clear than in the middle ear. Mammals have three middle ear bones: the malleus, incus, and stapes. All other tetrapods have only the stapes. The fossil record of synapsids shows how these two extra bones, once part of the jaw articulation, were freed from their jaw-bracing function to become integrated into the middle ear region, where they form part of the sound-detecting apparatus used in hearing.

The great dying

The Permian ended with the greatest mass extinction of all, with an estimated 95 percent of all marine and 70 percent of terrestrial species dying out after volcanic eruptions poisoned the oceans and atmosphere (see p.32).

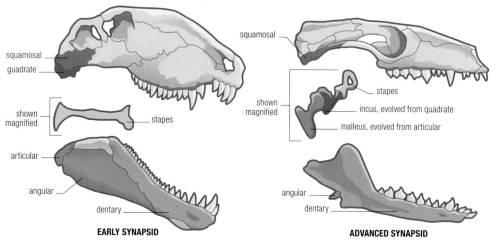

squamosal
quadrate
squamosal
stapes
incus, evolved from quadrate
malleus, evolved from articular
shown magnified
stapes
shown magnified
articular
angular
dentary
EARLY SYNAPSID
angular
dentary
ADVANCED SYNAPSID

THE EVOLUTION OF EAR BONES
Most tetrapods (far left) have only a single middle ear bone, the stapes, which articulates with the tympanum. The jaw hinges between the quadrate and articular. In advanced synapsids (including mammals, left), the jaw hinges between the dentary and squamosal, freeing the quadrate and articular to become incorporated into the middle ear.

GROUPS OVERVIEW

The Permian started with a fauna almost indistinguishable from that of the Late Carboniferous and ended with a synapsid-dominated fauna with lots of herbivores and fewer predators. The groups shown here, just four of many, highlight important innovations in vertebrate evolution during the Permian's 48 million years.

TEMNOSPONDYLS
During the Permian, the amphibious temnospondyls either became specialized for life on land or perfected life in water. Some, like *Eryops*, became quite alligatorlike semi-aquatic animals, whereas others, such as *Archegosaurus*, kept to a life in lakes. Fully terrestrial species include *Cacops*, whose back was protected by a double row of armor plates.

LEPOSPONDYLS
This group of small tetrapod amphibians occupied many lifestyles. The salamanderlike microsaurs wandered the land or became specialized burrowers. The snakelike lysorophians and adelospondylids had reduced limbs and an aquatic lifestyle. Nectrideans, like the boomerang-headed *Diplocaulus*, were also aquatic and were the largest of the group.

EDAPHOSAURIDS
These early synapsids, including the sail-backed *Edaphosaurus*, were herbivorous. Their teeth became short, closely packed, and blunt, for cropping and grinding. Their bellies became extended to host the bacteria in their guts necessary to digest plants. Eventually, herbivores became the largest part of the terrestrial vertebrate fauna.

CYNODONTS
These advanced synapsids arose in the Late Permian. Early cynodonts, such as *Procynosuchus*, show a number of innovations toward a mammal-like condition, including hair, skull construction, and limb placement. These and other skeletal changes indicate that the higher metabolic rate of mammals was in the process of being acquired.

SPINY SHARK
Acanthodians are also known as "spiny sharks," although they are not true sharks. *Acanthodes* had fewer spines than many of its relatives, but they would still have served to protect it.

Acanthodes

GROUP Acanthodians
DATE Early Carboniferous to Middle Permian
SIZE Usually 8in (20cm) long; a few 6½ft (2m) long
LOCATION Europe

Although a small, slender-bodied fish, *Acanthodes* was well protected from predators, with paired spines on the front edges of its pectoral and pelvic fins, as well as projecting from its dorsal and anal fins. More specific characteristics included a single dorsal fin set far down its back near the base of its sharklike tail and closely packed, roughly rectangular scales. It had no teeth, however. *Acanthodes* inhabited freshwater, where it lived as a fast-swimming filter-feeder.

Xenacanthus

GROUP Chondrichthyans
DATE Late Devonian to Late Triassic
SIZE 28in (70cm) long
LOCATION Europe, USA

Xenacanth sharks have been described mostly on the basis of isolated teeth found in marine and freshwater rocks from all over the world. As in all other shark groups, two pairs of fins projected from the belly region: the pectoral fins closest to the head, and the pelvic fins closest to the tail. These acted as hydrofoils, lifting the shark as it moved through the water. The male had a pair of rodlike claspers projecting backward from the pelvic fins, used to deposit sperm directly into the female's body during mating. The final fin on the underside of *Xenacanthus* was the anal fin, which was divided into two parts. Its tail was completely in line with the rest of the body, and the upper and lower lobes were about the same size.

HEAD SPINE
Xenacanthus's head spine is unusual in that it is attached to the back of the head instead of supporting the dorsal fin. It was probably a defensive structure and may well have had poison glands associated with it.

narrow skull

NARROW JAWS
With its narrow jawline and crocodilelike proportions, *Archegosaurus* was probably primarily a fish-eater, feeding on species such as *Acanthodes* in the waters of deep lakes.

crocodilelike teeth

Archegosaurus

GROUP Temnospondyls
DATE Early Permian
SIZE 5ft (1.5m) long
LOCATION Germany

Archegosaurus was discovered in 1847 and was one of the first Permian amphibians to be described by scientists. Its name means "founder of the reptile race," and it resembled a small, narrow-snouted crocodile. Almost all *Archegosaurus* fossils come from deposits that formed on the bed of a deep lake. Modern narrow-snouted crocodiles such as gharials are fish-eaters, and *Archegosaurus* probably hunted in the same way. Some archegosaur specimens have been found with *Acanthodes* bones inside their bodies. When *Archegosaurus* was first found, it was the earliest animal known to have legs (and toes); however, it shared many features with fish, which led some scientists to suggest it was a transition species. Today, *Archegosaurus* is regarded as an early amphibian.

Eryops

GROUP Temnospondyls
DATE Late Carboniferous to Early Permian
SIZE 6½ft (2m) long
LOCATION North America

Eryops is one of the best-known of the temnospondyl amphibians. A large number of specimens have been collected, and because the animal was large and robust, these specimens are remarkably well preserved. *Eryops* was also unusual among early amphibians in that the entire skeleton was ossified (made from bone), which is far more readily fossilized. Its contemporaries had more cartilage in their skeletons, and because these were not mineralized, their complete anatomy has not survived. *Eryops* first appeared in the Late Carboniferous of North America, but most specimens that have been found come from the Early Permian. *Eryops* was one of the top predators in its ecosystem. It resembled a short, fat alligator and almost certainly fed on fish and small amphibians. The long rectangular muzzle is the basis of its name, which means "drawn-out face." *Eryops* probably underwent a change of diet during the Permian. Early specimens have a muzzle shaped like a rounded shovel with small, almost marginal teeth suitable for seizing small fish. However, later *Eryops*, including the one shown here, had a longer, more rectangular snout with clumps of larger teeth at intervals, much like those found in modern crocodiles, which feed on larger animals caught at waterholes and in rivers.

KEY SITE
TEXAS RED BEDS

Since the 1870s, the major source of Early Permian amphibian and reptile fossils has been a series of formations known as the Texas Red Beds. They extend across northeast Texas into Oklahoma and represent flood deposits, pools, lake beds, and river accumulations on a tropical, coastal flood plain. The fauna they contain is rich and includes a range of aquatic, amphibious, and terrestrial animals, many of which have been found as complete skeletons. Fossils that have been found include the shark *Xenacanthus*; amphibians *Diplocaulus*, *Seymouria*, and *Eryops*; and the reptile *Captorhinus*.

MISSING RIBCAGE
Like modern salamanders, and unlike reptiles and mammals, the ribs of *Eryops* were short, straight rods that did not wrap around the chest to form a ribcage. The ribs were mainly for body support and not used for filling the lungs with air. Like salamanders, *Eryops* would have filled its huge mouth with air and pushed it back into the lungs.

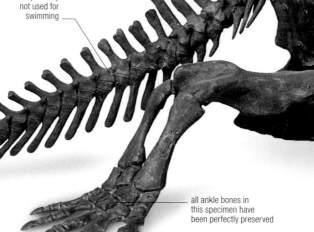

large sacral rib transmitted the body weight through to the pelvis and back legs

narrow tail probably not used for swimming

heavily built pelvis for supporting body weight

all ankle bones in this specimen have been perfectly preserved

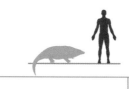

Sclerocephalus

GROUP Temnospondyls
DATE Early Permian
SIZE 5ft (1.5m) long
LOCATION Germany, Czech Republic

Sclerocephalus was a smaller relative of *Eryops* (above). It lived in large, shallow, tropical alpine lakes that developed between rows of mountains across what is now Central Europe. In some cases, the water became stagnant, and the large adults and small larvae of *Sclerocephalus* died together in large numbers. This specimen is a partly grown juvenile.

WELL PRESERVED
In this fossil, the outline of the body is seen as a black shape on the rock. When the animal died, it was buried quickly, along with the bacteria that were decomposing it. They were reduced to a black coal-like film, which solidified.

Seymouria

GROUP Stem amniotes
DATE Early Permian
SIZE 31in (80cm) long
LOCATION USA, Germany

Seymouria was a small carnivore that, for many years, was considered to be close to the amphibian–reptile transition. There was much debate as to whether it was a reptile or not until paleontologists noticed that its smaller relative, *Discosauriscus* (right), had external gills and lateral-line canals in its juvenile stage. This pointed to *Seymouria* being prereptile, and it is now considered to be a primitive amphibious tetrapod that shared with reptiles some of its adaptations to terrestrial life.

For many years, scientists thought *Seymouria* was a primitive reptile.

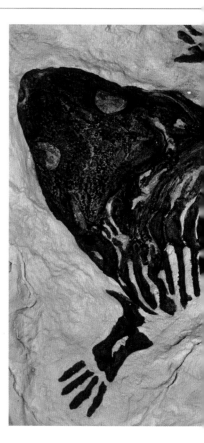

LAND LOVERS
Although *Seymouria* probably spent most of its time on land, females would almost certainly have returned to the water to lay their eggs.

Unusual among amphibians, the entire skeleton of *Eryops* was made of bone, which means that complete specimens have been preserved.

extra blades on some ribs helped it to breathe deeper and force air in and out of lungs

neck vertebra with no spine permitted up-and-down head movement

orbit would have contained a sideways-directed eye

massive U-shaped shoulder girdle for supporting large, heavy head

nostril, probably mainly used for smelling, not breathing

Diplocaulus

GROUP Nectridean
DATE Early Permian
SIZE 3ft 3in (1m) long
LOCATION USA, Morocco

Diplocaulus is one of the most common fossils found in the Texas Red Beds (left), but it has no living relatives. From the neck backward, it resembles a fat salamander, but its skull is a strange boomerang shape, with long, flattened "horns" on each side. One idea is that these horns acted like hydrofoils—by using them to manipulate the water current, *Diplocaulus* could gain lift or downward thrust, which enabled it to rise and sink with little effort.

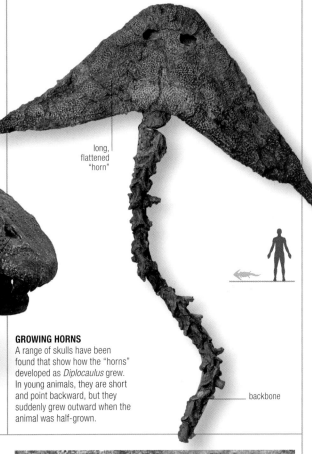

long, flattened "horn"

backbone

GROWING HORNS
A range of skulls have been found that show how the "horns" developed as *Diplocaulus* grew. In young animals, they are short and point backward, but they suddenly grew outward when the animal was half-grown.

pleurocentrum

Discosauriscus

GROUP Stem amniotes
DATE Early Permian
SIZE 16in (40cm) long
LOCATION Czech Republic, Germany, France

Discosauriscus was a smaller relative of *Seymouria* (left). Unlike its larger cousin, *Discosauriscus* seems to have been aquatic throughout its life, having external gills when it was young and lateral-line canals throughout its life. Most specimens of *Discosauriscus* have been found in the Czech Republic, in an area where the surface rocks represent a shallow Early Permian lake that became stagnant, killing all the fish and amphibians in it. Thousands of *Discosauriscus* have been collected from this area, making it one of the most common fossil amphibians in existence.

LITTLE-DISK LIZARD
Each vertebra has a small disklike structure called a pleurocentrum. These gave *Discosauriscus* its name, which means "little-disk lizard."

Orobates

GROUP Reptiliomorphs
DATE Early Permian
SIZE 3ft 3in (1m) long
LOCATION Germany

Orobates was an early, primitive vertebrate herbivore. It is known from a few complete skeletons found in an abandoned sandstone quarry near Bromacker, Germany. The fossils found in this quarry represent one of the first truly terrestrial reptile and amphibian communities found in the fossil record. *Orobates* means "mountain walker" for the rugged habitats in which it lived. *Orobates* is the most primitive member of the family Diadectidae. These herbivorous reptiles were large for their time. They had back teeth for chewing and grinding vegetation and longer front teeth for stripping foliage off branches.

broad hind feet

MOUNTAIN WALKER
With its broad feet and stocky limbs, *Orobates* is thought to have been able to cope with rugged terrain as it foraged for the vegetation that made up its diet.

Captorhinus

GROUP Eureptiles
DATE Early Permian
SIZE 19¹/₂in (50cm) long
LOCATION USA

Captorhinus was a primitive reptile with batteries of crushing teeth that were continuously replaced. It could have been an omnivore, feeding on smaller animals but also capable of grinding leaves or stripping twigs. Later Permian captorhinids were almost certainly plant-eaters.

pointed snout

LIZARDLIKE SKULL
The skull shape of *Captorhinus* was not unlike that of a modern lizard, which it also resembled in body type.

Eudibamus

GROUP Parareptiles
DATE Early Permian
SIZE 9³/₄in (25cm)
LOCATION Germany

Eudibamus was a small herbivore belonging to the bolosaurids. This group is noticeable for the unusual teeth with protrusions, or "heels," on the inner surface. When the mouth is closed, the heels on the upper and lower teeth meet, grinding plant material between them. *Eudibamus* itself represents the earliest known example of bipedal locomotion: walking on two legs. Its legs are extremely long relative to its arms, and the toes are elongated to increase stride length. The tail is also lengthened to improve balance. The morphology of the knee indicates that *Eudibamus*'s legs were held upright, much like the later dinosaurs, rather than having the sprawling stance of its nearest relatives. All these adaptations allowed *Eudibamus* to reach high speeds, and it may have adopted the bipedal posture when it needed to move fast to escape predators.

Mesosaurus

GROUP Parareptiles
DATE Early Permian
SIZE 3ft 3in (1m) long
LOCATION South Africa, South America

Mesosaurus was the first aquatic reptile that has been found. It represents a return of a small group of vertebrates to the water, about 20 million years after vertebrates adapted to life on land (amniotes) evolved in the Late Carboniferous (see p.155). *Mesosaurus* was a reptile, so it had waterproof, probably scaly, skin and long limbs. Because it had lungs instead of gills, it had to come to the surface to breathe. The tail was long and flat like a crocodile's, and the feet were large and could be used in swimming. Almost 200 long, needle-shaped teeth trapped small fish and crustaceans as *Mesosaurus* swept its long, narrow snout rapidly sideways through the water to capture prey.

crocodilelike tail

STREAMLINED BODY
With its long, streamlined body and powerful tail, *Mesosaurus* would have been a strong swimmer, yet its relatively small teeth meant that it probably survived on equally small prey.

long, narrow snout

BIOGRAPHY
ALFRED WEGENER

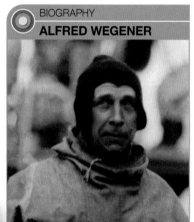

German meteorologist Alfred Wegener (1880–1930) first published the theory of continental drift in 1915. He observed that the coastline of West Africa fit neatly into South America's east coast, proposing that Earth's land masses were once joined in one vast supercontinent, Pangea, before separating about 300 million years ago. *Mesosaurus* fossils helped support his theories (see p.21).

Scutosaurus

GROUP Parareptiles
DATE Late Permian
SIZE 6¹/₂ft (2m) long
LOCATION Russia

Scutosaurus was a large herbivore belonging to the pareiasaur group. All pareiasaurs had long, thick ribs forming a barrel-shaped chest. The large torso contained an extensive digestive tract to break down tough plant material; blunt teeth were capable of cropping vegetation, not chewing it. Its skull was broad and flat, and short, down-pointing tusks hung from its lower jaw and the sides of its face.

bony scutes

PILLARLIKE LEGS
Although its name means "scute lizard" because of the small, bony plates embedded in its skin, *Scutosaurus*'s legs were not lizardlike, but nearly vertical, like pillars.

strong, pillarlike legs

Ascendonanus

GROUP Varanopids
DATE Early Permian
SIZE 15³/₄in (40cm)
LOCATION Germany

Ascendonanus was a close relative of *Varanops* (see opposite), although it was much smaller and likely fed on insects. It was found in the fossil forest at Chemnitz, Germany, where volcanic ash has preserved plants and animals in exceptional detail. Even the skin of *Ascendonanus* was preserved, with scales like a lizard. *Ascendonanus* is the earliest known arboreal (tree-climbing) vertebrate. Its long, slender, straight fingers with highly curved claws would have dug into branches—rather than clinging to them like a monkey—allowing *Ascendonanus* to climb vertically up large trees.

Varanops

GROUP Synapsids
DATE Early Permian
SIZE 4ft (1.2m) long
LOCATION USA, Russia, Southern Hemisphere

Compared to other pelycosaurs, *Varanops* and its relatives were active and agile, with long limbs and sharp, backward-curved teeth. They were voracious predators and also the most widely distributed of the primitive synapsids—varanopid remains have been found in Germany, Russia, South Africa, and North America, but *Varanops* itself is not known outside the USA. They outlasted even the primitive synapsids known as caseids, and *Elliotsmithia*—a relative of *Varanops* with a primitive, lizardlike body plan—would have been an unusual sight among the mammal-like therapsids of the Late Permian. In 2006, a skeleton of *Varanops* was reported to have bite marks, and scientists theorized that its remains had been chewed on by a scavenger before being buried and fossilized. The shape of the bite marks shows the scavenger to have been a large temnospondyl amphibian. This is the oldest evidence of scavenging among land vertebrates.

MIDSIZED HUNTER
Although not large compared to many dinosaur predators, *Varanops* was nonetheless roughly the size of some modern monitor lizards with similar predatory behavior.

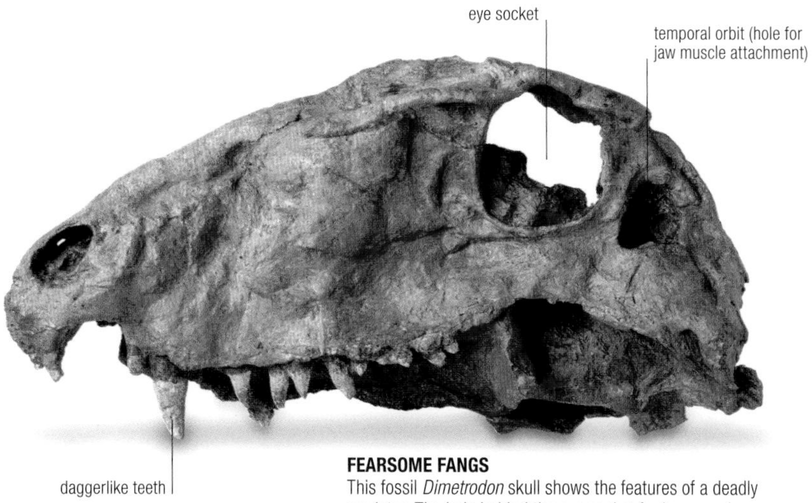

eye socket

temporal orbit (hole for
jaw muscle attachment)

back-sail

FEARSOME FANGS
This fossil *Dimetrodon* skull shows the features of a deadly
predator. The hole behind the eye socket for jaw muscle
attachment is a feature shared by mammals.

daggerlike teeth

This **successful predator**
is one of the **most
common fossils of its age**.

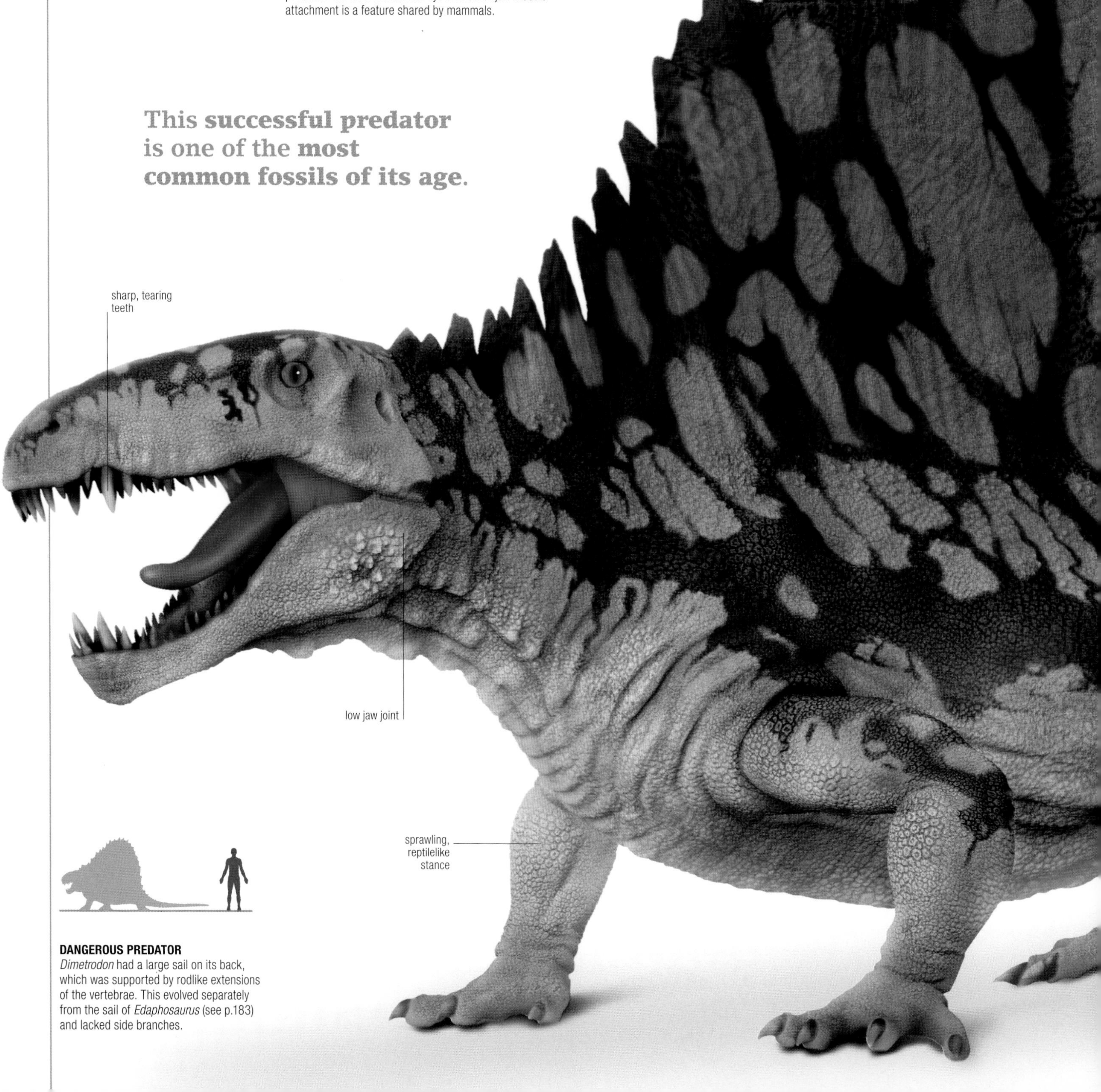

sharp, tearing
teeth

low jaw joint

sprawling,
reptilelike
stance

DANGEROUS PREDATOR
Dimetrodon had a large sail on its back,
which was supported by rodlike extensions
of the vertebrae. This evolved separately
from the sail of *Edaphosaurus* (see p.183)
and lacked side branches.

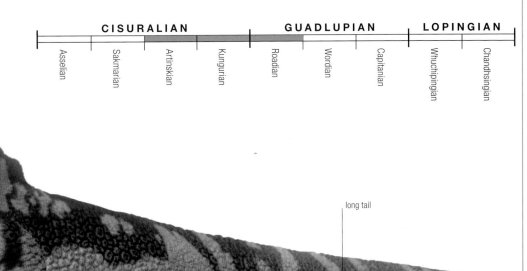

rodlike extensions of
vertebrae support sail

Dimetrodon

GROUP Synapsids
DATE Early Permian
SIZE 10½ft (3.2m) long
LOCATION Germany, USA

Dimetrodon is often included in dinosaur books, and thus commonly mistaken for a dinosaur. It is an early synapsid, and in fact more closely related to mammals, with which it shares some anatomical features, such as a single hole in the skull behind the eye. It lived in the Early Permian, at least 40 million years before the first dinosaurs, and lacks dinosaur features, such as upright legs. *Dimetrodon* was among the largest carnivorous pelycosaurs (see panel, right) and was the most fearsome predator of its time. Its high skull and relatively short snout gave it a powerful bite. *Dimetrodon* means "two-sized tooth" and refers to the two sizes of teeth in the jaw. The two caninelike fangs in the upper jaw were roughly located in the position of canine teeth of modern mammals, such as dogs and humans, and their roots extended through the bone all the way up the side of the face. Although they are not as striking as the saber-teeth of big cats, they were powerful and deadly. In all, *Dimetrodon* had 80 large, pointed teeth, which show that it was specialized for killing other large land vertebrates.

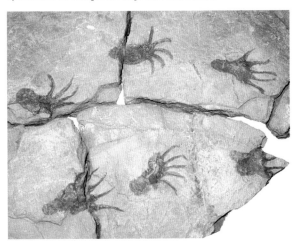

ANATOMY
SAIL-BACKED REPTILE

Many early synapsids, or pelycosaurs, are informally known as "sail reptiles," although not all pelycosaurs had sails. *Dimetrodon*'s sail helped increase its body surface area so that it could warm up more quickly in the sun. The cold-blooded *Dimetrodon* would have been sluggish in the morning before basking, and the sail may have helped it warm up before its prey did.

MAKING TRACKS
These five-toed tracks may have been made by *Dimetrodon*—which was one of the most common animals in its ecosystem. However, it is difficult to assign tracks to their makers because body fossils are almost never found with the footprints.

CISURALIAN				GUADLUPIAN			LOPINGIAN	
Asselian	Sakmarian	Artinskian	Kungurian	Roadian	Wordian	Capitanian	Whuchipingian	Chandhsingian

long tail

Moschops

GROUP Synapsids
DATE Middle Permian
SIZE 8¼ft (2.5m) long
LOCATION South Africa

Moschops was a thick-skulled herbivore with stout legs and thick ribs that formed a large, barrel-shaped chest. It had a very large gut for digesting tough plant matter. Overall, *Moschops* was a heavily built animal and full-grown adults probably had little to fear from the predators of the day. The dinocephalians, which included *Moschops*, were one of the earliest therapsid groups to evolve and were very common in the first part of the Late Permian. They diversified to exploit many of the ecological roles formerly carried out by pelycosaurs (see p.181). Carnivorous and herbivorous dinocephalians are known, but all of them have peculiar front teeth. Unlike those of other animals, the upper incisors do not overlap the lower incisors, but interlock with them, allowing effective nipping.

(see p.181)

ANATOMY
THICK SKULL

The skull of *Moschops* was astonishingly thick. In a 12in- (30cm-) long skull, the bone of the top surface is more than 4in (10cm) thick. This was an adaptation to allow ritualized head-butting, possibly in contests between males competing for mates, as in modern-day bighorn sheep. The neck joint and vertebrae were adapted to transmit the force of the blow without resulting in whiplash. One old theory was that thickening of the skull occurred due to overactive pituitary glands. This was thought to lead to blindness because the space for the eye was reduced. Today, skull thickening is recognized as an adaptation to social behavior.

EARLY HERDS
Moschops may have formed small social groups or herds. Indeed, discoveries have been made of several individuals fossilized together; the thick skull is probably an adaptation for social behavior, such as ritualized combat.

Eothyris

GROUP Synapsids
DATE Early Permian
SIZE Skull 2¼in (6cm) long
LOCATION USA

The mysterious *Eothyris* is only known from a single broad, flat skull described in 1937. It is distinctive due to two large, fanglike

PROMINENT FANGS
Eothyris was a small, lizardlike, early synapsid with peculiar double fangs in the upper jaw.

teeth on each side of the upper jaw. The reason for this peculiar arrangement is not known. The remaining teeth were also sharply pointed, so *Eothyris* was a carnivore, although the prey must have been small because *Eothyris* was shorter than 12in (30cm). It is possible that it hunted insects and other vertebrates smaller than itself. Surprisingly, *Eothyris* is related to the giant herbivore *Cotylorhynchus* (see below).

The short, broad **skull** of *Eothyris* gave it a **rapid, snapping bite**.

hypothetical reptilelike body

short legs

double fangs

Cotylorhynchus

GROUP Synapsids
DATE Early Permian
SIZE 13ft (4m) long
LOCATION USA

Cotylorhynchus was the largest of the caseids, a specialized group of herbivorous pelycosaurs. It had the smallest skull relative to its body size of any synapsid. *Cotylorhynchus* had leaf-shaped teeth with serrated tips designed to crop and tear tough plant leaves, but most of the digestive work was done by the enormous gut, held within a barrel-shaped ribcage. Caseids were highly successful and survived longer than most other pelycosaurs. They were still living in the Late Permian when therapsids had become dominant.

Edaphosaurus

GROUP Synapsids
DATE Late Carboniferous to Early Permian
SIZE 10¾ft (3.3m) long
LOCATION Czech and Slovak Republics, Germany, USA

Edaphosaurus was a herbivorous pelycosaur with numerous blunt teeth that formed crushing plates. The teeth were not just around the jaws but also on the roof of the mouth and sides of the lower jawbone. Although it was herbivorous, *Edaphosaurus* was not related to *Cotylorhynchus* (see left). One of the most distinctive features of *Edaphosaurus* is the "sail" crest on its back. Unlike the sail of *Dimetrodon* (see p.181), small prongs branched off from the main rods that supported the sail, giving them the appearance of twigs. Despite this

difference, the sails of *Dimetrodon* and *Edaphosaurus* probably evolved for the same purpose— trapping heat while basking. This meant that *Edaphosaurus* could warm its body quickly in the morning sun and become active before many predators.

BACK SAIL
The neural spines of *Edaphosaurus*'s backbone extended upward. The spines were covered in skin, producing the characteristic "sail" on its back. The spines also had smaller prongs branching off.

canine tusk

Robertia

GROUP Synapsids
DATE Middle Permian
SIZE 16½in (42cm) long
LOCATION South Africa

The dicynodonts were an extraordinarily diverse group of specialized herbivores that survived for longer than all other therapsid groups except the cynodonts. Dicynodonts continued into the Late Triassic, while cynodonts evolved into today's mammals. The earliest dicynodont known from good

NEARLY TOOTHLESS
Apart from the canines, *Robertia* lacked teeth. Instead, it had a horny beak, like that of a turtle. The skeleton was adapted to dig for food.

fossils is *Robertia*. It was relatively small, about the size of a house cat. Despite its early appearance, it shows recognizable dicynodont features. For example, it has two large canine tusks, although the other teeth are absent. It is likely that the canine tusks were used for digging.

Procynosuchus

GROUP Synapsids
DATE Late Permian
SIZE 19½in (50cm) long
LOCATION South Africa, Zambia

Procynosuchus is one of the earliest cynodonts, a group of therapsids that gave rise to mammals. It shows the first signs of mammal-like cheek teeth, differentiated into premolars and molars. Other therapsids, and most reptiles, have only one kind of tooth in the side of the jaw. Despite its advanced teeth, the body was still primitive.

semisprawling legs

PRIMITIVE BODY
Procynosuchus had a long tail and a more reptilian, semisprawling stance.

Pelanomodon

GROUP Synapsids
DATE Late Permian
SIZE 3ft 3in (1m) long
LOCATION South Africa

Dicynodonts such as *Pelanomodon* were stocky, piglike animals with shorter tails than most other early synapsids. Their jaw mechanism was unusual, with a sliding jaw joint and jaw-closing muscles that were located behind the jaw and attached around the rim of the skull opening behind the eye (temporal fenestra). In addition to chewing food by moving the jaw up and down, like humans and many other animals, dicynodonts could also move the jaw backward with each chewing stroke. This produced a powerful slicing motion that pulverized the tough plant material on which they fed. This adaptation must have been effective because dicynodonts were one of the most successful groups of herbivores in Earth's history. At about 15½in (40cm) long, *Diictodon* (see far right) was a small dicynodont. It made corkscrew-shaped burrows over 19½in (50cm) deep that have also been found fossilized. These become wider as they coil downward and end in an expanded chamber. They are remarkably similar to those made 220 million years later by the primitive beaver *Palaeocastor* (see p.384). *Pelanomodon* was a more typical, nonburrowing dicynodont, but may have dug in the ground for nutritious roots and tubers. The remains of some dicynodonts are so common that they are used by geologists to indicate the age of the rocks around them. Dating rocks based on the fossils they contain is known as biostratigraphy (see pp.39 and 44). Both *Pelanomodon* and *Diictodon* occur in the *Cistecephalus* Assemblage Zone, a layer of rock named after a particularly abundant dicynodont. Because *Cistecephalus*, *Diictodon*, and *Pelanomodon* were around for only a few million years, the rock strata in which they are discovered is easily assigned. Slightly older rocks belong in the *Tropidostoma* Assemblage Zone and younger rocks are in the *Dicynodon* Assemblage Zone.

Dicynodont fossils are so abundant that they are used to date the rock in which they are found.

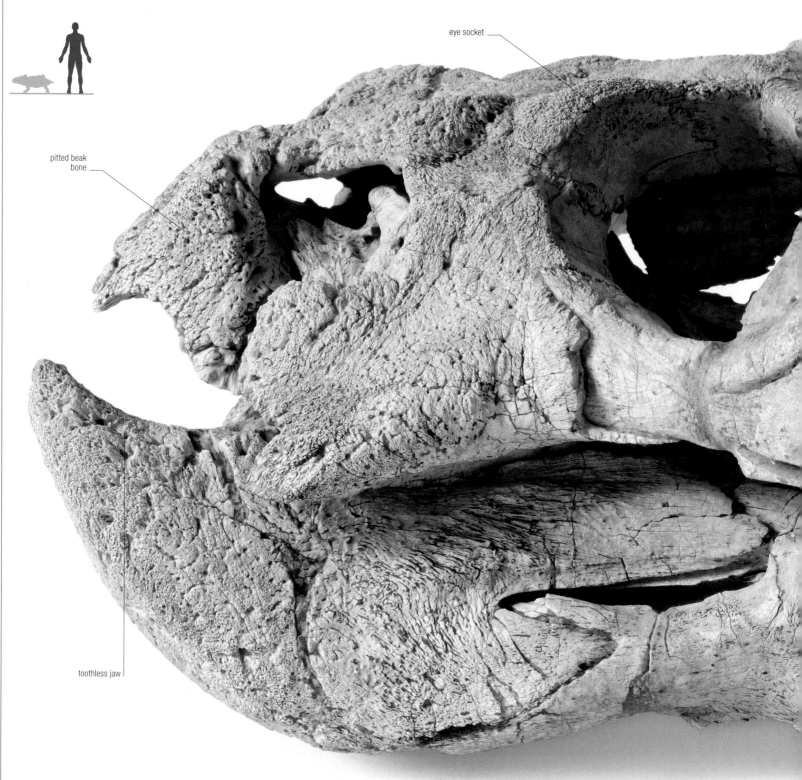

eye socket

pitted beak bone

toothless jaw

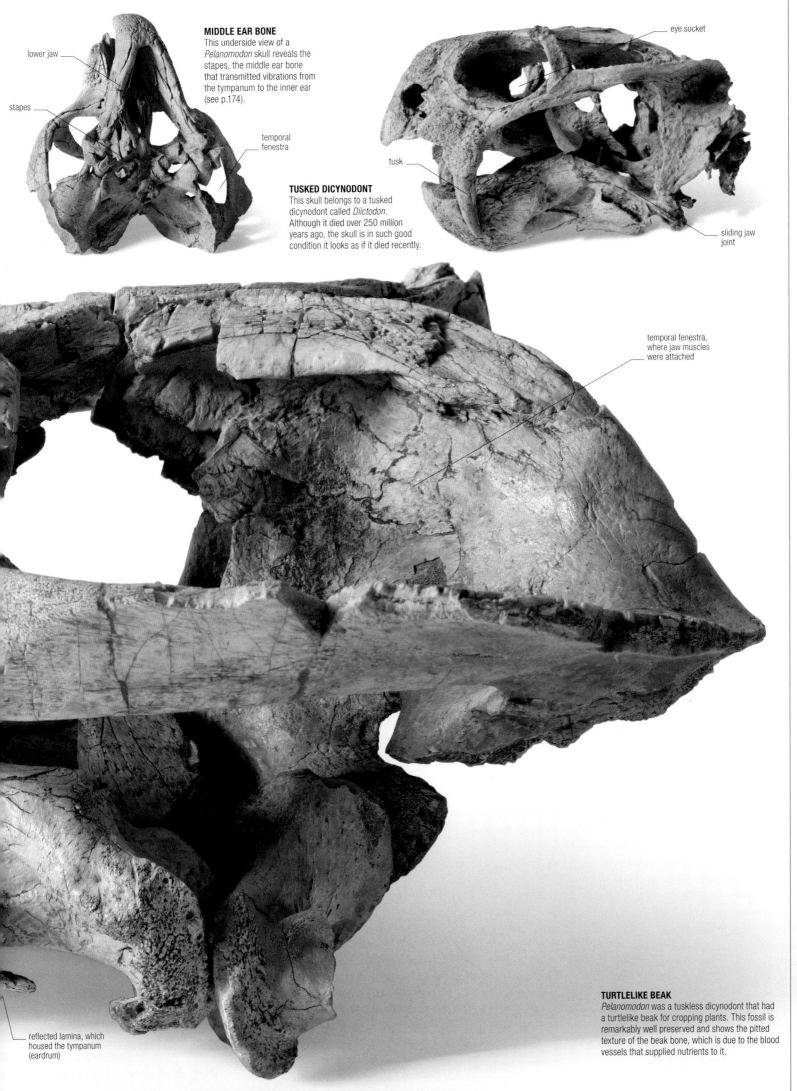

MIDDLE EAR BONE
This underside view of a *Pelanomodon* skull reveals the stapes, the middle ear bone that transmitted vibrations from the tympanum to the inner ear (see p.174).

lower jaw

stapes

temporal fenestra

eye socket

TUSKED DICYNODONT
This skull belongs to a tusked dicynodont called *Diictodon*. Although it died over 250 million years ago, the skull is in such good condition it looks as if it died recently.

tusk

sliding jaw joint

temporal fenestra, where jaw muscles were attached

reflected lamina, which housed the tympanum (eardrum)

TURTLELIKE BEAK
Pelanomodon was a tuskless dicynodont that had a turtlelike beak for cropping plants. This fossil is remarkably well preserved and shows the pitted texture of the beak bone, which is due to the blood vessels that supplied nutrients to it.

TRIASSIC

Triassic

The Triassic Period was the time of recovery after the devastation of the end-Permian faunal and floral crisis. This was also a period of very high global temperatures. Dinosaurs and mammals evolved into the recovering ecosystems on land, and were able to spread rapidly. In the sea, modern corals appeared, while sea urchins began to diversify after having almost become extinct at the end of the Permian.

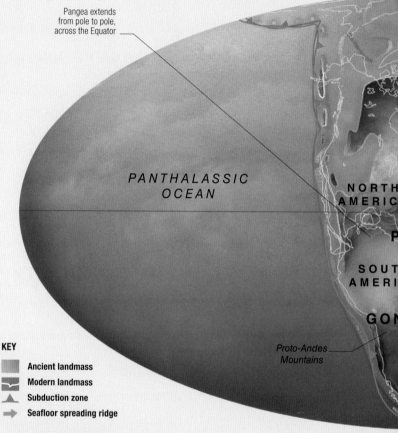

TRIASSIC SEDIMENTARY ROCKS
Newark Group sedimentary rocks form outcrops along the North American east coast stretching from South Carolina to Nova Scotia. Red beds laid down in streams alternate with thinner black deposits from lake floors.

Oceans and continents

The world's land masses were still combined in the supercontinent of Pangea, which extended from pole to pole. This reached its maximum extent around the time of the Middle to Late Triassic boundary when, due to lower sea levels, Earth's land area was greater. Throughout the Triassic, Pangea was moving steadily northward and, as a consequence, Siberia moved to the North Pole. A narrow sea, which had separated Europe and Kazakhstania during the Permian closed, causing further mountain-building in the Urals during the Early Triassic. As Pangea moved north, it also rotated counterclockwise, taking North and South China in a northerly direction. Meanwhile, the islands of Cimmeria moved north, crossing the Equator, widening the rift between it and the southeastern part of Pangea, and expanding the Tethys Ocean. The Paleo-Tethys Ocean began to contract as Cimmeria spread north at a faster pace than the eastern extension of Pangea that is now North China. Pangea, comprising all the component parts of Gondwana, still extended close to the South Pole. By the end of the Triassic, the world map was changing. Pangea began to break up, a process that was to continue through the rest of the Mesozoic into the Cenozoic. Triassic life on land and in the sea is preserved in some spectacular fossil localities. Early Triassic land vertebrates are found in the Urals in Russia and the Karoo Basin, South Africa, while the San Giorgio beds in Switzerland and Italy contain beautifully preserved aquatic reptiles and ammonites from the Middle Triassic. The agile dinosaur *Coelophysis* would have lived on river floodplains in the Late Triassic.

TRIASSIC SALT DEPOSITS
These salt deposits in Cheshire, UK, now mined for table salt, were formed by intense evaporation in shallow coastal lagoons in the hot Late Triassic climate.

KEY
- Ancient landmass
- Modern landmass
- Subduction zone
- Seafloor spreading ridge

LATE TRIASSIC WORLD MAP
Pangea reached its maximum extent around the Middle to Late Triassic boundary. By the end of the Triassic, it was beginning to break up.

Pangea extends from pole to pole, across the Equator

PANTHALASSIC OCEAN

NORTH AMERICA

P

SOUTH AMERICA

GON

Proto-Andes Mountains

Gondwana begins to fragment, separating the southern tip of South America from southern Africa and western Antarctica

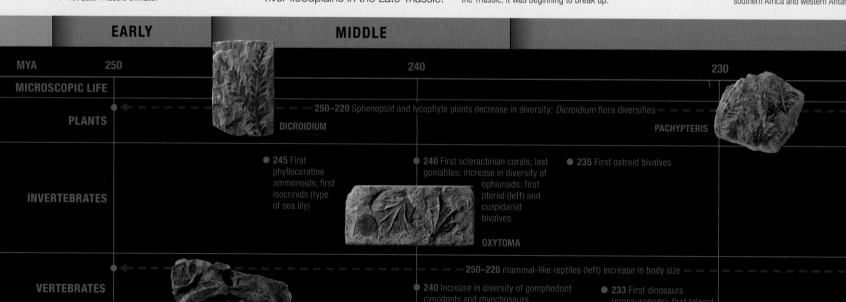

	EARLY	MIDDLE	
MYA	250	240	230
MICROSCOPIC LIFE			
PLANTS	250–220 Sphenopsid and lycophyte plants decrease in diversity; *Dicroidium* flora diversifies		
	DICROIDIUM		PACHYPTERIS
INVERTEBRATES	245 First phylloceratine ammonoids; first isocrinids (type of sea lily)	240 First scleractinian corals; last goniatites; increase in diversity of ophiuroids; first pteriid (left) and cuspidariid bivalves	235 First ostreid bivalves
		OXYTOMA	
VERTEBRATES		250–220 mammal-like reptiles (left) increase in body size	
		240 Increase in diversity of gomphodont cynodonts and rhynchosaurs	233 First dinosaurs (prosauropods); first teleost (ray-finned bony) fish
	CYNOGNATHUS		

SCALE-LEAF CONIFER
The vegetation in the Triassic was adapted to dry conditions. The flora included scale-leaved conifers, similar to this Lawson cypress, and cuticle-thickened seed ferns.

Climate

After the end of the Permian Period, rapid global warming in the Early Triassic produced one of the hottest periods in Earth's history. Some estimates put subtropical marine surface temperatures as high as 100°F (38°C). During the Triassic, the large size of the supercontinent Pangea gave rise to extreme climatic zonation, with some areas becoming very dry, whereas others were influenced by monsoonal conditions. One consequence of this zonation was the separation of plants into distinct northern and southern realms. However, this endemism (belonging to a particular place) was reduced through the Triassic. Climates were very warm, even in high latitudes, and both poles were ice free and were warm even in winter. The presence of coal deposits in high northern and high southern latitudes suggests that these regions were wetter than lower latitudes, probably being warm temperate. However, much of the continent was dry, especially in lower latitudes, and the arid-adapted nature of the vegetation reflects this. In the Early Triassic, arid areas extended to relatively high latitudes, perhaps as high as 50 degrees both north and south of the Equator. During the Triassic the equatorial tropical belt grew and extended from Cimmeria northwest across the European and western Siberian parts of Pangea. As it did so, the arid belts narrowed, the northern and southern warm temperate belts expanding into lower latitudes, especially in the Southern Hemisphere.

MONTE SAN GIORGIO FAUNA
Fossils found preserved in this pyramid-shaped mountain, on Lake Lugano in Canton Ticino, Switzerland, are the best-known record of life in a lagoon during the Middle Triassic.

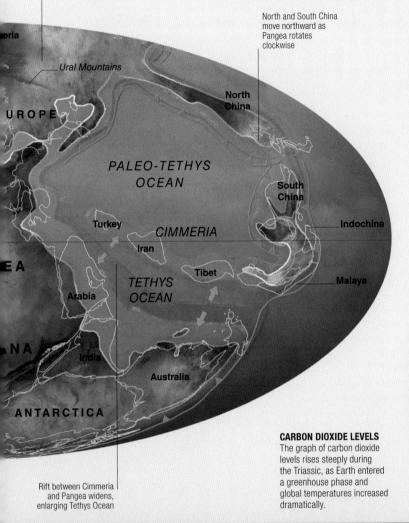

Europe and Kazakhstania join, as Siberia moves to the North Pole

Ural Mountains

North and South China move northward as Pangea rotates clockwise

EUROPE

North China

PALEO-TETHYS OCEAN

South China

Turkey

CIMMERIA

Indochina

Iran

Tibet

TETHYS OCEAN

Malaya

Arabia

India

Australia

ANTARCTICA

Rift between Cimmeria and Pangea widens, enlarging Tethys Ocean

CARBON DIOXIDE LEVELS
The graph of carbon dioxide levels rises steeply during the Triassic, as Earth entered a greenhouse phase and global temperatures increased dramatically.

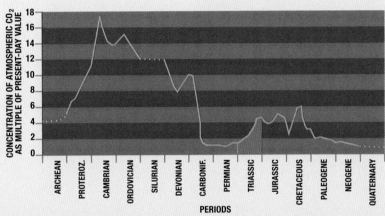

CONCENTRATION OF ATMOSPHERIC CO₂ AS MULTIPLE OF PRESENT-DAY VALUE

ARCHEAN | PROTEROZ. | CAMBRIAN | ORDOVICIAN | SILURIAN | DEVONIAN | CARBONIF. | PERMIAN | TRIASSIC | JURASSIC | CRETACEOUS | PALEOGENE | NEOGENE | QUATERNARY

PERIODS

LATE

220   210   200   MYA

225 First coccolithophores (single-celled marine planktonic algae) — MICROSCOPIC LIFE

225 Cycads, bennettitaleans, and conifers increase in diversity   ● **215** Some plant extinctions — PLANTS

225 First cardiid bivalves — CARDINIA

220–200 Increase in diversity of echinoids

200 Last ceratitid ammonoids ●

● **220** First flies   ● **215** Last trepostome bryozoans; last orthocerid nautiloids (right); first pentacrinitids   ● **210** First thraciid bivalves — INVERTEBRATES

CENOCERAS   CERATITES

220–200 Ornithischian dinosaur diversity increases; diversity of mammal-like reptiles decreases; gastric processing herbivorous reptiles diversify

● **215** First mammals, e.g. *Eozostrodon*; some vertebrate extinctions

200 Major extinction of ● terrestrial vertebrates and many large amphibians — VERTEBRATES

TRIASSIC PLANTS

The transition from the Permian to Triassic saw the demise of glossopterids, as well as other more ancient groups of plants from the Paleozoic. However, the Triassic brought further modernization of the flora, as well as the appearance and diversification of new groups of ferns and seed plants.

Relatively little is known about the plants of the Early Triassic, but rich floras of the Late Triassic—for example, from Greenland and South Africa—contain new kinds of ferns and many new groups of seed plants. Much research needs to be done to better understand these ancient seed plants, including clarifying how they relate to each other and their possible living relatives.

PETRIFIED FOREST
Massive fossil logs in the Petrified Forest, Arizona, which have been almost entirely replaced by silica, clearly show the great size attained by some of the ancient conifers of the Triassic.

New seed plants

Many new kinds of seed plants come into the fossil record for the first time during the Triassic. Studies of the Molteno flora of South Africa show a range of different reproductive structures for these plants. Many of them are difficult to understand in terms of the living plant groups to which they may be closely related. Especially prominent among the Triassic seed plants are corystosperms, which are particularly diverse in the Southern Hemisphere. Bennettitaleans are another important group of seed plants to appear for the first time in the Triassic.

Their leaves are superficially similar to those of living cycads, but the reproductive structures of the two groups are quite different. Those of cycads are simple cones, while those of bennettitaleans are often flowerlike. Some bennettitalean "flowers" produced only pollen or seeds, and a few were bisexual.

Modern seed plants and ferns

Cycads and ferns continued to diversify through the Triassic. This period also sees the first reliable evidence of the ginkgo lineage. Coniferlike leafy shoots are common in Triassic floras. Some of these plants are clearly related to living conifers, but the relationships of others are much less certain. Various groups of plants during the Triassic may have had coniferlike foliage but very different kinds of reproductive structures.

NORFOLK ISLAND PINE
The living conifer commonly known as the Norfolk Island pine (*Araucaria heterophylla*) has leafy shoots that are similar to those of early conifers.

GROUPS OVERVIEW

Triassic plants looked quite different from those of the Permian. Conifers were still very prominent during the Triassic, but they were joined by different kinds of seed plants that often had more or less pinnate leaves. During the Triassic, ferns also began to take on a more modern aspect.

FERNS
The Triassic saw a new wave of fern evolution. Earlier forms from the Paleozoic, such as the royal ferns and marattialeans, were joined for the first time by families of so-called primitive leptosporangiate ferns, such as Dipteridaceae.

CONIFERS
The more ancient conifers of the Permian were replaced by more diverse conifers during the Triassic. These voltzialean conifers show the next steps in the evolution of modern conifer cones from the more lax reproductive structures that were seen in the late Paleozoic.

CYCADS
Cycads survived the extinction that occurred at the end of the Permian, and by the Late Triassic, they were diverse and also geographically widespread. They are important in Jurassic vegetation from all over the world.

OTHER SEED PLANTS
The seed-plant group that is most characteristic of the Triassic is the corystosperms. They are especially important in the Southern Hemisphere. Other seed-plant groups that appeared for the first time during the Triassic include ginkgos, bennettitaleans, and caytonialeans.

Muscites

GROUP Bryophytes
DATE Carboniferous to Neogene
SIZE 3in (7cm) across
LOCATION Worldwide

Muscites is a name established for fossilized leafy moss shoots for which precise systematic affiliation is uncertain. *Muscites brickiae* was discovered in Middle to Late Triassic rocks of Kyrgyzstan, Central Asia, along with *Thalloid* liverworts. The moss has a helical arrangement of small, narrow leaves that taper to a point at each end. Like many modern mosses,

Muscites brickiae did not have any internal vessels for conducting water. As a result, the plant was restricted to damp habitats. Most plant fossils are formed by the sporophyte, or spore-producing, phase in the life cycle. The intervening phase, known as the gametophyte, is typically much smaller and is rarely found as a fossil. In mosses and liverworts, this situation is reversed. The fossil shown here represents *Muscites*'s dominant gametophyte form. The smaller sporophyte grew on the female gametophyte, where it released spores. The spores grew into new gametophyte individuals.

MUSCITES BRICKIAE
The entire plant was only a few cells thick and formed a flat thallus covering the ground. A thallus is a simple plant body without an obvious stem or tip.

Pleuromeia

GROUP Lycophytes
DATE Triassic
SIZE 6½ft (2m) high
LOCATION Worldwide

At the beginning of the Triassic, Earth's vegetation was emerging from the greatest of all mass extinctions. Many plants died out, but *Pleuromeia* benefited from the catastrophic change. Fossils show that it grew right across the world, appearing in a wide variety of different habitats left vacant by the disappearance of competing plants. *Pleuromeia* belonged to the lycophytes, a group that included the giant clubmosses that formed the coal swamps of the Carboniferous period (see p.138). Although *Pleuromeia* was a tree, it was built on a smaller scale. It had a single, unbranched trunk, topped by a tuft of grasslike leaves. Its root system consisted of four bulb-shaped lobes connected to rootlets that fanned out through the soil. *Pleuromeia* reproduced with spores from cones. Some species produced several cones, but many had just one on top of the stem.

Fossils show that *Pleuromeia* grew right across the world in the Triassic.

GROWTH SCARS
As it grew, *Pleuromeia*'s oldest leaves were discarded, leaving a ruff of broken bases. These bases became smooth leaf scars that covered the length of the trunk.

Thaumatopteris

GROUP Ferns
DATE Late Triassic to Early Jurassic
SIZE 3ft 3in (1m) high
LOCATION Worldwide

Also known as *Dictyophyllum*, this common fossil plant is often used as a geological marker, helping identify layers of rock that were formed at a given point in time. Some *Thaumatopteris* species straddle the Triassic–Jurassic boundary, but many fossils appear only at the point where the Jurassic period began. *Thaumatopteris* can be identified by the complex network of veins in its long fronds. The veins divide and rejoin, forming a mesh on the leaf surface.

network of veins

central rib

THAUMATOPTERIS SCHENKII
The mesh of veins seen on *Thaumatopteris*'s leaves is unusual for ferns and is more like the anatomy of flowering plants.

Dioonitocarpidium

GROUP Cycads
DATE Late Triassic
SIZE Cone scale 2in (5cm) long
LOCATION Worldwide

The earliest fossils of cycads date back to the Permian period, but they became more common at the dawning of the Triassic. Typically shaped like squat palm trees, cycads such as *Dioonitocarpidium* reproduce by growing cones. Individual plants are either male or female. The cones of male cycads release pollen grains, which fertilize the female cones. The seeds then develop inside the female cone.

DIOONITOCARPIDIUM LILIENSTERNII
Dioonitocarpidium fossils are the scales from female cones. They have finely divided tips and often show traces of seeds.

Dicroidium

GROUP Corystosperms
DATE Triassic
SIZE 13–98ft (4–30m) high
LOCATION Southern Hemisphere

Dicroidium belonged to a group of plants called the corystosperms, which were seed plants of uncertain relationship found mainly in the Southern Hemisphere. They had fernlike leaves but produced seeds instead of spores. *Dicroidium* and other corystosperms evolved in Gondwana, which was the southern landmass of the supercontinent of Pangea during the Triassic (see p.188). A fully grown *Dicroidium* would have been beyond the reach of ground-based herbivores, but all corystosperms would have been browsed by animals at some stage in their lives. These browsers included the Triassic dicynodont *Lystrosaurus* (see p.212), whose fossils have been found in the same rocks as corystosperm remains in Antarctica.

paired leaflets

network of veins

woody central stalk

DICROIDIUM SP.
Like all corystosperms, *Dicroidium*'s foliage was made up of pairs of leaflets growing on either side of a central stalk. The stalks joined at Y-shaped forks, and a network of veins ran through the leaves.

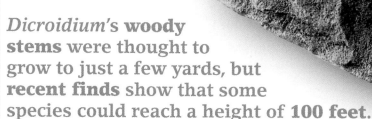

Dicroidium's **woody stems** were thought to grow to just a few yards, but **recent finds** show that some species could reach a height of **100 feet**.

Baiera

GROUP Ginkgos
DATE Triassic to Jurassic
SIZE Leaf up to 6in (15cm) long
LOCATION Worldwide

Ginkgos are not only living fossils—they form part of an ancient group of flowerless seed plants called the gymnosperms, which also include conifers and cycads. They have a fossil history stretching back more than 200 million years. During this time, evolution has created countless variations on the theme of fan-shaped leaves. With its highly divided leaves, *Baiera* looks very different from its living relative, the well-known *Ginkgo* or maidenhair tree (see p.39).

BAIERA MUNSTERIANA
In living ginkgos, the fanlike leaf is almost complete, apart from a central notch, but in *Baiera*, it was split into separate ribs.

Stachyotaxus

GROUP Conifers
DATE Late Triassic to Early Jurassic
SIZE 33ft (10m) high
LOCATION Northern Hemisphere

A shrub or a small tree, *Stachyotaxus* was one of the most abundant conifers during the Late Triassic, but died out along with its relatives in the Jurassic. It was evergreen and had narrow leaves arranged in opposing ranks. Their narrow width made the leaves resistant to drought, in much the same way as the needles of today's conifers. *Stachyotaxus* produced seeds using cones. Pollen from male cones was spread by the wind and fertilized the female seed-bearing cones. The pollen grains were spherical and lacked the sacs or winglike scales that pine pollen uses to drift in the wind.

cone scales

STACHYOTAXUS SEPTENTRIONALIS
The female cones of *Stachyotaxus* were about 4in (10cm) long. Nothing is known about the structure of the male cone.

Voltzia

GROUP Conifers
DATE Triassic
SIZE 16½ft (5m) high
LOCATION Worldwide

The Triassic was a time of great change for conifers. In this period, most of the earlier conifers of the Paleozoic were dying out, and new ones emerged and diversified into the conifers that survive until the present day. The small tree or shrub *Voltzia* belonged to one of the early, now-extinct groups, but the structure of its cones shows that it had links to today's forms. *Voltzia*'s male cone had pollen cases arranged in a spiral around the central core. The seed-producing scales of the female cones were fused together. This was a continuation of the evolutionary trend in which cones developed originally from dwarf shoots. The *Voltzia* conifer is named after the Voltzia sandstone of northeast France, where it is a common find.

scaly stem

VOLTZIA COBURGENSIS
Voltzia coburgensis had short scales that hugged the stems, although some other species also had sprays of needles a couple inches long.

TRIASSIC
INVERTEBRATES

The great Late Permian mass extinction cleared the field for many new kinds of invertebrate marine organisms—extinctions are both destructive and constructive. But the Triassic was itself terminated by yet another mass extinction, the fourth during the Phanerozoic.

Recovery from Permian extinctions

It took several million years for marine life to recover fully from the Permian catastrophe. Among the first postextinction life forms were stromatolites—it was like a return to the Precambrian. Although Paleozoic corals, trilobites, and most brachiopod groups were gone forever, some brachiopods, along with ammonoids, bivalves, gastropods, and other marine invertebrates, had survived in refuges around the world. New kinds of invertebrates would soon join them.

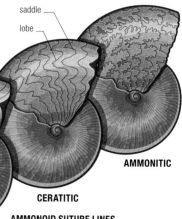

TRIASSIC OYSTER
This is a fossilized Triassic oyster cemented to a rock surface. These specialized bivalves arose during the Triassic, and similar species of living oyster remain very important today.

Inception of modern faunas

By the Early Triassic, most elements of the Cambrian fauna had vanished, and the Paleozoic fauna had been severely damaged, especially with the loss of so many groups of brachiopods, whose role as dominant seafloor filter feeders was taken over by bivalves. It was during the Triassic that the modern evolutionary fauna, which had begun in a small way during the Ordovician, became dominant. Paleozoic echinoids were replaced by new versions, and scleractinian corals, derived independently from soft-bodied sea anemones, appeared. Ceratitid ammonoids flourished, but these, like many other marine animals, died out during the important mass extinctions toward the end of the Triassic. It seems that there were several extinction events during the later Triassic, although their causes are not well understood. Possible causes are climatic change and falling sea levels, leading to habitat loss.

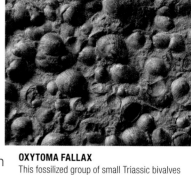

OXYTOMA FALLAX
This fossilized group of small Triassic bivalves consists of a single species. Such monospecific groups are typical of stressed environments.

saddle
lobe

AMMONITIC
CERATITIC
GONIATITIC
NAUTILOID

AMMONOID SUTURE LINES
Sutures mark where the partitions of the internal chambers (septa) met the inside of the shell. The suture lines became more complex as ammonoids evolved, from nautiloids through to ammonites.

GROUPS OVERVIEW

Groups of invertebrates that persisted through the Permian mass extinction into the Triassic continued to do well, with some groups of brachiopods and gastropods surviving to the present day. Bivalves flourished in the turbulent Triassic environments because they were able to adapt relatively quickly to the difficult, fluctuating conditions.

BRACHIOPODS
Following the Permian catastrophe, only a few brachiopod groups, chiefly terebratulides and rhynchonellides, survived to the present day. Strophomenides died out at the end of the Triassic.

GASTROPODS
The earliest gastropods are Cambrian in age, and they could be considered part of the Cambrian evolutionary fauna. They did well after the Permian mass extinction, persisted through the Triassic, and are an important component of the marine fauna today.

CERATITIDS
These are typical Triassic ammonoids. The suture line has forward-directed semicircular "saddles," and backward-pointing frilled "lobes" (see above). Ceratitids died out at the end of the Triassic, and only a few ammonoid groups continued into the Jurassic.

BIVALVES
These became the important seafloor shelled filter-feeders after the Permian, taking over from brachiopods like ships that pass in the night. Some Triassic bivalves are tolerant of changing salinity, evidently a factor in their survival in stressed conditions.

Arcticopora

GROUP Bryozoans
DATE Triassic
SIZE Colony 1/16in (2mm) across
LOCATION Europe, Asia, North America

This tiny bryozoan colony was made up of tubular structures called autozooecia, which housed many soft-bodied individuals (autozooids). Each autozooid fed using a lophophore—a ring of tentacles around the mouth. After feeding, the lophophores were withdrawn into the autozooecia for protection. In the outer parts of the colony, the autozooecia had thickened walls and were partitioned with diaphragms. The colony walls contained calcite rods, which projected like spines above the surface of the colony.

ARCTICOPORA CHRISTIEI
This specimen is a cross-section of an upright colony, showing the tubular autozooecia, which housed the feeding zooids.

Monophyllites

GROUP Cephalopods
DATE Middle to Late Triassic
SIZE 2¾–4½in (7–11cm) across
LOCATION Worldwide

Monophyllites was one of the earliest members of the phylloceratid ammonoids. This group evolved in the Early Triassic from *Meekoceras* (see below right) and survived to the end of the Cretaceous, changing little during this time. Phylloceratids are characterized by the leaf-shaped terminations of the forward-facing folds (saddles) of the suture lines. The suture lines mark where the walls (septa) of the internal chambers join the inside of the shell. Like most phylloceratids, *Monophyllites* has a disk-shaped form, with slightly overlapping whorls. Distribution patterns show that this group preferred open seas to more shallow water.

MONOPHYLLITES SPHAEROPHYLLUS
Like most other phylloceratids, *Monophyllites* has no pattern on the outside of its shell, except for strong growth lines, which are S-shaped.

rounded whorl

leaf-shaped terminations

S-shaped growth line

Coenothyris

GROUP Brachiopods
DATE Triassic
SIZE 3/8–3/4in (1–2cm) long
LOCATION Europe, Middle East

This small brachiopod had perforations in its upper shell that in life housed small bristles. The function of the bristles may have been sensory or to aid respiration. Inside the smaller (brachial) valve, a loop of calcite would have supported the lophophore, which was the animal's feeding organ. There is a small opening in the larger (pedicle) valve, through which a muscular stalk, the pedicle, would have extended to attach the animal to the seabed. A thickened ring extends around the inside of the pedicle opening. The outer edge of the smaller valve is gently folded outward; the larger valve has a corresponding infold. The main pattern on the shell is a series of growth lines, which are crossed by fine radial ribs. Some specimens of this genus have been found with bands of color running across the valves—a feature rarely seen in fossils.

larger valve
pedicle opening
smaller valve

COENOTHYRIS VULGARIS
Both valves are convex, but the larger valve curves outward slightly more than the smaller valve.

rounded venter

strong radial ribs

Preflorianites

GROUP Cephalopods
DATE Early Triassic
SIZE 3/4–2in (2–5cm) across
LOCATION Worldwide

Preflorianites was a ceratitid ammonoid from the Early Triassic. It had a simple ceratitic suture, meaning that the folded joints between the internal walls (septa) and the shell have a serrated pattern where they face away from the shell's living chamber or aperture. The whorls (coils) that make up the spiral shell only overlap each other slightly. Shells like this, where most of each whorl is visible, are described as evolute. The whorls are rounded in cross-section and converge into a rounded venter, which is the outer curve of the last whorl.

PREFLORIANITES TOULAI
Strong radial ribs cross all of the shell's whorls, but are most visible on the inner whorls. The ribs do not reach all the way around to the venter.

Meekoceras

GROUP Cephalopods
DATE Early Triassic
SIZE 1¼–4in (3–10cm) across
LOCATION Worldwide

Meekoceras was a smooth ceratitid ammonite. The genus was discovered in the western USA in 1879 by C.A. White. It is usually fairly tightly coiled and has a small umbilicus—the navel-like structure at the center of the shell. Such coiling is said to be involute. As the animal matures, the coiling becomes less tight and the umbilicus widens. The sides of the shell are slightly convex, and the ventral (outer) side of the shell is usually flat.

flat ventral side

MEEKOCERAS GRACILITATUS
The shell is ornamented by very fine growth lines; as is usual in mollusks, each one of these lines probably represents a day's growth.

Ceratites

GROUP Cephalopods
DATE Middle Triassic
SIZE 2¾–6in (7–15cm) across
LOCATION France, Germany, Spain, Italy, Romania

Ceratites gives its name to the Ceratitina—a major suborder of ammonoids that originated during the Middle Permian. The great majority of Triassic ammonoids belonged to this suborder, and they became a very successful group that flourished until their extinction at the end of the Triassic. The suture lines, which were strongly folded on *Ceratites,* marked the point where the internal walls (septa) dividing the shell met the inside of the shell. The folds that faced toward the shell opening (saddles) were simple, smooth curves, while those facing backward (lobes) were finely serrated. A suture line with this combination is known as "ceratitic." The last two suture lines on the specimen shown below are closer together than the others, which shows that this ammonoid was fully grown. This specimen's large size suggests that it was probably female.

CERATITES NODOSUS
This fossil is an internal mold. It formed from sediment filling the shell and shows the detailed features of the interior wall.

strong ribbing

smooth saddle

suture line

serrated lobes

ANATOMY
SUTURE LINES

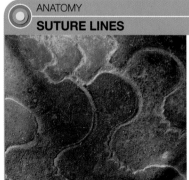

As a shelled cephalopod grows, it produces a new septum behind its body that is cemented to the inside of the shell. The shell is strengthened by the folding of the junction of each septum with the shell, so that the length of the cementation (the suture line) increases. Suture lines (which are only visible on internal molds) become more elaborate from the Late Paleozoic goniatites, through the Permian and Triassic ceratites, to the Jurassic and Cretaceous ammonites.

Arcomya

GROUP Bivalves
DATE Middle Triassic to Late Cretaceous
SIZE ⅜–1¼in (1–3cm) across
LOCATION Worldwide

The two valves, or shells, of *Arcomya* were elongated and the two beaks lay very close together. The hinge lacked interlocking teeth, and there would have been a strong external ligament to open and close the valves. *Arcomya* lived buried deeply in the seabed, with two siphons that extended up to the sea floor emerging from its permanently gaping posterior end. One siphon sucked in oxygenated water that also carried food particles, while the other expelled deoxygenated water and waste.

ARCOMYA SP.
Both valves show a series of fine growth lines. The hinge, which enabled the two valves to open and close, is also visible.

hinge

fine growth lines

Pustulifer

GROUP Gastropods
DATE Middle to Late Triassic
SIZE 1½–9½in (3.5–24cm) long
LOCATION Peru

This gastropod had a very tall and narrow shell that is referred to as "turreted." The patterns on the outside of the shell consisted of two rows of bumps, or tubercles, that were arranged in two spirals. One row ran around the upper margin of each whorl and the other around the lower margin. Irregular growth lines ran across the whorls, in some cases connecting the tubercles on either side of the whorl margins to the whorl sutures, the line where the whorls meet. This genus is coiled dextrally, meaning that the shell opening is on the right when the point (spire) of the shell is facing upward. However, the side of the shell with the aperture is damaged in this specimen, so the opposite side is shown. There would have also been a short inhalant canal, for water intake, at the mouth of the shell.

> This gastropod has a **very tall and narrow shell that is referred to as "turreted."**

PUSTULIFER SP.
This narrow and acutely angled specimen shows very little overlap of one whorl by the succeeding one.

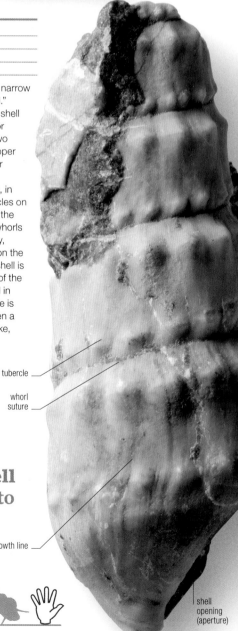

tubercle

whorl suture

growth line

shell opening (aperture)

FEEDING FANS
Encrinus extended each of its 10 arms to form a feeding fan to trap plankton that drifted close enough. If threatened by a predator, it could close its arms tightly together.

Encrinus

GROUP	Echinoderms
DATE	Middle Triassic
SIZE	Cup 1 1/2–2 1/4 in (4–6cm) long
LOCATION	Europe

Encrinus was a crinoid with a large cup consisting of three circlets of plates. The cup displays five-fold symmetry and the base is slightly concave. It was topped by 10 arms, which would have borne many smaller branches called pinnules. When the arms and pinnules were splayed, they formed a feeding fan. In life, the mouth was hidden beneath a leathery dome, called the tegmen, that formed the roof of the cup. Particles of food were passed along the arms via food grooves that led to further food grooves below the tegmen; these were lined with tiny hairlike structures called cilia that led to the mouth. The stem is circular in cross-section and features larger, regularly spaced bonelike structures (ossicles) between more frequent, smaller ones. *Encrinus* is the last survivor of a very large group of crinoids that can be traced back to the Early Ordovician. Most of the members of this group died out in the end-Permian extinction.

cup, with 10 closed arms

small ossicle

large ossicle

ENCRINUS LILIIFORMIS
The long, cylindrical stem was attached to the seabed. The evenly spaced enlarged ossicles would have provided structural support.

TRIASSIC
VERTEBRATES

The Triassic was a critical time in vertebrate evolution. It is regarded as the birth of modern ecosystems, as successful living groups such as mammals, turtles, and archosaurs arose in the aftermath of the devastating Permo–Triassic extinction, which killed off many primitive reptile and amphibian groups.

The list of vertebrates that arose during the Triassic is staggering: dinosaurs, crocodylomorphs, pterosaurs, mammals, turtles, ichthyosaurs, and frogs. The Triassic, which stretched from 250 to 200 million years ago, was a time of transition. Archaic groups that dominated Permian ecosystems had faded into extinction, and the arid landscape of Triassic Pangea was the perfect breeding ground for entirely new groups of vertebrates, many of which remain successful today.

THE GREAT SURVIVOR
The stocky, piglike *Lystrosaurus* looks fairly ordinary, but this early mammal cousin thrived across the globe in the Early Triassic. It is a classic example of a "disaster taxon."

Extinction recovery

The Permo–Triassic extinction was the greatest catastrophe in Earth's history—up to 95 percent of all species became extinct after volcanic eruptions poisoned the atmosphere. Among the departed were several groups of terrestrial (land-living) vertebrates, most notably many synapsids (see p.155), insect-eating reptiles, and fish-eating amphibians. In the aftermath, many of the world's ecosystems were barren. The evolutionary clock had been reset and new groups had the opportunity to evolve, expand, and dominate.

Within the first few million years of the extinction, ecosystems remained in disarray and a few vertebrates spread across the globe. These so-called "disaster taxa," such as *Lystrosaurus* (see left), were well equipped to withstand a toxic world because they were ecological generalists that could withstand a variety of climates. However, as the atmosphere returned to normal, ecosystems stabilized and major groups like the dinosaurs appeared.

The ruling reptiles

The archosaurs, or "ruling reptiles," include living birds and crocodilians, along with a range of extinct groups restricted to the Mesozoic, such as the dinosaurs. Archosaurs originated approximately 245 million years ago and quickly spread across

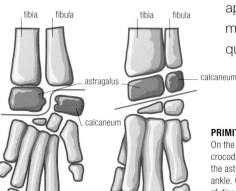

PRIMITIVE ANKLES
On the left is the crurotarsal ankle of crocodiles, in which there is rotation between the astragalus and calcaneum bones of the ankle. On the right is the mesotarsal ankle of dinosaurs, pterosaurs, and birds, in which the ankle bones together form a hinge that rotates against the foot.

GROUPS OVERVIEW

Many important groups of vertebrates originated in the Triassic, such as dinosaurs, crocodylomorphs, and other reptile groups. Other groups, such as the cynodonts and temnospondyl amphibians, became amazingly diverse after holding on and enduring the Permo–Triassic extinction.

TEMNOSPONDYLS
One of the largest and most important groups of early amphibians, temnospondyls originated during the Carboniferous but flourished during the Triassic. They lived across the globe, ranged in size from tiny to enormous, and prospered both on land and in the water. The largest temnospondyls, such as the German *Mastodonsaurus*, were up to 13ft (4m) in length.

RHYNCHOSAURS
Rhynchosaurs were some of the most unusual reptiles of the Triassic. These barrel-gutted herbivores thrived for only a short period of time, becoming extinct about 220 million years ago, but they were exceptionally abundant and were the primary large herbivores in many ecosystems. They sheared plants with their beaks and several rows of teeth on the roof of the mouth.

PHYTOSAURS
One of the most familiar of the crocodile-line archosaurs, phytosaurs prospered during the Late Triassic. These long-snouted and semiaquatic predators resemble living crocodiles and probably had a similar lifestyle, hunting for fish and small reptiles around the water's edge. However, this resemblance is only superficial and is a prime case of convergent evolution.

AETOSAURS
Like the phytosaurs, aetosaurs were a subgroup of crocodile-line archosaurs that were exceptionally common during the last 30 million years of the Late Triassic but died out at the Triassic–Jurassic extinction. Aetosaurs were heavily armored and resembled giant military tanks. Most aetosaurs were herbivorous and fed on low-growing plants, but some may have eaten meat.

Birds actually evolved from the "lizard-hipped" theropods rather than the "bird-hipped" dinosaurs.

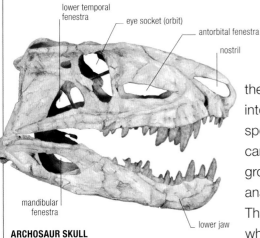

lower temporal fenestra
eye socket (orbit)
antorbital fenestra
nostril
mandibular fenestra
lower jaw

ARCHOSAUR SKULL
Postosuchus was a large predator. Its skull has a number of archosaur features, such as the antorbital fenestra and mandibular fenestra (skull openings).

the globe, diversifying into an array of different species, which generally can be divided into two groups based on the anatomy of the ankle. The birdlike archosaurs, which include dinosaurs, have a straight hinge joint between the ankle bones and the foot (a mesotarsal ankle), which allows for rapid locomotion (see left). However, the crocodile-line archosaurs have a ball-and-socket joint between the astragalus and calcaneum bones of the ankle (a crurotarsal ankle). This joint allows the two bones to rotate against each other, which offered a degree of maneuverability but prevented most crocodile-line archosaurs from attaining high speed.

Dinosaurs

Dinosaurs are the most familiar of the extinct archosaur groups and are closely related to the flying pterosaurs. The earliest-known dinosaurs appeared at the beginning of the Late Triassic, approximately 230 million years ago. Dinosaurs quickly

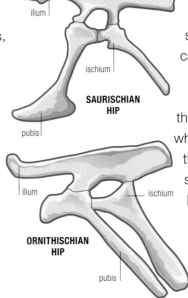

ilium

ischium

SAURISCHIAN HIP

pubis

ilium

ischium

ORNITHISCHIAN HIP

pubis

TWO TYPES OF HIP
In the saurischian hip of theropods and sauropodomorphs (top), the pubis points forward. In the ornithischian hip, the pubis points backward and lies against the ischium. These hips define the two major groups of dinosaurs.

diversified into an array of different body forms, but it took them much longer to multiply into a large number of species and to become exceptionally abundant in individual ecosystems.

The most primitive dinosaurs, such as *Herrerasaurus* and *Eoraptor*, were sleek, bipedal predators boasting an arsenal of serrated teeth and sharp claws. From these small carnivorous ancestors, dinosaurs proceeded to divide into two major groups: the saurischians and the ornithischians. The saurischians, which include the predatory theropods and the long-necked sauropodomorphs, possess a "lizard hip," in which the pubis bone points forward. The ornithischians, or "bird-hipped" dinosaurs, have a modified pelvis in which the pubis points backward, as in living birds. However, in one of the great ironies of paleontology, birds actually evolved from the "lizard-hipped" theropods.

RAUISUCHIANS

The rauisuchians, yet another group of crocodile-line archosaurs, were the keystone predators of most Triassic terrestrial ecosystems. Over 25 species have been found and include giant quadrupedal hunters, sleek bipedal omnivores, and lumbering beasts with deep sails on their backs. They probably occupied the large-predator niche later filled by theropods.

SPHENOSUCHIANS

Although they barely look like their later relatives, the sphenosuchians were some of the earliest and most primitive members of the crocodylomorph group. Sphenosuchians, which flourished during the Late Triassic and Early Jurassic, were sleek predators that walked upright (some could walk on two legs), could run fast, and resembled greyhounds in their overall proportions.

THEROPODS

Perhaps the most recognizable group of dinosaurs, the theropods include predators such as *Tyrannosaurus*, *Allosaurus*, and *Velociraptor*. Theropods first evolved in the Triassic but did not evolve into giant sizes until later, during the Jurassic. Most Triassic theropods, such as *Coelophysis*, were only several feet long and hunted small prey in the shadow of the giant rauisuchians.

CYNODONTS

Cynodonts are a large group that includes true mammals. The first cynodonts evolved in the Permian, but many groups prospered during the Triassic. Their characteristic mammalian features included hair, a large brain, and upright posture. Many were small, but some were enormous and filled the large-herbivore niche before the evolution of sauropodomorph dinosaurs.

Saurichthys

GROUP Actinopterygians
DATE Early to Late Triassic
SIZE 3ft 3in (1m) long
LOCATION All continents except Antarctica

Saurichthys, meaning "lizard fish," was a fast, open-water hunter, similar to present-day barracuda. This primitive ray-finned fish was a top predator. Its streamlined shape, posteriorly positioned pelvic fins, and forked tail can be clearly seen in this fossil. Whether it hunted in groups, like barracuda, or laid in wait for its prey is not known. One of its closest modern relatives is the sturgeon.

LONG SNOUT
Its long snout made up over 50 percent of the total head length and its narrow jaws were lined with sharp, conical teeth.

Mastodonsaurus

GROUP Temnospondyls
DATE Middle to Late Triassic
SIZE 20ft (6m) long
LOCATION Europe, Russia

Mastodonsaurus was the first early amphibian to be recognized and described, in 1828. It is far larger than any other fossil amphibian ever found. Although the skull is crocodilelike in shape and very large, it is more like a frog's in its construction. It was assumed at first that *Mastodonsaurus* had a squat, rounded body like a frog. It was only when bones from the body were found that it became apparent that the creature resembled a thickset crocodile with a gigantic head and no neck.

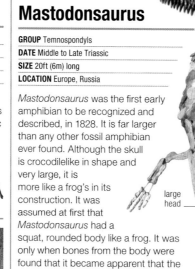

large head

LARGEST AMPHIBIAN
At 20ft (6m) in length, *Mastodonsaurus* is the largest amphibian to have ever lived. Fossil finds in coastal lagoons suggest it was saltwater tolerant.

Metoposaurus

GROUP Temnospondyls
DATE Late Triassic
SIZE 6½ft (2m) long
LOCATION Europe, India, North America

The metoposaurs were a family of amphibians that were widespread for a short time in the Late Triassic. This was when the supercontinent Pangea was just beginning to split up and was crisscrossed by rift valleys filled with lakes and rivers. *Metoposaurus* spread along these waterways, which explains why they have been found in many parts of the world. By the end of the Triassic, metoposaurs had been replaced by early crocodiles.

SKILLFUL SWIMMERS
Metoposaurus probably used its tail to propel itself through water, as well as undulating its whole body like a crocodile.

Odontochelys

GROUP Turtles
DATE Late Triassic
SIZE 15½in (40cm) long
LOCATION China

Turtles have a classic body plan that is not seen in any other group. This includes a short skull, a tiny tail, and a thick shell that surrounds the body. The mystery of how the turtle shell evolved has been clarified by *Odontochelys*, the oldest turtle yet discovered that shows the shell. It had a fully formed plastron (the lower part of the shell) but lacked a carapace (the upper part). This shows that the shell began as armor on the belly, with the carapace forming much later as the ribs became thicker and broader and fused with bony plates (osteoderms) inside the skin.

ANCIENT TURTLE
Odontochelys is the oldest-known turtle with a shell. It lived in nearshore waters, suggesting that turtles evolved in the sea.

Diphydontosaurus

GROUP Lepidosaurs
DATE Late Triassic
SIZE 4in (10cm) long
LOCATION British Isles, Italy

A nearly complete Late Triassic ecosystem has been reconstructed from the rich fossil deposits of the Bristol Channel region that lies between England and Wales. About 205 million years ago, this area was home to a labyrinth of small caves that formed after the thick limestone bedrock was eroded by acidic rainwater. The remains of a menagerie of small animals have been found in these dark caverns, having been drowned and washed inside by monsoon floods. They ranged in size from small sauropodomorph dinosaurs, which were up to 9¾ft (3m) long, to minute sphenodontian reptiles just 4in (10cm) long. The most common sphenodontian was *Diphydontosaurus*. Once a diverse group, the tuatara of New Zealand is now the only living member. *Diphydontosaurus* was far smaller than the tuatara, which may have been an adaptation for hunting down its insect prey in crevices.

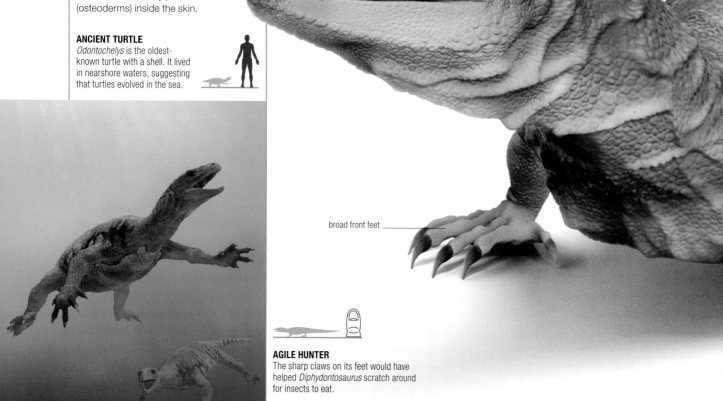

muscular neck

scaly skin

broad front feet

AGILE HUNTER
The sharp claws on its feet would have helped *Diphydontosaurus* scratch around for insects to eat.

sharp claws

Lariosaurus

GROUP Sauropterygians
DATE Middle to Late Triassic
SIZE 19½–28in (50–70cm) long
LOCATION Italy

Nothosaurs filled the same general niche that seals do today. Like seals, these reptiles primarily lived in the water, but they also came ashore to rest. *Lariosaurus* was one of the smallest known nothosaurs. Its front feet were webbed flippers and, uniquely, its hind feet retained five individual toes. This suggests that *Lariosaurus* was more active on land than other nothosaurs, and it may have hunted small shoreline animals as well as fish.

WEBBED FEET
Like all nothosaurs, *Lariosaurus* had webbed feet and an elongated neck. In the water, it used its long neck to lunge after fish, catching them with its needle-sharp teeth.

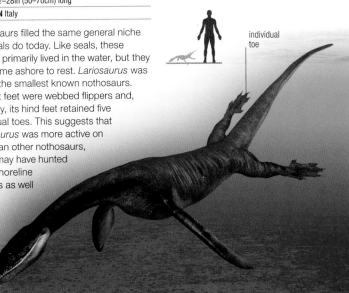

individual toe

long tail

The **only living sphenodontian** reptile is the **tuatara** from **New Zealand.**

sprawling hind limb

Placodus

GROUP Sauropterygians
DATE Middle Triassic
SIZE 6½–9¾ft (2–3m) long
LOCATION Germany

Living alongside nothosaurs in the warm, shallow seas that covered Europe in the Middle Triassic were some peculiar reptiles called placodonts. The most common and best-studied, placodont is *Placodus*. It had all the characteristic features of the placodonts: a large, barrel-chested body; webbed hands and feet that formed paddles; and a long, deep tail. The hands, feet, and tail were specially adapted for swimming and indicate that *Placodus* was probably an accomplished swimmer despite its bulky frame. It is also likely that *Placodus*, like many other marine reptiles, spent time out of the water feeding and mating, although it was probably a clumsy creature on land. Placodont skulls, including that of *Placodus*, look quite unusual. They feature distinct protruding teeth at the front of the jaw and a series of smooth, slablike teeth on the roof of the mouth. This strange collection of teeth is unlike anything found in any other reptiles. It is thought the front teeth were used for spearing fish and the ones in the roof of the mouth were for crushing the hard shells of mollusks.

PERFECT TEETH
The bizarre Triassic reptile *Placodus* foraged in the shallow waters along nearshore reefs looking for mollusks and fish to eat. Its unusual teeth were perfectly suited for dealing with its prey.

Pachypleurosaurus

GROUP Sauropterygians
DATE Middle Triassic
SIZE 12–15½in (30–40cm) long
LOCATION Italy, Switzerland

There are two subgroups of nothosaur: nothosaurids and pachypleurosaurids. The most characteristic pachypleurosaurid is *Pachypleurosaurus*. Its body was long and narrow with an elongated neck, and its limbs acted as robust paddles and for steering and stability in the water. This agile reptile swam by laterally undulating its slim yet muscular torso.

long, narrow torso

FAST MOVER
The small and gracile *Pachypleurosaurus* used its muscular body to dart quickly through the water while chasing after small prey.

Mixosaurus

GROUP Ichthyosaurs
DATE Middle Triassic
SIZE Up to 3ft 3in (1m) long
LOCATION North America, Europe, Asia

One of the oldest-known members of the ichthyosaur group, *Mixosaurus* lived right across the globe during the Middle Triassic. It had the characteristic ichthyosaur body plan (the word ichthyosaur means "fish lizard"), which resembles the dolphins of today. Most notably, the snout was long and slender, the torso was streamlined, the limbs were modified into paddles, the tail was deep, and there was a dorsal fin for stability in the water.

Mixosaurus was undoubtedly a fast swimmer and used its speed to surge through schools of fish, which made up the bulk of its diet. Compared to most other ichthyosaurs, *Mixosaurus* was small, with most individuals falling short of 3ft 3in (1m) in length; however, some later cousins reached more than four times that size. Not only was *Mixosaurus* one of the smallest ichthyosaurs, but it was also among the most primitive. In fact, *Mixosaurus* was so named because George Baur, the German paleontologist who first studied it, thought it showed a "mixture" of primitive and advanced characteristics.

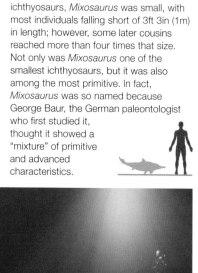

Procolophon

GROUP Primitive amniotes
DATE Early Triassic
SIZE 12–14in (30–35cm) long
LOCATION South Africa, Antarctica

The small frame of *Procolophon* and other procolophonid reptiles belies their ability to survive global catastrophes. Although unspectacular at first glance, these sluggish reptiles were one of the few groups to endure the devastating Permo–Triassic extinction. After surviving the most destructive global die-off in Earth's history, procolophonids evolved into a new array of species. One of these was *Procolophon*, the namesake of the group.

Procolophon was no larger than a small iguana. It had a wide body and small tail, and its broad, triangular skull is similar to that of many burrowing reptiles living today. Perhaps procolophonids survived the extinction by hiding away in burrows, sequestering themselves from intense bouts of heat and cold and showers of acid rain.

arm

FORELIMBS
The forelimbs of *Procolophon* were small and sleek, but they may have been strong enough to dig burrows.

broad skull

backward-facing cheek spike

large eye socket

BURROW-DWELLER
The wide, triangular skull and squat body of *Procolophon* is similar to that of many modern lizards, such as the gila monster of North America—another reptile that often seeks shelter in burrows.

Hyperodapedon

GROUP Rhynchosaurs
DATE Late Triassic
SIZE 4–5ft (1.2–1.5m) long
LOCATION Scotland, India, Brazil, Argentina

Rhynchosaurs were just one of many groups of herbivorous reptiles that lived during the Late Triassic. They were also among the main land-dwelling herbivores that existed around the globe immediately prior to the origination of dinosaurs, as well as during the first few million years of dinosaur evolution. One of the best-studied rhynchosaurs is *Hyperodapedon*, a species represented by over 35 skeletons from 230-million-year-old rocks discovered in Elgin, Scotland. Further species of *Hyperodapedon* are known from several other continents, showing it lived worldwide.

The skull of *Hyperodapedon* was broad and triangular when viewed from above. Its sharp "beak," composed of a pair of tusks, had razor-sharp edges that formed a scissorlike mechanism designed to shear through vegetation. Farther back in the jaw, rows of teeth at the sides of the snout and on the roof of the mouth ground down food before it was swallowed. The body of *Hyperodapedon* was short and stocky, and it may have used its limbs to dig small holes, which would have enabled this barrel-chested herbivore to access nutritious tubers and roots. Rhynchosaurs became extinct during the Late Triassic. It is thought that their disappearance allowed dinosaurs to replace them as the dominant land-based herbivores.

tail may have dragged on ground

Tanystropheus

GROUP Tanystrophids
DATE Middle Triassic
SIZE 18–21½ft (5.5–6.5m) long
LOCATION Europe and the Middle East

Tanystropheus must rank among the most bizarre reptiles of all time, instantly recognizable by its outsized, noodlelike neck that was longer than its body and tail combined. This extraordinary neck was more than 9¾ft (3m) long—about twice the height of an average man—yet *Tanystropheus* possessed only 10 greatly elongated cervical vertebrae to support it.

A semiaquatic animal that fed primarily on fish, *Tanystropheus* could capture fish in the murky waters of the seashore by rapidly extending its neck. This strategy allowed it to take fish while its body remained completely still, in some cases while it stood on dry land.

ANATOMY
PLANT-EATER

Rhynchosaurs were common, midsized herbivores that filled the top plant-eating niche in many ecosystems. Rhynchosaurs such as *Hyperodapedon* had an extraordinary set of anatomical adaptations that facilitated their plant-vacuuming lifestyle. The rhynchosaur skull has paired tusks (the "beak") at the front, and the roof of the mouth is covered with hundreds of teeth, which were constantly replaced throughout life in a conveyor-belt fashion. The tusks cropped and sheared plants, while the tooth fields ground and crushed the plant matter before it was swallowed.

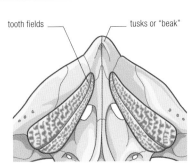

tooth fields

tusks or "beak"

BEAKED LIZARD

A large nasal cavity suggests that *Hyperodapedon* had a good sense of smell, and large eye sockets indicate good vision. This squat, rather heavy-set lizard also had claws on its back legs that were suitable for "scratch digging," possibly for food. It also had a sharp-edged "beak," which may have helped the lizard scrape up food to its powerful tongue.

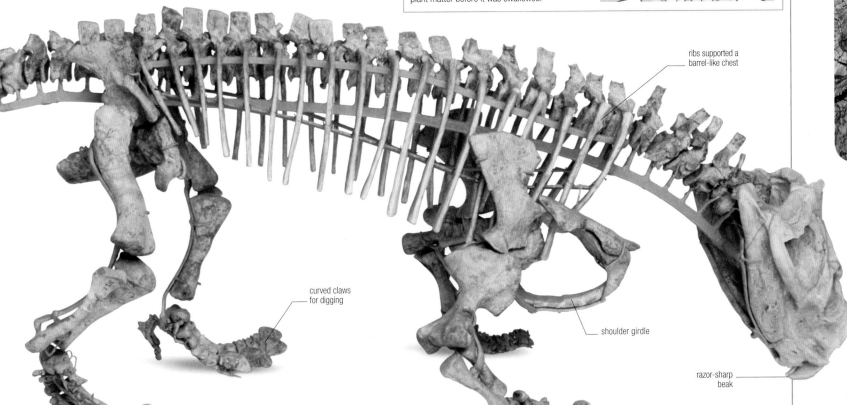

ribs supported a barrel-like chest

curved claws for digging

shoulder girdle

razor-sharp beak

Proterosuchus

GROUP Proterosuchids
DATE Early Triassic
SIZE 3¼–6½ft (1–2m) long
LOCATION South Africa

Following the Permo–Triassic mass extinction, entirely new groups of organisms emerged. One of these was *Proterosuchus*, a Komodo-dragon-sized predator that is a distant cousin of the archosaurs—the large group that contains birds, crocodiles, and dinosaurs.

The skull of *Proterosuchus* was long and narrow, with enlarged eye sockets and numerous small, curved teeth. The body was also long and thin, with a flexible, tapering tail. It is likely that *Proterosuchus* alternated between living on land and in the water, giving it unparalleled access to a range of prey. Although quite small even by today's standards, *Proterosuchus* was among the largest carnivores of its day.

HOOKED MOUTH

A hook-shaped mouth and curved teeth gave prey little chance of escape when *Proterosuchus* struck.

stout legs aided walking on land

Euparkeria

Group Primitive archosaurs
Date Early Triassic
Size 28in (70cm) long
Location South Africa

Euparkeria was one of many reptile species that lived during the first few million years of the Triassic in what is now South Africa. This small carnivore had a mouth full of sharp teeth and was covered in scaly skin embedded with bony plates (osteoderms) along its back and tail. A close cousin of the archosaurs, it showed many characteristics that developed more fully in dinosaurs and other archosaurs. Longer back legs allowed it to run or stand upright at times, which gave it the life-saving advantage of speed and agility, and its skull had a large opening in front of the eye that housed an enlarged space or sinus, which points to a good sense of smell.

osteoderms along tail

MEAGER DEFENSES
Despite its sharp teeth and fearsome appearance, *Euparkeria* lived in a world of many predators. Aside from speed, sharp claws on its thumbs were its only defense.

needle-sharp teeth

shorter front legs

powerful hind legs

KEY SITE

KAROO SITE

Many spectacular fossils have been found in the Karoo Desert of South Africa. The footprints being cleaned below were made by a synapsid, a mammal-like reptile that preyed on *Euparkeria*. Rocks deposited during the Permian, Triassic, and Jurassic in this large inland basin, filled with lakes and rivers, have been explored for centuries.

Gualosuchus

GROUP Proterochampsids
DATE Middle Triassic
SIZE 3¼–6½ft (1–2m) long
LOCATION Argentina

Gualosuchus lived around 235 million years ago alongside an array of dicynodonts, cynodonts, and close dinosaur cousins such as *Marasuchus* and *Lagerpeton*. The skull of *Gualosuchus* was long and very broad when viewed from the top, with large eye sockets and a long snout filled with sharp, curved teeth. *Gualosuchus* lived in and around the margins of the seas and rivers and probably hunted fish, along with any small vertebrates that ventured too close to the shore.

Parasuchus

GROUP Phytosaurs
DATE Late Triassic
SIZE 6½ft (2m) long
LOCATION Worldwide

A crocodilelike body characterizes a group of fossil vertebrates known as phytosaurs. These reptiles, some of which grew to sizes close to that of *Tyrannosaurus rex* (see pp.302–303), lived for about 25 million years during the Late Triassic. One of the most studied is *Parasuchus*; complete skeletons have been found in India, as well as fragmentary fossils all across the world. Its name means "similar to crocodiles," which fits this sprawling, fish-eating reptile.

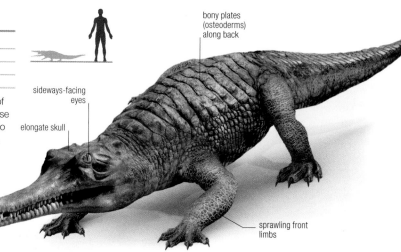

bony plates (osteoderms) along back

sideways-facing eyes

elongate skull

sprawling front limbs

Stagonolepis

GROUP Aetosaurs
DATE Late Triassic
SIZE 9¾ft (3m) long
LOCATION Scotland, Poland

Aetosaurs were exceptionally abundant from 225 to 200 million years ago. One of the most recognizable is *Stagonolepis*. Like all aetosaurs, this animal shielded its body with bony armor plates (osteoderms). The skull of *Stagonolepis* was short and deep and ended in a shovel-like snout that was useful for uprooting tubers or slurping bugs from the mud. *Stagonolepis* is a common fossil find in the Late Triassic rocks of Elgin, Scotland.

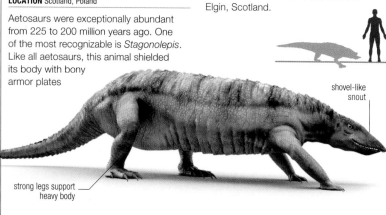

shovel-like snout

strong legs support heavy body

Ornithosuchus

GROUP Primitive archosaurs
DATE Late Triassic
SIZE 3¼–6½ft (1–2m) long
LOCATION Scotland

Ornithosuchus may look like a dinosaur, but its skeleton suggests that it is a crocodile-line archosaur related to aetosaurs. Among the largest predators in its ecological niche 230 million years ago, when the first dinosaurs were evolving, *Ornithosuchus* spent most of its time on four legs, but had the ability to walk or run on two. Its sharp teeth indicate that it was a carnivore.

dinosaurlike head

bony plates on back

forelegs smaller than hind legs

Postosuchus

GROUP Rauisuchians
DATE Late Triassic
SIZE 9¾–15ft (3–4.6m) long
LOCATION USA

Postosuchus is one of the best known and most studied members of the rauisuchians, a large group of Triassic archosaurs. Like the ornithosuchians to which *Ornithosuchus* belongs (see opposite), this group is part of the same lineage as modern crocodiles and included some of the largest predators of the day. *Postosuchus* grew to about 15ft (4.6m) in length and may have weighed as much as 1,500lb (680kg). In many ways, it resembles the large theropod dinosaurs of the Jurassic and Cretaceous periods—which is why it was first described as a primitive ancestor of *Tyrannosaurus rex* (see pp.302–303). Indeed, the large skull

and banana-shaped cutting teeth do resemble those of the more well-known carnivorous dinosaurs. However, its ankle anatomy and the presence of bony plates (osteoderms) down the back strongly support placing *Postosuchus* among the crocodile line of archosaurs.

Although *Postosuchus* is not a true dinosaur, it did live alongside some of the oldest theropod dinosaurs, which it probably also ate. Perhaps unexpectedly, it was much larger than these early theropods and was clearly the main predator in North America during the Late Triassic.

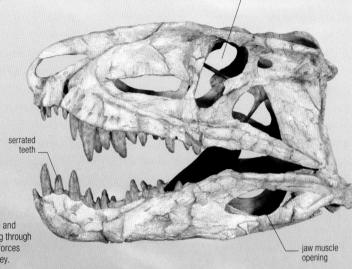

eye socket

serrated teeth

jaw muscle opening

POWERFUL JAWS
The skull of *Postosuchus* is large and strong, a perfect design for tearing through flesh and withstanding the high forces associated with tackling large prey.

FORMIDABLE PREDATOR
With its sharp, serrated teeth; huge body; and powerful jaws, *Postosuchus* resembled a smaller *Tyrannosaurus* but was unable to move as swiftly.

Effigia

GROUP Rauisuchians
DATE Late Triassic
SIZE 6½–9¾ ft (2–3m) long
LOCATION USA

One of the weirdest-looking creatures of the Late Triassic world was *Effigia*. This archosaur was a sleek, fast-running omnivore. It lived alongside early dinosaurs such as *Coelophysis* and *Chindesaurus*, as well as large rauisuchian predators. *Effigia* greatly resembles a theropod dinosaur in its overall body plan. It was bipedal (meaning it could walk on two legs), had large attachment sites on its pelvis for strong leg muscles, possessed a long tail for balance, and had holes in its neck vertebrae for air sacs. In fact, when its close cousin *Shuvosaurus* was discovered in 1994, it was described as a Late Triassic "ostrich dinosaur," due to its birdlike, toothless skull. *Effigia* has a similar skull, capped with a beak at its front, which was perfect for cracking seeds, grinding nuts, and possibly shearing vegetation and killing small vertebrates. However, despite their resemblances to theropods, *Effigia* and *Shuvosaurus* are clearly members of the crocodile line because they have a crocodilian ankle, as well as many other crocodilian features. This is a prime example of convergence: distantly related organisms independently evolving the same body shape in response to a similar shared lifestyle.

The word *Effigia* means **"ghost,"** a reference to the **Ghost Ranch** area of New Mexico where it was **discovered.**

MISTAKEN IDENTITY
Although it lived at least 80 million years before true ostrich dinosaurs, *Effigia okeeffeae* was originally thought to belong to this group. In addition to referring to the area in which it was found, the "ghost" in its name is also a nod to the fact that it was effectively "invisible" to science for decades because the fossil was collected in the 1940s but wasn't examined until 2006. The species name honors American artist Georgia O'Keeffe, who lived close to Ghost Ranch.

SIMILAR BODY PLAN

Effigia is a member of the crocodile-line archosaur group, but greatly resembles a theropod dinosaur. In fact, scientists once thought that a close relative, *Shuvosaurus*, was a member of the ornithomimosaur ("ostrich dinosaur") group. Like these theropods, such as *Struthiomimus* (shown here), *Effigia* was a sleek and fast animal that walked upright on two legs. It had claws on its hands and a brain specialized for intelligence and keen senses. *Effigia* also resembles these dinosaurs in that it had no teeth in its jaws, which instead were covered in a keratinous beak.

toothless, beaked skull

long tail

clawed hands

long legs

Lotosaurus

GROUP Rauisuchians
DATE Middle Triassic
SIZE 5–8¼ft (1.5–2.5m) long
LOCATION China

A barrel-chested quadruped with an array of fanlike spines on its neck and back, *Lotosaurus* lived about 240 million years ago, when the world's continents were joined together. This allowed it to migrate around the globe. The *Lotosaurus* skull was lightweight, lacked teeth, and had a sharp beak. The elongated spines on its vertebrae—more than three times as long as an individual vertebra is deep—may have been used to attract mates or regulate body temperature.

fanlike spines

sharp beak

STRANGE SPINES
The long spines of the vertebrae of *Lotosaurus* are an odd feature also seen in a wide variety of unrelated reptiles.

Terrestrisuchus

GROUP Crocodylomorphs
DATE Late Triassic
SIZE 30in–3ft 3in (75cm–1m) long
LOCATION British Isles

Only an adult *Terrestrisuchus* approached 3ft 3in (1m) in length, and most weighed less than 33lb (15kg). The skull was lightweight and fragile, and many limb and hand bones were pencil-thin. However, this tiny carnivore

was one of the oldest relatives of the present-day crocodile, although it had a very different lifestyle to the modern reptile. *Terrestrisuchus* was a small, slender, and agile carnivore that walked with its body raised off the ground. It was able to sprint at high speeds on all four legs. In many ways, sphenosuchians resemble small theropod dinosaurs, but, like living crocodiles, they had bony plates (osteoderms) in their skin and elongated bones in their wrists.

tail well off the ground

osteoderms along back

elongated wrists

Herrerasaurus

GROUP Theropods
DATE Late Triassic
SIZE 9¾–20ft (3–6m) long
LOCATION Argentina

Herrerasaurus is one of the oldest, most primitive dinosaurs yet discovered. This small predator lived about 228 million years ago, and shared its ecosystem with an array

of primitive reptiles that would eventually be supplanted by the dinosaurs. Even at first glance, *Herrerasaurus* resembles a carnivorous dinosaur. It walked on two legs, boasted an array of sharp claws, and had jaws studded with numerous serrated teeth: it was clearly a brawny predator that could outrun and overpower its prey. It was named in honor of its discoverer Victorino Herrera, an Argentinian rancher.

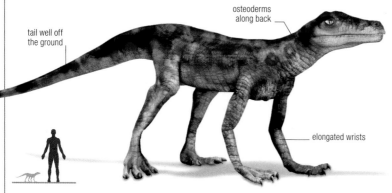

carnivorous jaws

large back legs for bipedalism

clawed toes

Eudimorphodon

GROUP Pterosaurs
DATE Late Triassic
SIZE 3ft 3in (1m) long
LOCATION Italy, Greenland

Eudimorphodon is one of the oldest and most primitive pterosaurs, and its fossilized remains are common in the folded rocks of the mountains of northern Italy. It was clearly a pterosaur, possessing all of the characteristics of this group of flying reptiles, including a greatly enlarged fourth finger that supported the wing membrane. *Eudimorphodon* was a relatively small pterosaur, weighing about 22lb (10kg). This weight pales in comparison to the monstrous proportions of its later relatives, such as *Quetzalcoatlus* (see p.293). *Eudimorphodon* is also distinguished from most other pterosaurs by its complex teeth. Whereas most pterosaurs had simple

teeth—or even lacked them altogether—*Eudimorphodon* boasted a mouthful of bizarre teeth, each with several small points called cusps. These would have been perfect for grabbing and grinding prey and may have been an adaptation for eating large fish.

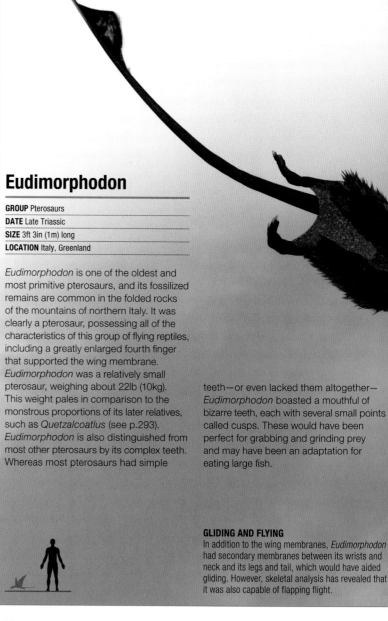

GLIDING AND FLYING
In addition to the wing membranes, *Eudimorphodon* had secondary membranes between its wrists and neck and its legs and tail, which would have aided gliding. However, skeletal analysis has revealed that it was also capable of flapping flight.

Eoraptor

GROUP Theropods
DATE Late Triassic
SIZE 3ft 3in (1m) long
LOCATION Argentina

During an expedition to Argentina in the late 1980s, a scientist named Paul Sereno and his team came across a curious fossil. No larger than a small human child, this skeleton was clearly a dinosaur, with an open hip

socket, an enlarged muscle attachment on the upper arm bone (humerus), and other characteristic features. In many ways, it resembled a carnivorous theropod because of its sharp teeth, killer claws, and the ability to walk on two hind legs (bipedal gait). Sereno and his colleagues named this dinosaur *Eoraptor*, one of the most primitive dinosaurs ever found. It lived about 228 million years ago.

sharp claws

elongated finger supported wing membrane

cusped teeth for grinding prey

Chindesaurus

GROUP Theropods

DATE Late Triassic

SIZE 6½–7½ft (2–2.3m) long

LOCATION USA

Chindesaurus is a puzzling dinosaur found in 210- to 220-million-year-old rocks in the American Southwest. It is thought to have been one of the foremost predators in western North America during the Late Triassic. Unfortunately, *Chindesaurus* is known only from incomplete skeletons, and no skull bones have been found. By contrast, the Argentinian species *Herrerasaurus* (see opposite) is the best-known dinosaur from the same period, with a completely preserved skeleton that has been studied in great detail. Similarities between the pelvic bones and hind limbs of *Chindesaurus* and *Herrerasaurus* show that they were probably close cousins. Because all the world's continents were joined together during the Late Triassic, it is likely that close relatives of *Herrerasaurus* lived across the globe.

claws on back feet

long back legs

Staurikosaurus

GROUP Theropods

DATE Late Triassic

SIZE 6½ft (2m) long

LOCATION Brazil

Barely larger than a human toddler, *Staurikosaurus* raced across the lowland plains of South America during the first few million years of the reign of dinosaurs. Very little is known about this nimble hunter, since only one fossil skeleton has been discovered. However, it is clear that *Staurikosaurus* was an agile and dangerous predator and resembled a theropod dinosaur in its general body shape. It is difficult to determine whether it was a true theropod or a more primitive dinosaur that simply looked like a theropod because of shared carnivorous lifestyles. However, it is likely that *Staurikosaurus* is a close cousin of *Herrerasaurus* (see opposite). Unlike *Herrerasaurus*, however, *Staurikosaurus* was a much smaller, slighter, and lighter animal, and as a result was probably able to run much faster than its heavier relative. When it was found and described in 1970, *Staurikosaurus* was one of only a handful of dinosaurs known from South America. In fact, its name means "Lizard of the Southern Cross"—a reference to the constellation that can only be seen in the Southern Hemisphere. Today, dozens of species of dinosaurs have been named from South America.

long tail

large hind legs for bipedalism

SMALL BUT DEADLY
Although small for a dinosaur, *Eoraptor* was large compared to many of the close cousins of dinosaurs that were alive at the time.

BIOGRAPHY
ROBERT BAKKER

Dr. Robert Bakker (b.1945) is one of the most familiar paleontologists of the past half-century. As a student in the 1970s, he was the first scientist to argue seriously that dinosaurs, such as *Eoraptor*, were smart, agile, energetic animals—not the slow, sluggish creatures often portrayed at the time.

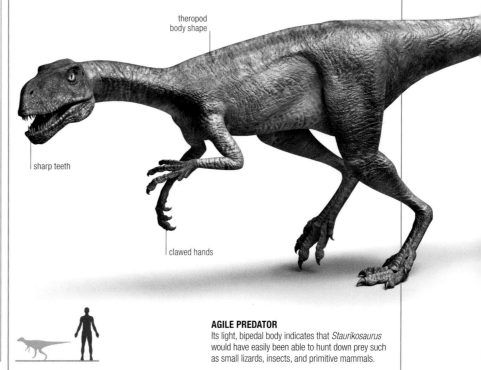

theropod body shape

sharp teeth

clawed hands

AGILE PREDATOR
Its light, bipedal body indicates that *Staurikosaurus* would have easily been able to hunt down prey such as small lizards, insects, and primitive mammals.

Liliensternus

GROUP Theropods
DATE Late Triassic
SIZE 16½–20ft (5–6m) long
LOCATION Germany

Liliensternus was a large, ferocious predator that lived in central Europe 210 million years ago; in fact, it was the largest land predator of its day. Its long hind legs allowed for great speed, its short arms ended in sharp-clawed hands, and its jaws were studded with serrated teeth. *Liliensternus* feasted on large prosauropod herbivores, including *Plateosaurus* (see opposite) and *Efraasia*, which inhabited wet, humid lowlands during the Late Triassic. Unfortunately, only two skeletons have been found, so it is difficult to know whether or not *Liliensternus* traveled in packs.

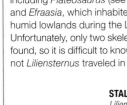

STALKING PREY
Liliensternus may have stalked its prosauropod prey, waiting for an opportunity to strike.

Gojirasaurus

GROUP Theropods
DATE Late Triassic
SIZE 16½–23ft (5–7m) long
LOCATION USA

Gojirasaurus prowled across the arid scrublands of North America 210 million years ago, and is named after the Japanese movie monster Godzilla (*Gojira* in Japanese). It seems strange that a 23ft- (7m-) long predator is named after the colossal monster of popular culture. However, during its time in the Late Triassic, most predatory dinosaurs were small; *Gojirasaurus* stood out as the largest, strongest, and fiercest hunter of the American southwest. Unfortunately, only a few bones of *Gojirasaurus* have been found, and scientists have much to learn about this carnivore.

deep tail

strong neck

muscular forelimbs

Saturnalia

GROUP Sauropodomorphs
DATE Late Triassic
SIZE Up to 6½ft (2m) long
LOCATION Brazil

This small herbivore lived 225 million years ago. Adults barely grew to 6½ft (2m) long and weighed around 26–33lb (12–15kg). A tiny, primitive cousin of later giants such as *Brachiosaurus* (see p.255), *Saturnalia* lived alongside a curious mixture of early dinosaurs and more primitive reptiles such as rhynchosaurs and dicynodonts.

SOCKET-TOOTHED LIZARD
Thecodontosaurus means "socket-toothed lizard," a name that reflects this plant-eater's blunt, leaf-shaped teeth with serrated edges.

shortish neck

Thecodontosaurus

GROUP Sauropodomorphs
DATE Late Triassic
SIZE 3¼–9¾ft (1–3m) long
LOCATION British Isles

About 205 million years ago, much of the Bristol Channel area of western England was filled with a network of caves. An entire ecosystem of small creatures inhabited these caverns, including some of the earliest mammals and many types of lizards. Dinosaurs also lived here, including the common, primitive sauropodomorph *Thecodontosaurus*. This skinny herbivore, which grew to no more than 9¾ft (3m) in length and weighed between 44–66lb (20–30kg), took refuge in the caves when threatened by predators. Usually, this was an effective strategy, but every so often a cave collapsed, entombing *Thecodontosaurus* and other animals that were inside. Today, the fossilized remains of these animals are common discoveries around Bristol and in South Wales.

five-toed feet

five-fingered hands

Plateosaurus

GROUP Sauropodomorphs
DATE Late Triassic
SIZE 20–33ft (6–10m) long
LOCATION Germany, Switzerland, Norway, Greenland

The prosauropod *Plateosaurus* is one of the best understood dinosaurs. Most dinosaurs are known from only a few scraps of bone, or at most, a single skeleton, but *Plateosaurus* is represented by more than 50 complete skeletons. Most of these have come from the Late Triassic basins of Germany, which were deposited 220 million years ago by a series of large rivers. Other specimens have turned up in the glacier-covered rocks of Greenland and even in drill cores from the bottom of the North Sea. These specimens have allowed scientists to study this dinosaur in remarkable detail. *Plateosaurus* was one of the largest prosauropods, and some individuals were 33ft (10m) long and weighed 1,540lb (700kg). Its long mouth was crammed full of leaf-shaped teeth, and these were covered in rough bumps (denticles) that were perfect for grinding vegetation. It is likely that *Plateosaurus* could walk on either two or four legs, although recent research suggests that this animal's arms and hands were not very suitable for locomotion and instead were used mainly for collecting food. Its name means "flat lizard."

> *Plateosaurus* could **increase** or **decrease** its **growth rate**, depending on the **season**.

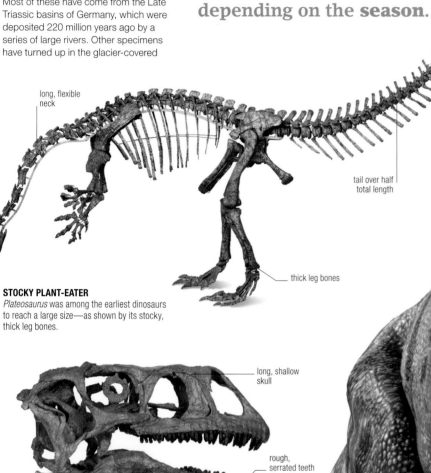

long, flexible neck

tail over half total length

thick leg bones

long neck for high browsing

STOCKY PLANT-EATER
Plateosaurus was among the earliest dinosaurs to reach a large size—as shown by its stocky, thick leg bones.

long, shallow skull

rough, serrated teeth

SMALL BUT EFFECTIVE
The skull of *Plateosaurus* was relatively small and long, yet its jaws hosted numerous rough, serrated teeth that were ideally designed for grinding and tearing the plants that made up its diet.

clawed hands

weight-bearing hind legs

HIGH FEEDER
Plateosaurus's sturdy, weight-bearing bone structure allowed it to stand upright, which in turn enabled it to feed off plants out of reach of other herbivorous dinosaurs and reptiles.

Eocursor

GROUP Ornithischians
DATE Late Triassic
SIZE 3ft 3in (1m) long
LOCATION South Africa

One of the few known Triassic ornithischians, *Eocursor* was a small and extremely fast herbivore from what is now South Africa. It had most of the major features of ornithischians, including a

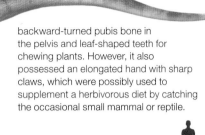

backward-turned pubis bone in the pelvis and leaf-shaped teeth for chewing plants. However, it also possessed an elongated hand with sharp claws, which were possibly used to supplement a herbivorous diet by catching the occasional small mammal or reptile.

Lystrosaurus

GROUP Synapsids
DATE Late Permian to Early Triassic
SIZE 3ft 3in (1m) long
LOCATION Africa, Asia, Antarctica

This barrel-chested, piglike herbivore lived across half the world and survived the violent Permo-Triassic mass extinction that wiped out up to 95 percent of all species. *Lystrosaurus* had a wide body; a slow, sprawling gait; and an enormous skull with a horny beak and two tusklike canine teeth. The canines were probably used for display or defense, and the beak was a perfect tool for shearing through plant stems and branches. It probably weighed less than 220lb (100kg).

horny beak

tusklike canine tooth

Placerias

GROUP Synapsids
DATE Late Triassic
SIZE 6½–11½ft (2–3.5m) long
LOCATION USA

Dicynodonts, reptilelike cousins of mammals, were some of the most successful herbivores of the Late Permian to Middle Triassic periods. One of the

last surviving members of this once-great group was *Placerias*. Resembling an ancient hippopotamus, *Placerias* reached weights of as much as 4,400lb (2,000kg). It lived approximately 220 million years ago in southwestern USA and was the largest herbivore in the region. Although it lived

alongside early carnivorous dinosaurs, *Placerias* was such a successful plant-eater that it was much more abundant than rival herbivores. Its skull was ornamented with two enormous tusks that were probably used for social display. Fossil discoveries suggest *Placerias* traveled in large herds.

short, almost stubby tail

powerful hind legs

Kannemeyeria

Group Synapsids

Date Early to Middle Triassic

Size 9¾ft (3m) long

Location South Africa, China, India, Russia

Only a few groups of therapsids survived into the Triassic Period, including the dicynodonts. The rhinoceros-sized *Kannemeyeria* was descended from one of these and had large canine tusks but otherwise lacked teeth. Instead, it possessed a beak for cropping food, and horny plates lined its mouth, just as in modern turtles. Tough plant matter was pulped by the forward and backward closing motion of the jaws. This body plan for herbivores was clearly successful, because it remained largely unchanged for the next 50 million years in successive dicynodonts after *Kannemeyeria*.

Thrinaxodon

GROUP Synapsids

DATE Early Triassic

SIZE 17½in (45cm) long

LOCATION South Africa, Antarctica

Thrinaxodon, a cat-sized predator, was the most common cynodont of the Early Triassic. Like mammals, it had a distinct waist in front of its hips. In addition, its broad ribs interlocked with each other, stiffening its body and making it difficult to breathe via rib movement alone. This suggests that *Thrinaxodon* used a diaphragm to breathe (a more efficient method) and points to an evolving mammal-like metabolism and similar activity levels. *Thrinaxodon*'s limbs were held almost under its body, as in mammals, and this further supports the suggestion that *Thrinaxodon* was an active animal capable of running.

MAMMAL-LIKE MOTION

Unlike in earlier synapsids, where the limbs were held out to the side in a lizardlike posture, *Thrinaxodon*'s legs were located almost underneath its body.

badgerlike jaw

waist area

legs held close
to body

4,400lb The maximum
weight of the hippolike
dicynodont *Placerias*.

Cynognathus

Group Synapsids

Date Early to Middle Triassic

Size 6ft (1.8m) long

Location South Africa, Antarctica, Argentina

Cynognathus means "dog-jaw," and this mammal-like reptile had a large canine tooth on either side of its upper and lower jaws. A single bone, the dentary, made up most of its lower jaw, a feature more common in modern mammals than reptiles (see p.174). *Cynognathus* was probably an extreme carnivore and ate nothing but meat. It possessed bladelike teeth behind its canine teeth for slicing flesh. Its skull also had a wide area for holding large jaw muscles, suggesting that *Cynognathus* possessed a powerful bite.

A JAW FOR HEARING

The bones that form the jaw joint in cynodonts are the same as two of the middle ear bones in mammals. The small size of these jawbones in *Cynognathus* means that they were probably already used for hearing.

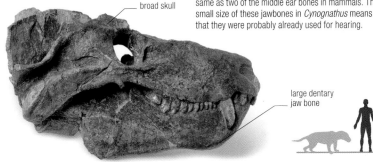

broad skull

large dentary
jaw bone

Morganucodon

GROUP Synapsids

DATE Late Triassic to Early Jurassic

SIZE 3½in (9cm) long

LOCATION Europe, China, USA

Morganucodon, one of the earliest true mammals, was described in 1949 from thousands of teeth, jaws, and fragments of bone from a South Wales quarry. A small, shrew-sized insect-eater, *Morganucodon* was probably nocturnal and had a good sense of smell. Like the most advanced cynodonts, *Morganucodon* had a double jaw joint—the new mammalian joint lying alongside the old reptilian one. The reduced size of the old jaw joint bones indicates more acute hearing in *Morganucodon*. Many early mammals retained this condition, and fossils show that the fully enclosed middle ear and single jaw joint of modern mammals evolved independently in monotremes (see p.337) on the one hand, and the marsupials and placental mammals on the other.

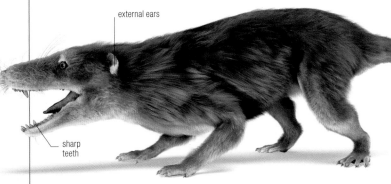

external ears

sharp
teeth

ridged area
behind eye

turtlelike beak for
tearing vegetation

ornamental
tusks

narrow lower
jaw

sturdy forelegs

"REPTILIAN HIPPO"

Placerias has often been called a "reptilian hippo" because of its similar body shape and weight. Over 40 *Placerias* skeletons have been unearthed in Petrified Forest National Park, Arizona.

JURASSIC

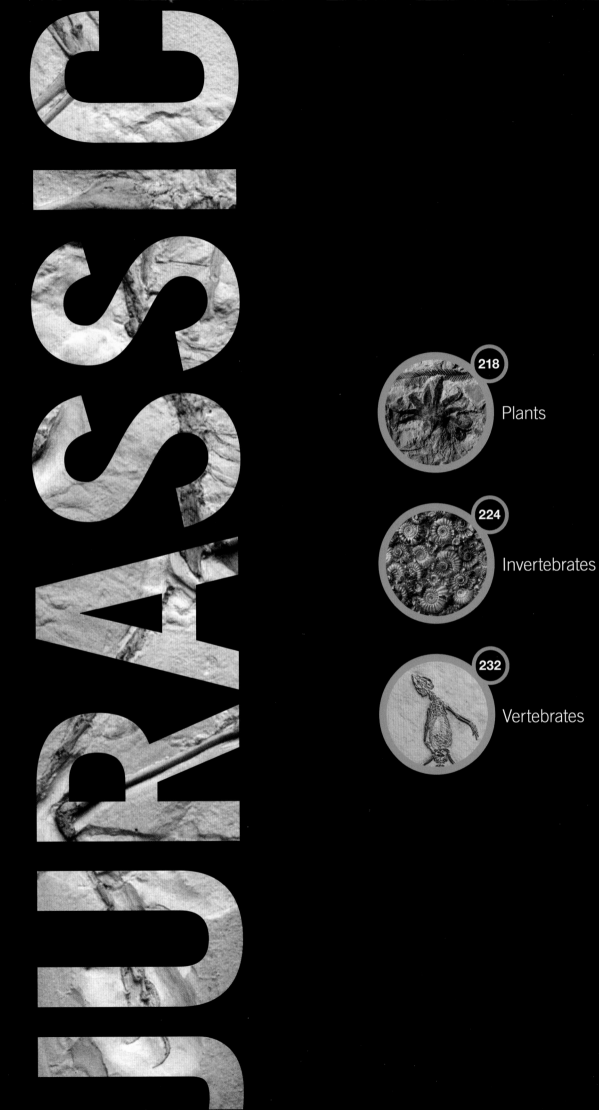

Jurassic

The word Jurassic conjures up a picture of life on a large scale, with dinosaurs roaming the land, huge reptiles surging across the sea, and flying reptiles soaring in the air. The oceans teemed with new predators, including ammonites, belemnites, and a range of shell-crushing fish. On a much smaller scale, the plankton were evolving and changing ocean chemistry.

Oceans and continents

Plate movement continued to reshape continents and widen oceans during the Jurassic period. The separation of the northern and southern parts of the supercontinent Pangea that began during the Triassic continued during the Early Jurassic, enlarging the Tethys Ocean. This ran in an east–west direction, which had a significant effect on ocean flora and fauna, and on the world's climate. Ocean currents here flowed east to west. This allowed the east–west radiation of animals and plants, with many groups spreading mainly eastward. Fossils found in limestones in Western Australia are therefore remarkably similar to those found on the southern coast of England. In the Middle Jurassic, another ocean, the Proto-Atlantic, began to open as America moved north-westward. During the later Jurassic, Laurasia was separated further from Pangea by the northward movement of

North America away from northwest Africa. This resulted in the widening of the western side of the Tethys Ocean. As Pangea continued to be torn apart by rifting and subduction (see p.20), shelf areas around the continents expanded and shallow-water environments become globally widespread. In the oceans, new types of plankton emerged, with calcareous and siliceous skeletons, which formed limestone and silica-rich sediments. Massive limestone rocks form the Jura Mountains on the border of France and Switzerland, after which the Jurassic period is named.

ZION NATIONAL PARK
Vast sand dunes accumulated in the western interior of North America throughout the Jurassic, creating the dramatic red formations exposed today in Zion National Park in Utah.

DINOSAUR TRACKWAYS
The parallel trackways of Jurassic sauropods on an ancient lakeshore are preserved in sedimentary rocks in the Morrison Formation, in southeast Colorado.

DURDLE DOOR
The spectacular arch at Durdle Door on England's coast was eroded into limestone that has been pushed up and subjected to the same tectonic events that produced the European Alps.

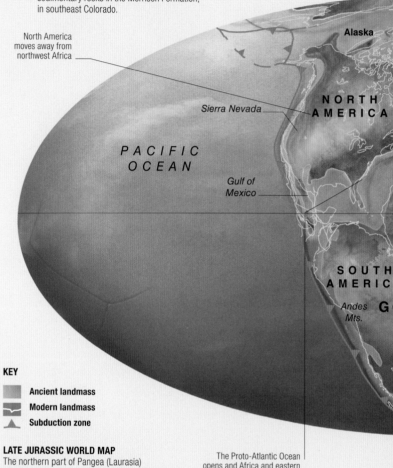

North America moves away from northwest Africa

Alaska

L

NORTH AMERICA

Sierra Nevada

PACIFIC OCEAN

Gulf of Mexico

SOUTH AMERICA

Andes Mts.

G C

KEY
Ancient landmass
Modern landmass
Subduction zone

LATE JURASSIC WORLD MAP
The northern part of Pangea (Laurasia) and the southern part (Gondwana) continued to separate, widening the Tethys Ocean.

The Proto-Atlantic Ocean opens and Africa and eastern North America rift apart, driven by seafloor spreading

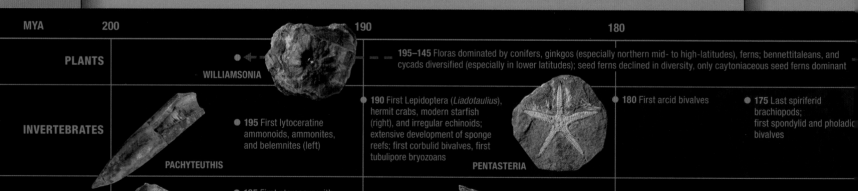

EARLY

MYA	200		190		180	
PLANTS		WILLIAMSONIA	195–145 Floras dominated by conifers, ginkgos (especially northern mid- to high-latitudes), ferns; bennettitaleans; and cycads diversified (especially in lower latitudes); seed ferns declined in diversity, only caytoniaceous seed ferns dominant			
INVERTEBRATES		195 First lytoceratine ammonoids, ammonites, and belemnites (left) PACHYTEUTHIS	190 First Lepidoptera (*Liadotaulius*), hermit crabs, modern starfish (right), and irregular echinoids; extensive development of sponge reefs; first corbulid bivalves, first tubulipore bryozoans PENTASTERIA		180 First arcid bivalves	175 Last spiriferid brachiopods; first spondylid and pholadic bivalves
VERTEBRATES		195 First pterosaur with specialized feeding (*Dorygnathus*); first sauropod dinosaurs; diversification of small ornithischian dinosaurs (heterodontosaurids (left), fabrosaurids, and scelidosaurids) HETERODONTOSAURUS	190 First pliosaurs LIOPLEURODON			

GINKGO TREE
One species of the genus *Gingko* survives today and is popularly known as the maidenhair tree. During the Jurassic, this plant diversified and spread through Laurasia.

South-central Asia assembles

Ural Mts.

Amurian Seaway

North China

South China

Southeast Asia

Turkey · Iran · Tibet

Indochina

Tethyan Trench

TETHYS OCEAN

Arabia

India

Australia

ARCTICA

Tethys Ocean separates northern continents from Gondwana

Climate

Global climates were warm and humid, with temperatures of up to 86°F (30°C) until the Middle Jurassic. Subtropical climates may have extended as far north as 60 degrees latitude. Low latitudes, including present-day southwestern North America, South Africa, and South America were more arid. Middle latitudes, including present-day Western Australia, India, and the southern tip of South America, were seasonally arid. High-latitude temperatures in Australia during the mid-Jurassic probably averaged about 64–59°F (15–18°C). Higher latitudes were wetter in both the Northern and Southern Hemispheres, with lush forests, demonstrated by the occurrence of coal deposits. In Laurasia these areas became drier during the Jurassic due to breakdown of monsoonal circulation that had been well established in the Early Jurassic. Global temperatures declined during the Late Jurassic, with relatively cool oceanic temperatures, measuring about 68°F (20°C). Pronounced seasonality, where temperatures swing from extremes of heat in summer to very cold in winter, was a feature of the Late Jurassic. At this time an arid equatorial belt spread across the Equator, up to about 45–50 degrees latitude, a narrow belt of seasonally wet climates extended to 60 degrees, and wet temperate conditions existed toward the poles. In the wetter regions, forests that consisted primarily of conifers, laid down extensive coal deposits. Mid- to high-latitude floras were dominated by *Ginkgo*-like plants and ferns, as well as conifers. Benettitaleans (cycadlike plants) and cycads were the main plants in low latitudes.

TEMPERATE RAIN FOREST
Temperate forests flourished in moist conditions in both hemispheres. Conifers predominated, but there were also ferns, like these in South Australia.

CARBON DIOXIDE LEVELS
Studies have shown that carbon dioxide levels were fairly high at the Triassic–Jurassic boundary. They dip and increase again with the warm greenhouse conditions that existed during the Jurassic.

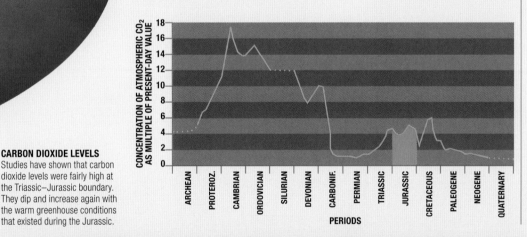

CONCENTRATION OF ATMOSPHERIC CO_2 AS MULTIPLE OF PRESENT-DAY VALUE

18 16 14 12 10 8 6 4 2 0

ARCHEAN · PROTEROZ. · CAMBRIAN · ORDOVICIAN · SILURIAN · DEVONIAN · CARBONIF. · PERMIAN · TRIASSIC · JURASSIC · CRETACEOUS · PALEOGENE · NEOGENE · QUATERNARY

PERIODS

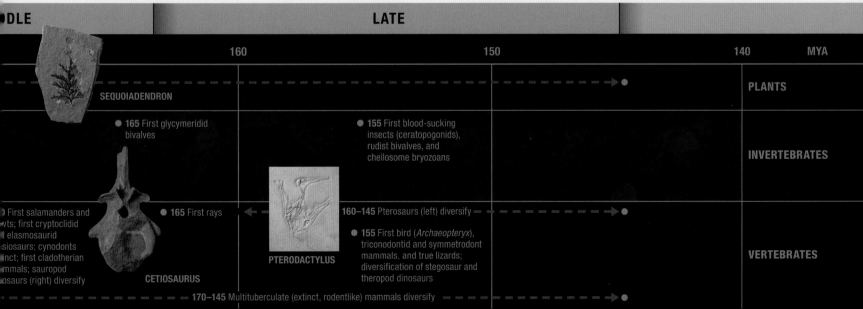

DLE

LATE

160 · 150 · 140 · MYA

PLANTS

SEQUOIADENDRON

● 165 First glycymerid bivalves

● 155 First blood-sucking insects (ceratopogonids), rudist bivalves, and cheilosome bryozoans

INVERTEBRATES

First salamanders and wts; first cryptoclidid elasmosaurid siosaurs; cynodonts nct; first cladotherian mmals; sauropod osaurs (right) diversify

● 165 First rays

160–145 Pterosaurs (left) diversify

● 155 First bird (*Archaeopteryx*), triconodontid and symmetrodont mammals, and true lizards; diversification of stegosaur and theropod dinosaurs

PTERODACTYLUS

CETIOSAURUS

VERTEBRATES

170–145 Multituberculate (extinct, rodentlike) mammals diversify

JURASSIC
PLANTS

During the Jurassic, the new kinds of plants that developed in the Triassic continued to diversify. Some of the best information on Jurassic plants comes from studies of the rich and well-preserved Middle Jurassic flora found in Yorkshire, in northern England.

Jurassic plants are known from many parts of the world, but fossils from several localities in northern England have told us much of what we know about Jurassic plants. This so-called Yorkshire Jurassic flora has been studied since the earliest days of paleontology, and attempts to reconstruct extinct plants from the different dispersed parts preserved in Yorkshire flora have been especially successful.

Before flowering plants

Ferns and conifers are common in many floras from Yorkshire, but cycadlike leaves are also prominent. Some of these leaves were certainly produced by true cycads, but some are bennettitalean leaves, or may belong to other kinds of plants that are distantly related to cycads. Bennettitaleans, for example, may be closer to living Gnetales than to other groups of living seed plants. Another interesting group of extinct Jurassic seed plants are the caytonialeans. They have leaves with four leaflets, all of which have net venation similar to that of

CLASSOPOLLIS POLLEN
The characteristic *Classopollis* pollen of the fossil conifer family Cheirolepidiaceae is one of the most common components of fossil pollen and spore floras in the later Jurassic and Early Cretaceous.

glossopterids. Caytonialeans were originally thought to be related to angiosperms, but later research showed that they lacked the key reproductive specializations characteristic of this group.

Plants and insects

The evolution of land plants and insects had been closely linked since the Devonian, but during the Mesozoic, several orders of insects still living, especially flies and beetles, diversified rapidly. Many of these insects probably fed on plants or on decaying plant matter, but some may also have been involved in pollination. By the Jurassic, it is probable that insects were already involved in the transfer of pollen from one plant to another, bringing a new dimension to plant reproduction and a new potential for specialization. Insects play an important role in the pollination of living

FOSSIL BEETLE
The diversification of beetles was well underway by the Jurassic. It is possible that some species may have contributed to the pollination of seed plants during this period.

cycads and Gnetales. There were also Jurassic bennettitaleans with both pollen-producing and ovule-producing parts. Such bisexual "flowers" were probably insect pollinated.

GROUPS OVERVIEW

The major groups of plants that flourished during the Jurassic were very similar to those of the Late Triassic. Ferns and a variety of seed plants were the dominant groups. Conifers, cycads, and bennettitaleans were the main groups of seed plants, along with a few other smaller lineages, such as caytonialeans and czekanowskialeans.

FERNS
The dominant fern families of the Triassic, Jurassic, and Early Cretaceous were lineages that diverged very early from the main line of fern evolution. They were unusual compared to most modern ferns in that they flourished in open habitats. These ancient ferns lived in similar habitats to those occupied by grasses today.

CONIFERS
Conifers are important in most Jurassic floras and in general were more similar to the living families of conifers than to the voltzialeans from the Triassic. Living groups of conifers that can be recognized in Jurassic fossils include forms related to the swamp cypress and yew families. The monkey-puzzle family is also widely represented.

CYCADS
Cycads were diverse during the Jurassic, but the best-understood fossil cycads had pollen and seed cones very similar to those of living forms. However, the fossil leaves were much smaller than those of living cycads, suggesting that the stems of these ancient forms were slender and perhaps more highly branched.

GINKGOS
Many different kinds of *Ginkgo*-like plants are known from the Jurassic, and during this interval, the group reached its maximum diversity. Especially well understood are several different *Ginkgo*-like plants from the Jurassic of China. Some of these differ very little from the single living species *Ginkgo biloba*.

Selaginella

GROUP Lycophytes
DATE Carboniferous to present
SIZE 4in (10cm) high
LOCATION Worldwide

These small plants look like mosses, but they have two types of simple leaf—large and small—that are arranged in four rows along the stem. More importantly, they produce spores from spore-producing structures (sporangia), grouped into small cones, rather than from simple spore capsules as true mosses do. The oldest-known *Selaginella* remains are over 300 million years old, making it a plant with one of the longest-known fossil records.

SELAGINELLA ZEILLERI
This fossilized leafy shoot of *Selaginella zeilleri* shows the rows of small and large leaves that were attached to the stem.

⊙ LIVING RELATIVE
SELAGINELLA

Modern *Selaginella* are usually either small or scrambling plants, but occasionally they grow on the branches of trees. About 750 species are alive today. Most are found in tropical and warm temperate regions, but some are found in cooler climates.

rachis — fertile pinnule — stem — sterile pinnule

Coniopteris

GROUP Ferns
DATE Triassic to Neogene
SIZE Frond 3ft 3in (1m) long
LOCATION Worldwide

This is one of the most diverse and widespread genera of Jurassic ferns. Some fronds were vegetative, with leaflets (pinnules) that had broad lobes to collect light for photosynthesis. Other fronds, with more slender leaflets, were fertile, with small clusters of spore-producing sacs (sporangia) at the end of the leaflets. The detailed structure of the sporangia shows that these fossil ferns were related to the modern-day tree fern *Dicksonia*. At least some of the Mesozoic *Coniopteris* fronds also came from tree ferns, although others may have grown on plants that were much smaller.

CONIOPTERIS HYMENOPHYLLOIDES
Broken fragments of the dicksoniaceous fern *Coniopteris* include remains of stem and rachis and leaflets from both fertile and sterile fronds.

Dictyophyllum

GROUP Ferns
DATE Late Triassic to Middle Jurassic
SIZE Frond 8–12in (20–30cm) wide
LOCATION Worldwide

Just like members of the Matoniaceae, such as *Weichselia* (see p.274), *Dictyophyllum* had leaf segments (pinnae) that radiated from the end of a stalk, which was attached to an underground creeping stem (rhizome). However, this fern is distinctive in the way in which the veins formed a meshwork across the leaflets (pinnules). The veins, and the structure of the spore-producing bodies (sporangia), suggest that *Dictyophyllum* belonged to the family Dipteridaceae. Today, this family only grows in Southeast Asia. During Jurassic times, *Dictyophyllum* was widespread, but it declined and eventually became extinct later in the Jurassic.

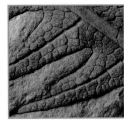

LEAF VEINS
Each triangular leaflet has a central, thick midvein. On either side of the midvein, there are lateral veins that create a polygonal mesh.

DICTYOPHYLLUM NILSSONII
This partial leaf of a Jurassic fern has characteristic meshed veins. Ferns with this type of mesh veins often preferred drier habitats.

Equisetites

GROUP Sphenophytes
DATE Late Carboniferous to present
SIZE 8¼ft (2.5m) tall
LOCATION Worldwide

Fossil casts of *Equisetites* stems, with clear ribs running along their length and more widely spaced "cross-bars" (nodes), are frequently found in the Jurassic. They are particularly common in rocks that formed along riverbanks or the shores of lakes, where the plants probably grew at the water's edge. They look very similar to the stems of living *Equisetum*, but many are larger than any living species.

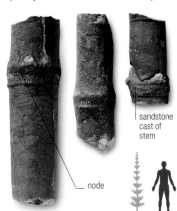

sandstone cast of stem — node

⊙ LIVING RELATIVE
HORSETAILS

Horsetails (*Equisetum*) remain a widespread and common group of plants today. They spread easily using underground stems called rhizomes. Their scaly leaves are arranged in whorls around the stem at characteristic nodes. Despite a lengthy and varied fossil record, only about 20 species survive today.

Cladophlebis

GROUP Ferns
DATE Triassic to Cretaceous
SIZE Frond 3ft 3in (1m) long
LOCATION Worldwide

Both extinct and modern-day members of the fern family Osmundaceae, to which *Cladophlebis* belongs, have nonfertile vegetative leaves and fertile leaves that have spore-bearing sacs (sporangia)—as seen in the living royal fern (*Osmunda regalis*). These two types of frond can look quite different, and their fossils are often found separately. Paleontologists therefore give them different genus names: *Cladophlebis* for the vegetative fronds with distinctive, triangular leaflets, and *Todites* for the fertile fronds.

CLADOPHLEBIS FUKIENSIS
This fossil from China shows part of a typical sterile frond of an osmundaceous fern. Its small triangular leaflets never bore sporangia.

fertile pinnule

TODITES SP.
The leaflets of fertile leaves, such as those seen on this fossilized leaf fragment from the Jurassic, often have clusters of spore-bearing sacs on their undersides.

Klukia

GROUP Ferns

DATE Late Triassic to Early Cretaceous

SIZE Frond 12–20in (30–50cm) long

LOCATION Europe, Central Asia, Japan

The distinctive, small, tongue-shaped leaflets (pinnules) of *Klukia* ferns are common in Jurassic floras. The fertile leaves are particularly distinctive. They have quite large spore-bearing sacs (sporangia) that are borne singly on the underside of the leaflets. This is quite different from most other Jurassic ferns, where the sporangia occur in a series of clusters (sori) on the pinnules. Members of this genus are most closely related to the modern-day climbing ferns of the family Schizaeaceae, which grow today in tropical and subtropical areas.

> ### *Klukia's* **distinctive pinnules** are often found in **Jurassic floras.**

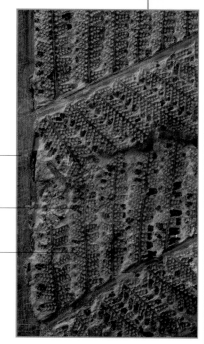

rachis

tongue-shaped leaflet

sporangia

KLUKIA EXILIS
The small, circular marks on this fossil are the distinctive sporangia that are borne on the undersides of the pinnules.

Pseudoctenis

GROUP Cycads

DATE Late Permian to Cretaceous

SIZE Leaf 3ft 3in (1m) long

LOCATION Worldwide

Leaves that are divided into a series of elongate leaflets (pinnae) and attached to a central axis (rachis) are characteristic of many Mesozoic fossil floras. They are similar to the leaves of modern-day cycads and some, but not all (see *Zamites*, p.222), are actually cycad remains. Various genera have been identified based on the shape and veining of the leaflets and how they are attached to the rachis. *Pseudoctenis* leaves were produced by true cycads. They have unbranched veins and leaflets that were attached to the sides of the rachis. Although fossilized examples are not as abundant as some other cycad leaf fossils, such as *Nilssonia*, they are very widespread. *Pseudoctenis* plants occurred in both tropical and temperate floras. Examples have even been found in Permian rocks, making them among the oldest-known cycad fossils.

rachis

PSEUDOCTENIS HERRIESII
This fossil of a Jurassic cycad leaf clearly shows a thick central rachis, with slender pinnae on either side. Each pinna has numerous fine veins running along its length.

Caytonia

GROUP Caytonialeans

DATE Late Triassic to Early Cretaceous

SIZE Cone 2in (5cm) long

LOCATION Europe, North America, Central Asia

Caytonia is the seed-bearing organ of a group of extinct plants called *Caytoniales*, which were abundant in subtropical areas during the Jurassic. Several seeds were enclosed in the protective, helmet-shaped, seed-bearing structures (cupules) that were arranged in two rows on either side of a central axis (rachis). The enclosure of the seeds within the cupules superficially resembles how today's flowering plants bear their seeds. The leaves produced by the plant also had features in common with flowering plants, particularly their mesh venation. For many years, it was thought that *Caytonia* might have been a direct ancestor of the flowering plants, although this idea has now been rejected.

leaflet

CAYTONIA SP.
Seed-bearing cupules are attached on either side of the main axis. Remains of several of the seeds that were borne in each cupule are just visible.

cupule

rachis

SAGENOPTERIS NILSSONIANA
The plants that produced *Caytonia* seed-bearing structures had characteristic leaves, called *Sagenopteris*, with four leaflets and mesh veins. (One leaflet is missing from this example.) The leaves are far more commonly found than the cones and give a better idea of the distribution of these plants.

Androstrobus

GROUP Cycads

DATE Late Triassic to Early Cretaceous

SIZE Cone 2in (5cm) long

LOCATION Europe, Siberia

Androstrobus is the name given to pollen-producing cones that are found together with cycad leaves, such as *Pseudoctenis* (see above); they are thought to be part of the same plant. Their spirally arranged structure is thought to be made by modified leaves (sporophylls), each of which has an upturned scale at the end. The scales overlap each other to provide protection for the numerous pollen sacs (sporangia) on the underside of the sporophylls.

ANDROSTROBUS PICEOIDES
These are two pollen-producing cones of a Jurassic cycad. Each cone has spirally arranged, bractlike scales, with pollen-bearing sacs on the lower surface.

bractlike scale

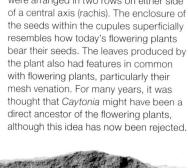

Williamsonia

GROUP Bennettitaleans
DATE Early Jurassic to Late Cretaceous
SIZE Flower 4in (10cm) long
LOCATION Worldwide

The bennettitaleans were a characteristic group of Jurassic plants. Many species had a stocky trunk, up to 6¹/₂ft (2m) high, with cycadlike leaves and distinctive flowerlike reproductive structures. A particular type of seed-producing flower, known as *Williamsonia*, had a central dome-shaped structure (receptacle) to which numerous stalked seeds, separated by scales, were attached. The entire flower was surrounded by a protective layer of bracts. Often, these bracts are the only part that has been fossilized. Several aspects of the complex structure of *Williamsonia* suggest that these plants may have been related to the ancestors of flowering plants. However, the exact relationship between bennettitaleans and flowering plants is still unclear.

WILLIAMSONIA GIGAS
This fossil, preserved in an ironstone nodule, shows the ring of protective bracts that surrounded the bennettitalean cone—seen from the underside.

SCALY TRUNKS
The plants that bore *Williamsonia* flowers had stocky trunks with the flowers and leaves borne at the top. The scaly surface of the trunk was created by the bases of dead leaves.

Zamites

GROUP Bennettitaleans
DATE Late Triassic to Late Cretaceous
SIZE Leaf 20in (50cm) long
LOCATION Worldwide

These commonly found plant fossils look very similar to the leaves of modern-day cycads. However, the detailed cellular structure of *Zamites*, particularly the small breathing pores (stomata) on the underside of the leaves, is very different. It is now known that these cycadlike fossils are from plants that also bore the flowerlike reproductive structures that are characteristic of all the bennettitaleans. In fact, the majority of cycadlike leaves found in Jurassic floras are likely to be a type of bennettitalean rather than true cycads. The genus *Zamites* is a good indicator of warm Jurassic climates because it occurred only in tropical or subtropical conditions.

pinna

leaf rachis (stem)

ZAMITES GIGAS
This fossil shows typical bennettitalean leaves and is from the Jurassic plant beds exposed on the Yorkshire coast in England.

Weltrichia

GROUP Bennettitaleans
DATE Jurassic
SIZE Cone 4in (10cm) across
LOCATION Worldwide

The bennettitalean plants that bore *Williamsonia* (see p.221) are also thought to have borne pollen-producing flowers that are known as *Weltrichia*. These flowers were made up of groups of long, leaflike bracts that had pollen sacs on their upper surface. In some species, there were also small, stalked bodies on the upper surface that may have secreted a nectarlike substance to attract pollinating insects. The bracts radiated from a cup-shaped base (receptacle), giving it a flowerlike appearance. *Weltrichia* has never been found directly attached to a plant, so it is hard to know exactly how the flowers were borne. They are, however, almost always found associated with *Williamsonia*, as well as with their leaves, known as *Ptilophyllum*, and there is little doubt that they belonged to the same plants.

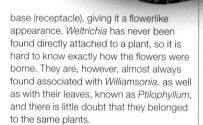

Ptilophyllum leaf

pollen sacs

WELTRICHIA SPECTABILIS
This is an example of an almost complete male flower showing the bracts and pollen-bearing sacs. Also present is part of the leaf (known as *Ptilophyllum*) of the same plant.

bract

Czekanowskia

GROUP Czekanowskialeans
DATE Late Triassic to Early Cretaceous
SIZE Leaf 8in (20cm) long
LOCATION Europe, Greenland, North America, Central Asia, Siberia, China

This enigmatic genus was once thought to be related to *Ginkgo*, but it is now classified in its own extinct group: the czekanowskialeans. The leaves look superficially like clusters of pine needles, except for the fact that they fork several times. They are fused at the base to form a short shoot. The plants that bore these leaves produced seeds in loose, conelike structures called *Leptostrobus*. Czekanowskialeans mainly seem to have favored warm or hot, humid conditions.

CZEKANOWSKIA ANGUSTIFOLIA
This fossil shows a short shoot with a group of slender leaves that are fused together at the base.

slender leaf

base of shoot

LEPTOSTROBUS LUNDBLADIAE
These cones, which were produced by the plant that had *Czekanowskia* leaves, were made up of a series of small, two-valved capsules arranged around a thin axis. When mature, the capsules would open up to reveal up to five seeds.

Ginkgo

GROUP Ginkgos
DATE Late Triassic to present
SIZE 164ft (50m) tall
LOCATION Worldwide

Identifiable by their distinctive leaves, ginkgos were widespread and abundant in Jurassic times, but went into decline during the Late Cretaceous period. Today, there is just one living species, *Ginkgo biloba* (the maidenhair tree), which is only found growing in the wild in the mountains of China. This species is probably one of the best examples of a "living fossil" in the entire plant kingdom. Just like its fossilized cousins, the beautiful fan-shaped leaves with many fine veins are easy to recognize. Many different species of fossilized ginkgos have been identified by the shape and number of the lobes on the leaves and the shape of the leaf as a whole. Sometimes they are found together with small seeds that are attached to the ends of stalks. These seeds and seed stalks are very similar to those produced by the living *Ginkgo*. In other species, there were more complex clusters of seeds that are known as *Karkenia*.

GINKGO SP.
This beautiful fossil from Afghanistan shows the fan-shaped leaves and radiating veins that are characteristic of most ginkgoalean plants. It is not known what sort of seeds this particular species produced, so it is not clear exactly how closely related it is to today's *Ginkgo biloba*. Nevertheless, paleontologists still usually refer to such leaves as *Ginkgo*.

SCIENCE

GINKGO BIOLOGY

Ginkgo seeds have a fleshy outer covering and a hard inner layer. They germinate better if the outer fleshy layer is removed. In prehistoric times, dinosaurs or ancient mammals may have been attracted to these succulent fruits, eaten them, and eventually dispersed the seeds in their droppings.

fan-shaped leaf

leaf stalk

fleshy outer layer

hard inner layer

seed

seed, attached to axis

branch

KARKENIA CYLINDRICA
The cone above is from Iran. It occurs with *Ginkgo*-like leaves and contains numerous seeds that are attached to a central axis.

fan-shaped leaf

leaf stalk

Podozamites

GROUP Conifers
DATE Late Triassic to Late Cretaceous
SIZE Leaf 3¼in (8cm) long
LOCATION Worldwide

The leaves of most conifers are either short and scaly, or slender needles; both types have a single vein along their length. However, some Jurassic conifers, such as *Podozamites*, had broader leaves with several veins running along their length. It was originally thought that *Podozamites* leaves belonged to cycads, but studies of the structure of their breathing pores (stomata), on the underside of leaves, suggest that they were conifers.

unusually wide leaf

PODOZAMITES DISTANS
This specimen is part of a fossilized shoot of this unusual conifer. The leaves are wide, and each leaf has several veins.

Elatides

GROUP Conifers
DATE Middle Jurassic to Early Cretaceous
SIZE 98ft (30m) tall
LOCATION Europe, Canada, Siberia, Central Asia, China

The remains of conifer shoots and cones are common in Jurassic floras and can often be assigned to the same families as living conifers. *Elatides*, for example, has seed cones that are very similar to the living "China-fir" (*Cunninghamia*), in that they both had three to five small seeds attached to each seed-producing scale. However, *Cunninghamia* has much larger and straighter leaves than the shoots that bore these Jurassic cones. *Cunninghamia* is found growing today in China and Vietnam, in conditions that are probably similar to the warm climates *Elatides* seems to have favored in the Jurassic.

JURASSIC
INVERTEBRATES

Following the late Triassic extinctions, marine invertebrate life proliferated and diversified greatly as new ecosystems arose in the warm seas. An overall greenhouse climate prevailed, and on land there were rich forests, and also dinosaurs, flying reptiles, and the first birds.

Flourishing marine life

Typical of the Jurassic are the ammonites—these were the last group of ammonoids to evolve, and they displayed a characteristic complex suture (see p.194). They spread out and diversified rapidly, genera or whole groups succeeding one another, often with a turnover rate of less than a million years. Belemnites were important in the Jurassic. Bivalves, corals, gastropods, echinoderms, bryozoans, and scleractinian corals spread out, and both terebratulid (see right) and rhynchonellid brachiopods became locally abundant. Lobsterlike crustaceans became more common in some habitats. In places, large sponge reefs grew, although there were no true coral reefs yet. The presence of large fish and marine reptiles gave a new dimension to Jurassic marine life, in contrast to that of the Carboniferous, for example. In the face of this increased predation, many groups of echinoids and bivalves adopted the strategy of burrowing into the sediment for protection.

LYME REGIS BEACH
Along this shore, remains of Lower Jurassic ammonites are abundant; the large specimen here has been eroded to show its crushed septa.

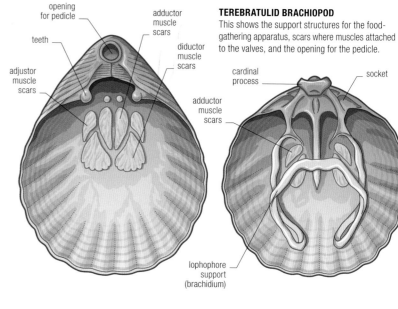

TEREBRATULID BRACHIOPOD
This shows the support structures for the food-gathering apparatus, scars where muscles attached to the valves, and the opening for the pedicle.

opening for pedicle

teeth

adjustor muscle scars

adductor muscle scars

diductor muscle scars

cardinal process

socket

adductor muscle scars

lophophore support (brachidium)

A biological arms race?

In all animals with voluminous and edible flesh, such as mollusks, the threat of predation has had a significant effect on their evolution. In times following a mass extinction, such as the Jurassic, the food-catching ability of predators increased dramatically. But this was matched in the prey by increasingly effective antipredator devices. Cephalopods, for example, depend on escape or avoidance, but sessile mollusks developed other strategies, which have enabled them to survive very well.

GROUPS OVERVIEW

Many invertebrate groups flourished during the Jurassic. Ammonites appeared at the beginning of the period, and new forms of crustaceans also began to appear. Scleractinian corals continued to evolve and were very abundant in places. Brachiopods were common, and fossils of the terebratulid type are often found in Jurassic limestone.

SCLERACTINIAN CORALS
These arose in the Triassic after the demise of the Paleozoic groups, and some genera survive today. The main dividing partitions (septa) in their skeletons are laid down on six lines, not four as in the rugose corals, and are radially symmetrical. They originated from a group of soft-bodied sea anemones.

TEREBRATULID BRACHIOPODS
Brachiopods were common at this time, although they had lost their Paleozoic importance. Terebratulid brachiopods (see also above) formed one of the two main groups of calcareous-shelled brachiopods in the Jurassic and were widespread. They were attached to the seafloor by a fleshy stalk called a pedicle.

AMMONITES
The Jurassic was the heyday of these rapidly evolving coiled cephalopods, and their complex sutures are readily distinguished from those of other ammonoids. Their variegated form attests to adaptation to many environments. Ammonites are more valuable for studying rock strata than any other kind of invertebrate fossil.

CRUSTACEANS
Shrimplike crustaceans were present in the Carboniferous, but by the Jurassic, the first lobsterlike forms had originated, many being predatory. The descendants of some of these Jurassic crustaceans inhabit the deep sea to the present day.

Peronidella

GROUP Sponges
DATE Carboniferous to Cretaceous
SIZE 2–4¹/₂in (5–12cm) across
LOCATION Worldwide

Peronidella was a calcareous sponge, characterized by a skeleton composed of calcium carbonate spicules (in the form of calcite) that were needlelike and fused together. The rigid structure created by the spicules was cylindrical, and in most species, it was also branched. In life, this sponge would have fed and breathed by drawing seawater in through pores in the outer body wall, which then passed into its central cavity. As the seawater passed through the thick body walls of the sponge, oxygen was absorbed and food particles were filtered out. With the help of many small hairlike structures called cilia that helped maintain the current, the water then passed out through an opening (osculum) along with any waste material that needed to be excreted.

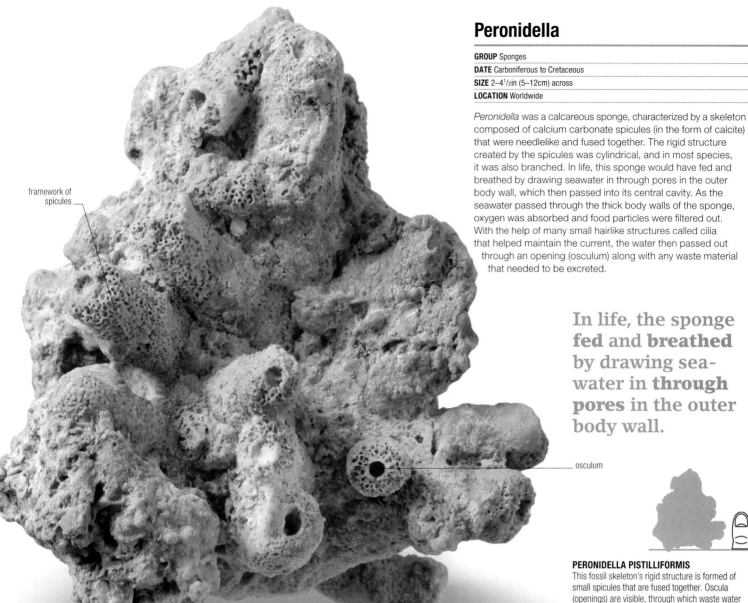

framework of spicules

osculum

> In life, the sponge **fed** and **breathed** by drawing sea-water in **through pores** in the outer body wall.

PERONIDELLA PISTILLIFORMIS
This fossil skeleton's rigid structure is formed of small spicules that are fused together. Oscula (openings) are visible, through which waste water passed out of the sponge.

Thecosmilia

GROUP Anthozoans
DATE Middle Triassic to Cretaceous
SIZE Corallites ¹/₂–1¹/₄in (1.5–3cm) across
LOCATION Worldwide

Thecosmilia was a scleractinian coral that consisted of numerous loosely packed, cylindrical individual polyp skeletons (corallites). Extra corallites were formed by a process called asexual budding, in which offspring of the soft-bodied polyps grew out of the body of the parent. This process created more branches and increased the width of the colony. The outer wall (epitheca) of some of the fossilized corallites shows growth increments, but this has often worn away.

THECOSMILIA SP.
Numerous thin walls (septa) divided the inside of each individual corallite.

septa

growth increments

Isastrea

GROUP Anthozoans
DATE Middle to Late Cretaceous
SIZE Colony up to 3ft 3in (1m) across
LOCATION Europe, Africa, North America

Isastrea was a colonial coral that belonged to a group called the hexacorals, so called because of the hexagonal shape of their polyp skeletons (corallites). *Isastrea*'s corallites were closely packed together, and the walls of the adjacent corallites were fused together—an arrangement referred to as cerioid. The insides of the corallites were divided up by tightly spaced, slender septa. *Isastrea* had several cycles of these septa arranged radially, each in multiples of six. It was a hermatypic coral, which meant that it formed reefs. Hermatypic corals need warm, clear, shallow water and live in symbiotic relationships with microscopic algae, which use carbon dioxide produced by the coral to photosynthesize and provide oxygen for the coral. *Isastrea* may have been able to withstand slightly lower temperatures because it occurs farther north than other hermatypic corals.

ISASTREA SP.
This specimen clearly shows the polygonal corallites, most of which are hexagonal. The striking radial septa give it a distinctive, star-shaped appearance.

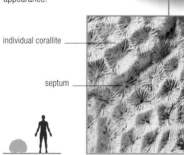

individual corallite

septum

> *Isastrea* was a **hermatypic coral**, living in **large coral reefs,** often with **dozens** of **different species.**

BERENICEA
Each autozooecium developed
a long tube (peristome) at the
entrance to the chamber.

Berenicea

GROUP Bryozoans
DATE Jurassic
SIZE Colony ¹/₂in (1.5cm) across
LOCATION Worldwide

"Berenicea" is the informal name for a type
of cyclostome bryozoan that can only be
identified more precisely if brood chambers
(gynozooecia), where bryozoan larvae
developed, are present. "Berenicea"
formed a thin, encrusting colony that was
roughly circular in shape. The calcified
chamber (autozooecia) that contained the
feeding soft-bodied animals (autozooids)
grew from the center of the colony. These
chambers were closely packed and in
contact with each other.

Spiriferina

GROUP Brachiopods
DATE Carboniferous to Early Jurassic
SIZE 1–2¹/₄in (2.5–5.5cm) long
LOCATION Europe, North America

Early Jurassic forms of *Spiriferina*, such as
the specimen shown below, were the last
survivors of the order Spiriferida, which
first appeared in Silurian times. A long,
straight hinge-line connected the valves,
which were strongly patterned with four
to seven well-marked ribs that were
more prominent toward the outer edge.
A strong fold in the center of the smaller
(brachial) valve corresponded with an
equally strong infold on the larger (pedicle)
valve. Examination under a magnifying
glass reveals that *Spiriferina*'s shell was
perforated by minute pores (punctae).

growth
increment

pedicle
opening

SPIRIFERINA WALCOTTI
This *Spiriferina* shell has some well-marked growth
increments and a pedicle opening, from which its
attaching stalk would have emerged.

Homeorhynchia

GROUP Brachiopods
DATE Early to Middle Jurassic
SIZE ³/₈–1in (1–2.5cm) wide
LOCATION Europe

This unusually shaped brachiopod belonged to the rhynchonellids.
Other members of this group usually had strongly ribbed shells
(valves), but *Homeorhynchia* was characterized by its smooth
shell surfaces. The brachial valve and its pointed end, or beak,
were more prominent than the pedicle valve, which had a small
opening for the pedicle, a fleshy stalk that attached the organism
to the seabed. The center of the brachial valve had a very
strong fold that corresponded with an equally strong infold that
ran down the center of the pedicle valve. The two valves rose
up in a sharp angle to the point at which they met along the
junction line (commissure).

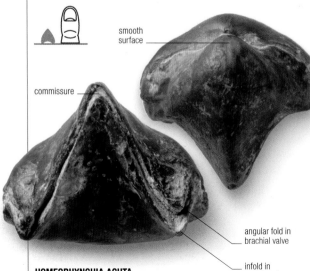

smooth
surface

commissure

angular fold in
brachial valve

infold in
pedicle valve

HOMEORHYNCHIA ACUTA
This species had a distinctive angular fold
on the outer edge of the brachial valve, which
corresponded with an infold in the pedicle valve.

Eutrephoceras

GROUP Cephalopods
DATE Middle Jurassic to Early Neogene
SIZE 4¹/₂–12in (12–30cm) wide
LOCATION Worldwide

Eutrephoceras was a tightly coiled
nautiloid whose outer whorls almost
enveloped its inner whorls. It had a
small and deep central part to the shell
(umbilicus) and is therefore described as
involute. The shell was almost spherical, and
the whorl sections were kidney-shaped. Inside the
shell, the siphuncle, a tube of tissue that connected
the inner chambers, was small in diameter and
positioned roughly in the center. The suture
lines where the septa joined the shell were simple
and fairly straight, but otherwise, the shell
was quite plain, apart from some fine
growth lines. The shell opening had
a broad recess (sinus) on the inner
side, through which a funnel
would have protruded
when *Eutrephoceras*
was swimming.

narrow
umbilicus

rounded
outer margin

suture line

EUTREPHOCERAS SP.
This internal mold clearly shows the thick,
straight suture lines and the rounded outer
margin. Most of the body chamber is missing.

Dactylioceras

GROUP Cephalopods
DATE Early Jurassic
SIZE 2¼–3¼in (6–8.5cm) across
LOCATION Europe, Iran, North Africa, Arctic, Japan, Indonesia, Chile, Argentina

Dactylioceras was a heavily ribbed ammonite, with whorls that barely overlapped each other. This produced a wide and shallow central part of the shell, or umbilicus. Ammonites with this kind of whorl and umbilicus configuration are referred to as evolute. The shell had sharp, radial ribs that were dense on the inner whorls and gradually became more widely spaced on the outer whorls. On the very outer edge of the shell, many of the ribs split into two, while others remained simple and unbranched. In cross-section, these whorls appear almost circular. Ammonite fossils are abundant in some parts of northwest Europe. In medieval times, people believed that these rocks were snakes that had been turned to stone and called them snakestones.

KEY SITE
THE JURASSIC COAST

The Jurassic Coast is situated on the south coast of England. It stretches from Exmouth in Devon to beyond Swanage in Dorset and comprises 95 miles (153km) of coastline that documents 185 million years of history, from the Triassic all the way through to the Lower Paleogene periods. The sections through the Jurassic rocks are famous for their beautifully preserved fossils. In 2001, the Jurassic Coast was made a World Heritage Site.

In medieval Europe, people thought ammonites were fossilized snakes.

carved snake's head

wide umbilicus

sharp radial rib

DACTYLIOCERAS SP.
In the 19th century, a snake head was sometimes carved onto an ammonite so that it could be sold as a fossil snake.

Gryphaea

GROUP Bivalves
DATE Late Triassic to Late Jurassic
SIZE 1–6in (2.5–15cm) high
LOCATION Worldwide

Gryphaea was an asymmetrical member of the oyster family. Its left valve was large and outwardly curved, providing stability on the seabed. The right valve was flat or slightly concave and acted as a lid. A ligament controlled the opening of the shell, while a single, strong adductor muscle closed the valves by contracting.

GRYPHAEA ARCUATA
Due to its odd shape, *G. arcuata* is called the "devil's toenail."

Cylindroteuthis

GROUP Cephalopods
DATE Early Jurassic to Early Cretaceous
SIZE 4–8½in (10–22cm) long
LOCATION Europe, Africa, North America, New Zealand

Cylindroteuthis was a large belemnite and is a common fossil in many Jurassic localities. The part of a belemnite most frequently preserved is its pointed guard, which was made of calcite and acted as a stiffening structure for the animal's squidlike body. Some fossils show traces of blood vessels on the outside of the guard, which indicate that it was an internal feature. A conical depression in the guard's blunt end housed the phragmocone, a kind of internal, chambered shell that helped *Cylindroteuthis* regulate its buoyancy. Well-preserved examples show some soft body parts, including the body, 10 arms, and an ink sac similar to those of modern squid.

CYLINDROTEUTHIS PUZOSIANA
This specimen shows the animal's long, tapered guard, which is attached at its blunt end (the alveolus) to the aragonite hragmocone. These hard parts were internal features.

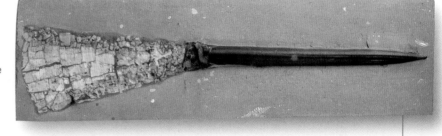

The pointed guard is the most commonly preserved part of a belemnite. It gave support to the animal's squidlike body.

SQUIDLIKE SWIMMER
With its ink sac, soft body, and 10 arms, *Cylindroteuthis* was very much a squidlike creature that used its elongated guard as a buoyancy device, as well as for support.

Pleuromya

GROUP Bivalves
DATE Middle Triassic to Early Cretaceous
SIZE ³⁄₄–2³⁄₄in (2–7cm) long
LOCATION Worldwide

Pleuromya was a medium-sized bivalve with two valves of equal size. At the rear end was a permanent gape through which two siphons protruded. *Pleuromya* lived buried in seabed sediment, drawing in food through its inhalant siphon and expelling waste through its exhalant siphon.

PLEUROMYA JURASSI
The outside of most *Pleuromya* shells have fine growth lines, but some species have concentric ribs.

fine growth lines

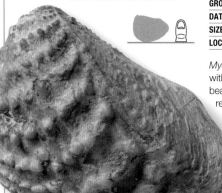

M. CLAVELLATA
The main surface of *Myophorella*'s shell is marked with tubercles.

Myophorella

GROUP Bivalves
DATE Early Jurassic to Early Cretaceous
SIZE 1¹⁄₂–4in (4–10cm) long
LOCATION Worldwide

Myophorella was a wedge-shaped bivalve with a distinct shell pattern. Behind its two beaks lay the short ligament that was responsible for opening the valves. The ligament ran along a flat, lozenge-shaped section on both valves, which was smoother than the rest of the shell, and had three rows of small bumps (tubercles) on each valve. On the main surface of the shell were rows of much larger, irregular tubercles.

Modiolus

GROUP Bivalves
DATE Devonian to present
SIZE ³⁄₈–4¹⁄₂in (1–12cm) long
LOCATION Worldwide

Modiolus has adapted to a range of environments and survives today. The beaks of its shells are to the front, and the shell surface has fine growth lines on it; in some species, these become heavier ribs. It attaches to the seabed by a byssus—horny threads often called a "beard" on mussels.

MODIOLUS BIPARTITUS
The shells of *Modiolus* get progressively wider the farther back from the beak they go.

Pleurotomaria

GROUP Gastropods
DATE Early Jurassic to Early Cretaceous
SIZE ³⁄₄–3in (2–7.5cm) high
LOCATION Worldwide

Pleurotomaria was a conical gastropod. The center of its shell was smooth, but it developed a strong ornament consisting of combinations of spiral and radial patterns. This specimen shows distinct bumps (tubercles) where the radial pattern crosses the spiral one. Around the side of the whorl is the slit band, a groove that continues around the shell to a slit in its opening that separates the waste-carrying exhalant current from the inhalant current. The animal could withdraw into its shell and close off the opening with a horny lid called an operculum.

smooth juvenile shell

tubercle

PLEUROTOMARIA ANGLICA
Tubercles are clearly visible on this specimen of *P. anglica*.

Pentacrinites

GROUP Echinoderms
DATE Jurassic
SIZE Arms up to 31in (80cm) across
LOCATION Europe

The crinoid *Pentacrinites* is often found in Jurassic rocks in the form of isolated pentagonal stem segments called ossicles. The small cup is often hidden by the abundant side branches, called cirri, that grow from the stems. The cup was made of circles of regularly arranged plates and roofed by a domed membrane called the tegmen, which incorporated a number of small calcite plates. Long arms grew from the arm plates at the top of the cup, dividing many times and producing hundreds of densely packed branches. The whole crown may have been as much as 31in (80cm) across. Some specimens have stems over 3ft 3in (1m) long. *Pentacrinites* is often found with fossil wood, so it has been suggested that this crinoid attached itself to floating wood and died when the wood became waterlogged and sank—a mode of life known as pseudoplanktonic.

densely packed arm branches

small, virtually hidden cup

PENTACRINITES SP.
With its hundreds of densely packed arm branches, as well as its association with fossil wood, *Pentacrinites* resembles a type of beautiful, fibrous plant more than an invertebrate animal.

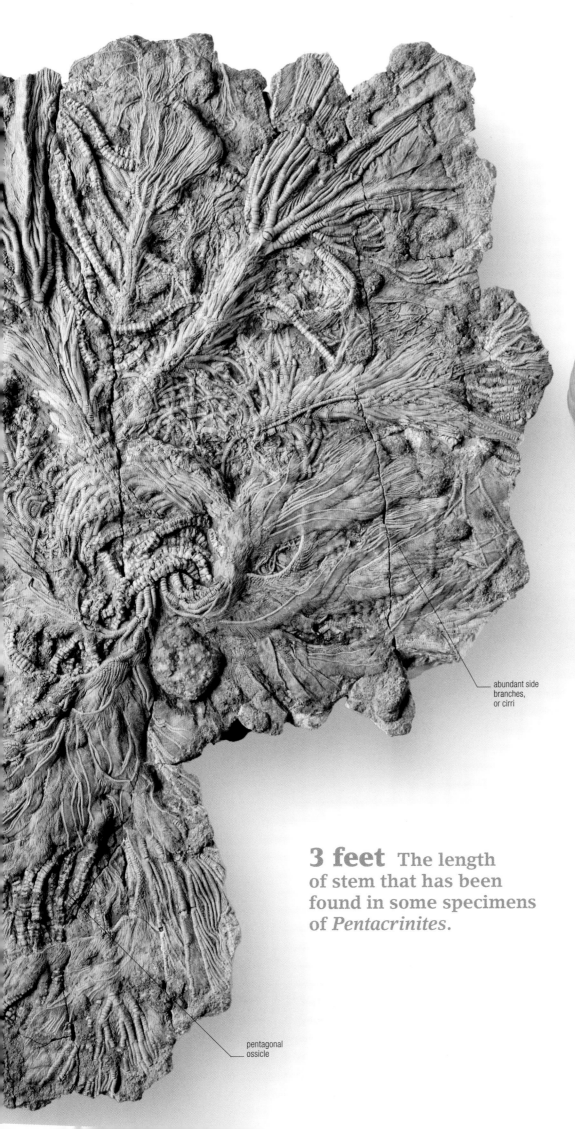

3 feet The length of stem that has been found in some specimens of *Pentacrinites*.

abundant side branches, or cirri

pentagonal ossicle

Apiocrinites

GROUP Echinoderms

DATE Middle to Late Jurassic

SIZE 12in (30cm) high, with stem; cup 1¼in (3cm) across

LOCATION Europe, Asia

Apiocrinites was a crinoid with a large cup. Its long stem was made up of individual segments called ossicles, and the cup was made up of two rings, each with five plates. Above these were rows of arm plates, from which arose 10 arms used for gathering food. Each arm had one or two branches, which housed a groove along which food particles were transferred to the mouth within the cup. A leathery dome (tegmen) covered the oral surface. *Apiocrinites* attached itself to hard surfaces in clear, warm waters.

arm plate

APIOCRINITES ELEGANS
This specimen shows the large cup and circular stem. Half of what seems to be the cup is really solid stem ossicles.

Pentasteria

GROUP Echinoderms

DATE Lower Jurassic to Lower Paleogene

SIZE Up to 4½in (12cm) across

LOCATION Europe

arm

PENTASTERIA COTTESWOLDIAE
Pentasteria had five arms of roughly equal length.

Pentasteria was a typical Jurassic asteroid or starfish, and in most respects, it differed little from starfish alive today. Like them, it had a double row of tube feet along the underside of each arm, which it used to move along with, and its mouth was positioned on its underside in the center of its body. As its name suggests, *Pentasteria* had five arms, each of similar length, that projected outward like the points of a star. Unlike many modern starfish, *Pentasteria* lacked suction disks on its tube feet, so it was unable to use them to pry open closed bivalve shells.

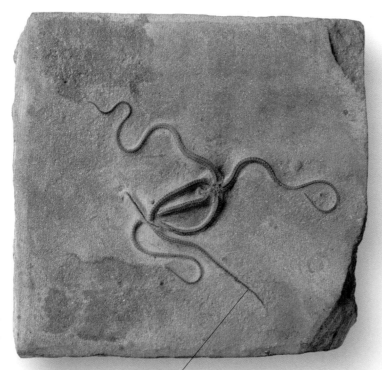

Palaeocoma

GROUP Echinoderms
DATE Early Jurassic
SIZE 2–4in (5–10cm) across
LOCATION Europe

Like all brittlestars, *Palaeocoma*'s body openings were located on the lower surface of its flat, disklike body. These included a central, star-shaped mouth that also served as the anus, the opening into the water vascular system, and five pairs of gill slits

tentaclelike arm

PALAEOCOMA EGERTONI
Brittlestars are delicate and decay quickly. Only a swift burial in sand preserved this specimen intact.

found on either side of each arm near the disk's outer edge. In addition to being used for respiration, the gill slits acted as release points for eggs or sperm. Food grooves with tiny, muscular tube feet ran along the underside of the arms and led to the mouth. The tube feet brought food toward the mouth and helped the organism to move.

Clypeus

GROUP Echinoderms
DATE Middle to Late Jurassic
SIZE 2–4½in (5–12cm) across
LOCATION Europe, Africa

Clypeus was a large, flattened sea urchin. On its upper surface, the ambulacra were shaped like petals; in the center lay the

apical disk, with four genital plates containing pores for the release of eggs or sperm. Its anal opening was in a groove at the rear of the upper surface. The mouth was located in the center of the lower surface; the ambulacra on this side had large pores, and a single tube foot protruded through each pair of these. The tube feet were used for respiration, feeding, and locomotion, including burrowing.

petal-shaped ambulacrum

CLYPEUS PLOTI
Large *Clypeus* fossils such as this are common in parts of the English Midlands, where they are often known as "poundstones" due to their uniform size and weight.

Hemicidaris

GROUP Echinoderms
DATE Middle Jurassic to Late Cretaceous
SIZE Including spines, 8in (20cm) diameter, without spines, ¾–1½in (2–4cm) diameter
LOCATION England

Shaped like a flattened sphere, *Hemicidaris* was a medium-sized sea urchin. The most noticeable feature of this species are the large bumps, called tubercles, that ran along the body. In life, these tubercles carried large, tapering spines up to 3¼in (8cm) long. The sockets found at the end of the spines fitted onto the ball-shaped ends of the tubercles, and the muscles that moved the spines were attached to

the relatively smooth area around these tubercle end points. The mouth of *Hemicidaris* was positioned at the center of the animal's lower surface, and the outer margin of the membrane that housed it had 10 notches that corresponded to the position of external gills. The anal opening was located almost at the center at the top of the shell. Five genital plates with pores surrounded it; one of these five plates also served as a pressure sensor to the water vascular system. Between and outside these were five smaller ocular plates that bore light-sensitive tube feet. *Hemicidaris* would have lived on firm seabeds and used the sticky tube feet on its lower surface for locomotion, as well as for feeding.

secondary tubercles

large primary tubercle

HEMICIDARIS INTERMEDIA
With its combination of large, spine-bearing primary tubercles and smaller secondary tubercles, *H. intermedia* resembles a jeweled and studded Christmas ornament.

ANATOMY
BALL-AND-SOCKET JOINTS

Ball-and-socket joints are common in vertebrate skeletons. Echinoids have evolved a similar type of articulation for their spines. This is most readily seen in those with large tubercles on their shells; these are the "ball" part of the ball-and-socket joint. The smooth plate area surrounding the ball serves as a muscle-attachment area; muscles stretch from here to the base of the spine around the socket part of the joint, allowing controlled movement.

Libellulium

GROUP Arthropods
DATE Jurassic
SIZE Wingspan up to 5½in (14cm)
LOCATION Europe

Libellulium was a large, prehistoric dragonfly. Like modern dragonflies, its head had large, bulging eyes located at the front. Its thorax was short, expanding slightly immediately behind the head. The abdomen was composed of seven or eight long, narrow segments, with visible divisions between them. The limbs, not seen on this specimen, were situated at the front end of the thorax. The long, narrow, powerful paired wings originated from the thorax at the front end of the animal. The forewings were a little narrower and slightly longer than the hind wings, which slightly overlapped them. Some of the fine venation of the wings, including major and secondary veins, can readily be seen. Dragonflies (*Odonata*) have a long geological history—the earliest appeared during the Carboniferous period.

LIVING RELATIVE
GIANT DRAGONFLY

Libellulium belonged to the same family of large living dragonfly species such as *Petalura gigantea*, commonly known as the giant dragonfly, which is found in New South Wales, Australia. Despite a wingspan of just under 5½in (14cm), it is actually a fairly poor flyer and rarely moves far from the area in which it emerged.

LIBELLULIUM SP.
With its two sets of wings; long, segmented body; and large head (not well preserved here), *Libellulium* differs little from modern dragonflies.

Mesolimulus

GROUP Arthropods
DATE Late Jurassic
SIZE Up to 3¼–3½in (8–9cm) long, excluding telson
LOCATION Germany

The prehistoric horseshoe crab *Mesolimulus* was very similar to today's species. The large headshield was horseshoe shaped, and its eyes small and compound. The abdomen was covered by a single, semioval plate with a narrow, raised area flanked by raised lobes. Its broad marginal area was flat and adorned on either side by seven long, pointed spines. A long, sharp-pointed barb or telson arose from the back of the abdomen.

telson ____

MESOLIMULUS WALCHII
The long, pointed telson of *Mesolimulus* is almost longer than the rest of the animal's exoskeleton, as can be seen in this specimen.

LIVING RELATIVE
HORSESHOE CRAB

The name "horseshoe crab" is derived from the shape of these animals' headshields; however, they are unrelated to true crabs. *Mesolimulus*, like other extinct forms, was smaller and flatter than its living relative *Limulus*, which is shown here, but they inhabited similar shallow marine environments. *Limulus* swims upside down, and this is likely to have been the case with its extinct cousins. *Limulus* is commonly found on the Atlantic coast of North America and in the Gulf of Mexico, and its lineage has a long pedigree. The oldest fossil horseshoe crabs have been found in rocks dating from the Permian period.

Eryma

GROUP Arthropods
DATE Early Jurassic to Late Cretaceous
SIZE 1½in (3.5cm) long
LOCATION Europe, E. Africa, Indonesia, North America

Eryma was a lobster. Its hard, upper shell (carapace) at the body's front end had a broadly oval outline and a prominent, backward-curved groove in the area between the front limbs or chelipeds, which ended in large, lozenge-shaped pincers. Each pincer had an inner, moveable finger and an outer, rigidly attached one. There were four more long, slender, delicate walking limbs. Behind the carapace, the abdomen was narrower and composed of five broad segments. It terminated in the tail fan, made up of five, radiating flat lobes.

ERYMA LEPTODACTYLINA
This specimen, from limestone quarried in southern Germany's Solnhofen region, clearly shows the large claws and five-lobed tail fan.

JURASSIC
VERTEBRATES

Dinosaurs began to dominate in the Jurassic. Although dinosaurs originated about 30 million years earlier, it was only after the Triassic—Jurassic extinction of 200 million years ago that they became the top land vertebrates across the world. Meanwhile, in the seas, giant reptilian predators flourished.

The major subgroups of dinosaurs—theropods, sauropodomorphs, and ornithischians—originated during the Late Triassic. However, for the first 30 million years of their history, dinosaurs were only supporting actors on the world stage and in many ecosystems were outnumbered and outmuscled by crurotarsans (crocodile-line archosaurs). Very suddenly, about 200 million years ago, all crurotarsans except for the crocodylomorphs disappeared in the midst of a global environmental meltdown, as volcanoes spewed toxic gas into the atmosphere and temperatures skyrocketed. This extinction was one of the "big five" mass extinctions in Earth's history, and without it dinosaurs might have never risen to dominance.

Crocodylomorphs

Crocodylomorpha is the technical term for living crocodiles and alligators and their closest fossil kin. This remarkably successful group is the only surviving remnant of the great crurotarsan radiation of the Late Triassic, when crocodile-line archosaurs such as rauisuchians, phytosaurs, and aetosaurs dominated ecosystems around the globe. True crocodylomorphs evolved in the Late Triassic alongside their crurotarsan cousins. These proto-crocodiles looked nothing like the crocodiles of today, but instead were small, slender animals that walked upright and could run at high speeds. These animals, called

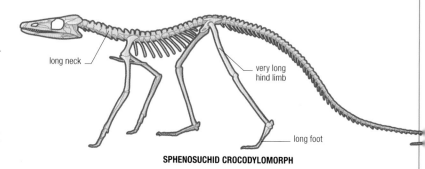

long neck
very long hind limb
long foot

SPHENOSUCHID CROCODYLOMORPH

elongated skull
paddlelike forelimb
deep tail provides thrust while swimming

METRIORHYNCHID CROCODYLOMORPH

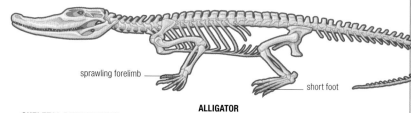

sprawling forelimb
short foot

ALLIGATOR

SKELETAL DIFFERENCES
Jurassic crocodylomorphs were a varied group that explored a wide range of skeletal designs. The earliest crocodylomorphs, the sphenosuchians, were small and graceful predators adapted for running. Metriorhynchids were colossal predators that lived in the ocean, with skeletons fine-tuned for their marine habitat. Modern alligators are slow, bulky creatures that lurk near the shore and ambush unsuspecting prey.

GROUPS OVERVIEW

The Jurassic was a time of reptilian dominance. Dinosaurs were the preeminent land vertebrates, and reptilian megapredators, such as plesiosaurs, ichthyosaurs, and metriorhynchids, ruled the oceans. However, other groups, such as mammals and amphibians, multiplied in the shadow of the dinosaurs.

TURTLES
Turtles are among the most unusual reptilian groups, and little is known about the early history of these animals. The oldest turtles come from the Middle Triassic, but these creatures only had a shell on their bellies. Some of the best-known primitive turtles lived in the Jurassic, such as *Kayentachelys* from southwestern USA, and had fully formed shells.

PLESIOSAURS
The long-necked, pot-bellied plesiosaurs look a bit like a sea-dwelling sauropod dinosaur. But unlike the peaceful, plant-eating sauropods, plesiosaurs were the most fearsome predators in most Jurassic oceans. Some of these creatures were over 49ft (15m) long (much longer than *Tyrannosaurus*) and could tackle the largest prey that the ocean had to offer.

ICHTHYOSAURS
The finned, streamlined ichthyosaurs resemble dolphins and tuna but were actually ocean-living reptiles. Ichthyosaurs were enormously successful during the entire Mesozoic and fed on a variety of prey, from squid to large fish. The biggest ichthyosaurs were over 66–69ft (20–21m) long, making them the largest marine reptiles that have ever existed.

METRIORHYNCHIDS
Metriorhynchids are among the most unusual crocodylomorphs. Unlike most crocs, which are sluggish creatures that lurk at the water's edge, metriorhynchids spent their entire lives in the open ocean. They were among the largest and fiercest predators of their day and thrived worldwide before being replaced by plesiosaurs later in the Jurassic and Cretaceous.

50 tons The weight of *Brachiosaurus*, one of the largest animals ever to walk the Earth.

sphenosuchians, persisted until deep into the Jurassic. However, during the Jurassic, crocodylomorphs continued to diversify and many specialized subgroups evolved. These included not only sprawling species that lurk at the water's edge, such as living crocodiles, but also an unusual group called metriorhynchids that lived entirely in the open ocean.

The age of giants

The Middle and Late Jurassic are often referred to as the "age of dinosaur giants." This fitting description pays homage to the evolution of several remarkable groups of colossal dinosaurs, which dominated terrestrial ecosystems across the planet. Foremost among these giants were the long-necked and plant-guzzling sauropods, such as *Brachiosaurus* and *Diplodocus*. The Late Jurassic was the peak of sauropod evolution.

FEARSOME PREDATOR
The powerful hunter *Allosaurus* was the most feared dinosaur of the Late Jurassic. It lived across North America and in Portugal and used its size and strength to take down large prey, such as the sauropod giants.

MASSIVE PROPORTIONS
The Late-Jurassic *Brachiosaurus* is probably the most familiar of all of the long-necked, barrel-chested sauropods. It was one of the largest animals ever to walk the Earth. It approached lengths of 82ft (25m) and weighed as much as 50 tons.

At no other time were these thundering beasts as abundant and diverse. As many as 25 different sauropod species lived alongside each other in the Late Jurassic of North America, and today their fossils are common discoveries in the famous Morrison Formation (see p.42). Other colossal sauropods are known from Africa, China, and Portugal. Some of these creatures were among the largest animals ever to walk the Earth, and the ground literally quaked underneath their footsteps. Living alongside these herbivores, and probably preying on them, were many monstrous theropod carnivores. The poster child for Jurassic theropods is *Allosaurus*, a hatchet-skulled hunter that may have reached lengths of 39ft (12m) and weights of 4,000lb (1,800kg). Other nightmarish beasts lived alongside *Allosaurus*, including *Ceratosaurus* and *Torvosaurus*. Truly, the Jurassic was an age of giants.

PTEROSAURS

Often mistaken for dinosaurs, pterosaurs instead form their own unique subgroup of archosaurs. Pterosaurs arose during the Late Triassic and were the first vertebrate group to evolve flight. Most Triassic and Jurassic pterosaurs were small and fed on fish, but some truly colossal species, such as *Quetzalcoatlus*, lived in the Cretaceous.

TETANURANS

The tetanurans are one of the most important and successful subgroups of theropod dinosaurs. Most Late Triassic and Early Jurassic theropods were small and had primitive skeletons, but in the Middle Jurassic, the much larger and more diverse tetanurans arose, including *Megalosaurus*, the "great lizard." Birds are a member of this subgroup.

SAUROPODOMORPHS

The sauropodomorph subgroup contains some of the most distinctive dinosaurs, such as the familiar long-necked herbivores *Brachiosaurus* and *Diplodocus*. Triassic and Early Jurassic sauropodomorphs were pretty small and had unspecialized skeletons, but giant, quadrapedal, earth-shaking species with long, sinuous necks evolved in the Middle Jurassic.

ORNITHISCHIANS

Along with sauropodomorphs and theropods, ornithischians are one of the three major subgroups of dinosaurs. Ornithischians were rare in the Late Triassic, but became more common, and evolved into a diverse array of body forms, in the Jurassic. These include the plated stegosaurs, the armored ankylosaurs, and the herbivorous iguanodontians.

MAMMALS

Mammals originated in the Late Triassic, around the same time as the dinosaurs. However, the furry mammals remained in the shadow of their giant reptilian contemporaries for most of the Mesozoic. But just because mammals were small does not mean they were unimportant. Mammals evolved into a range of different body types and ecological niches during the Jurassic.

Hybodus

GROUP Chondrichthyans

DATE Late Permian to Late Cretaceous

SIZE 6½ft (2m) long

LOCATION Europe, North America, Asia, Africa

The skeleton of *Hybodus* was very robust and more densely calcified than that of modern sharks. The jaws were solid and armed with rows of teeth that were continuously being replaced, as in modern sharks. The teeth measured up to ¾in (2cm) long and varied in shape—some had a series of upright spikes or cusps, while others were low and stubby. In contrast to living sharks, in which the mouth opens on the underside of the head, the mouth of *Hybodus* opened at the end of the snout. In addition to pelvic claspers, used to insert sperm directly into the female, a male *Hybodus* also had specialized spines in the soft tissues of its head. These may have helped the male grip the female during mating. One or two pairs of these spines were situated just behind the eye.

STREAMLINED SHARK
The two dorsal fins were supported by long, ribbed fin spines that were embedded in the soft tissues. These probably helped the animal cut through the water more efficiently.

Ischyodus

GROUP Chondrichthyans

DATE Middle Jurassic to Miocene

SIZE 5ft (1.5m) long

LOCATION Europe, North America, Kazakhstan, Australia, New Zealand, Antarctica

Fish similar to *Ischyodus* are today restricted to the deep oceans, but in its time, *Ischyodus* was a common bottom-dweller. It had a mainly cartilaginous skeleton, large eyes, pectoral (side) fins, and a long, whiplike tail. The first dorsal fin had a prominent spine that could be raised and was probably used for defense. Instead of teeth, *Ischyodus* had dental plates that often protruded from its mouth. This is where it gets its common name, "rabbitfish." There were two pairs of plates in the upper jaw and one pair in the lower jaw. They were used to crush mollusk shells and crustaceans. Dental plates are the most commonly preserved part of this fish and are used to distinguish different species.

HEAD GEAR
Males had a curious head appendage, called a tenaculum. The tip was covered in sharp, toothlike structures and may have played a part during mating.

Leedsichthys

GROUP Actinopterygians

DATE Middle to Late Jurassic

SIZE 26–55ft (8–17m) long

LOCATION Europe, Chile

Possibly the largest fish that ever lived, *Leedsichthys* was first discovered near Peterborough, England, in 1886. Despite its formidable size, *Leedsichthys* was a harmless filter-feeder. Its huge mouth could take in large volumes of water in a single gulp. It then sifted microorganisms from the sea water with gill rakers, or mesh plates at the back of its mouth, just as

GENTLE GIANT
A thrust from its powerful 16½ft (5m) tail was the only defense *Leedsichthys* had against attackers.

basking sharks and baleen whales do today. *Leedsichthys* is classified with the pachycormids: fish that have partly calcified skeletons. Because cartilage does not fossilize well, the lack of calcification means that many of its bones have not been preserved and others are so thin that they are easily crushed. Teeth, however, are more robust than bones, and it has been possible to estimate that *Leedsichthys* had more than 40,000 of them.

Leptolepides

GROUP Actinopterygians
DATE Late Jurassic to Early Cretaceous
SIZE up to 4in (10cm) long
LOCATION Europe, North America

Although one of the commonest fossil fish, specimens of *Leptolepides* are often damaged, making them difficult to study. This is possibly the reason that it was not recognized as being distinct from the very similar *Leptolepis* until 1974. *Leptolepides* was a primitive bony fish that lived in shoals in shallow-water lagoons. It had a somewhat elongated body with a depth about one-sixth of its length. The single dorsal fin was relatively long, and the tail fin was deeply forked. Its teeth were delicate—those present on the parasphenoid, a bone in the roof of the mouth, being a feature that distinguishes it from *Leptolepis*. Other diagnostic features include wide sensory canals in the head region and the structure of the skeleton supporting the tail.

SAFETY IN NUMBERS
Like the modern herring, this relatively small fish lived in large shoals for protection. It was predated by large fish and aquatic reptiles.

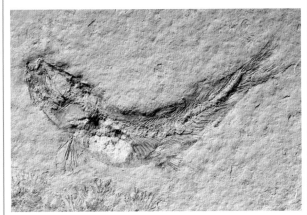

Prosalirus

GROUP Amphibians
DATE Early Jurassic
SIZE 2¼in (6cm) long
LOCATION USA

Only three fossilized skeletons of this small frog have ever been found. They were discovered in Arizona in the 1990s, and are the earliest frog skeletons to show the full set of jumping adaptations that characterize frogs today. These include long hip bones, long hind leg bones, and long ankle bones. Earlier ancestral frogs had been found in Madagascar and Poland, but these examples still had tails and did not have jumping adaptations, so they would not have been capable of froglike leaping movements. Frogs began to diversify long after *Prosalirus* had died out, so it is not directly related to any one modern family. All the fossils were found in one small area, which probably represented the last puddle of water on a dried-up flood plain.

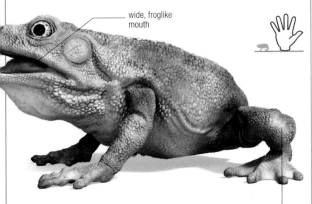

wide, froglike mouth

JUMPING FROG
This species had hind legs adapted for jumping. The name *Prosalirus* comes from the Latin verb prosalire, which means "to leap forward."

long hind legs

Eocaecilia

GROUP Amphibians
DATE Early Jurassic
SIZE 7in (18cm) long
LOCATION USA

This small, burrowing amphibian was discovered in Arizona. It had a small head with heavy internal armor, a very long body, four tiny limbs, and a short tail. The skull of *Eocaecilia* is very like those of the group of living amphibians called caecilians, or "worm amphibians." *Eocaecilia* is believed to be the earliest known member of this group. Modern-day caecilians are burrowing creatures that have lost their limbs altogether. Although *Eocaecilia* still had limbs, its heavily built skull suggests that it used its head to plow through the ground like modern-day caecilians.

WORM AMPHIBIAN
Eocaecilia had a very elongated body, with around 50 vertebrae in the trunk section.

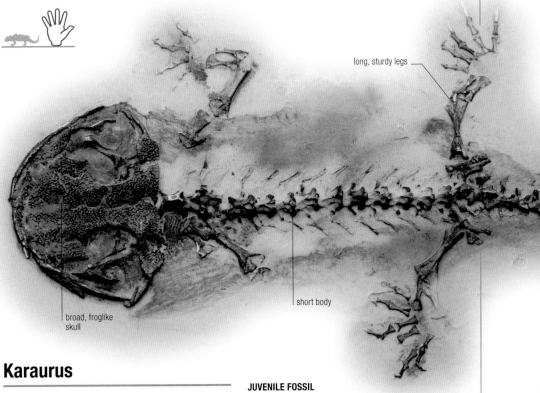

long, sturdy legs

short body

broad, froglike skull

Karaurus

GROUP Amphibians
DATE Late Jurassic
SIZE 7½in (19cm) long
LOCATION Kazakhstan

The only fossil of *Karaurus* was discovered in the 1970s in a fossil lake bed in Kazakhstan. For some years, it was the earliest salamander known. The name *Karaurus* comes from a combination of "Kara" (after the Karatau Mountains where it was found) and "Urus," meaning tail. *Karaurus* was a medium-sized salamander and was very heavily built with a broad, froglike skull, a short body, and large legs. It was more primitive in many of its skeletal features than any living salamander, which has led scientists to believe that modern-day salamanders evolved from an animal that was more advanced than *Karaurus*. Since *Karaurus* was discovered, other Jurassic salamanders have been found in China, Europe, and North America, which suggests that they were widespread across the northern continents.

JUVENILE FOSSIL
In the only known skeleton of *Karaurus*, the limb bones are only partly developed, which suggests that this was a juvenile animal that would have grown larger.

SHINY SCALES
This beautifully fossilized *Lepidotes* shows many of
the characteristic features of this fish, in particular
its thick, diamond-shaped scales. In life, each scale
had an outer layer of ganoine, an enamel-like
substance that reflected light and gave the fish a
shiny appearance. Rows of peglike teeth, once used
for crushing shellfish, are also clearly visible.

Kayentachelys

GROUP Turtles
DATE Early Jurassic
SIZE 10–14in (25–35cm) long
LOCATION USA

Kayentachelys was one of the oldest turtles and an important link between primitive Triassic turtles, such as *Odontochelys* (see p.200), and modern turtles. Like all modern turtles, *Kayentachelys* was sheathed in a boxlike shell comprised of a carapace on top and a plastron underneath. Its skull was short and wide and was capped with a sharp beak. *Kayentachelys* lived in southwestern North America alongside dinosaurs such as the theropod *Dilophosaurus* (see p.246) and early crocodylomorphs, such as *Protosuchus* (see p.242). The presence of large dinosaurs, crocodylomorphs, and turtles gave this ecosystem a modern character in comparison with ecosystems of the Middle to Late Triassic.

KEY SITE

THE KAYENTA FORMATION

The rusty red sandstones and shales of the Kayenta Formation cover broad swathes of western USA, including many of the most famous national parks, such as Grand Canyon and Zion. In addition to providing stunning desert scenery, the Kayenta Formation has yielded many important fossils from the Early Jurassic period. During this time, about 190 million years ago, a varied fauna of early dinosaurs, mammals, crocodiles, and turtles thrived in the arid conditions of a region that at this time was located in the supercontinent Pangea. Many famous fossils, including *Kayentachelys*, have been found in the Kayenta Formation.

SMALL TURTLE
Kayentachelys was a small animal no more than 14in (35cm) long and weighing only a couple of pounds. Its ecological habits are questionable, but it probably spent a lot of time on land.

Pleurosaurus

GROUP Lepidosaurs
DATE Late Jurassic
SIZE 19½–28in (50–70cm) long
LOCATION Germany

New Zealand's tuatara is the only living species representing the once-diverse reptile order known as Sphenodontia. Perhaps the most unusual of these animals, *Pleurosaurus* was slender and streamlined, with a long tail. Its hands and feet were modified into paddles, enabling its maneuvers as it chased small fish.

SOLNHOFEN LIMESTONE
Pleurosaurus was found in the same German rock unit in which the first bird, *Archaeopteryx* (see p.252), was discovered.

Plesiosaurus

GROUP Sauropterygians
DATE Early Jurassic
SIZE 9¾–16½ft (3–5m) long
LOCATION British Isles, Germany

Plesiosaurus was a large marine reptile that lived in the shallow seas of what is now Europe during the Early Jurassic, about 190 million years ago. In many ways, it looked like an aquatic version of a dinosaur. However, it was only distantly related to the dinosaurs and was more closely related to lizards and snakes. *Plesiosaurus* was a typical plesiosaur, with a stocky torso, four large flippers, an extremely long neck, and a tiny skull filled with many small teeth. This combination of features allowed *Plesiosaurus* to dominate its marine ecosystem. It was a highly successful predator of fish, squid, and other relatively small, fast-moving prey. *Plesiosaurus* was a fast and powerful swimmer that could strike at schools of fish with great speed using its muscular neck.

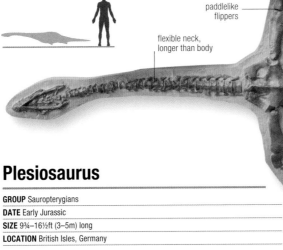

paddlelike flippers

rigid spine

digits elongated by additional bones

flexible neck, longer than body

small head

weak tail not used for propulsion

FIRST FOUND
Plesiosaurus was one of the first ancient reptiles to be discovered. It was found by Mary Anning (see p.241) and described in 1821 by William Conybeare.

COMMON FOSSILS
Plesiosaurus fossils are common in the so-called Liassic rocks of southern England. These are well exposed along the coast of Dorset, particularly around the town of Lyme Regis.

POWERFUL PADDLES
Liopleurodon used its four huge paddles to propel itself through the water after prey. The size of the paddles suggests that it was a powerful swimmer capable of high speeds in short spurts.

rib positioned in neck

CERVICAL RIB
The cervical ribs (or neck ribs) of plesiosaurs were short and robust, which allowed for the attachment of strong muscles and great mobility.

Liopleurodon

GROUP Sauropterygians
DATE Middle to Late Jurassic
SIZE 23–33ft (7–10m) long
LOCATION British Isles, France, Russia, Germany

Liopleurodon was the apex predator of Europe's seas during the Middle to Late Jurassic. This monstrous reptile reached lengths of 33ft (10m), making it larger than a modern-day killer whale. *Liopleurodon* was a plesiosaur, a stout-chested carnivorous marine reptile. Most plesiosaurs were long-necked and fed on fish. However, *Liopleurodon* was one of the pliosaurids, a subset of fearsome, predatory plesiosaurs with shorter necks and larger heads. Like other pliosaurids, *Liopleurodon* used its powerful jaws, which were up to 5ft (1.5m) long and studded with conical teeth, to spear marine reptiles and large fish. Although hefty, its body was streamlined, and it used its four paddlelike limbs to drive itself through the water. *Liopleurodon* was an amazingly successful animal: it existed for around 10 million years and inhabited a wide belt of seas across all of ancient Europe.

FOSSIL VERTEBRA
Pliosaurids were massive animals. Their spines were made up of vertebrae the size of dinner plates.

Ichthyosaurus

GROUP Ichthyosaurs
DATE Early Jurassic
SIZE 6½ft (2m) long
LOCATION British Isles, Belgium, Germany

At first glance, *Ichthyosaurus* looks rather like a dolphin or a fish, but it was a reptile closely related to lizards. Its streamlined body was designed for speed, with hands and feet modified into flippers, a dorsal fin, and a tall, paddlelike tail. *Ichthyosaurus*'s skull was adapted for hunting fast-moving, slippery prey—it was elongated and studded with rows of sharp teeth.

Stenopterygius

GROUP Ichthyosaur
DATE Early to Middle Jurassic
SIZE 6½–13ft (2–4m) long
LOCATION British Isles, France, Germany

Stenopterygius was a close relative of *Ichthyosaurus* (see p.239). Like its more famous cousin, *Stenopterygius* was a dolphinlike reptile that hunted schools of fish in the warm, shallow seas of Jurassic Europe. It grew to a slightly larger size than *Ichthyosaurus* but had a smaller skull, which was perfectly adapted for quickly striking, capturing, killing, and eating fish.

Stenopterygius's snout had the streamlined profile of a missile or submarine, and it could blast through the water like a projectile, rapidly striking fish with little or no warning. This creature's tail was also adapted for speed. The vertebral column in the tail bent downward and supported a deep tail fin, similar to those seen in modern sharks. This fin shape would have powered *Stenopterygius*'s large body as it raced through the water in search of its next meal. One of the most interesting *Stenopterygius* fossils shows a mother giving birth—proving that ichthyosaurs, like the dolphins of today, bore live young, and that the young were born tail-first.

60mph The estimated maximum speed reached by *Stenopterygius*, based on comparison with modern-day tunas.

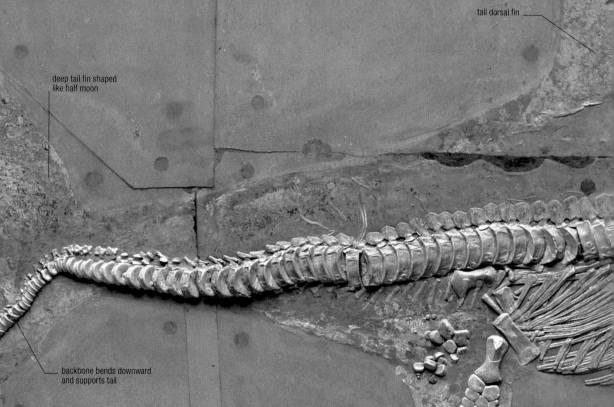

tall dorsal fin

deep tail fin shaped like half moon

backbone bends downward and supports tail

short rear limb

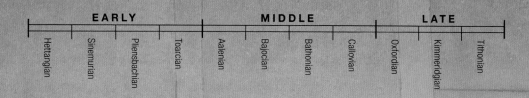

	EARLY			MIDDLE				LATE		
Hettangian	Sinemurian	Pliensbachian	Toarcian	Aalenian	Bajocian	Bathonian	Callovian	Oxfordian	Kimmeridgian	Tithonian

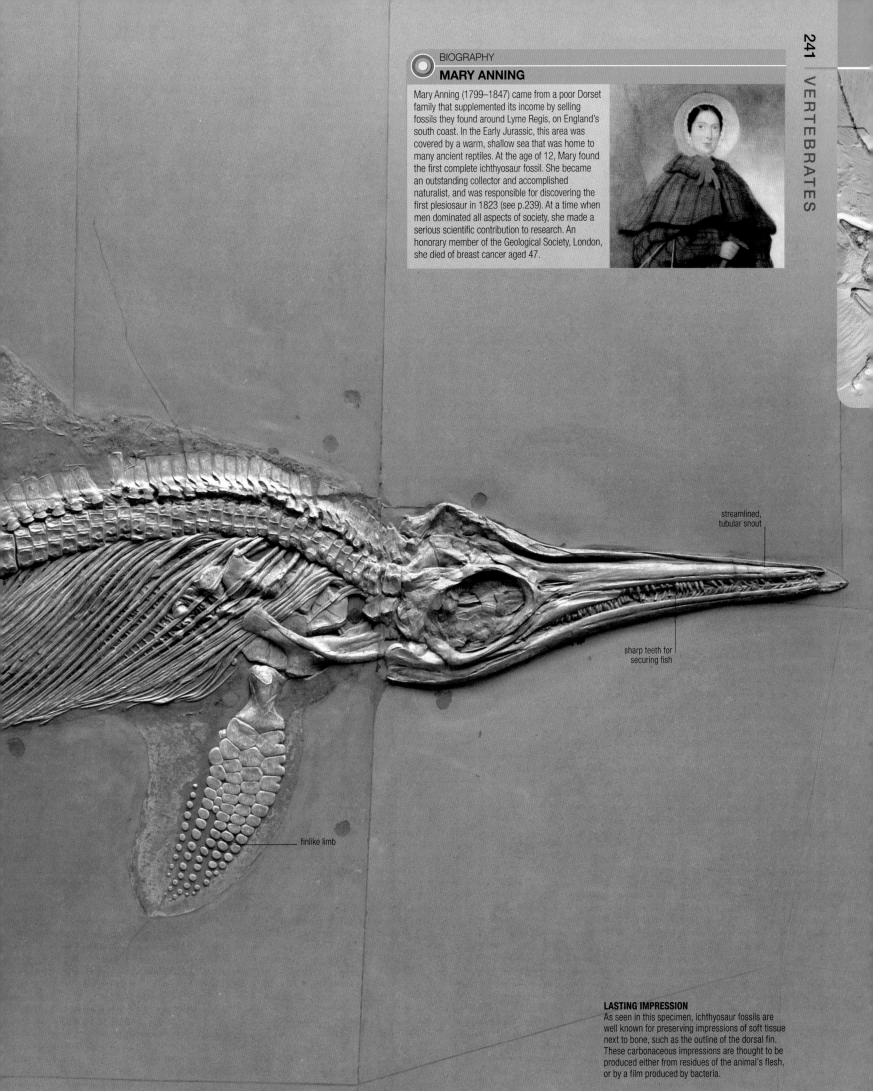

BIOGRAPHY
MARY ANNING

Mary Anning (1799–1847) came from a poor Dorset family that supplemented its income by selling fossils they found around Lyme Regis, on England's south coast. In the Early Jurassic, this area was covered by a warm, shallow sea that was home to many ancient reptiles. At the age of 12, Mary found the first complete ichthyosaur fossil. She became an outstanding collector and accomplished naturalist, and was responsible for discovering the first plesiosaur in 1823 (see p.239). At a time when men dominated all aspects of society, she made a serious scientific contribution to research. An honorary member of the Geological Society, London, she died of breast cancer aged 47.

streamlined, tubular snout

sharp teeth for securing fish

finlike limb

LASTING IMPRESSION
As seen in this specimen, ichthyosaur fossils are well known for preserving impressions of soft tissue next to bone, such as the outline of the dorsal fin. These carbonaceous impressions are thought to be produced either from residues of the animal's flesh, or by a film produced by bacteria.

Protosuchus

GROUP Crocodylomorphs

DATE Early Jurassic

SIZE 3ft 3in (1m) long

LOCATION Worldwide

Protosuchus is one of the earliest and most primitive relatives of living crocodiles. However, in many ways *Protosuchus* was very different from living crocodiles. It walked with an almost upright stance, its limbs were long and slender, and its feet were thin and capped with claws. These features suggest that *Protosuchus* was a strong runner and adapted for hunting on land rather than in the water. However, like living crocodiles, *Protosuchus* had a wide skull, powerful jaw muscles, and thick, conical teeth. It is no surprise that *Protosuchus* possessed a mixture of primitive and more modern physical features because it lived toward the beginning of crocodylomorph evolutionary history. *Protosuchus* was first found in Arizona but more recently its fossils have turned up in many other places across the world.

thick, powerful tail

back covered in bony, platelike scales

wide skull with powerful jaws

long, slender legs

thin feet adapted for running

SMALL STATURE
At about 3ft 3in (1m) long, and weighing only 65–100lb (30–45kg), *Protosuchus* was a small animal that resembled a juvenile crocodile.

tooth sockets

narrow snout

STENEOSAURUS SKULLS
These recognizably crocodilian *Steneosaurus* skulls were found in Oxfordshire, England.

Steneosaurus

GROUP Crocodylomorphs

DATE Early Jurassic to Early Cretaceous

SIZE 3¼–13ft (1–4m) long

LOCATION Europe, Africa

Steneosaurus was unlike most modern crocodiles. It did not lurk around the edge of the water, but rather was a competent swimmer that could venture into estuaries and coastal waters to hunt fish. However, *Steneosaurus* was not fully aquatic. It still had a bony coat of armor and its limbs were not modified into flippers. This meant that it could hunt on land if necessary.

Sphenosuchus

GROUP Crocodylomorphs

DATE Early Jurassic

SIZE 3¼–5ft (1–1.5m) long

LOCATION South Africa

Around the same time that *Protosuchus* was running across the sand dunes of Arizona (see above), another crocodilian relative called *Sphenosuchus* was sprinting through the humid landscape of southern Africa. Although these two early crocodylomorphs lived at about the same time, they were quite different in anatomy. *Protosuchus* was a quick runner that stood fairly upright, but *Sphenosuchus* had an even more upright stance and would have been much faster. It may have even reared up on two legs at times, and, like its close cousin *Terrestrisuchus*, it greatly resembled theropod dinosaurs in its skeletal anatomy and lifestyle.

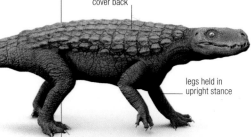

bony plates cover back

legs held in upright stance

STRANGE CROCODILE
A fast runner with an upright stance, *Sphenosuchus* was a strange early crocodile with a lifestyle very different from that of modern crocodiles.

Dakosaurus

GROUP Crocodylomorphs
DATE Late Jurassic to Early Cretaceous
SIZE 13–16½ft (4–5m) long
LOCATION Worldwide

Dakosaurus was the largest and fiercest member of a group of marine reptiles distantly related to modern crocodiles. This behemoth was much larger than the other marine reptiles that, along with fish, formed most of its diet. It was well adapted to life in the open oceans, with a streamlined body and paddlelike limbs. However, unlike many of its relatives, which had tubular snouts, *Dakosaurus* had a deep skull that resembled those of predatory dinosaurs such as *Tyrannosaurus*. Its teeth were also similar to those of theropods, as well as modern killer whales. They were large, narrow, and covered with an array of sharp serrations, perfect for cutting flesh.

MONSTROUS SKULL
Dakosaurus had a monstrous skull full of huge teeth. One South American skull specimen measuring more than 3ft 3in (1m) long has been nicknamed "Godzilla" because of its size.

Geosaurus

GROUP Crocodylomorphs
DATE Late Jurassic to Early Cretaceous
SIZE 6½–9¾ft (2–3m) long
LOCATION Europe, North America, the Caribbean

Like *Dakosaurus* (above), *Geosaurus* is a member of the metriorhynchids, a group of distinctive crocodylomorphs that lived in the open oceans. Unlike its land-living relatives, *Geosaurus* had lost its protective coat of bony scales, making its body lighter and more streamlined for swimming. In addition, its hands and feet had evolved into flippers, its body was thin and elongated, and its tail had a deep fin. There is even evidence that *Geosaurus*'s skull housed a large salt gland—an organ present in some marine creatures that enables them to drink salty sea water and eat aquatic prey without dehydrating. A successful predator, it used its sharp teeth to plunder schools of fish and squid.

> **Geosaurus**, meaning "**earth lizard**," is a strange name for a creature that lived in the open oceans.

MEDIUM-SIZED
Geosaurus was a medium-sized metriorhynchid, reaching up to 9¾ft (3m) in length. It had a narrow, tapering snout full of fairly small, sharp teeth.

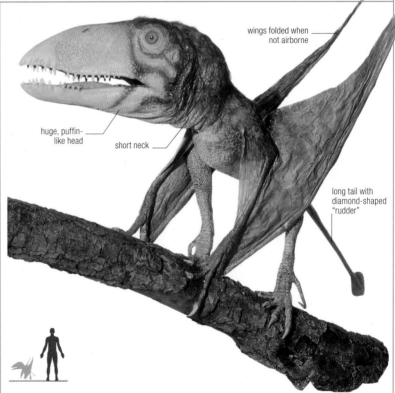

wings folded when not airborne

huge, puffin-like head

short neck

long tail with diamond-shaped "rudder"

Dimorphodon

GROUP Pterosaurs
DATE Early Jurassic
SIZE 3ft 3in (1m) long
LOCATION British Isles

A relatively small pterosaur, *Dimorphodon* was only about 3ft 3in (1m) long but had a skull that was about 10in (25cm) long. Its body was surprisingly light, and its weight-saving adaptations for flight included hollow limb bones and a skull reduced to little more than a scaffolding around its large eye and face.

ON ALL FOURS
When not in the air, *Dimorphodon*, like other pterosaurs, would have walked—probably rather clumsily—on four legs and taken off from a crouched, quadrupedal stance.

Dimorphodon lived along the coasts of ancient seas and probably used its puffinlike skull and fanglike teeth to capture fish and rip them apart. This pterosaur would probably have caught fish at the surface by lunging for them while skimming low over the water. It is not thought to have entered the water or dived.

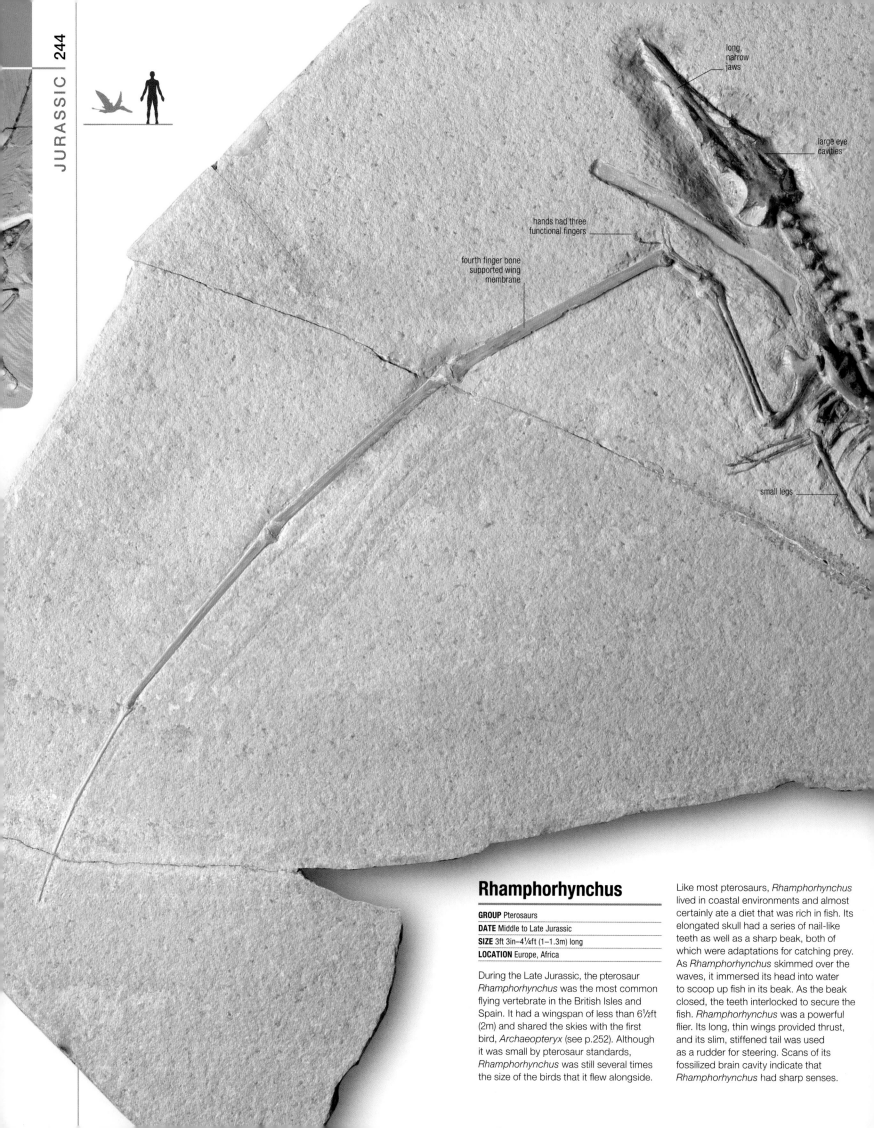

long, narrow jaws

large eye cavities

hands had three functional fingers

fourth finger bone supported wing membrane

small legs

Rhamphorhynchus

GROUP Pterosaurs

DATE Middle to Late Jurassic

SIZE 3ft 3in–4¼ft (1–1.3m) long

LOCATION Europe, Africa

During the Late Jurassic, the pterosaur *Rhamphorhynchus* was the most common flying vertebrate in the British Isles and Spain. It had a wingspan of less than 6½ft (2m) and shared the skies with the first bird, *Archaeopteryx* (see p.252). Although it was small by pterosaur standards, *Rhamphorhynchus* was still several times the size of the birds that it flew alongside.

Like most pterosaurs, *Rhamphorhynchus* lived in coastal environments and almost certainly ate a diet that was rich in fish. Its elongated skull had a series of nail-like teeth as well as a sharp beak, both of which were adaptations for catching prey. As *Rhamphorhynchus* skimmed over the waves, it immersed its head into water to scoop up fish in its beak. As the beak closed, the teeth interlocked to secure the fish. *Rhamphorhynchus* was a powerful flier. Its long, thin wings provided thrust, and its slim, stiffened tail was used as a rudder for steering. Scans of its fossilized brain cavity indicate that *Rhamphorhynchus* had sharp senses.

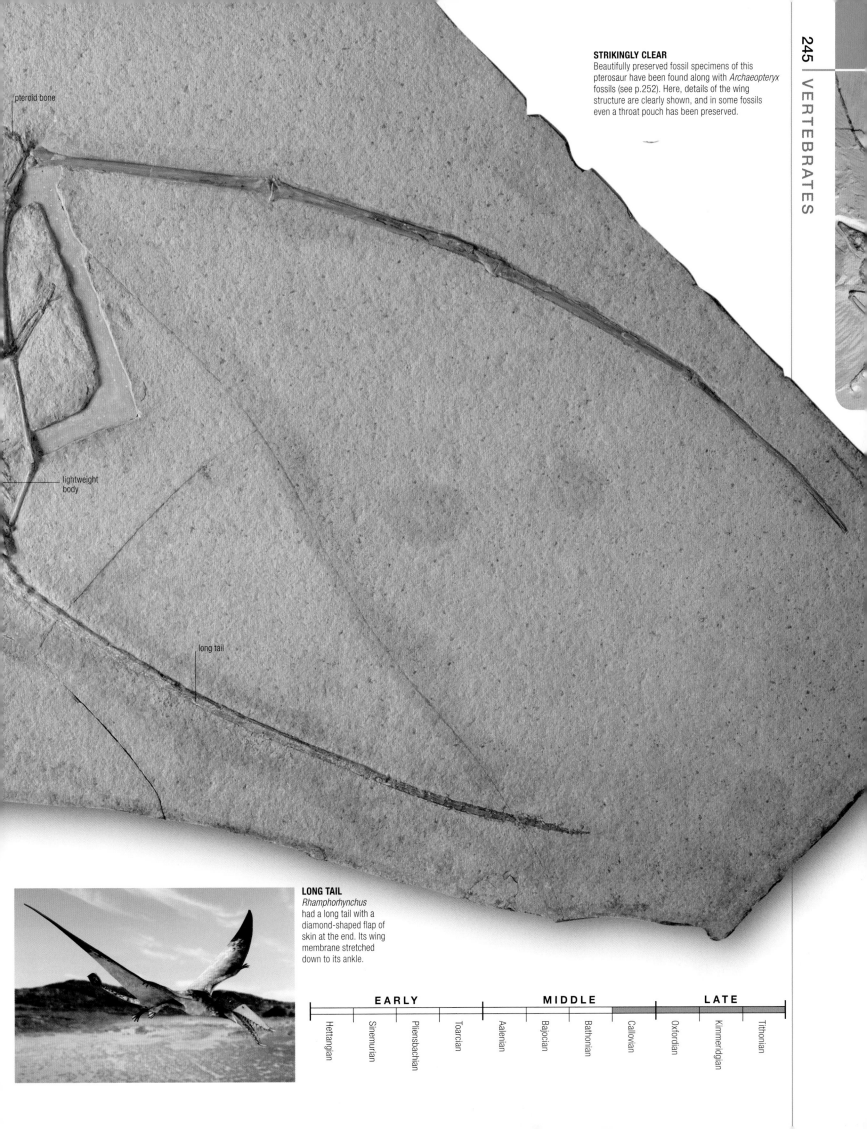

pteroid bone

STRIKINGLY CLEAR
Beautifully preserved fossil specimens of this pterosaur have been found along with *Archaeopteryx* fossils (see p.252). Here, details of the wing structure are clearly shown, and in some fossils even a throat pouch has been preserved.

lightweight body

long tail

LONG TAIL
Rhamphorhynchus had a long tail with a diamond-shaped flap of skin at the end. Its wing membrane stretched down to its ankle.

EARLY				MIDDLE			LATE			
Hettangian	Sinemurian	Pliensbachian	Toarcian	Aalenian	Bajocian	Bathonian	Callovian	Oxfordian	Kimmeridgian	Tithonian

elongated fourth digit supports
wing membrane

small crest made of
skinlike tissue

wing
membrane

CRESTED HEAD
Recent fossil finds have shown that
Pterodactylus had a small crest on its
head that was probably used for display.

four-toed feet

long, slim legs

Pterodactylus

GROUP Pterosaurs

DATE Late Jurassic

SIZE 3ft 3in (1m) long

LOCATION Germany

Pterodactylus, meaning "wing finger," is
one of the best-known pterosaurs. Many
complete skeletons have been found,
from small juveniles to large adults.
Pterodactylus was one of the first
pterodactyloid pterosaurs. With shorter
tails and longer necks, pterodactyloids
were better adapted to flight than their
earlier relatives. Unlike most other
pterodactyloids, *Pterodactylus* was
relatively small, with a wingspan of about
3ft 3in (1m). However, there were a number
of different *Pterodactylus* species, and it
can be difficult to distinguish the young of
a large species from the adults of a small
species. *Pterodactylus* lived in the coastal
areas of an inland sea that covered
southern Germany in the Late Jurassic.
It hunted small fish using its long jaws,
which were full of small teeth.

thick knobbly crest

relatively long
neck

Dilophosaurus

GROUP Theropods

DATE Early Jurassic

SIZE 20ft (6m) long

LOCATION USA

One of the few known theropods from the Early Jurassic,
Dilophosaurus was also one of the most striking. It was slender
and lightly built, but it is best known for its remarkable skull.
This had two parallel, platelike crests arranged along the upper
surface of the snout. No one knows why *Dilophosaurus* had these
crests, but the most popular idea is that they were used for sexual
displays. Inside its mouth, there was a prominent notch between
the tip of the upper jaw and the rest of the teeth, and this may
have helped *Dilophosaurus* to grab small prey. It was originally
described as a new species of *Megalosaurus* (see opposite)
and not assigned its own genus until 1970.

bony, semicircular crests

mouth packed
with large teeth

flexible tail

four-fingered
hands

Monolophosaurus

GROUP Theropods

DATE Middle Jurassic

SIZE 20ft (6m) long

LOCATION China

Many large theropods had crests, bumps,
ridges, or horns on their heads, but one of
the biggest and certainly oddest theropod
skulls belonged to *Monolophosaurus*. This
predatory dinosaur had a very thick and
knobbly crest on its head. Rather than being
attached to the top of the skull, as in many
theropods, the crest was integrated into the
whole head. It is hard to say how the crest
functioned. It is very thick and strong at
the top but thin and hollow lower down,
suggesting that as a whole it was weak
and could not resist heavy forces. Only one
specimen of *Monolophosaurus* is known,
and some parts of this are missing—
including the entire tail—so we do not know
a great deal about this animal. However, it
was certainly a top predator and would have
been a capable killer.

long, slim
legs

TWO-RIDGED LIZARD
Dilophosaurus, meaning "two-ridged lizard,"
was named for the striking pair of bony, but
thin and fragile, crests that adorned its head.

three long,
forward-facing,
clawed toes

relatively long
arms with three
clawed fingers

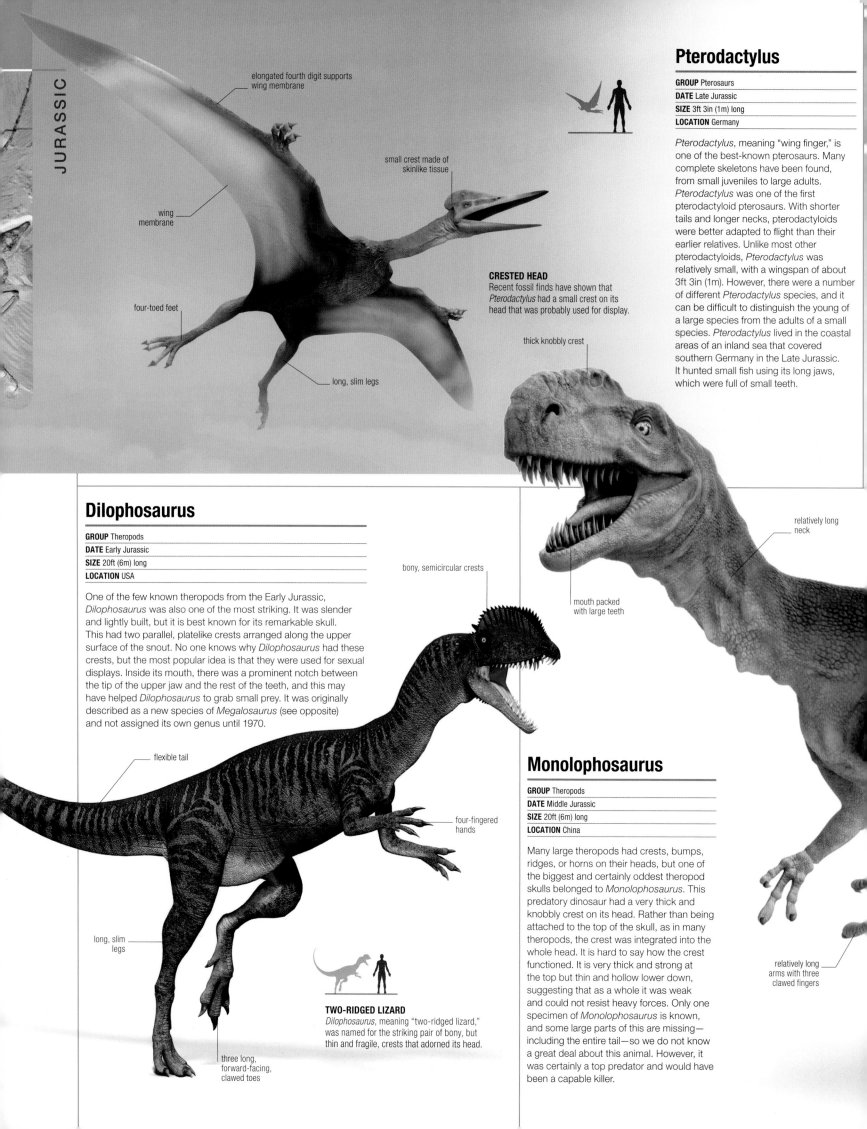

Cryolophosaurus

GROUP Theropods
DATE Early Jurassic
SIZE 21½ft (6.5m) long
LOCATION Antarctica

Discovered in the Transantarctic Mountains during the 1990s, *Cryolophosaurus* was clearly a top predator. Fragments of ankylosaurs, ornithopods, and theropods had previously been found in Antarctica, but *Cryolophosaurus* is currently the only impressive dinosaur specimen known from the continent. *Cryolophosaurus* had a very unusual bony crest above its eyes. This was a thin, sheetlike structure that curved upward and forward over the top of the skull and had parallel lines decorating the front and back. Perhaps a close relative of *Dilophosaurus* (see opposite), *Cryolophosaurus* had a shallow skull and long, slender proportions.

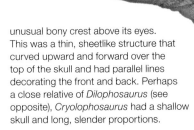

shallow skull

unusual, sheetlike crest

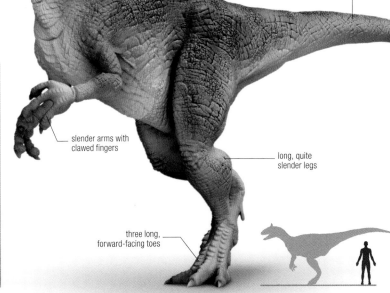

FORMIDABLE SIZE
With an estimated weight of more than 770lb (350kg), *Cryolophosaurus* is one of the largest Early Jurassic theropods. The only known specimen was not fully mature; adults were probably even larger than this.

long, stiff tail

slender arms with clawed fingers

long, quite slender legs

three long, forward-facing toes

SCIENCE
ANTARCTIC FIELDWORK

Cryolophosaurus is one of several dinosaurs that have recently been discovered in the rocks of the Transantarctic Mountains, Antarctica. Although covered by ice and snow today, the rocks on Antarctica surely contain many fossils of species that have yet to be discovered. Fieldwork in Antarctica presents special complications, because people and machinery must be able to withstand the cold, strong winds, and dangerous terrain. Furthermore, the rock containing *Cryolophosaurus* was particularly hard, so proper preparation of its bones could only be carried out in the laboratory.

This dinosaur's name means **"single-crested lizard."** Despite the similar name, it is not a very close relative of *Dilophosaurus*, the **"double-crested lizard."**

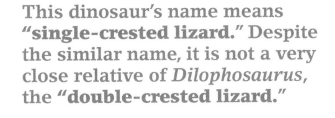

stiff tail outstretched for balance

long, powerful legs

three forward-facing toes

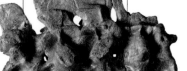

AGILE PREDATOR
A medium-sized theropod, *Monolophosaurus* had long, powerful legs and was probably a fairly agile predator that would have been a threat to other dinosaurs.

Megalosaurus

GROUP Theropods
DATE Middle Jurassic
SIZE 26ft (8m) long
LOCATION England

Megalosaurus, "great lizard," was described in 1824 after several large reptile bones were discovered in Oxfordshire, England. This makes it the very first dinosaur to be recognized by science. Because the fragmentary remains were superficially similar to the bones of large, living lizards, *Megalosaurus* was originally imagined to be a gigantic, four-footed reptile similar to modern monitor lizards. Today we know that *Megalosaurus*, like other large theropods, was a two-legged predator with short arms. Despite the length of time since it was first discovered, *Megalosaurus* remains enigmatic due to the poor fossil record.

imprint left by three-toed foot

row of vertebrae

FOSSILIZED FOOTPRINTS
This is one of a group of fossilized *Megalosaurus* footprints found in a quarry near Oxford, England, in 1997.

SACRAL VERTEBRAE
These vertebrae were among the first *Megalosaurus* fossils to be found. They were identified by geologist William Buckland in 1824.

Gasosaurus

GROUP Theropods
DATE Middle Jurassic
SIZE 11½ft (3.5m) long
LOCATION China

The bones of this carnivore were discovered by chance after the rocks they were encased in were blown apart with dynamite. As a result, they are not in the best condition, and many are missing. Nevertheless, *Gasosaurus* is an important and interesting animal because it comes from the Middle Jurassic, a time from which we have few dinosaur fossils. *Gasosaurus* reveals information about the evolution of some of the earlier predatory dinosaurs. It was one of the larger predators of its time but was too small to threaten many of the giant herbivores that were its contemporaries.

Gasosaurus **takes its name** from the fact that it was found by a company **searching for natural gas.**

bulky body

large jaws with sharp teeth

long, muscular tail

three forward-facing, clawed toes

TYPICAL THEROPOD
This species is thought to have had the typical theropod body shape, with a large head, long legs, and a thick tail. The few fossil fragments include arm, pelvis, and leg bones, but no skull.

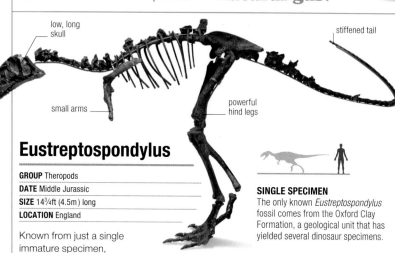

low, long skull

stiffened tail

small arms

powerful hind legs

Eustreptospondylus

GROUP Theropods
DATE Middle Jurassic
SIZE 14¾ft (4.5m) long
LOCATION England

SINGLE SPECIMEN
The only known *Eustreptospondylus* fossil comes from the Oxford Clay Formation, a geological unit that has yielded several dinosaur specimens.

Known from just a single immature specimen, *Eustreptospondylus* is one of Europe's best-preserved large theropods. It had a low, crestless skull, and a shallow notch in the upper jaw—close to the front of the snout—gave it curved margins to the edges of its mouth. The lower jaw was long and slender and had a thickened and deepened tip. These features are like those developed to an extreme in the Cretaceous spinosaurids, which had skulls like those of modern crocodiles. Indeed, some experts have suggested that *Eustreptospondylus* may have foraged on shorelines for carcasses and marine life.

Ceratosaurus

GROUP Theropods
DATE Late Jurassic
SIZE 20ft (6m) long
LOCATION USA, Portugal

Ceratosaurus was a formidable and frightening predator. Its name—"horned reptile"—is a reference to its rounded nasal horn, but it also had tall triangular or rounded horns in front of its eyes. It had a large, deep skull that was well suited for a life of preying on large animals. Unique among theropods, *Ceratosaurus* had a continuous row of flat, bony plates known as scutes running along its neck, back, and tail, which was deep but narrow.

very long upper teeth

four fingers; fourth lacks claw

long foot with reduced back toe

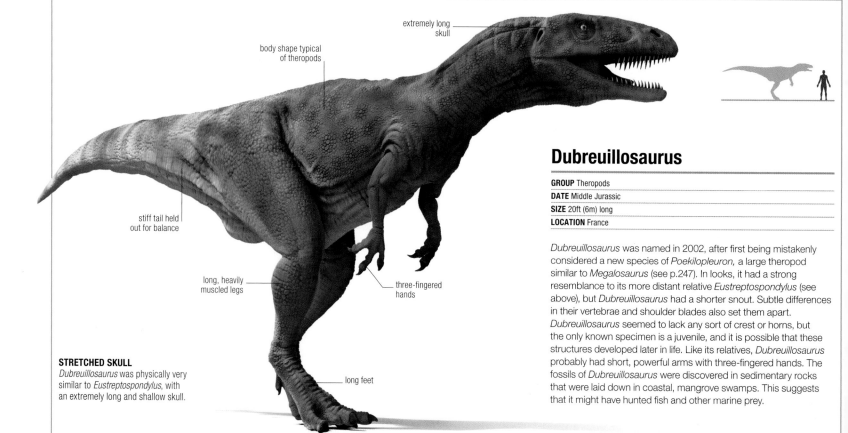

extremely long skull

body shape typical of theropods

Dubreuillosaurus

GROUP Theropods
DATE Middle Jurassic
SIZE 20ft (6m) long
LOCATION France

Dubreuillosaurus was named in 2002, after first being mistakenly considered a new species of *Poekilopleuron,* a large theropod similar to *Megalosaurus* (see p.247). In looks, it had a strong resemblance to its more distant relative *Eustreptospondylus* (see above), but *Dubreuillosaurus* had a shorter snout. Subtle differences in their vertebrae and shoulder blades also set them apart. *Dubreuillosaurus* seemed to lack any sort of crest or horns, but the only known specimen is a juvenile, and it is possible that these structures developed later in life. Like its relatives, *Dubreuillosaurus* probably had short, powerful arms with three-fingered hands. The fossils of *Dubreuillosaurus* were discovered in sedimentary rocks that were laid down in coastal, mangrove swamps. This suggests that it might have hunted fish and other marine prey.

stiff tail held out for balance

long, heavily muscled legs

three-fingered hands

long feet

STRETCHED SKULL
Dubreuillosaurus was physically very similar to *Eustreptospondylus*, with an extremely long and shallow skull.

Allosaurus

GROUP Theropods
DATE Late Jurassic
SIZE 28ft (8.5m) long
LOCATION USA, Portugal

One of the most abundant of all Jurassic theropods, *Allosaurus* is known from many specimens discovered in the Morrison Formation in western USA. A formidable predator, its large skull and jaws held sharp, serrated teeth (see panel), and the three large claws on its hands may have been used to grip the sides of prey. Its powerful jaws and claws, combined with its considerable size, have led most scientists to imagine *Allosaurus* as a predator of stegosaurs, ornithopods, and perhaps sauropods. Evidence for the fact that *Allosaurus* fed on these large plant-eating dinosaurs comes from tooth marks preserved on their bones, although like all predators, it would also have scavenged on dead dinosaurs as well. One *Allosaurus* specimen has a hole in one of its tail vertebrae exactly matching the size and shape of a stegosaur tail spike. On this occasion, it seems that the herbivore succeeded in causing a major injury to the predator.

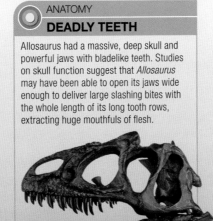

ANATOMY
DEADLY TEETH

Allosaurus had a massive, deep skull and powerful jaws with bladelike teeth. Studies on skull function suggest that *Allosaurus* may have been able to open its jaws wide enough to deliver large slashing bites with the whole length of its long tooth rows, extracting huge mouthfuls of flesh.

massive skull lightened by large openings

tail held outstretched for balance

powerful limbs armed with long claws

RICH FOSSIL RECORD
Allosaurus fossils are particularly abundant in the Morrison Formation, a large tract of sedimentary rock that has proved a rich fossil-hunting ground.

FAMILIAR PREDATOR
Allosaurus had low, parallel ridges running along the top of its snout, and distinctive triangular horns were present in front of its eyes. After *Tyrannosaurus* (see pp.302–303), *Allosaurus* is probably the best-known predatory dinosaur.

Sinraptor

GROUP Theropods
DATE Late Jurassic
SIZE 30ft (9m) long
LOCATION China

Sinraptor means "Chinese hunter," and so far all of its fossils have been excavated in that country. It was a large predator that closely resembled its more famous relative, *Allosaurus* (see p.249). Several specimens are known from different parts of China, and more work

is needed to establish exactly which specimens belong to which species. Some paleontologists think that several animals named as separate species of *Sinraptor* are actually different enough to be placed in their own genus. One thing that is known for certain is that *Sinraptor* occasionally fought with others of its kind. Scientists have found tooth marks on the skull and jaw of one fossil that appear to have been made by another *Sinraptor*.

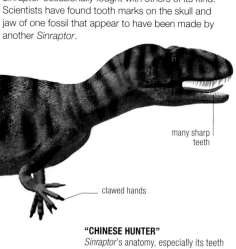

strong, thick tail

long, powerful legs

clawed hands

many sharp teeth

"CHINESE HUNTER"
Sinraptor's anatomy, especially its teeth and claws, indicates that it was a formidable hunter—probably the largest predator in Asia during the Late Jurassic.

Proceratosaurus

GROUP Theropods
DATE Middle Jurassic
SIZE 6½ft (2m) long
LOCATION England

Discovered in Gloucestershire, England, *Proceratosaurus* is known from only one well-preserved skull. At the tip of its snout, the base of a crest is preserved. Because the top of the skull is missing, the shape of this crest is not very well known, but it is thought to have been similar to the crest of *Guanlong* (see below). Alternatively, it may have had a long crest that extended the whole length of its skull.

base of crest

MISTAKEN IDENTITY
Due to its nasal horn, *Proceratosaurus* was named to reflect a close relationship to *Ceratosaurus*, but this relationship was later disproved. It is now recognized as a primitive tyrannosaur, like *Guanlong*.

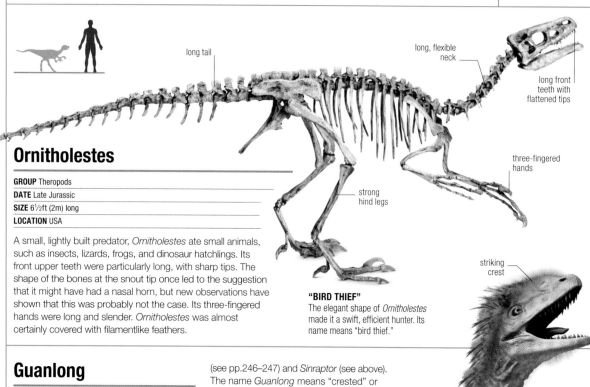

long tail

long, flexible neck

long front teeth with flattened tips

three-fingered hands

strong hind legs

Ornitholestes

GROUP Theropods
DATE Late Jurassic
SIZE 6½ft (2m) long
LOCATION USA

A small, lightly built predator, *Ornitholestes* ate small animals, such as insects, lizards, frogs, and dinosaur hatchlings. Its front upper teeth were particularly long, with sharp tips. The shape of the bones at the snout tip once led to the suggestion that it might have had a nasal horn, but new observations have shown that this was probably not the case. Its three-fingered hands were long and slender. *Ornitholestes* was almost certainly covered with filamentlike feathers.

"BIRD THIEF"
The elegant shape of *Ornitholestes* made it a swift, efficient hunter. Its name means "bird thief."

striking crest

Guanlong

GROUP Theropods
DATE Late Jurassic
SIZE 8¼ft (2.5m) long
LOCATION China

Guanlong is an early form of tyrannosaur, and although it may not look much like the famous *Tyrannosaurus* (see pp.302–303), it has much in common with its fiercer cousin. Among other things, the fused nasal bones on top of the snout and the shape of the hips show that there are evolutionary ties between the two species. *Guanlong* was a relatively small and lightly built predator that would have hunted other small dinosaurs. It was found in the fossil beds of western China, where it may have competed for prey against larger and more ferocious contemporaries, such as *Monolophosaurus*

(see pp.246–247) and *Sinraptor* (see above). The name *Guanlong* means "crested" or "crowned dragon" in Chinese, and the large, thin crest on its head—thought primarily to be a display structure— was made of bone.

Tyrannosaurus-style hips

talons on three-toed feet

three-fingered hands

Guanlong's **striking crest** ran from its nose to the back of its head and was probably used mainly for **display**.

PRIMITIVE TYRANNOSAURUS
Although small in comparison, *Guanlong* shared enough characteristics with *Tyrannosaurus* for scientists to be confident that the two were related.

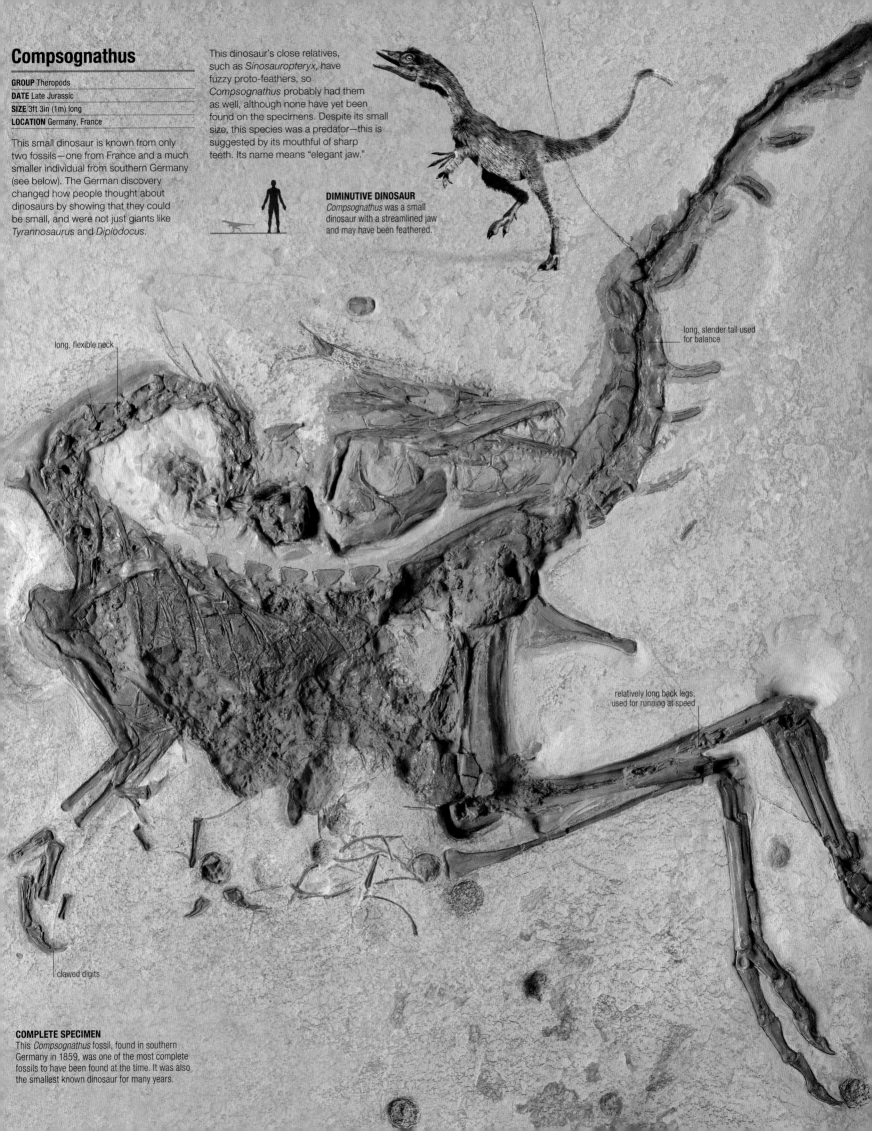

Compsognathus

GROUP Theropods
DATE Late Jurassic
SIZE 3ft 3in (1m) long
LOCATION Germany, France

This small dinosaur is known from only two fossils—one from France and a much smaller individual from southern Germany (see below). The German discovery changed how people thought about dinosaurs by showing that they could be small, and were not just giants like *Tyrannosaurus* and *Diplodocus*.

This dinosaur's close relatives, such as *Sinosauropteryx*, have fuzzy proto-feathers, so *Compsognathus* probably had them as well, although none have yet been found on the specimens. Despite its small size, this species was a predator—this is suggested by its mouthful of sharp teeth. Its name means "elegant jaw."

DIMINUTIVE DINOSAUR
Compsognathus was a small dinosaur with a streamlined jaw and may have been feathered.

long, flexible neck

long, slender tail used for balance

relatively long back legs, used for running at speed

clawed digits

COMPLETE SPECIMEN
This *Compsognathus* fossil, found in southern Germany in 1859, was one of the most complete fossils to have been found at the time. It was also the smallest known dinosaur for many years.

fingers

feather impressions

skull contained true teeth

bony tail

SPECTACULAR FOSSIL
The most spectacular *Archaeopteryx* fossil, and the third of its kind to be found, the Berlin specimen exhibits strikingly clear feather impressions, including those of flight feathers on the wings.

lightly built, feathered body

elongated hands

FEATHERED FLYER
In addition to its wing and tail feathers, *Archaeopteryx* also possessed long feathers on its hind limbs, and its neck and body were fully feathered, too.

feathered hind legs

four-toed feet

Archaeopteryx

GROUP	Theropods
DATE	Late Jurassic
SIZE	12in (30cm) long
LOCATION	Germany

Archaeopteryx, meaning "ancient wing," was discovered in the Jurassic limestone of Solnhofen in Germany in 1859. Perhaps the most famous fossil in the world, the first specimen—now known as the London Specimen—was disarticulated and incomplete. At least 11 specimens are now known, and some are amazingly well preserved. Like modern birds, *Archaeopteryx* had long wing and tail feathers, and this is why it has always been regarded as the first bird. However, we now know that complex feathers and other birdlike features were widespread among the meat-eating dinosaurs known as maniraptorans. While *Archaeopteryx* remains recognized as one of the earliest ancestors of modern birds, many maniraptorans were almost as birdlike.

Archaeopteryx was found two years after Charles Darwin published his *On the Origin of Species* and provided powerful support for the theory of evolution by natural selection.

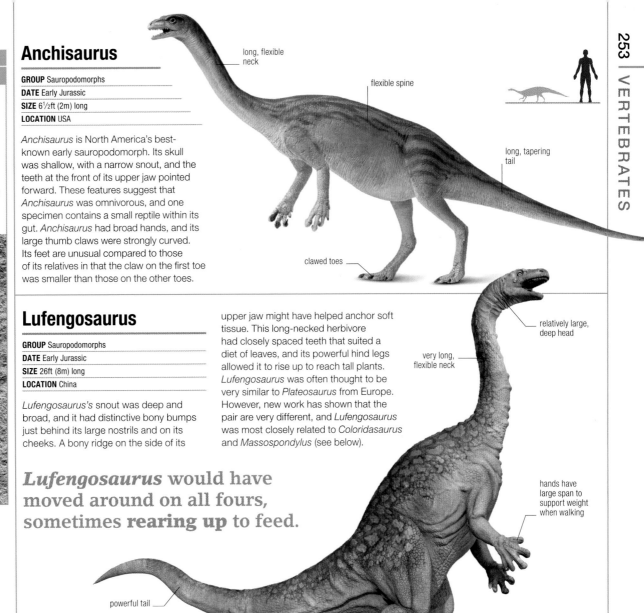

Anchisaurus

GROUP Sauropodomorphs
DATE Early Jurassic
SIZE 6½ft (2m) long
LOCATION USA

Anchisaurus is North America's best-known early sauropodomorph. Its skull was shallow, with a narrow snout, and the teeth at the front of its upper jaw pointed forward. These features suggest that *Anchisaurus* was omnivorous, and one specimen contains a small reptile within its gut. *Anchisaurus* had broad hands, and its large thumb claws were strongly curved. Its feet are unusual compared to those of its relatives in that the claw on the first toe was smaller than those on the other toes.

long, flexible neck

flexible spine

long, tapering tail

clawed toes

Lufengosaurus

GROUP Sauropodomorphs
DATE Early Jurassic
SIZE 26ft (8m) long
LOCATION China

Lufengosaurus's snout was deep and broad, and it had distinctive bony bumps just behind its large nostrils and on its cheeks. A bony ridge on the side of its upper jaw might have helped anchor soft tissue. This long-necked herbivore had closely spaced teeth that suited a diet of leaves, and its powerful hind legs allowed it to rise up to reach tall plants. *Lufengosaurus* was often thought to be very similar to *Plateosaurus* from Europe. However, new work has shown that the pair are very different, and *Lufengosaurus* was most closely related to *Coloridasaurus* and *Massospondylus* (see below).

relatively large, deep head

very long, flexible neck

hands have large span to support weight when walking

Lufengosaurus would have moved around on all fours, sometimes **rearing up** to feed.

powerful tail

long, clawed toes

three-fingered claws

feathered wings for effective flight

long, flexible neck

bulky body

EMBRYO FOSSIL
This *Massospondylus* embryo skeleton, measuring 6in (15cm) long, was found preserved inside its egg.

Massospondylus

GROUP Sauropodomorphs
DATE Early Jurassic
SIZE 16½ft (5m) long
LOCATION South Africa

Massospondylus is known from several complete skeletons and skulls and even some eggs containing embryos. It was a medium-sized, broad-skulled sauropodomorph with large eye sockets. Its broad, five-fingered hands had particularly large, curved thumb claws. At the front of its ribcage, it had two separate clavicles that met each other in a V-shape similar to the wishbone in birds. It had been thought that *Massospondylus* was able to walk on all fours. However, recent work on the mechanics of its short forearms suggests that it moved around all of the time on its hind legs.

large, curved thumb claw

long, muscular hind legs

long tail with thin, whiplash end

Vulcanodon

GROUP Sauropodomorphs
DATE Early Jurassic
SIZE 23ft (7m) long
LOCATION Zimbabwe

Vulcanodon was an early sauropod known from a single, partial skeleton without a skull. However, like other early sauropods it probably had a deep, blunt-snouted skull with leaf-shaped teeth. *Vulcanodon*'s feet were short, and large claws were present, at least on the inside toes. In contrast to later sauropods, the bones of its first toe were quite long. Two of the claws on each foot were broad and nail-like, not deep and narrow as they were in other sauropods. *Vulcanodon* was originally thought to have been a close relative of the sauropods, but not actually a sauropod in the true sense. However, it has now been established that it was indeed a true sauropod. One of its closest relatives is *Tazoudasaurus*, and both animals have been united in the family Vulcanodontidae.

nostrils positioned high on nose

long, slim neck

bulky, sloping body

relatively long front legs

short feet

EARLY SAUROPOD
An early sauropod, *Vulcanodon* was a specialized quadruped with long, columnlike limbs. The first skeleton was found in sandstone sandwiched between two layers of volcanic rock—after which it is named.

Barapasaurus

GROUP Sauropodomorphs
DATE Early Jurassic
SIZE 59ft (18m) long
LOCATION India

Barapasaurus was originally named for just a sacrum (part of the backbone), but a large number of additional remains are now also thought to belong to it. Its skull is unknown, although isolated teeth have been found. These are broad at the tip but have a narrower base, and there are coarse serrations either side of the crown. *Barapasaurus* had particularly slender limbs for a sauropod. Its neck vertebrae were long but its body vertebrae were quite compressed. The shape of its vertebrae is unique among sauropods, leading some experts to suggest that it was an unusual side-branch in sauropod evolution.

short, deep head

neck made up of long vertebrae

heavy, bulky body

relatively slender legs

SHORT, DEEP HEAD
Although fossils of most parts of *Barapasaurus*'s body have been found, exceptions are the skull and feet. Paleontologists think that its head was relatively short and deep, like skulls from other primitive sauropods.

Shunosaurus

GROUP Sauropodomorphs
DATE Middle Jurassic
SIZE 32ft (10m) long
LOCATION China

Shunosaurus is the most abundant dinosaur in a fossil assemblage that has been named after it—the "*Shunosaurus* fauna." *Shunosaurus*'s neck was quite short compared with that of many other sauropods, but it does appear to have been quite flexible. Its skull was deep seen from the side but narrow looked at from above. *Shunosaurus* had more teeth than most other sauropods, with 25 or 26 being present in each half of the lower jaw.

relatively short, flexible neck

ALMOST COMPLETE
Shunosaurus is known from many specimens including several almost complete skeletons.

This dinosaur had the **longest neck** of any known animal. One specimen had a neck more than **43ft (13m) long.**

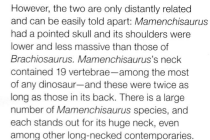

tiny head

enormously elongated neck

back slopes from shoulders

long front legs

Mamenchisaurus

GROUP Sauropodomorphs
DATE Late Jurassic
SIZE 85ft (26m) long
LOCATION China

At first glance, one could mistake the front half of *Mamenchisaurus* with *Brachiosaurus* (see opposite), as they both have domed foreheads and enormously long necks. Both dinosaurs certainly used their necks to reach for food.

CLUBLESS TAIL
Older reconstructions and illustrations of *Mamenchisaurus* sometimes show a club on the tail. This in now known to be a mistake.

However, the two are only distantly related and can be easily told apart: *Mamenchisaurus* had a pointed skull and its shoulders were lower and less massive than those of *Brachiosaurus*. *Mamenchisaurus*'s neck contained 19 vertebrae—among the most of any dinosaur—and these were twice as long as those in its back. There is a large number of *Mamenchisaurus* species, and each stands out for its huge neck, even among other long-necked contemporaries.

Brachiosaurus

GROUP	Sauropodomorphs
DATE	Late Jurassic
SIZE	75ft (23m)
LOCATION	USA, Tanzania

Unlike most sauropods, *Brachiosaurus* had very long front legs and a tall, erect neck. These features gave *Brachiosaurus* a huge reach. Without needing to stretch, a large individual could probably reach more than 49ft (15m) up into trees to feed. Despite what some early illustrations show, this herbivore probably could not rear up on its hind legs. Most of its weight was carried on its huge front limbs, so it would have been hard to take its full weight on its hind legs and stay balanced. In any case, *Brachiosaurus* was so tall it probably had little need to rear up. It could have reached higher to feed than almost any other dinosaur.

Brachiosaurus had a large skull, even when compared to similarly immense sauropods. It had a distinctive bar of bone in the middle of its forehead that created a large bulge on the top of its head. Inside the skull, this bar separates the two openings of the nostrils. It was once thought that the nostrils were large holes located high up on the skull, but recent research suggests that they were in fact relatively small and located closer to the front of the head.

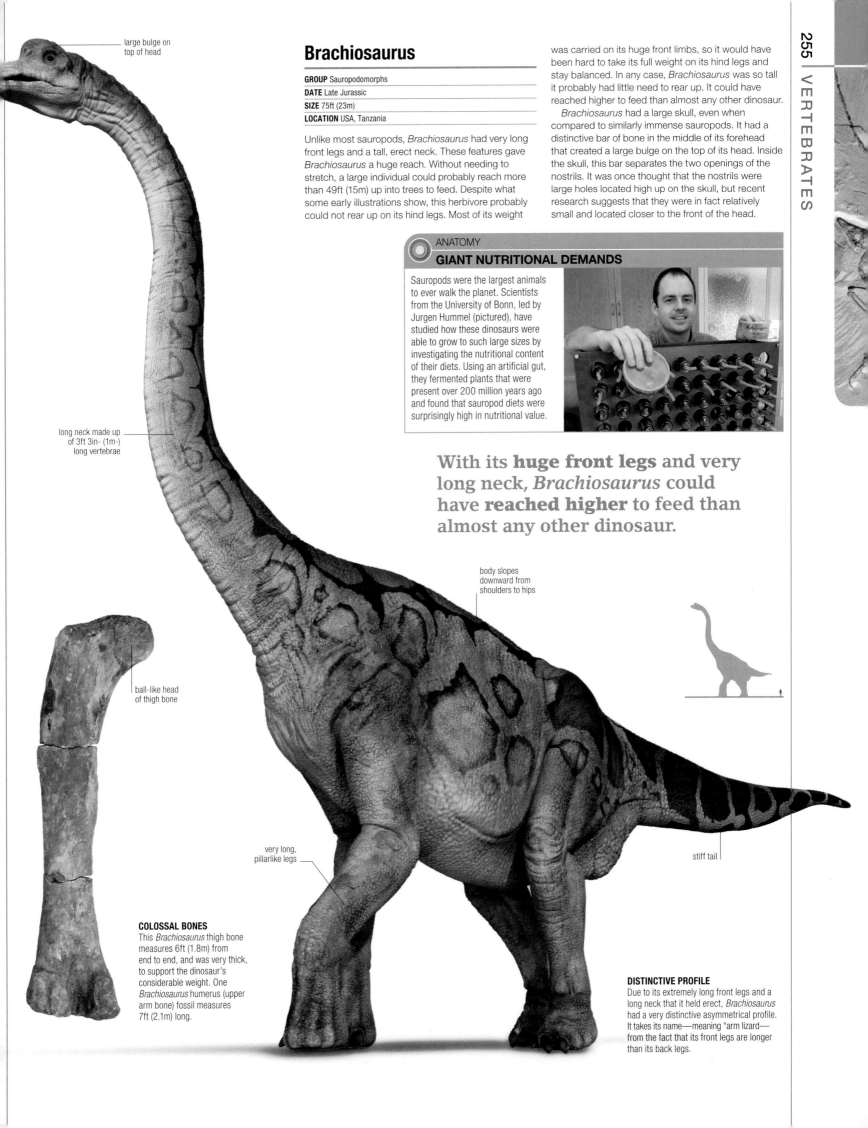

large bulge on top of head

long neck made up of 3ft 3in- (1m-) long vertebrae

ball-like head of thigh bone

body slopes downward from shoulders to hips

very long, pillarlike legs

stiff tail

ANATOMY
GIANT NUTRITIONAL DEMANDS

Sauropods were the largest animals to ever walk the planet. Scientists from the University of Bonn, led by Jurgen Hummel (pictured), have studied how these dinosaurs were able to grow to such large sizes by investigating the nutritional content of their diets. Using an artificial gut, they fermented plants that were present over 200 million years ago and found that sauropod diets were surprisingly high in nutritional value.

With its **huge front legs** and very long neck, *Brachiosaurus* could have **reached higher** to feed than almost any other dinosaur.

COLOSSAL BONES
This *Brachiosaurus* thigh bone measures 6ft (1.8m) from end to end, and was very thick, to support the dinosaur's considerable weight. One *Brachiosaurus* humerus (upper arm bone) fossil measures 7ft (2.1m) long.

DISTINCTIVE PROFILE
Due to its extremely long front legs and a long neck that it held erect, *Brachiosaurus* had a very distinctive asymmetrical profile. It takes its name—meaning "arm lizard—from the fact that its front legs are longer than its back legs.

huge eye
sockets

very large nasal
chambers

small
braincase

long, flat lower jaw

ELONGATED NECK
Diplodocus's neck contained 15 massive
vertebrae and made up a considerable
proportion of its total length. Whether the
neck was held vertically or horizontally
has been a subject of much debate.

Diplodocus

GROUP	Sauropodomorphs
DATE	Late Jurassic
SIZE	82ft (25m) long
LOCATION	USA

One of the best known of all dinosaurs,
Diplodocus was like other diplodocids in
having 15 neck vertebrae, proportionally
short forelimbs, and a whiplike tip
to its tail. Seen from above, its skull
is rectangular and ends with a broad,
squared-off mouth. *Diplodocus* had

triangular spines along its back, and this
was possibly the case in all sauropods.
Studies of tooth wear suggest that
Diplodocus used a feeding strategy
known as unilateral branch stripping:
a branch was gripped between its
peglike teeth, its head was pulled
sharply upward or downward, and
as a result, either the upper or lower
tooth row stripped the foliage off the
branch. Two species of *Diplodocus* are
currently recognized: *D. carnegii* and *D.
hallorum*, also known as *Seismosaurus*.

DOUBLE BEAM
The long tail consisted of up to 80
vertebrae. The name *Diplodocus*,
meaning "double beam," comes from
the presence of chevron-shaped
bones under the tail vertebrae.

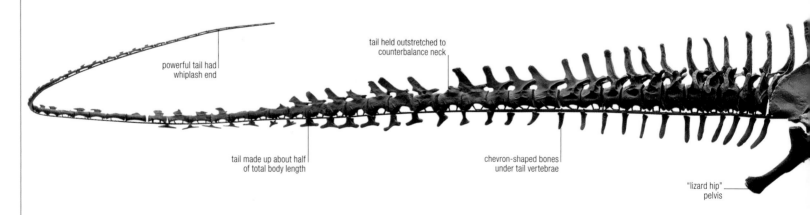

powerful tail had
whiplash end

tail held outstretched to
counterbalance neck

tail made up about half
of total body length

chevron-shaped bones
under tail vertebrae

"lizard hip"
pelvis

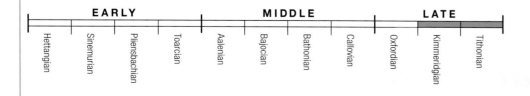

EARLY				MIDDLE				LATE		
Hettangian	Sinemurian	Pliensbachian	Toarcian	Aalenian	Bajocian	Bathonian	Callovian	Oxfordian	Kimmeridgian	Tithonian

IMPRESSIVE PROPORTIONS
Diplodocus was one of the longest of all
dinosaurs, possibly reaching up to 98ft (30m).
Its tail, which made up about half its total length,
counterbalanced an extremely long neck.

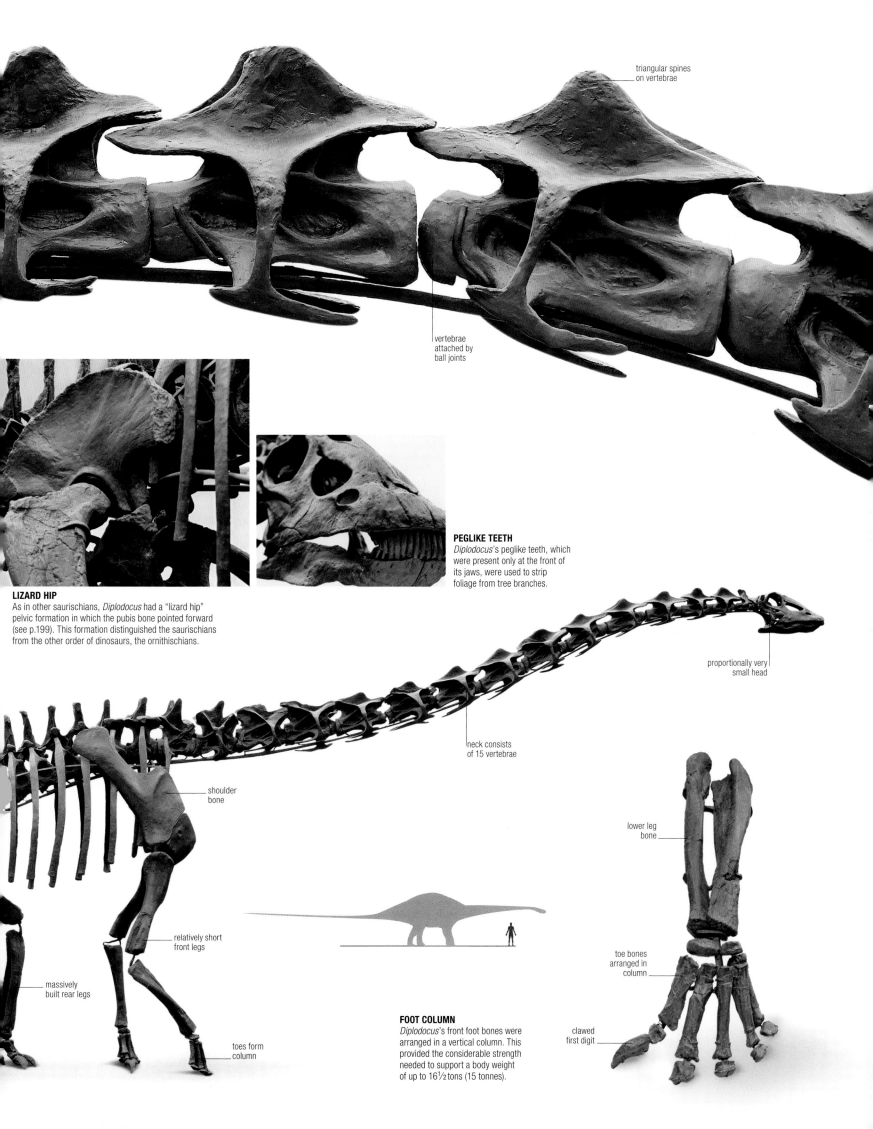

triangular spines
on vertebrae

vertebrae
attached by
ball joints

PEGLIKE TEETH
Diplodocus's peglike teeth, which
were present only at the front of
its jaws, were used to strip
foliage from tree branches.

LIZARD HIP
As in other saurischians, *Diplodocus* had a "lizard hip"
pelvic formation in which the pubis bone pointed forward
(see p.199). This formation distinguished the saurischians
from the other order of dinosaurs, the ornithischians.

proportionally very
small head

neck consists
of 15 vertebrae

shoulder
bone

lower leg
bone

relatively short
front legs

massively
built rear legs

toes form
column

toe bones
arranged in
column

FOOT COLUMN
Diplodocus's front foot bones were
arranged in a vertical column. This
provided the considerable strength
needed to support a body weight
of up to 16½ tons (15 tonnes).

clawed
first digit

Barosaurus

GROUP Sauropodomorphs
DATE Late Jurassic
SIZE 92ft (28m) long
LOCATION USA

Barosaurus was similar to *Diplodocus* (see p.256) in many ways. Among other features, it shared distinctive hollows in its tail vertebrae and almost identical limbs. They differed primarily in the length of their neck vertebrae. *Barosaurus* probably had 16, which is one more than other diplodocids, and they were about a third longer than those of *Diplodocus*. This suggests that *Barosaurus* probably had a larger feeding range than its shorter relative. For some time, fossils found in Tanzania were thought to belong to *Barosaurus*. It has now been confirmed that they are two different, but very similar, diplodocids: *Tornieria* and *Australodocus*.

long, slender neck

short, deep chest cavity

whiplike tail tip

huge muscles at the tail base

stout, columnlike forelimbs

LONG REACH
Thanks to its particularly long neck, *Barosaurus* was able to reach higher into trees and access more food than its close relatives.

robust neck

box-shaped head

Apatosaurus

GROUP Sauropodomorphs
DATE Late Jurassic
SIZE 75ft (23m) long
LOCATION USA

Together with "Brontosaurus," *Apatosaurus* formed the diplodocid subfamily Apatosaurinae. Like other diplodocids, its tail had large muscles at the base, while the tail tip was slender and whiplike. Many old restorations of *Apatosaurus* give it a box-shaped skull, but a true *Apatosaurus* skull, as described in 1978, shows that its head was long and rectangular—much like that of *Diplodocus*, but broader. In general, all apatosaurines had thicker legs and were stockier and more heavily constructed than other diplodocids.

HEAVYWEIGHT
The apatosaurine subfamily included several species of *Apatosaurus* and its close relative "Brontosaurus." A large *Apatosaurus* weighed as much as four elephants.

Dicraeosaurus

GROUP Sauropodomorphs
DATE Late Jurassic
SIZE 39ft (12m) long
LOCATION Tanzania

Dicraeosaurus is a member of a group of diplodocoid sauropods called the dicraeosaurids. These dinosaurs were small compared to other sauropods and had relatively short necks. The neck contained 12 unusually short vertebrae, so it could probably browse vegetation only from ground level to a height of around 9¾ft (3m).

relatively short neck

bony ridge along back

Camarasaurus

GROUP Sauropodomorphs
DATE Late Jurassic
SIZE 59ft (18m) long
LOCATION USA

Camarasaurus is the most common North American sauropod. Its neck was broad, but not as long as the necks of many sauropods. This indicates that it probably fed on vegetation growing at various heights. It had a broad skull with short, strong jaws and stout, spoon-shaped teeth that could have enabled it to feed on even the coarsest vegetation. *Camarasaurus* is one of the most primitive members of the macronarians, the sauropod group that also includes the brachiosaurs and titanosaurs. Like many macronarians, *Camarasaurus*'s skull had particularly large nostril openings. Its vertebrae also contained large openings that housed air sacs connected to its lungs. It is from these chambers that the dinosaur gets its name: "chambered lizard."

sturdy neck

massive, heavy body

thick, strong legs

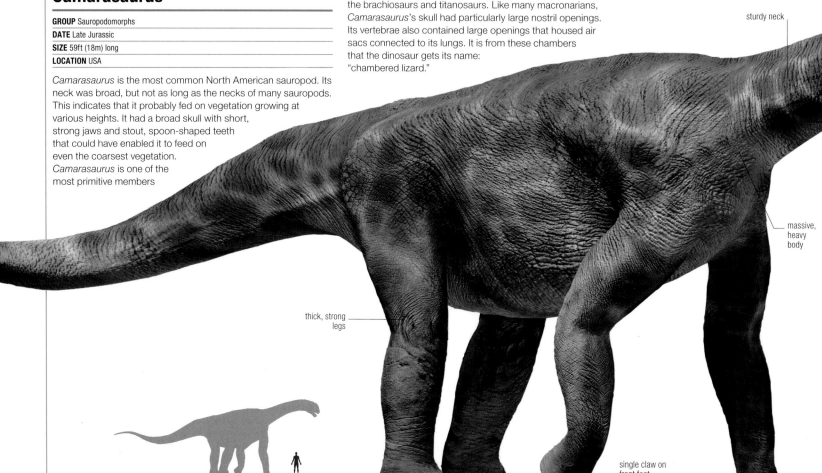

single claw on front feet

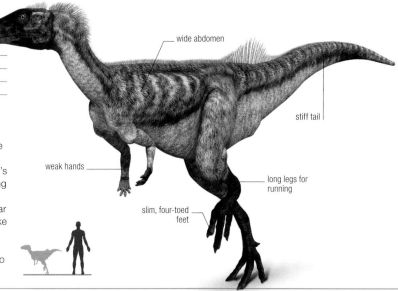

Scelidosaurus

GROUP Ornithischians
DATE Early Jurassic
SIZE 9¾ft (3m) long
LOCATION England

Discovered in the 1850s, *Scelidosaurus* was one of the first dinosaurs to be found as a complete skeleton. This original find is still one of the best preserved European dinosaurs. It had stout limbs and walked on all fours. Rows of oval armor plates ran along its neck, body, and limbs, and it had hooked spikes down the sides of its limbs. Because all the fossils have been found in marine rocks, it has been suggested that this dinosaur may have been an island-dweller or lived on the coast.

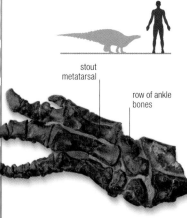

stout metatarsal

row of ankle bones

DINOSAUR TOES
Scelidosaurus had four toes, each tipped with a blunt claw. Its toes were longer than those of the ankylosaurs and stegosaurs.

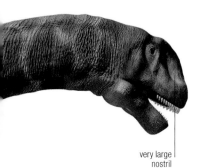

very large nostril

The name "chambered lizard" comes from the large air sacs that were housed in the vertebrae of *Camarasaurus*.

ABUNDANT SAUROPOD
Camarasaurus was a robust, broad-skulled sauropod with spoon-shaped teeth. It was the most abundant North American sauropod and is better understood than any other dinosaur of this type.

Lesothosaurus

GROUP Ornithischians
DATE Early Jurassic
SIZE 3ft 3in (1m) long
LOCATION Lesotho

Lesothosaurus is one of the earliest members of the ornithischians. It had long, slender hind limbs and small forelimbs with hands that would not have been able to grasp properly. Like all ornithischians, the tips of *Lesothosaurus*'s upper and lower jaws were horny, forming a beaklike structure. Behind the beak, leaf-shaped teeth lined the jaws, and near the front of the upper jaw were 12 fanglike teeth. Analysis of its teeth have shown that *Lesothosaurus* sliced up plant material with its beak and was not able to chew its food.

wide abdomen

stiff tail

weak hands

long legs for running

slim, four-toed feet

Heterodontosaurus

GROUP Ornithischians
DATE Early Jurassic
SIZE 3ft 3in (1m) long
LOCATION South Africa

Heterodontosaurus is a member of a small group of peculiar ornithischians called heterodontosaurids. Unlike virtually all other ornithischians, these animals had long, grasping hands with strongly curved claws. *Heterodontosaurus*, meaning "different-toothed lizard," had three different tooth shapes. Small, incisorlike teeth were present at the front of the upper jaw, and blunter, chisel-shaped teeth were positioned farther back. Most obvious were the large, fanglike teeth in both the upper and lower jaws. In addition, as in all ornithischians, it had a beak at the front of the upper and lower jaws. *Heterodontosaurus* was probably herbivorous, but its strong jaws, large fangs, and grasping hands suggest that it may also have eaten small animals. Its long, slender hind limbs also suggest that it was a fast runner.

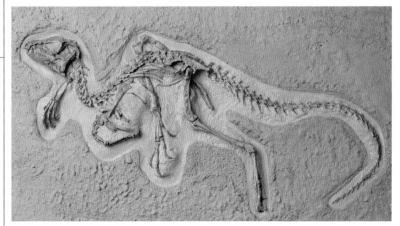

BIPEDAL
This excellent, articulated skeleton shows that the hind limbs were much longer than the arms, which suggests it moved around on its hind legs.

ANATOMY
WORN TEETH

The top surfaces of *Heterodontosaurus*'s teeth are heavily worn down—clearly, it ate hard, abrasive foods. Other dinosaurs that experienced heavy tooth wear had numerous replacement teeth: as a tooth became worn down, a new one emerged from the jaw. Surprisingly, replacement teeth are usually absent in *Heterodontosaurus*, showing that tooth replacement occurred only very rarely.

Scutellosaurus

GROUP Ornithischians
DATE Early Jurassic
SIZE 3ft 3in (1m) long
LOCATION USA

Scutellosaurus is one of the oldest and most primitive thyreophorans, the ornithischian group that later included the armored ankylosaurs and plated stegosaurs. It was lightly built and probably capable of walking on its hind legs. Like other thyreophorans, *Scutellosaurus* had rows of armor plates along its body and tail. These formed parallel rows, with as many as five rows on each side. It also had a double row of scutes, or external plates, running neck to tail.

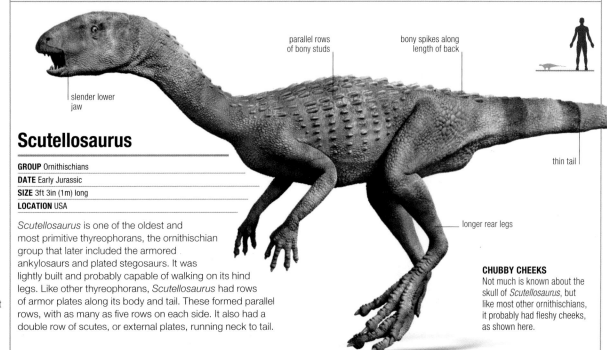

slender lower jaw

parallel rows of bony studs

bony spikes along length of back

thin tail

longer rear legs

CHUBBY CHEEKS
Not much is known about the skull of *Scutellosaurus*, but like most other ornithischians, it probably had fleshy cheeks, as shown here.

Stegosaurus

GROUP Ornithischians
DATE Late Jurassic
SIZE 30ft (9m) long
LOCATION USA, Portugal

Stegosaurus—the "plated lizard"—is the best-known stegosaur, and was the first member of that group to be named. Its bizarre, diamond-shaped plates have been the cause of much disagreement, but recent work has shown that they were arranged in two staggered rows that ran along the neck, back, and tail. In most other stegosaurs, the plates were paired, so *Stegosaurus* was unusual in that respect. It has often been suggested that the plates were used in defense or in controlling body temperature. However, their position on the body makes a defensive role unlikely, while a role in temperature regulation also seems unlikely because the plates have the same anatomy as the armor plates that covered the bodies of other dinosaurs. The most likely role for the plates is that they were used in display. Small, rounded bones called ossicles covered the throat region of *Stegosaurus*, and two pairs of long spikes that projected from the tip of the tail were almost certainly used in defense.

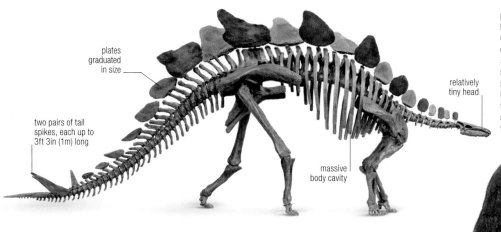

plates graduated in size

two pairs of tail spikes, each up to 3ft 3in (1m) long

relatively tiny head

massive body cavity

The bizarre, **diamond-shaped plates** along the back of *Stegosaurus* were probably **used in display** rather than defense.

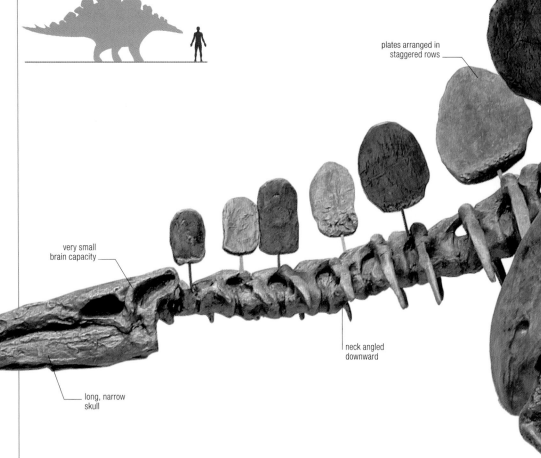

plates arranged in staggered rows

very small brain capacity

neck angled downward

long, narrow skull

relatively short forelegs enabled browsing

EARLY				MIDDLE				LATE		
Hettangian	Sinemurian	Pliensbachian	Toarcian	Aalenian	Bajocian	Bathonian	Callovian	Oxfordian	Kimmeridgian	Tithonian

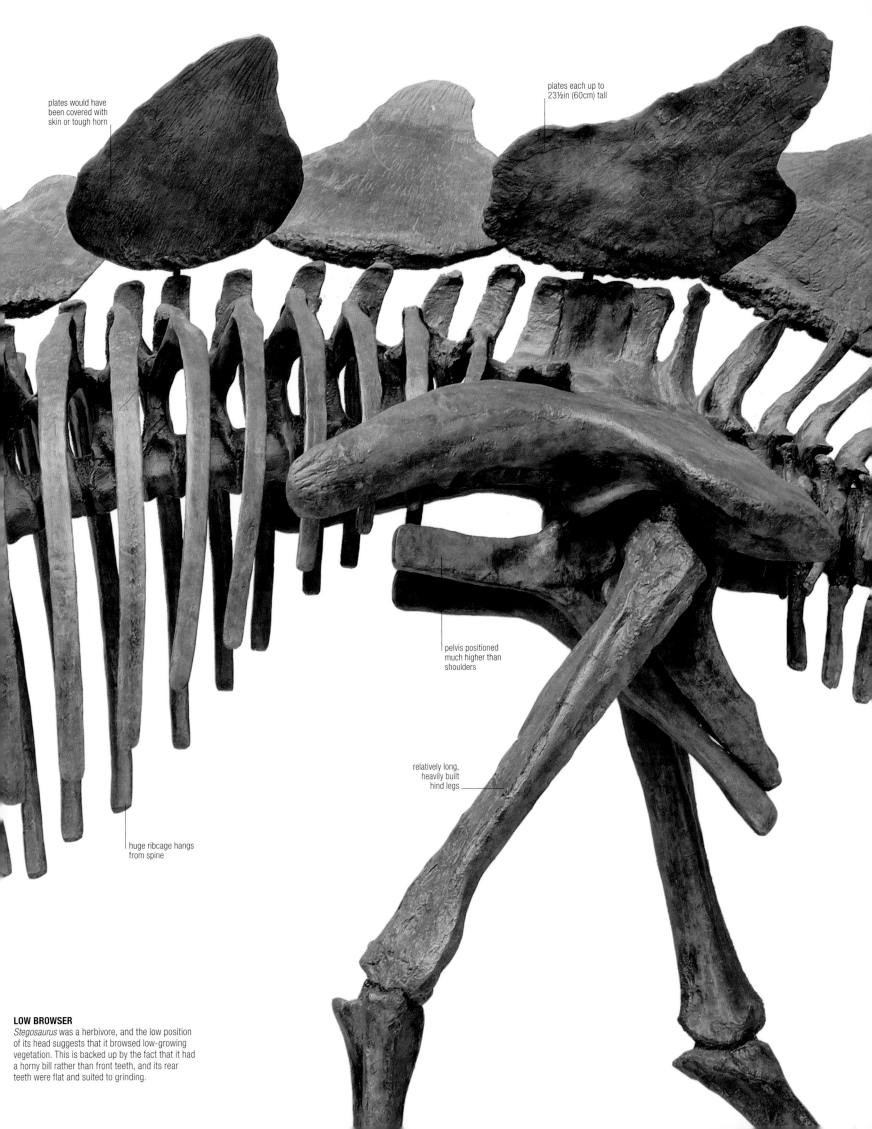

plates would have
been covered with
skin or tough horn

plates each up to
23½in (60cm) tall

pelvis positioned
much higher than
shoulders

relatively long,
heavily built
hind legs

huge ribcage hangs
from spine

LOW BROWSER
Stegosaurus was a herbivore, and the low position
of its head suggests that it browsed low-growing
vegetation. This is backed up by the fact that it had
a horny bill rather than front teeth, and its rear
teeth were flat and suited to grinding.

Huayangosaurus

GROUP Ornithischians
DATE Middle Jurassic
SIZE 13ft (4m) long
LOCATION China

Huayangosaurus is one of the more primitive members of Stegosauria. It differed from the more advanced members of this group in having teeth at the front of the upper jaw. Its hips were also different from those of later stegosaurs, as was its snout—other stegosaurs had long, slim snouts, but the skull of *Huayangosaurus* was relatively short and broad. In some specimens, small horns were present above the eyes. However, these horns may also be absent in some *Huayangosaurus* skulls, so it is possible that they only appeared in adulthood, or were restricted to males or females. A large spike was present on each shoulder. Such spikes were typical for stegosaurs; the species that lacked them were unusual. In addition to the plates and spikes that ran along the backbone and tail of *Huayangosaurus*, several large armor plates were also arranged along both sides of the body for protection.

Huayangosaurus was **small** for a **stegosaur**.

KEY SITE

DASHANPU FORMATION

Huayangosaurus was discovered at Dashanpu Quarry in Sichuan Province, China. This site includes both Middle and Upper Jurassic sediments and is one of the most famous dinosaur-bearing locations in the world. Over 8,000 specimens have been collected from the site and the discoveries have revolutionized our understanding of dinosaur diversity and evolution. In addition to *Huayangosaurus*, Dashanpu Quarry has yielded numerous sauropods, theropods, and small, bipedal ornithischians. Plesiosaurs, pterosaurs, and other fossil reptiles have also been discovered. Many of the fossils are carcasses that collected over the centuries in a gigantic lake.

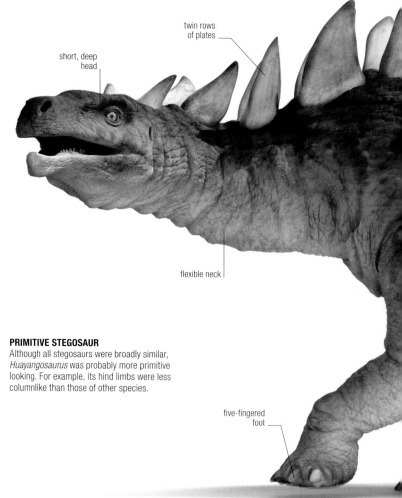

twin rows of plates

short, deep head

flexible neck

PRIMITIVE STEGOSAUR
Although all stegosaurs were broadly similar, *Huayangosaurus* was probably more primitive looking. For example, its hind limbs were less columnlike than those of other species.

five-fingered foot

The group of **spikes** at the end of a **stegosaur tail** has been nicknamed the **thagomizer**.

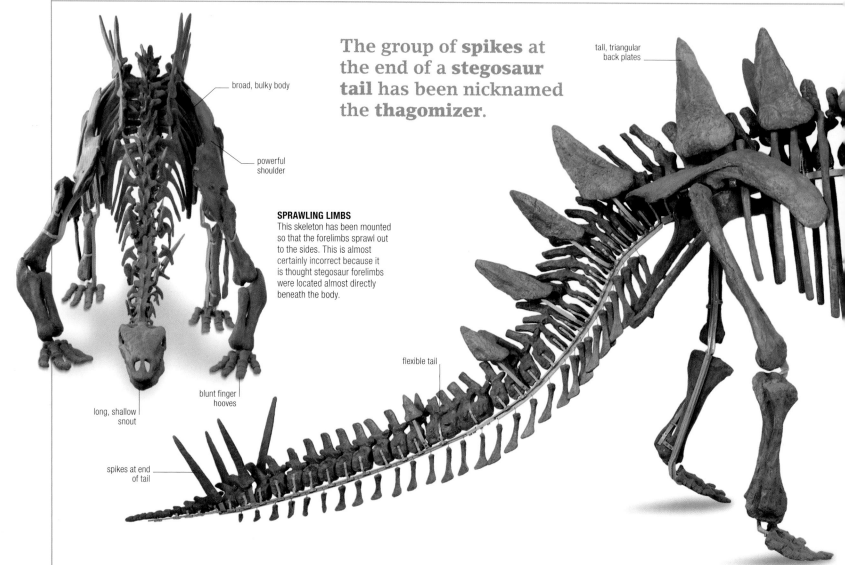

broad, bulky body

powerful shoulder

tall, triangular back plates

SPRAWLING LIMBS
This skeleton has been mounted so that the forelimbs sprawl out to the sides. This is almost certainly incorrect because it is thought stegosaur forelimbs were located almost directly beneath the body.

long, shallow snout

blunt finger hooves

flexible tail

spikes at end of tail

conical tail
plate

shoulder spike

large tail
muscles

tail spike

short foot

Tuojiangosaurus

GROUP Ornithischians

DATE Late Jurassic

SIZE 23ft (7m) long

LOCATION China

Tuojiangosaurus is one of the most complete stegosaurs ever discovered. It is known from a good partial skeleton, as well as the fragmentary remains of several other individuals. *Tuojiangosaurus* has often been illustrated with a long, shallow skull, similar to that of *Stegosaurus*, but unfortunately too few skull bones have been found to show exactly what the skull was really like. We do know, however, that some of the bones on the top of the skull were particularly broad. Spines were present at the end of the tail and armor plates lined its neck and back. As seems to have been the case for all stegosaurs except *Stegosaurus*, these plates were arranged in pairs. Those located over the hips were narrow and spinelike, whereas those from the neck and front part of the back were broader and flattened. Recent studies indicate that *Tuojiangosaurus* may be closely related to the African stegosaur *Paranthodon*, which is known from only a partial skull, and perhaps also to *Loricatosaurus*, found in England.

BIOGRAPHY

DONG ZHIMING

Tuojiangosaurus is one of almost 40 dinosaurs named and described by Dong Zhiming, one of China's most influential paleontologists. Initially working under the great pioneer of Chinese vertebrate paleontology Yang Zhong-jian (also known as C. C. Young), Dong was involved in the discovery of the Dashanpu Quarry in Sichuan Province in the 1970s, naming and describing many of the dinosaurs discovered there.

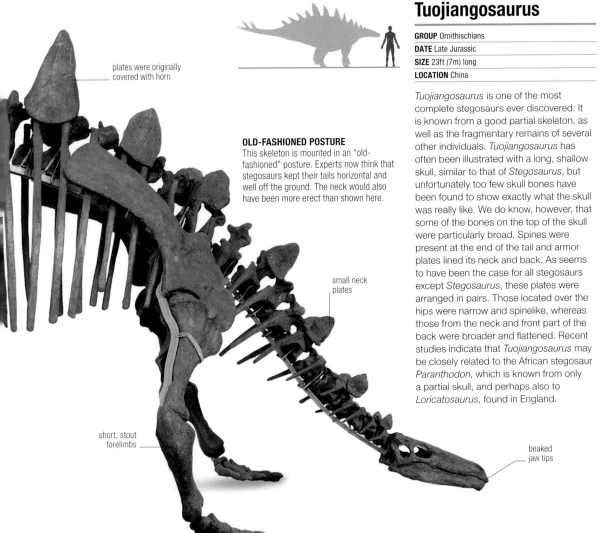

plates were originally
covered with horn

OLD-FASHIONED POSTURE
This skeleton is mounted in an "old-fashioned" posture. Experts now think that stegosaurs kept their tails horizontal and well off the ground. The neck would also have been more erect than shown here.

small neck
plates

short, stout
forelimbs

beaked
jaw tips

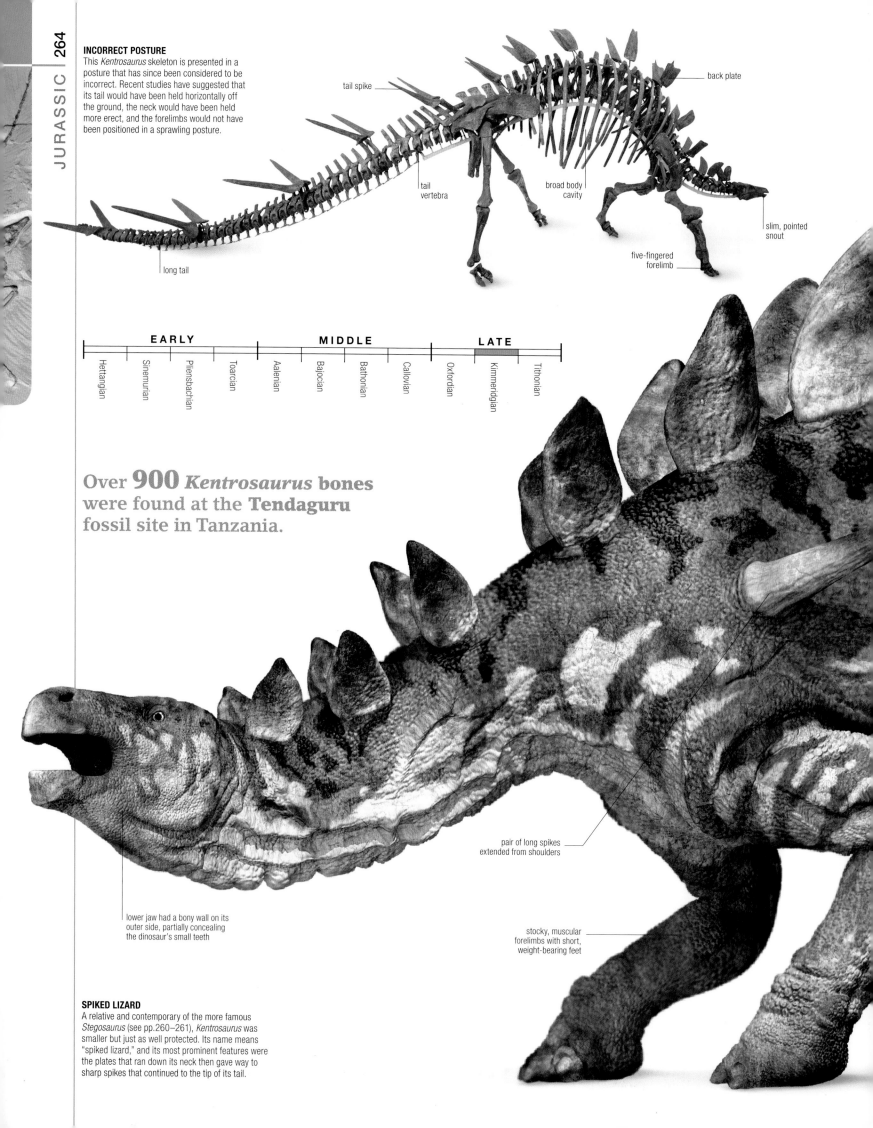

INCORRECT POSTURE
This *Kentrosaurus* skeleton is presented in a posture that has since been considered to be incorrect. Recent studies have suggested that its tail would have been held horizontally off the ground, the neck would have been held more erect, and the forelimbs would not have been positioned in a sprawling posture.

tail spike

back plate

tail vertebra

broad body cavity

slim, pointed snout

long tail

five-fingered forelimb

EARLY				MIDDLE				LATE		
Hettangian	Sinemurian	Pliensbachian	Toarcian	Aalenian	Bajocian	Bathonian	Callovian	Oxfordian	Kimmeridgian	Tithonian

Over **900** *Kentrosaurus* bones were found at the **Tendaguru** fossil site in Tanzania.

pair of long spikes extended from shoulders

lower jaw had a bony wall on its outer side, partially concealing the dinosaur's small teeth

stocky, muscular forelimbs with short, weight-bearing feet

SPIKED LIZARD
A relative and contemporary of the more famous *Stegosaurus* (see pp.260–261), *Kentrosaurus* was smaller but just as well protected. Its name means "spiked lizard," and its most prominent features were the plates that ran down its neck then gave way to sharp spikes that continued to the tip of its tail.

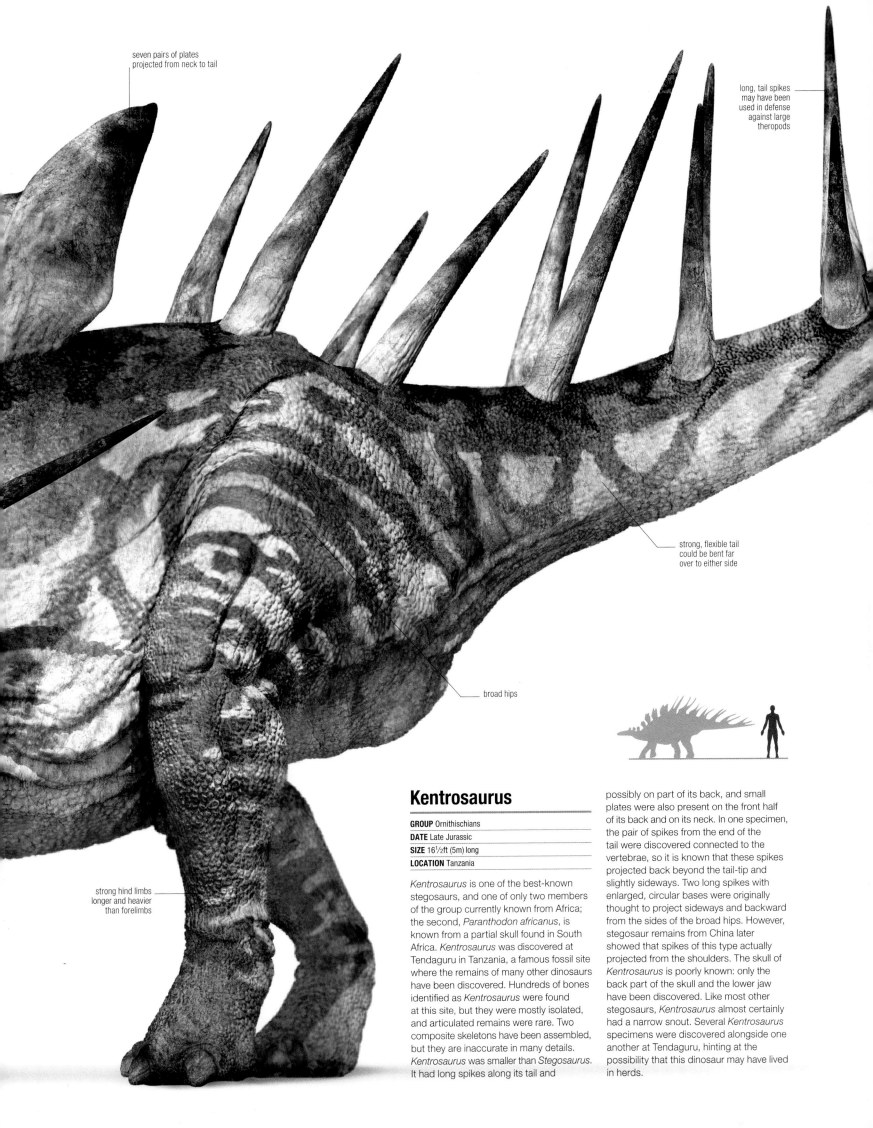

seven pairs of plates
projected from neck to tail

long, tail spikes
may have been
used in defense
against large
theropods

strong, flexible tail
could be bent far
over to either side

broad hips

strong hind limbs
longer and heavier
than forelimbs

Kentrosaurus

GROUP Ornithischians
DATE Late Jurassic
SIZE 16½ft (5m) long
LOCATION Tanzania

Kentrosaurus is one of the best-known stegosaurs, and one of only two members of the group currently known from Africa; the second, *Paranthodon africanus*, is known from a partial skull found in South Africa. *Kentrosaurus* was discovered at Tendaguru in Tanzania, a famous fossil site where the remains of many other dinosaurs have been discovered. Hundreds of bones identified as *Kentrosaurus* were found at this site, but they were mostly isolated, and articulated remains were rare. Two composite skeletons have been assembled, but they are inaccurate in many details. *Kentrosaurus* was smaller than *Stegosaurus*. It had long spikes along its tail and

possibly on part of its back, and small plates were also present on the front half of its back and on its neck. In one specimen, the pair of spikes from the end of the tail were discovered connected to the vertebrae, so it is known that these spikes projected back beyond the tail-tip and slightly sideways. Two long spikes with enlarged, circular bases were originally thought to project sideways and backward from the sides of the broad hips. However, stegosaur remains from China later showed that spikes of this type actually projected from the shoulders. The skull of *Kentrosaurus* is poorly known: only the back part of the skull and the lower jaw have been discovered. Like most other stegosaurs, *Kentrosaurus* almost certainly had a narrow snout. Several *Kentrosaurus* specimens were discovered alongside one another at Tendaguru, hinting at the possibility that this dinosaur may have lived in herds.

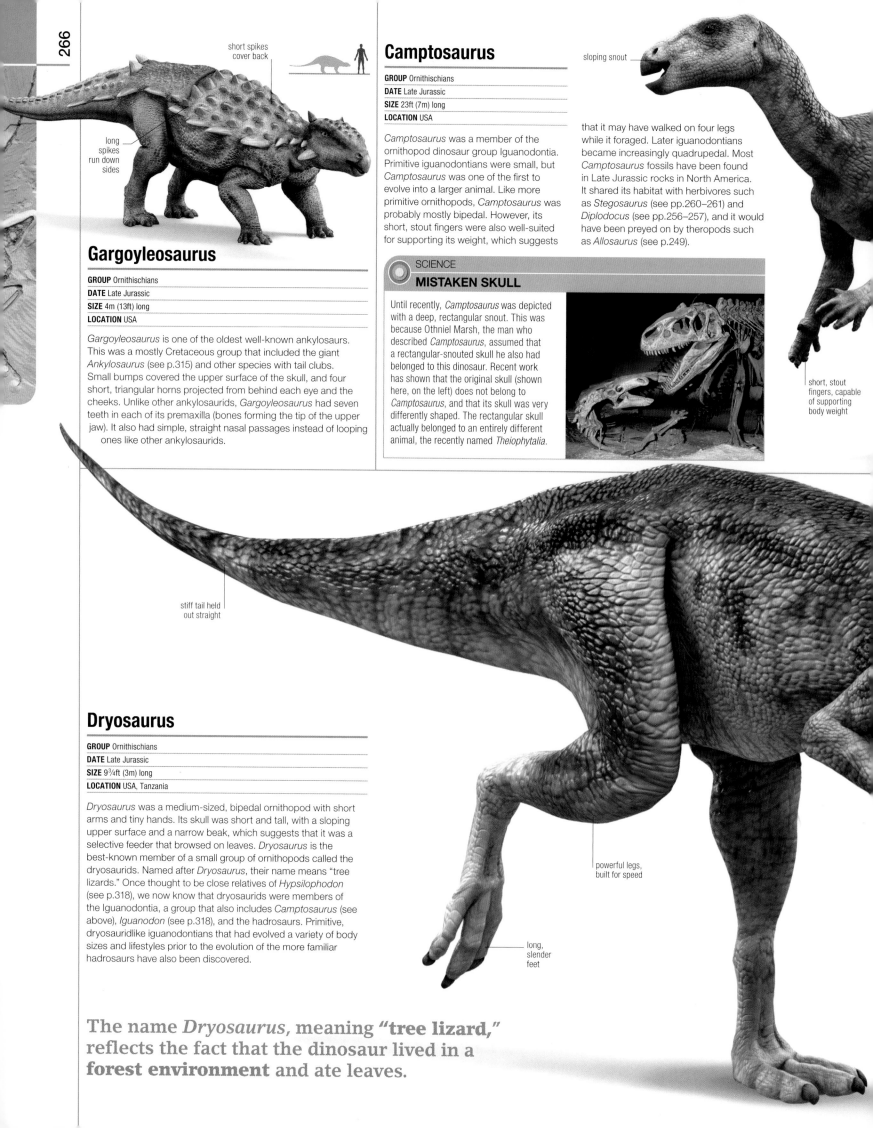

short spikes
cover back

long
spikes
run down
sides

Gargoyleosaurus

GROUP Ornithischians
DATE Late Jurassic
SIZE 4m (13ft) long
LOCATION USA

Gargoyleosaurus is one of the oldest well-known ankylosaurs. This was a mostly Cretaceous group that included the giant *Ankylosaurus* (see p.315) and other species with tail clubs. Small bumps covered the upper surface of the skull, and four short, triangular horns projected from behind each eye and the cheeks. Unlike other ankylosaurids, *Gargoyleosaurus* had seven teeth in each of its premaxilla (bones forming the tip of the upper jaw). It also had simple, straight nasal passages instead of looping ones like other ankylosaurids.

Camptosaurus

GROUP Ornithischians
DATE Late Jurassic
SIZE 23ft (7m) long
LOCATION USA

Camptosaurus was a member of the ornithopod dinosaur group Iguanodontia. Primitive iguanodontians were small, but *Camptosaurus* was one of the first to evolve into a larger animal. Like more primitive ornithopods, *Camptosaurus* was probably mostly bipedal. However, its short, stout fingers were also well-suited for supporting its weight, which suggests that it may have walked on four legs while it foraged. Later iguanodontians became increasingly quadrupedal. Most *Camptosaurus* fossils have been found in Late Jurassic rocks in North America. It shared its habitat with herbivores such as *Stegosaurus* (see pp.260–261) and *Diplodocus* (see pp.256–257), and it would have been preyed on by theropods such as *Allosaurus* (see p.249).

sloping snout

short, stout
fingers, capable
of supporting
body weight

SCIENCE
MISTAKEN SKULL

Until recently, *Camptosaurus* was depicted with a deep, rectangular snout. This was because Othniel Marsh, the man who described *Camptosaurus*, assumed that a rectangular-snouted skull he also had belonged to this dinosaur. Recent work has shown that the original skull (shown here, on the left) does not belong to *Camptosaurus*, and that its skull was very differently shaped. The rectangular skull actually belonged to an entirely different animal, the recently named *Theiophytalia*.

stiff tail held
out straight

Dryosaurus

GROUP Ornithischians
DATE Late Jurassic
SIZE 9¾ft (3m) long
LOCATION USA, Tanzania

Dryosaurus was a medium-sized, bipedal ornithopod with short arms and tiny hands. Its skull was short and tall, with a sloping upper surface and a narrow beak, which suggests that it was a selective feeder that browsed on leaves. *Dryosaurus* is the best-known member of a small group of ornithopods called the dryosaurids. Named after *Dryosaurus*, their name means "tree lizards." Once thought to be close relatives of *Hypsilophodon* (see p.318), we now know that dryosaurids were members of the Iguanodontia, a group that also includes *Camptosaurus* (see above), *Iguanodon* (see p.318), and the hadrosaurs. Primitive, dryosauridlike iguanodontians that had evolved a variety of body sizes and lifestyles prior to the evolution of the more familiar hadrosaurs have also been discovered.

powerful legs,
built for speed

long,
slender
feet

The name *Dryosaurus*, meaning **"tree lizard,"** reflects the fact that the dinosaur lived in a **forest environment** and ate leaves.

distinctive
arched back

LOW BROWSER
A low-browsing herbivore,
Camptosaurus had a distinctive
arched back, and arms long
enough to enable it to move on
all fours when feeding.

sturdy tail used
for balance

narrow beak
for browsing
on leaves

MISSING TOE
The first toe on the inner side of
Dryosaurus's foot (the hallux) was
missing, its feet were long and
slender, and its back legs were
powerful. These features suggest
that *Dryosaurus* was a fast runner.

tiny hands

CROSSING CONTINENTS

In 1919, remains were found in Tanzania that were similar to
Dryosaurus fossils later discovered in the USA—so similar that
the Tanzanian example was renamed as a second species of
Dryosaurus, suggesting that the American and African
continents were joined in the Late Jurassic. Research has shown
that it was, in fact, a different animal, which matches geological
evidence that America and Africa were not joined at this time.

Othnielosaurus

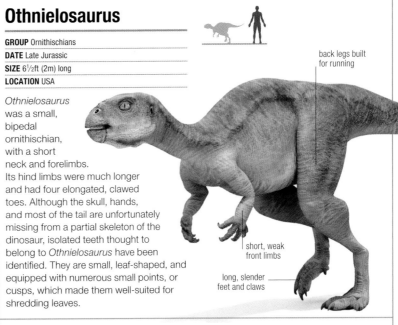

GROUP	Ornithischians
DATE	Late Jurassic
SIZE	6½ft (2m) long
LOCATION	USA

Othnielosaurus
was a small,
bipedal
ornithischian,
with a short
neck and forelimbs.
Its hind limbs were much longer
and had four elongated, clawed
toes. Although the skull, hands,
and most of the tail are unfortunately
missing from a partial skeleton of the
dinosaur, isolated teeth thought to
belong to *Othnielosaurus* have been
identified. They are small, leaf-shaped, and
equipped with numerous small points, or
cusps, which made them well-suited for
shredding leaves.

back legs built
for running

short, weak
front limbs

long, slender
feet and claws

Megazostrodon

GROUP	Synapsid
DATE	Early Jurassic
SIZE	4in (10cm) long
LOCATION	South Africa

Megazostrodon's name means "large
girdle tooth." Each cheek tooth possessed
short, triangular points, or cusps, arranged
in a line. They were simpler in shape to
those of later mammals, especially in the
way the cusps bit together, and were

probably used for cutting up insects.
The skeleton of *Megazostrodon* was
unspecialized for any particular lifestyle,
but it probably climbed, burrowed, and
ran, much like modern-day rats and
shrews. *Megazostrodon* shares several
features with a number of other Late
Triassic and Early Jurassic mammals.
These early mammals are grouped
together as Morganucodonta.

good eyesight aided
nocturnal lifestyle

hair-covered
body

Sinoconodon

GROUP	Synapsid
DATE	Early Jurassic
SIZE	12in (30cm) long
LOCATION	China

Sinoconodon is one of the largest primitive
mammals found to date. Several features
of its teeth and jaws were unusual—the
long gap between the canine and cheek

teeth, the robust jaw joint, and the stout,
strong chin. These features suggest that
Sinoconodon had a powerful bite and might
have preyed on large insects and small
reptiles. It has also been suggested that it
may have been related to *Megazostrodon*
(above) and *Morganucodon* (see p.213), but
recent studies have shown that
Sinoconodon was an even
more primitive mammal—
one of the earliest known.

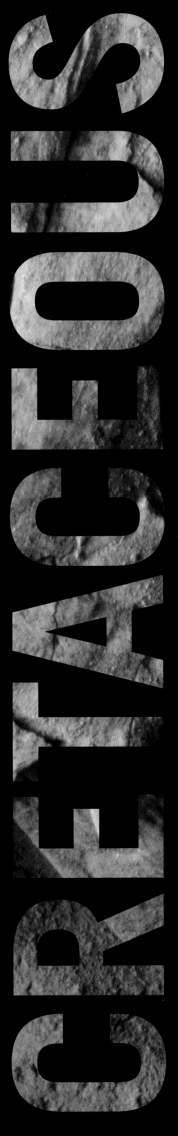

CRETACEOUS

Cretaceous

The Cretaceous period in Earth's history was generally a greenhouse world, with relatively high atmospheric carbon dioxide levels, high global temperatures, and sea levels about 650—980ft (200—300m) higher than today. Huge, shallow seas covered many of the present-day continents. It was a time when birds and flowering plants evolved and modern insect groups diversified.

Oceans and continents

The Cretaceous takes its name from its extensive chalk strata (from the Latin word *creta* for chalk), composed of the skeletons of countless tiny algae in shallow seas. It was a time when today's oceans were beginning to take shape. Pangea continued to disintegrate, with the South Atlantic Ocean opening up as Africa and South America rifted apart. Australia remained attached to Antarctica, but India separated from the western side of Australia during the Early Cretaceous and began to move in a relatively westerly direction. The Tethys Ocean contracted a little as the combined Eurasia, North China, South China, and Indochina blocks rotated clockwise, bringing Southeast Asia closer to the Equator. Sea levels were very high across the globe in Late Cretaceous times, flooding North America and creating a huge seaway that extended from the Gulf of Mexico to the newly forming Arctic Ocean. The North Atlantic, like its southern counterpart, was expanding, but only in the south. India parted from Madagascar, then rotated to move northward, beginning its collision with the Asian continent at the end of the Cretaceous. This initially resulted in the outpouring of vast amounts of magma, covering much of India with lava that solidified as a basalt, known as the Deccan Traps. Some experts suggest that the End-Permian and Cretaceous mass extinctions might have been caused by this type of excessive volcanic activity. Rock layers enriched with the rare element iridium indicate that Earth was also hit by asteroids, one of which left an enormous impact crater at Chicxulub, on Mexico's Yucatán Peninsula. This event is one of several proposed as causes for the mass extinction of species at the end of the Cretaceous.

MONUMENT ROCKS
The widespread carbonates and other sedimentary deposits make the Cretaceous rock record especially fine—for example, the rich marine fossils of Kansas's Smoky Hill Chalk Member.

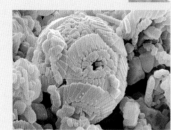

CALCAREOUS FOSSIL
This is a scanning electron micrograph of a coccolith, the fossilized skeleton of a coccolithophore (a single-celled marine alga).

WHITE CLIFFS OF DOVER
These famous chalk cliffs on England's south coast are composed almost entirely of coccoliths, deposited in Cretaceous seas.

KEY

▢	Ancient landmass
〜	Modern landmass
▲	Subduction zone
➡	Seafloor spreading ridge

LATE CRETACEOUS WORLD MAP
Pangea continued to disintegrate as Africa and South America parted. Oceans were rearranged, and Europe remained connected to North America.

ARC

Asian-Alaskan land bridge

NORTH AMERICA

Rocky Mountains

Gulf of Mexico

PACIFIC OCEAN

NORTH ATLANTIC

Proto-Caribbean Sea

SOUTH AMERICA

SOU ATLAN

Caribbean Sea opens up, moving North and South America apart

EARLY

MICROSCOPIC LIFE ● 140 Planktic foraminifers begin to diversify

PLANTS
● 125 First confirmed flowering plants
BETULITES

INVERTEBRATES
● 140 First venerid bivalves
● 135 First tellinid bivalves
● 125 First mactrid, teredenid, and donacid bivalves
TEREDINA
● 110 First limopsid, verticordiid, and thyasirid bivalves

140–100 Stegosaurs and high-browsing sauropod dinosaurs decrease in diversity; ornithopod and ankylosaur dinosaurs, and birds diversify

VERTEBRATES
● 140 First ceratopsian dinosaurs
● 130 First freshwater pelomedusid turtles
● 125 First enantiornithine birds
TRIONYX
● 115 First fossils of monotreme (egg-laying) mammals
● 110 First hesperornithiformes (diving birds)
HESPERORNIS (VERTEBRA)

MAGNOLIA TREE
Flowering plants (angiosperms), such as this magnolia, appeared in the Early Cretaceous and occupied new niches on the land. They co-evolved with pollinating insects, such as bees, and birds.

EURASIA

North China

South China

Indochina

Arabia

TETHYS OCEAN

...agascar

India

Australia

...RICA

ARCTICA

Eurasia, North China, South China, Indochina rotate clockwise

India separates from Madagascar and moves northward

Climate

The Cretaceous period had a variable climate, with oscillations between warm and cold about every 2 million years, and major swings in average temperature over a 20-million-year cycle. The Cretaceous is often described as a warm greenhouse world (see p.25), but this really only applies to part of it. At the outset, the Jurassic climate continued, with warm temperatures and no evidence for polar ice caps. Reptiles flourished, and flowering plants evolved and spread. However, by about 120 million years ago, global temperatures had reduced to the extent that upland areas of east Antarctica were covered by ephemeral ice sheets, and parts of Australia at latitudes above 65 degrees south were glaciated. High-latitude temperatures in Australia averaged only about 54°F (12°C). Over the next 30 million years, global temperatures increased—by 90 million years ago the Earth experienced one of its hottest periods. Globally, average surface temperatures were over 18°F (10°C) higher than today. Even in high latitudes temperatures were 72 to 82°F (22–28°C) and ocean temperatures in lower latitudes were as high as 97°F (36°C). The Late Cretaceous was much cooler, with ephemeral ice sheets developing on the South Pole. Average global temperatures dropped from their previous high of 88°F (31°C) to about 70°F (21°C) near the end of the Cretaceous.

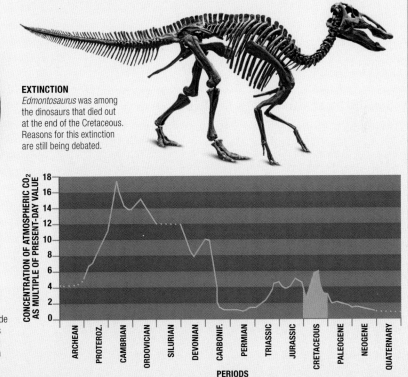

EXTINCTION
Edmontosaurus was among the dinosaurs that died out at the end of the Cretaceous. Reasons for this extinction are still being debated.

CARBON DIOXIDE LEVELS
The dip and rise in carbon dioxide reflects the changing climate as Earth moved from warm to cool to warm conditions. Gases from volcanic activity may have increased carbon dioxide levels.

CONCENTRATION OF ATMOSPHERIC CO₂ AS MULTIPLE OF PRESENT-DAY VALUE

ARCHEAN | PROTEROZ. | CAMBRIAN | ORDOVICIAN | SILURIAN | DEVONIAN | CARBONIF. | PERMIAN | TRIASSIC | JURASSIC | CRETACEOUS | PALEOGENE | NEOGENE | QUATERNARY

PERIODS

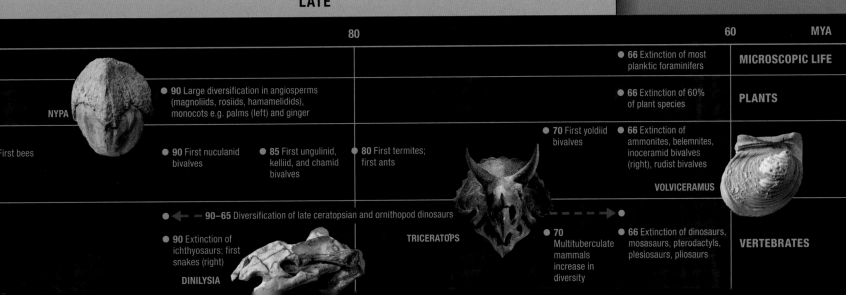

LATE

80   60   MYA

MICROSCOPIC LIFE
● **66** Extinction of most planktic foraminifers

PLANTS
● **90** Large diversification in angiosperms (magnoliids, rosiids, hamamelidids), monocots e.g. palms (left) and ginger

NYPA

● **66** Extinction of 60% of plant species

First bees

● **90** First nuculanid bivalves

● **85** First ungulinid, kelliid, and chamid bivalves

● **80** First termites; first ants

● **70** First yoldiid bivalves

● **66** Extinction of ammonites, belemnites, inoceramid bivalves (right), rudist bivalves

VOLVICERAMUS

●◄— **90–65** Diversification of late ceratopsian and ornithopod dinosaurs —►●

TRICERATOPS

● **90** Extinction of ichthyosaurs; first snakes (right)

DINILYSIA

● **70** Multituberculate mammals increase in diversity

● **66** Extinction of dinosaurs, mosasaurs, pterodactyls, plesiosaurs, pliosaurs

VERTEBRATES

CRETACEOUS PLANTS

Angiosperms, also known as flowering plants, made their first appearance during the Early Cretaceous. At first they were rare and not diverse, but steadily they became increasingly prominent, and by the end of the Cretaceous, they dominated the vegetation in most parts of the world.

Fossil floras from the earliest part of the Cretaceous consist mainly of ferns, conifers, cycads, and groups of extinct seed plants—they look very similar to fossil floras from the Jurassic. By about 130 million years ago, traces of the earliest flowering plants provide the first hints of evolutionary changes to come.

The earliest flowering plants

The earliest reliable fossils of flowering plants are pollen grains from the middle of the Early Cretaceous. They can be recognized as angiosperm pollen by their distinctive wall structure. Next, by about 125 million years ago, well-preserved fossils from eastern North America and Portugal record the first appearance of angiosperm leaves and the earliest fossil flowers. These flowers are all very small, but have the same basic structure seen in flowering plants today. Details of their pollen-producing structures, and of the organs in which the seeds develop, are also exactly like those of modern angiosperms, but different from the reproductive parts of other seed plants. Several of the earliest flowering plants from the Early Cretaceous were members of groups that are still alive today, such as the water lily and wintergreen families. The Chloranthaceae, an unusual family of living angiosperms, were especially prominent at this

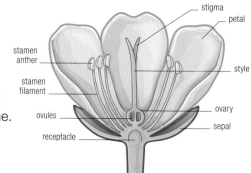

stamen anther
stamen filament
ovules
receptacle
stigma
petal
style
ovary
sepal

THE STRUCTURE OF A FLOWER
Typically, the flowers of angiosperms have a whorl of green, leaflike structures on the outside that protect the flower. In the center, the stamens, which produce the pollen, surround the carpels—the female part of the flower, including the ovary, style, and stigma. The carpels ultimately develop into the fruit, which contains the seeds.

time. There are also fossils of early monocots related to modern aroids. By about 100 million years ago, the number of modern angiosperm groups that can be recognized had increased further. They included plants very similar to living magnolia and bay laurel, as well as early eudicots related to modern plane trees, sacred lotus, and boxwood.

Toward modern vegetation

Flowering plants continued to diversify rapidly through the Late Cretaceous. By about 70 million years ago palms and gingers had evolved, which suggests that the variety of existing monocots had

DAKOTA SANDSTONE LEAF
The Middle Cretaceous Dakota Sandstone flora illustrates how much angiosperms had developed by about 100 million years ago. These fossils consist almost entirely of many different kinds of leaves from diverse angiosperms that grew on the banks of an ancient river system.

GROUPS OVERVIEW

The most obvious evolutionary change during the Cretaceous was the rise of flowering plants, but there was also expansion and decline of other groups. This resulted in ecosystems dominated by plants that are mostly familiar to us today. However, these ecosystems, populated by dinosaurs and other extinct reptiles, were still far from fully modern.

FERNS
The Cretaceous was a transitional time in the evolution of ferns. Ancient groups, such as Dipteridaceae, became less important, while Glecheniaceae initially flourished but later declined. By the end of the Cretaceous, new groups of ferns were starting to diversify into the new habitats created by angiosperms.

HORSETAILS
Horsetails were common in many Early Cretaceous floras and they have a continuous fossil record up to the present day. Some seem to have been larger than their modern counterparts, but they grew in similar habitats—along the banks of streams and around the edges of lakes.

CONIFERS
The Cretaceous was a time of major modernization in the history of conifers, resulting in many of the living groups seen today. But some groups of conifers declined at this time. Cheirolepidiaceae, which were very prominent in the Jurassic and Early Cretaceous, were extinct by the end of the Cretaceous.

CYCADS
Cycads are one of the groups that appear to become less important during the Cretaceous. Some may have suffered in competition with angiosperms. However, there is evidence of diversification among some groups of cycads that survived up to the present.

There are **several features** that distinguish all angiosperms from other seed plants, but most obvious is **the flower itself**.

WATER FERNS
Many groups of plants that have a long fossil history continued to diversify through the Cretaceous. Water ferns are one of several new groups of ferns that first appear at this time.

also increased. Among eudicots, early dogwoods and witch hazels appeared, along with many forms related to heathers. Plants related to modern oaks, walnuts, and hazels appeared in eastern North America and western Europe, and southern beeches are known from the Southern Hemisphere at this time. In parallel with the modernization of angiosperms, other groups of plants also showed important changes. Among ferns, archaic forms from earlier in the Mesozoic became less important, as did cycads, bennettitaleans, and a key family of extinct conifers. At the same time, new groups appeared and plant life as a whole began to look more modern. For example, relatives of modern pines are known from the Early Cretaceous and true pines became widespread in the Late Cretaceous.

Evolution of flowers

Several features distinguish all angiosperms from other seed plants, but most obvious is the flower itself. Recently, beautifully preserved fossil flowers have been discovered in many parts of the world from several different stages in the Cretaceous. They provide important new insights into the evolution of flowers. The earliest angiosperm flowers were small—typically just a quarter inch or less in diameter. Many were bisexual, but others were unisexual and produced only pollen or seeds. Very few had distinct petals. More typical showy petals appear later in the fossil record and are common in flowers from the Late Cretaceous. The structure of flowers became increasingly specialized in the Late Cretaceous. For example, petals became fused into a tube and the carpels were often fused into an ovary with a single pollen-receiving structure. Some of these changes probably indicate increasingly sophisticated

interactions with insect pollinators. Several key groups of pollinators, such as bees and modern kinds of moths and butterflies, are first known from the Late Cretaceous.

MAGNOLIA
Among the several different modern groups of flowering plants that can be recognized by the Middle Cretaceous, there are forms clearly similar to modern magnolia. By the end of the Cretaceous the diversification of the magnolia family and its relatives was well underway.

GINKGOS
Ginkgo-like plants were widespread in both the Northern and Southern Hemispheres during the Early Cretaceous. But by the end of the Cretaceous, they were less diverse and confined to the Northern Hemisphere. Ginkgo from the Early Paleogene is very similar to modern Ginkgo biloba.

GNETALES
Gnetales are a small group of living seed plants. Their distinctive pollen grains are first seen around the Permian–Triassic boundary. During the mid-Cretaceous, Gnetales underwent a major diversification. Fossil Gnetales similar to living Welwitschia (see p.275) and Ephedra (see p.397) are known from this period.

BENNETTITALEANS
Bennettitaleans are a group of extinct seed plants that appeared in the Triassic. They were important in Early Cretaceous floras, but died out in most areas by the Cretaceous–Paleogene boundary. New fossil evidence indicates that, in Australia, some survived into the Oligocene. In the Cretaceous, they are known mainly from cycadlike Cycadeoidea.

ANGIOSPERMS
Angiosperms radiated rapidly through the Cretaceous. Before the end of the Early Cretaceous, the three major groups of living flowering plant—eumagnoliids, monocots, and eudicots—existed. Today, eudicots include nearly three-quarters of all flowering plants, and monocots account for almost a quarter.

Nathorstiana

GROUP Lycophytes
DATE Cretaceous
SIZE Up to 1½in (4cm) long, ¾in (2cm) wide
LOCATION Germany

The fossil plant *Nathorstiana* is known only as a collection of molds and casts taken from a single location in Germany. Its root-bearing base grew downward, producing new roots and shedding older ones as it went. The root-growing tip was sunk in a central depression and covered with a membrane for protection. The radially symmetrical shape of the young bases divided into two and then four lobes as the plant matured; this maturation indicates age—not the very variable size shown by the fossils. *Nathorstiana*'s leaves and reproductive organs are unknown, but it is generally regarded as a lycophyte and part of a series of plants that gradually reduced in size between the Triassic *Pleuromeia* and the living *Isoetes*.

QUILLWORT

Isoetes consists of about 200 mainly aquatic or semiaquatic species. Narrow, quill-like leaves (hence the common name "quillwort") are attached to a rooted, cormlike stem. The leaves' swollen bases have a large sporangium containing either megaspores or microspores enclosed by a thin membrane called a vellum.

NATHORSTIANA SP.
This impression from Quelinburg in Germany shows the base of *Nathorstiana* with the lumps on the side marking the bases of the leaves.

Weichselia

GROUP Ferns
DATE Cretaceous
SIZE Leaf 5ft (1.5m) or more across
LOCATION Northern Hemisphere

Weichselia was a successful genus of treelike ferns that formed dense sward, rather like modern bracken. The plants had palmate leaves with 5 to 15 "fingers" or pinnae, each containing 14 to 20 spore-producing sporangia arranged in single rows on either side of the central rib. Middle Cretaceous climate change seems to have eliminated *Weichselia*. This treelike fern belonged to the family Matoniaceae, which died out throughout most of the world by the Late Cretaceous, perhaps out-competed by flowering plants. Today, only two living genera exist and are found in Malaysian Borneo.

DIPTERIS

Dipteris is considered to be a broadleaf fern. Its leaves show a characteristic known as dichotomous venation, in which the veins branch off from one another like the branches of a tree. With its unequal leaf lobes arranged around the top of a vertical stalk, *Dipteris* resembles a fan or umbrella.

sporangial clusters

central rib

WEICHSELIA RETICULATA
The fronds of *Weichselia* were similar to those of bracken.

sporangial cluster

SPORE PRODUCTION
This close-up shows the single rows of sporangial clusters on one side of the central rib. These clusters produced the plant's reproductive structures (spores).

Tempskya

GROUP Ferns
DATE Cretaceous
SIZE Up to 14¾ft (4.5m)
LOCATION Northern Hemisphere

This very unusual tree fern—placed in a family of its own—has been described from well-preserved petrified specimens found in many locations in the Northern Hemisphere. What appears to be a trunk is a false trunk, made up of a large number of stems surrounded by a thick mat of fibrous adventitious roots—that is, roots arising from a part of the plant other than a main root source. Sporelings grew vertically upward and branched or dichotomized repeatedly while giving off more adventitious roots that grew downward and bound the stems together. The result is that the base of the trunk was almost entirely composed of adventitious root fibers. Although no leaves have been found attached, the fossilized leaf bases show that they were produced from the sides of the stems, so the upper part of the trunk would have been covered with leaves, rather than having a single crown of leaves on top as in other forms of tree ferns.

TEMPSKYA
With its huge, fibrous false trunk and covering of leaves, *Tempskya* probably resembled a young cedar or redwood rather than a species of tree fern.

Salvinia

GROUP Ferns
DATE Cretaceous to present
SIZE Leaf ³⁄₁₆–³⁄₈in (5–10mm) long
LOCATION Tropical and subtropical regions worldwide

Salvinia is a living genus of small-leaved floating ferns. It has pairs of flat, oval, floating leaves attached to a rootless rhizome, and the leaves are covered with hairs to repel water. Living *Salvinia*, and the closely related genus *Azolla*, are two mainly tropical ferns that have fossil records dating from the Cretaceous period onward. Both can propagate easily by growth and division. *Salvinia* also has highly dissected, fertile leaves hanging downward into the water. They grow rounded, hard-coated sporocarps—specialized structures that produce and release spores. The sporocarps contain either a few female megasporangia, each with a single megaspore, or many male microsporangia, each with 64 microspores. Sporocarps enable the plants to survive periods of drought, only opening to release their spores when water returns. Individual megaspores are covered with a foamy layer that makes them buoyant, thereby allowing them to float on the water.

sporocarp

SALVINIA SP.
This specimen from Baden-Württemberg, Germany, shows a length of *Salvinia* rhizome (the dark lines) with many leaves and the reproductive sporocarps (the dark oval areas).

rhizome

hairs to repel water

SALVINIA FORMOSA
This magnified photograph of a fossilized *Salvinia* leaf shows the remains of the hairlike strands that grow from the tropical plant's leaves. These allow it to float on top of water.

Gleichenia

GROUP Ferns
DATE Cretaceous to present
SIZE Smallest leaflet about 1/16 in (2mm) long
LOCATION Tropical and subtropical regions worldwide

This fossil fern is branched, with short, rounded leaflets. It is identical to the tropical *Gleichenia* that is found growing today in open places and on the edges of forests. It has a creeping rhizome (a constantly growing underground stem) and fronds that grow enormous by continuously dividing their side branches. *Gleichenia* forms dense thickets by scrambling or climbing over other plants and is a genus of the primitive Gleicheniaceae family, whose members have simple rhizomes covered with hairs.

Its large sporangia are arranged in circles on the underside of the fronds. Gleicheniaceous ferns date from the Paleozoic, but they are particularly common in Cretaceous floras.

frond stem

sporangia on underside of leaflets

GLEICHENIA SP.
The leaves of *Gleichenia* could reach enormous sizes by repeated growth and branching. In contrast, its ultimate leaflets (pinnules) are small.

Gleichenia was much more **scarce** in the Cenozoic than in the Cretaceous, perhaps because of **competition** from **flowering plants**.

Cycadeoidea

GROUP Bennettitaleans
DATE Jurassic to Cretaceous
SIZE Stem up to 2in (5cm) across
LOCATION North America, Europe

These large, fossilized bennettitalean trunks are known from many places in North America and Europe. Most *Cycadeoidea* stems are short and barrel-shaped, with a dense covering of leaf bases. In life, they had a crown of pinnate leaves at the top of the stem.

They were seed plants, but were unlike any other group in having their reproductive organs arranged in complex flowerlike structures protected by stout, spirally arranged bracts. The size of these cones varied from species to species, and most species were bisexual. Their ovules were borne on a central swollen receptacle and separated by scales; the pollen organs surrounded the receptacle. It is likely that these flowers were self-pollinated, although it is possible that insects may have been involved.

AMONG FERNS
These slow-growing plants probably lived in open habitats, surrounded by many different species of ferns. The climate was probably strongly seasonal, and the vegetation may have burnt during the dry season, similar to many modern savannas.

One species of *Cycadeoidea* was **named from a stem** found as a burial object in an **Etruscan grave**.

Hausmannia

GROUP Ferns
DATE Triassic to Cretaceous
SIZE Leaf 2–3 1/4 in (5–8cm) long
LOCATION Northern Hemisphere

This group of ferns often had deeply lobed leaves with the major veins branching and running to the edge of the leaf-blade. Small sporangia were scattered across the lower, hairy surface of the leaves. *Hausmannia* is structurally very similar to the living tropical fern *Dipteris* and is therefore included in the family Dipteridaceae. Paleoecological studies of groups of fossils found together suggest that *Hausmannia* was a streamside dweller. It was also a pioneer species, like the living *Dipteris*, in the way it invaded open areas.

HAUSMANNIA DICHOTOMA
This leaf fragment shows the dividing veins. Between these veins, finer veins form polygonal meshes.

SPLITTING VEINS
This small fragment of a dipteridaceous fern leaf shows several veins branching and running to the edge of the leaf-blade.

Welwitschia

GROUP Gnetales
DATE Late Cretaceous to present
SIZE Leaf up to 30ft (9m) long
LOCATION Namibia

Welwitschia mirabilis is a highly specialized gymnosperm that grows in the Namib Desert in southwest Africa. It has a very short stem, tapering down into a long taproot, and two leaves that grow continually from their bases. Individual plants produce cones that have either pollen organs or ovules, and pollination is probably achieved through insects. Pollen similar to that from this genus and the related, living *Ephedra* is known from the Late Triassic onward. It is probable that *Welwitschia* and *Ephedra* diverged in the Late Cretaceous, especially as both *Welwitschia*-like and *Ephedra*-like fossils are known from this time.

Araucaria

GROUP Conifers
DATE Jurassic to present
SIZE Cone 1–1 3/4 in (2.5–4.5cm) long and 1–1 1/2 in (2.5–4cm) across
LOCATION South America

These large, conifer trees from the Jurassic of Argentina bore very characteristic female cones with numerous, spirally arranged cone scales attached to a central axis. The corresponding pollen cones are not known. Each fertile cone scale had an ovule and a smaller bract scale. Immature cones contained ovules with three-layered skin (integument). Pollination would have been by wind, as in the living members of the genus. Mature cones contained seeds with embryos, apparently in a dormant state; some seeds were loose within the cone. It also appears likely that the seeds were not shed until the cones fell to the ground, where they may have shattered on impact. The structure of the cone and the way that the seeds were dispersed is most similar to the living species *Araucaria bidwillii*, which is found in Southern Queensland, Australia.

cone scale

seed

central axis

woody ends of cone leaves

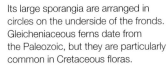

ARAUCARIA MIRABILIS
These anatomically perfectly preserved, silicified cones of *Araucaria mirabilis* come from the Cerro Cuadrado petrified forest in Patagonia, South America.

diamond pattern

silicified tissue

PERFECTLY PRESERVED
Preserved cones such as the ones shown here have been overcollected in the past for museums and private collectors, so it is now forbidden to collect and export them from Argentina without permission.

Brachyphyllum

GROUP Conifers
DATE Jurassic to Cretaceous
SIZE About 5/16in (8mm) wide
LOCATION Northern Hemisphere

Fossil conifer shoots that are covered with spirals of short, stubby, scalelike leaves are called *Brachyphyllum*. Pollen-producing cones were found on the ends of these shoots. These cones had many scales, and each scale had three pollen sacs on its underside. The leaves of *Brachyphyllum* possessed distinctive stomata—the respiratory pores that allow plants to "breathe." *Brachyphyllum* leaves show a number of features that indicate that they would have grown in arid environments.

Pityostrobus

GROUP Conifers
DATE Cretaceous
SIZE Cone 2–2¼in (5–6cm) long
LOCATION North America, Europe

Pityostrobus was a seed-bearing conifer cone. In *Pityostrobus*, the seed-bearing scales were positioned in the axils—the angle between a leaf stem (or the bract, in this case) and the main stem—of very small bracts. The seed-bearing (or ovuliferous) scales thinned toward their tips, unlike in modern pinecones. In mature cones, each ovuliferous scale had two winged seeds separated by a central ridge. The closest living relative to *Pityostrobus* is *Pinus*, the genus that contains pines and related conifers. However, *Pityostrobus* differs from

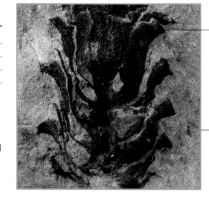

seed-bearing scale

Pinus in that it possessed separate bract and ovuliferous scales. In *Pinus* cones, the bract and ovuliferous scales were fused together and had a woody consistency.

PITYOSTROBUS DUNKERI
This carbonized cone from the Early Cretaceous is roughly three times longer than it is wide. In life, the cones would have been woody.

Drepanolepis

GROUP Conifers
DATE Cretaceous
SIZE 4in (10cm) long
LOCATION Northern Hemisphere

Drepanolepis was a genus of extinct conifers that produced seed-bearing shoots with a spiral arrangement of bracts. Each bract consisted of a single

sickle-shaped, winglike, seed-bearing scale with a single seed attached to its upper surface. The seeds themselves were curved, with their openings turned back toward the base of the scale. Each of these seed-bearing scales was positioned in the axil—the angle where the bract joined the stem—of a much longer bract. *Drepanolepis* was similar in structure to the Permian conifer *Ullmania*. It is also similar to the Araucariaceae, the

family to which the somewhat primitive-looking but living monkey-puzzle tree belongs. Some people think that *Drepanolepis* might be an evolutionary offshoot of early Araucariaceae. Like all conifers, *Drepanolepis* was a gymnosperm, meaning that its seeds were "naked"—as opposed to trees whose seeds are surrounded by fruit, which are known as angiosperms.

With its distinctive bracts and bract scales, *Drepanolepis* has sometimes been compared to cones of the Araucariaceae family.

large stem

DREPANOLEPIS SP.
The scales and ovules appear to depart at different angles from the axis. This is an illusion produced by the compression of the spiral during fossilization.

bract   axil

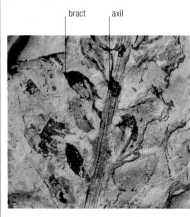

SPIRALING STRUCTURE
The shoots of *Drepanolepis* featured an arrangement of seed-bearing bracts that spiraled around a central stem.

leafy shoot

fertile shoot

Sequoia

GROUP Conifers
DATE Cretaceous to present
SIZE 230ft (70m) tall
LOCATION Northern Hemisphere

Sequoia is a genus of conifers of the bald cypress family. Also known as redwoods, they have both pollen and seed cones on the same tree. After pollination, the seed cones expand and become woody. The long, diamondlike shapes on the cone's surface consist of ovule-bearing scales (upper) and bract scales (lower). Such cones can remain closed for 25 years after falling from the trees and may not open until stimulated by fire.

ovule-bearing scale

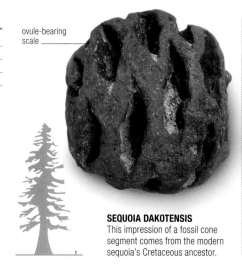

SEQUOIA DAKOTENSIS
This impression of a fossil cone segment comes from the modern sequoia's Cretaceous ancestor.

LIVING RELATIVE
REDWOODS

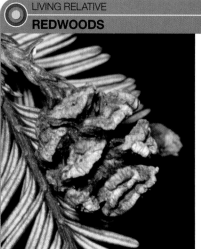

There are two species of evergreen conifer known as giant redwood. The Californian coast redwood is *Sequoia sempervirens* and the giant redwood (or *Wellingtonia*) is *Sequoiadendron giganteum*. *Sequoia* is confined to a narrow coastal strip running from the Canadian border to southern California. At up to 360ft (110m) high, these trees are the tallest in the world; some are over 2,000 years old. Their trunks can be up to 20ft (6m) in diameter, and their 12in- (30cm-) thick bark protects them from animal damage and forest fires. The leaves on the main shoots are needlelike and arranged into one plane, while the leaves on the cone-bearing shoots are scalelike and set in a spiral.

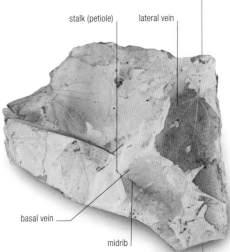

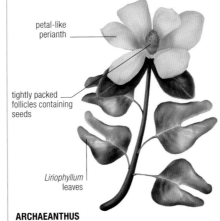

Geinitzia

GROUP Conifers
DATE Cretaceous
SIZE Shoot 5/16in (8mm) wide
LOCATION Northern Hemisphere

This conifer leafy shoot has long, outward-spreading, spirally arranged leaves. Each leaf is needle-shaped, as thick as it is broad, and merges below into a swollen leaf cushion. *Geinitzia*-like leaves are very similar to the young shoots of other conifers. Similar leafy shoots in the Permian are called *Walchia* and those in the Triassic are called *Voltzia*. According to some interpretations, *Geinitzia* is part of the extinct conifer family

Cheirolepidiaceae. Seed cones of two *Geinitzia* species are known from Belgium and Germany.

GEINITZIA FORMOSA
This fossilized example shows a small portion of a branching, fertile, leafy shoot with two lateral female cones. The ovals on the cones mark the ends of the ovule-bearing scales.

fertile, leafy shoot

Archaeanthus

GROUP Angiosperms
DATE Cretaceous
SIZE Cone 4in (10cm) long
LOCATION North America

This flower consisted of around 100 loosely packed, helically arranged, peapod-shaped follicles that were attached to a long, central axis. The flower was borne at the tip of a branch, with leaves in an alternate arrangement. Leaves associated with *Archaeanthus* had a prominent midrib and are called *Liriophyllum*. *Archaeanthus* flowers are very similar to those of modern Magnoliaceae, which points to *Archaeanthus* as the earliest member of the Magnolia family.

petal-like perianth

tightly packed follicles containing seeds

Liriophyllum leaves

ARCHAEANTHUS
With its large, magnolialike flower, petal-like perianth, and podlike fruits, *Archaeanthus* shows all the characteristics that are traditionally associated with insect pollination.

Mauldinia

GROUP Angiosperms
DATE Cretaceous
SIZE Leaf 4in (10cm) long
LOCATION Europe, North America, Central Asia

The leaves of *Mauldinia* were typically simple, elongated angiosperm leaves. Flowers were radially symmetrical, bisexual, and about 1/8in (3.5mm) in length. The flower parts were arranged in groups of

three. There were two whorls of three petals, surrounding three whorls of three stamens. The central ovary was rounded to triangular in cross-section and contained a single seed. Five individual flowers were grouped together on the upper surface of bilobed, scalelike structures. These scales were then spirally arranged on an elongated axis. The flower structure and leaves of *Mauldinia* are characteristic of the living angiosperm family Lauraceae (the laurels).

Araliopsoides

GROUP Angiosperms
DATE Cretaceous to present
SIZE 33ft (10m) high
LOCATION North America, Europe, Asia

The palmate leaves of *Araliopsoides* had three distinct lobes. The wide base of the leaf stalks (petioles) suggest that the leaves were deciduous and shed seasonally. The plants themselves were small, shrubby, and probably grew in warm temperate to subtropical deciduous forests in middle- and high-level northern latitudes during the Late Cretaceous. *Araliopsoides* was a forerunner of the maples that came later.

ARALIOPSOIDES CRETACEA
This impression of a trilobed leaf from the Dakota Sandstone shows beautiful preservation of leaf venation.

Archaefructus

GROUP Angiosperms
DATE Cretaceous
SIZE 4in (10cm) long
LOCATION China

Archaefructus is a herbaceous, aquatic angiosperm. It does not have petals or sepals, but it does have carpels and stamens. These are attached to an elongated stem with the staminate (pollen-producing) flowers below and pistillate (fruit-producing) flowers above. This ancient flower is similar in some respects to *Trithuria*, a peculiar living genus of Nymphaeales (waterlilies).

ARCHAEFRUCTUS LIAONINGENSIS
This fossil shows part of a reproductive axis with female carpels on short stalks. The oval outlines in the carpels are seeds.

Credneria

GROUP Angiosperms
DATE Cretaceous
SIZE Leaf 4in (10cm) long
LOCATION North America, Europe

Credneria is the name given to large angiosperm leaves produced by trees very similar to living plane trees. The leaves were originally attached to the shoots by stalks (petioles). Individual leaves were typically broadly oval or elliptical, with rounded bases. Their margins were smooth, and they had pointed or blunt

apices. Leaf venation was usually pinnate, with a main vein and lateral veins. During the Middle Cretaceous, the leaves are very common in many fossil floras from all over the Northern Hemisphere, where they have been given a range of different names, such as *Aralia* and *Araliopsoides* (see left). The extinct plane trees that produced *Credneria* were especially common on ancient riverbanks. Living plane trees grow in similar places today.

CREDNERIA ZENKERI
Although there is no plant material remaining, the detailed outlines of these two leaves and their intricate pattern of veins are clearly impressed into the sandstone in which they are preserved.

smaller veins make an angular pattern

central, main vein

lateral vein

Cercidiphyllum

GROUP Angiosperms
DATE Cretaceous to present
SIZE Leaf 2 1/4in (6cm) long
LOCATION Northern Hemisphere

The fossilized leaves of this Cretaceous plant, together with other fossils of its fruits and seeds, are very similar to those of living *Cercidiphyllum*. *Cercidiphyllum* is known today from two very similar species that grow in China and Japan. Both of these are commonly called katsura. Male and female flowers are produced on separate plants. The leaves are simple, ovate, and wider than they are long. They have a midrib and one or two pairs of basal veins that curve out toward the leaf apex. Fossil *Cercidiphyllum*-like seedlings from the Paleocene of Canada have been found in their original growth positions. This discovery suggests that this species grew on open floodplains during the Late Cretaceous and the Tertiary.

stalk (petiole)   lateral vein

basal vein

midrib

CERCIDIPHYLLUM SP.
This fossil, from Alberta in Canada, shows impressions of simple, ovate leaves related to the modern katsura tree (*Cercidiphyllum japonicum*).

CRETACEOUS
INVERTEBRATES

During the Cretaceous, the sea level stood higher than at any time in geological history. The continental shelves and adjacent lands were flooded, providing diverse marine habitats for invertebrate life. Yet this time of flourishing marine life ended in disaster—for terrestrial animals as well.

Diversification and extinction

Cretaceous marine invertebrate faunas were not unlike those of the Jurassic, with ammonoids, bivalves, gastropods, brachiopods, echinoids, crinoids, and bryozoans. Yet many ammonoids developed strange shapes, some being uncoiled or partially recoiled, or even twisted in a series of U-bends to form an unlikely tangle. The Late Cretaceous chalk of northern Europe is a source of beautifully preserved, although not especially common, marine invertebrates; this unique sediment is largely composed of calcareous plates of tiny algae (coccoliths). Marine life flourished, including sea-dwelling reptiles, while on land dinosaurs, flying reptiles, and birds proliferated. Yet all this was to change dramatically. At the very end of the Cretaceous, 66 million years ago, ammonites, belemnites, and plesiosaurs abruptly vanished, as did dinosaurs and pterosaurs. An asteroid hit Earth, creating shock waves, dust clouds, forest fires, and acid rain; the collision site has been identified in Yucatán, Mexico, buried by later sediment (see p.32). At the same time, huge volcanic eruptions in India created widespread devastation, and the effects of both together were catastrophic. Study of a fossil rain forest suggests that its recovery after this mass extinction took 1.5 million years.

NEOHIBOLITES MINIMUS
This small, slender mid-Cretaceous belemnite lived in vast numbers in the warm shelf-seas that existed during the Cretaceous, where it hunted small prey. It is a common fossil in Europe, where it is often found in clay deposits.

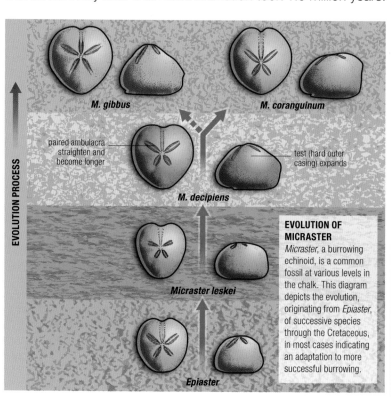

M. gibbus

M. coranguinum

paired ambulacra straighten and become longer

test (hard outer casing) expands

M. decipiens

EVOLUTION PROCESS

Micraster leskei

Epiaster

EVOLUTION OF MICRASTER
Micraster, a burrowing echinoid, is a common fossil at various levels in the chalk. This diagram depicts the evolution, originating from *Epiaster*, of successive species through the Cretaceous, in most cases indicating an adaptation to more successful burrowing.

GROUPS OVERVIEW

Many groups of invertebrates survived little changed from the Jurassic into the Cretaceous. Mollusks were abundant and diverse, as were cephalopods. Ammonites, belemnites, and nautiloids all evolved rapidly. Bivalves were abundant and became more diverse at low latitudes. Many new forms of echinoids also appeared.

AMMONITES
Cretaceous ammonites are of variable forms and include many differently shaped ones, such as helically coiled types, which look like gastropods. The helically coiled types may have lived like gastropods, creeping over the seafloor. Toward the end of the Cretaceous, ammonites declined, restricted to specific areas, before their final demise.

BELEMNITES
Belemnites are the internal shells of fossil squidlike cephalopods, but they look like elongated bullets and are unlike the shells of modern squids or cuttlefish. They were common in the Jurassic and Cretaceous, and fossils are often found as mass-accumulations, or "belemnite battlefields," which, as with modern squids, may represent mass mortality after spawning.

BIVALVES
A very unusual group of bivalves, known as rudists, evolved at this time and was confined to the Cretaceous. Their external appearance is so modified that they are hardly recognizable as bivalves at all. Usually one valve is long and conical and the other sits on top of it like a lid. They grew very large, and some species actually formed reefs.

ECHINOIDS
A trend toward deep burrowing, which began in the Jurassic, reached its greatest effectiveness in one group of Cretaceous echinoids, the spatangoids. Micraster (illustrated above) is one of these, and could burrow to many inches below the surface. Other unmodified echinoids, however, lived on the seafloor.

Ventriculites

GROUP Sponges
DATE Cretaceous
SIZE Up to 4½in (12cm) high
LOCATION Europe, North America

Ventriculites was a glass sponge, with a skeleton of siliceous needles (spicules) that were fused into a rigid network to support the sponge's body. Between them were pores that dotted the surface. The skeleton of this fossil is vase-shaped, and siliceous fibers anchored it to the seabed. In its center was a cavity called the cloaca. Food and oxygen were removed as water passed through the body walls before exiting through the top opening.

VENTRICULITES SP.
Ventriculites occurred in many different shapes. Some were vaselike, such as the one above, while others varied from squat cups to tall cylinders.

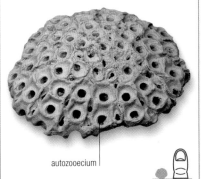

autozooecium

LUNULITES SP.
Lunulites colonies were highly organized systems. A soft-bodied autozooid lived inside each of the chambers, which were known as autozooecia.

Lunulites

GROUP Bryozoans
DATE Late Cretaceous to present
SIZE ³⁄₁₆–³⁄₈in (5–10mm) across
LOCATION Worldwide

Lunulites was a cheilostome bryozoan that formed very small, circular colonies. Its lower surface was concave, with radial grooves. The colony originated at the center, and radiating from this were rectangular to hexagonal autozooecia: chambers that housed soft-bodied, feeding autozooids—the animals that made up the colony. All autozooids possessed a lophophore or feeding structure consisting of a ring of tentacles surrounding the mouth, which could be withdrawn into the autozooecia for protection when not feeding. The autozooecia increased in size from the center and had wide openings. Modern relatives of *Lunulites* move across the seafloor by means of long, spiny appendages.

Crania

GROUP Brachiopods
DATE Cretaceous to present
SIZE Up to ½in (1.5cm) across
LOCATION Worldwide

Crania is a brachiopod that first appeared in the Cretaceous and still has relatives alive today. Scientists describe *Crania* as an inarticulate brachiopod because the two halves of its shell are not "articulated" or joined together, but are held together by muscles alone. Unlike most brachiopods, *Crania* has no pedicle. It is commonly found in chalk deposits with one half of its shell cemented to a larger organism, such as a mollusk or echinoderm. Because its shell halves are held together by soft tissue, the bottom half is often the only part preserved, the top half having been swept away before fossilization could take place.

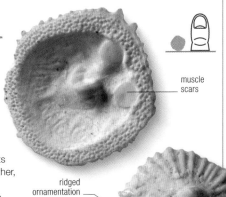

muscle scars

ridged ornamentation

CRANIA EGNABERGENSIS
Crania shells lack hinge teeth and show tiny perforations across the shell. Inside are scars where muscles were attached.

fine growth lines

Sellithyris

GROUP Brachiopods
DATE Cretaceous
SIZE Up to 1½in (3.5cm) long
LOCATION Europe

Sellithyris lived on the seabed, attached by a flexible stalk or pedicle that passed through an opening below the beak of the larger shell (the pedicle valve); the smaller shell is called the brachial valve. *Sellithyris* was almost pentagonal in outline. Its outer surface was smooth but showed growth lines. All over the surface are tiny perforations (punctae) that housed bristles.

SELLITHYRIS SP.
The outer edge of *Sellithyris*'s shell is strongly folded, and there is an inward fold at the center of the brachial valve, with two outward folds on either side.

Cymatoceras

GROUP Cephalopods
DATE Late Jurassic to Paleogene
SIZE Up to 12in (30cm) across
LOCATION Worldwide

Cymatoceras was a tightly coiled cephalopod mollusk related to the modern *Nautilus*. It had a broader shell than *Nautilus* and also had well-defined ornamental ribs, which were strongest on the outer part of the whorl and over the wide outer edge (venter). Like all cephalopods, *Cymatoceras* would have caught prey with its flexible tentacles. It would then have dismembered its prey using its horny beak.

strongly defined ribs

suture line

CYMATOCERAS SP.
On this internal mold, the distinction between the ribs and the suture lines is clear.

Deshayesites

GROUP Cephalopods
DATE Early Cretaceous
SIZE Up to 4in (10cm) across
LOCATION Europe, Greenland, Georgia

Deshayesites was an ammonite with a rather flattened, coiled shell. Like all ammonites, in life it would have had flexible, muscular tentacles around its mouth and emerging from its shell. *Deshayesites* has strong, sometimes branched, closely spaced ribs that have a slight forward bend in the middle, which points in the direction of the opening of the shell. The whorls themselves overlap each other very slightly, creating a wide, shallow depression, or umbilicus, at the center.

DESHAYESITES SP.
In between the longer main ribs are shorter secondary ones, which begin halfway up the whorls.

Mortoniceras

GROUP Cephalopods
DATE Early Cretaceous
SIZE Up to 12in (30cm) across
LOCATION Europe, Africa, India, North America, South America

Mortoniceras fossils are characterized by their square-section whorls and strong ornamentation, which is produced by a combination of ribs and tubercles. The whorls overlap only slightly, leading to a very shallow umbilicus—this coiling is described as moderately evolute. A pronounced ridge, known as the keel, is on the outer edge (venter). In some mature specimens, a large horn is visible at the opening on the shell. This has not been associated with just one sex in this genus.

prominent tubercles

MORTONICERAS ROSTRATUM
At the tops of the ribs on the outer whorl are prominent bumps, known as tubercles.

The **final growth stage** may have **prevented** the animal from actually **feeding**. It may be that, when it reached maturity, the adult **reproduced and then died**.

BUOYANCY CONTROL
Like other ammonites, *Scaphites* would have been able to alter the amount of liquid and gas in its chambered shell to regulate buoyancy.

Several **recently discovered fossils** are of Cretaceous dinosaurs that were **covered in feathers**. There is barely any distinction between **these dinosaurs** and early **true birds**.

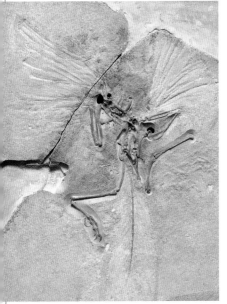

ARCHAEOPTERYX
There is no debate: *Archaeopteryx* is the most famous fossil of all time. Ten specimens of this, the oldest bird ever discovered, have been found in the limestones of Bavaria, Germany. Some of them preserve beautifully clear details of the feathers and skeleton.

members of strange and wholly extinct groups, and even early representatives of the lineage leading to modern birds. Clearly, birds were the dominant flying creatures of the Cretaceous, and after the dinosaur extinction, they continued to diversify.

The Mesozoic marine revolution

Marine ecosystems went through a remarkable period of change during the Mesozoic. This massive reorganization occurred on all levels, from primary producers all the way up to top predators. Large predators such as sharks and mosasaurs spread around the world, and bony fish became astonishingly common. The changes were not limited to vertebrates, and, in fact, many of the greatest changes involved invertebrates and microscopic organisms. The major modern groups of planktonic microorganisms—the primary producers that form the base of every oceanic food web—originated and diversified during this time. Meanwhile, archaic invertebrates

that were common during the Paleozoic—such as crinoids and brachiopods—were marginalized, and more modern groups such as clams, scallops, and heavily armored gastropods exploded in abundance. The reasons for these changes are complex but are likely to involve predation pressure from newly evolved giant predators (such as some sharks), as well as continental breakup and changes in sea level, and ocean chemistry that are poorly understood.

ANCIENT SHARK
The ancient shark *Hybodus* was a small predator, only about 6½ft (2m) long, that thrived for over 100 million years. It is closely related to the modern sharks but became extinct during the Late Cretaceous.

The end of the dinosaurs

The extinction of the dinosaurs has been one of the greatest mysteries of Earth's history. What could cause such a successful group to die off? Sixty-six million years ago, a massive asteroid struck the Yucatán Peninsula of Mexico (see p.32). Most scientists think that this impact set in motion a chain reaction of environmental disturbances, such as fallout of soot and ash, temperature fluctuations, and the collapse of all food pyramids.

ORNITHOPODS
The ornithopods were some of the most successful ornithischian dinosaurs. Among this group of large, bulky plant-eaters are the iguanodontids and the duck-billed hadrosaurs, such as *Maiasaura*, which were the most common herbivores in most Late Cretaceous ecosystems. Hadrosaurs and ceratopsians were probably the preferred prey of *Tyrannosaurus*.

CERATOPSIANS
Perhaps the most recognizable ornithischian subgroup is the ceratopsians, the horned-and-frilled herbivores that lumbered around on four legs. The three-horned *Triceratops* is the most familiar member of the group, but a variety of other species with even more bizarre horns also thundered across the North American plains during the Late Cretaceous.

AVES
Birds are theropods, and therefore represent a living subgroup of dinosaurs. The first bird, the beautifully feathered *Archaeopteyx*, is known from the Late Jurassic, about 150 million years ago. However, bird evolution was jump-started in the Cretaceous, as several modern groups evolved and a variety of strange birds, now totally extinct, dominated the skies.

MARSUPIALS
The marsupials, which carry their young in a pouch, are one of the major subgroups of living mammals. Modern marsupials include animals such as kangaroos and opossums. The first marsupials must have evolved in the Jurassic, but their fossils become common in the Cretaceous. They were more geographically widespread than the marsupials of today.

EUTHERIANS
The other major subgroup of living mammals is the placentals, or eutherians, whose young develop inside the mother's womb. The oldest unquestionable eutherian is the tiny *Eomaia* from the Early Cretaceous of China. Throughout the rest of the Cretaceous, eutherians became larger and more common, and today live across the globe.

Hoplopteryx

GROUP Actinopterygians
DATE Late Cretaceous
SIZE 10¹/₂in (27cm)
LOCATION North America, Europe, North Africa, SW. Asia

Hoplopteryx is an extinct relative of the modern slimeheads (see panel, below). Its remains have been found in chalk deposits, which suggests it inhabited a shallow-water environment. *Hoplopteryx* had large eyes, an upturned mouth, and jaws lined with small teeth. It was a deep-bodied fish with small pectoral fins and pelvic fins almost directly below them on the underside. The single anal fin had a straight rear edge, and the tail was deeply forked with equal-sized lobes. *Hoplopteryx* displayed an advance over other fish in the form of a series of bones called uroneurals in the upper lobe of the tail. These bones strengthened and supported the fin rays. Uroneurals in the tail fin allow fish to swim more powerfully, and their presence is considered a major advance in fish evolution. Fish with these skeletal elements are called teleosts.

Special tail fin bones, called uroneurals, gave *Hoplopteryx* an evolutionary advantage over other fish of the time.

 LIVING RELATIVE

SLENDER ROUGHY

A deep-sea reef fish from New Zealand, *Optivus elongatus* is commonly known as the "slender roughy." The roughys belong to a family known as "slimeheads" because of the numerous mucous canals in their heads. During the day, it hides in crevices and cracks in the reef, but at night, it emerges and hunts for prey, which it catches in its extensible jaws and swallows whole.

large eyes on the side of the head provided good vision for hunting prey

wide mouth and jaws lined with tiny teeth

operculum covered the gills

angular dorsal fin supported
by nine unjointed fin rays
increased the height of the fish

part of the powerful,
forked tail fin

deep, flat body

small pelvic fin

small pectoral fin

EARLY

LATE

Berriasian

Valanginian

Hauterivian

Barremian

Aptian

Albian

Cenomanian

Turonian

Coniacian

Santonian

Campanian

Maastrichtian

Xiphactinus

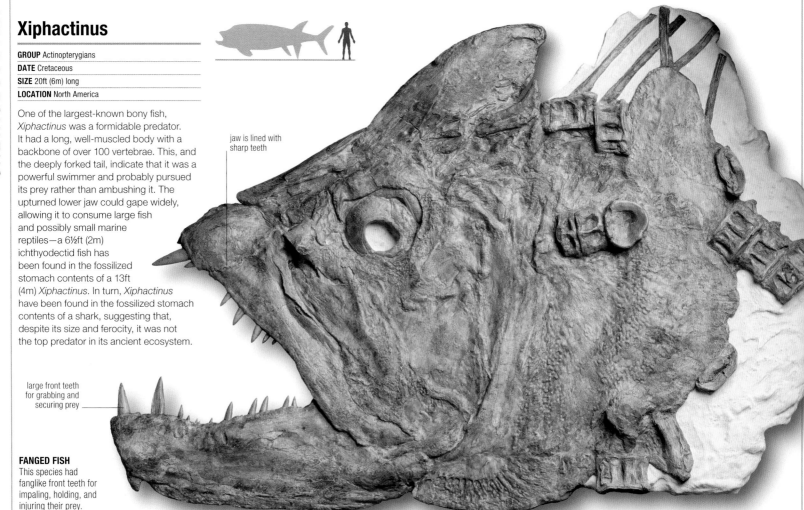

GROUP Actinopterygians

DATE Cretaceous

SIZE 20ft (6m) long

LOCATION North America

One of the largest-known bony fish, *Xiphactinus* was a formidable predator. It had a long, well-muscled body with a backbone of over 100 vertebrae. This, and the deeply forked tail, indicate that it was a powerful swimmer and probably pursued its prey rather than ambushing it. The upturned lower jaw could gape widely, allowing it to consume large fish and possibly small marine reptiles—a 6½ft (2m) ichthyodectid fish has been found in the fossilized stomach contents of a 13ft (4m) *Xiphactinus*. In turn, *Xiphactinus* have been found in the fossilized stomach contents of a shark, suggesting that, despite its size and ferocity, it was not the top predator in its ancient ecosystem.

jaw is lined with sharp teeth

large front teeth for grabbing and securing prey

FANGED FISH
This species had fanglike front teeth for impaling, holding, and injuring their prey.

Squalicorax

GROUP Chondrichthyans

DATE Cretaceous

SIZE 16½ft (5m) long

LOCATION Europe, North America, South America, Africa, Near East, India, Japan, Australia, Russia

Squalicorax means "crow shark." Like *Cretoxyrhina*, it is an extinct member of the mackerel shark group. It had a typical shark body shape, and its teeth were similar in outline to those of modern *Carcharhinus* species, such as the bull shark. It was a top predator and probably fed on mosasaurs, turtles, and fish. Isolated teeth have been found alongside a skeleton of *Cretoxyrhina*, which suggests it may also have scavenged on the carcasses of its larger cousin.

THORNY TOOTH
This tooth has a roughly rectangular root supporting a thornlike crown, with a single, serrated, bladelike cusp.

Lepisosteus

GROUP Actinopterygians

DATE Eocene to present

SIZE 30in (75cm) long

LOCATION North and Central America, Cuba

Lepisosteus (gar pikes) first appeared about 55 million years ago. Today they are found in freshwater habitats in North and Central America and Cuba. They are similar in appearance to lizardfish, and it is possible they evolved in parallel to suit particular niches that existed in both saltwater (lizardfish) and freshwater (gar pike) environments. Modern gars, such as *L. oculatus*, have changed little from their Early Cretaceous ancestors, which makes them "living fossils."

DIAMOND SCALES
Lepisosteus had a long body with dorsal and anal fins situated far down the back. It had a long snout; jaws lined with small, sharp teeth; and heavy, interlocking, diamond-shaped scales over its body.

GAR PIKES

There are seven existing species of gar pikes, but they have changed little from their fossil ancestors. They are easily recognized by their long snouts with nostrils at the tip and their body scales with an enamel-like covering. All species are found in freshwater and brackish water, except *Lepisosteus platostomus*, which is only found in the freshwater drainage systems in North America.

Beelzebufo

GROUP Amphibians

DATE Late Cretaceous

SIZE 15½in (40cm) long

LOCATION Madagascar

Beelzebufo ("devil toad") was only discovered in 2008. Its most remarkable feature is its size, which is far larger than any other known frog or toad, living or fossil. It would have coexisted with the last dinosaurs and been large enough to eat their newly hatched young. *Beelzebufo* is a close relative of the living South American horned frogs that ambush their prey, using their huge mouth and the large, fanglike spikes of bone in their jaws to grab small animals as they pass. Its resemblance to modern South American frogs rather than living Madagascan frogs is yet more support for a past connection between South America, India, and Madagascar in the Cretaceous.

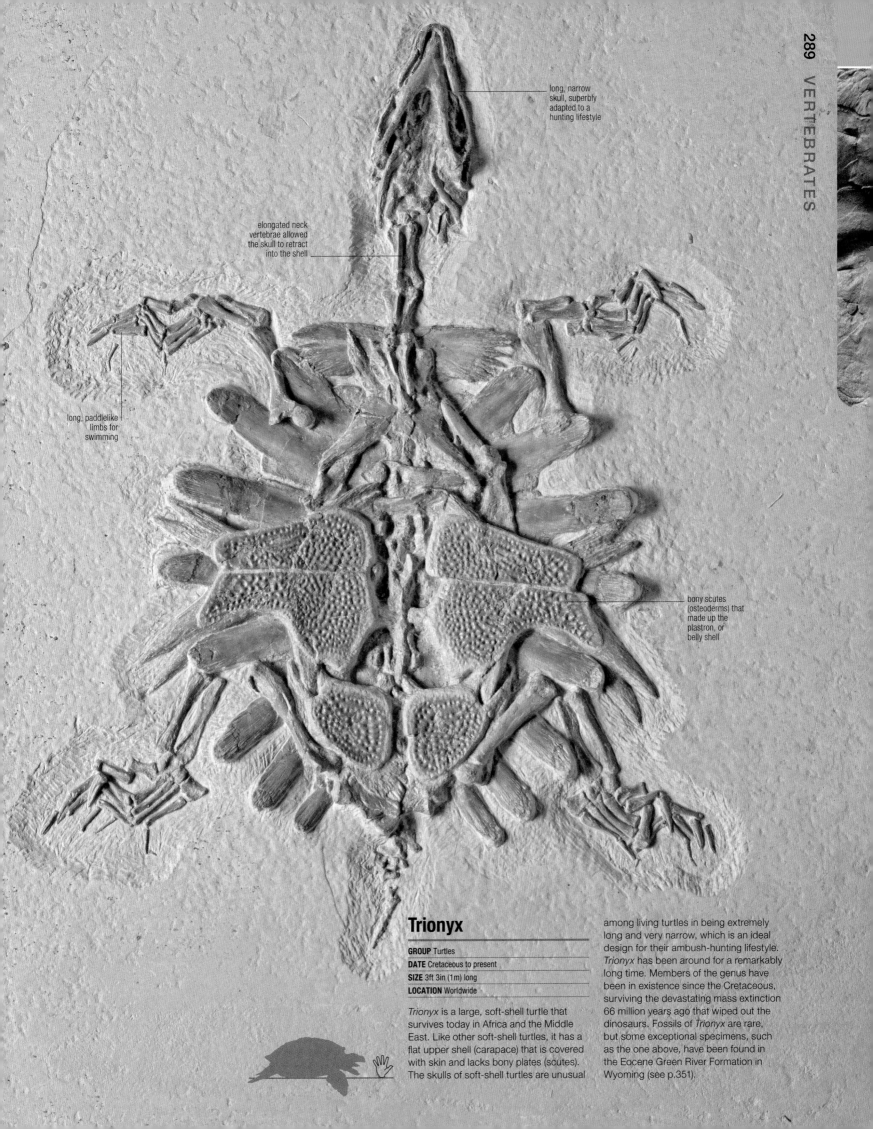

long, narrow skull, superbly adapted to a hunting lifestyle

elongated neck vertebrae allowed the skull to retract into the shell

long, paddlelike limbs for swimming

bony scutes (osteoderms) that made up the plastron, or belly shell

Trionyx

GROUP Turtles

DATE Cretaceous to present

SIZE 3ft 3in (1m) long

LOCATION Worldwide

Trionyx is a large, soft-shell turtle that survives today in Africa and the Middle East. Like other soft-shell turtles, it has a flat upper shell (carapace) that is covered with skin and lacks bony plates (scutes). The skulls of soft-shell turtles are unusual among living turtles in being extremely long and very narrow, which is an ideal design for their ambush-hunting lifestyle. *Trionyx* has been around for a remarkably long time. Members of the genus have been in existence since the Cretaceous, surviving the devastating mass extinction 66 million years ago that wiped out the dinosaurs. Fossils of *Trionyx* are rare, but some exceptional specimens, such as the one above, have been found in the Eocene Green River Formation in Wyoming (see p.351).

Protostega

GROUP Turtles
DATE Late Cretaceous
SIZE 9¾ft (3m) long
LOCATION USA

During the Late Cretaceous, North America was almost cut in two by a vast inland sea that stretched from the Arctic Ocean all the way to the present-day Gulf of Mexico. A strange menagerie of prehistoric life inhabited these warm waters, including enormous fish, toothed birds, and a colossal turtle called *Protostega*. It was one of the largest turtles ever known, reaching lengths of around 9¾ft (3m) and weights over 500lb (226kg), but it was only the second largest turtle of the Cretaceous, after *Archelon*. *Protostega* was fully marine and rarely ventured onto land. Its limbs functioned as effective paddles and its thick shell was relatively light and streamlined so that it could travel quickly through the water. The skull of *Protostega* was characteristic of a turtle: it was short and wide, lacked teeth on its snout, and sported a sharp beak.

FAVORITE FOOD
Although *Protostega* surely fed on fish, it is likely that it targeted jellyfish, squid, and other soft creatures as its preferred prey.

Elasmosaurus

GROUP Sauropterygians
DATE Late Cretaceous
SIZE 30ft (9m) long
LOCATION USA

Elasmosaurus was one of the last plesiosaurs. More than half its total body length was made up of the neck, which contained 71 vertebrae—more than any other animal that has ever existed. This was a source of great confusion for Edward Drinker Cope, who described the first specimen in 1869, because he placed the head on the end of the tail, which he had misidentified as the neck. *Elasmosaurus* had a relatively small head and probably ambushed fish by maneuvering its neck. Its long, narrow teeth would have been perfect for piercing and trapping small, soft prey.

LONGEST NECK
Elasmosaurus belonged to a subgroup of plesiosaurians known as the plesiosauroids. All of them possessed small heads and very long necks, although none of them had a neck as long as *Elasmosaurus*.

BIOGRAPHY
EDWARD DRINKER COPE

Born in Philadelphia, Edward Drinker Cope (1840–1897) was a paleontologist, herpetologist, and evolutionist. He led numerous natural history surveys and fossil hunting expeditions across western North America and described more than 1,000 vertebrate species, including a number of dinosaurs. Even so, Cope is most famous for his intense rivalry with fellow paleontologist Othniel Marsh, which has been dubbed the "Bone Wars."

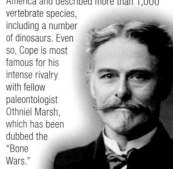

Mosasaurus

GROUP Lepidosaurs
DATE Late Cretaceous
SIZE 49ft (15m) long
LOCATION USA, Belgium, Japan, Netherlands, New Zealand, Morocco, Turkey

For about 20 million years during the Late Cretaceous, the oceans teemed with one of the most spectacular groups of predators to ever evolve: the mosasaurs. These were gigantic relatives of today's lizards and snakes that were adapted for life in the sea. *Mosasaurus* was a voracious, crocodilelike hunter that swam by undulating its long body. As a result, it was incapable of swimming fast over long distances, but it could accelerate very rapidly when required. It probably lived in the well-lit surface waters of the oceans, hunting slower-moving prey. Its bite marks have been found on ammonites and on the shells of large turtles, suggesting that it was capable of catching sizable prey. The first *Mosasaurus* skull was discovered in 1774 in Maastricht, Holland, in a limestone quarry.

FEARSOME LIZARD
Mosasaurs, including *Mosasaurus*, were some of the largest and most fearsome lizards in Earth's history. However, they survived for only a brief period of time—they first appeared in the Late Cretaceous and were wiped out along with the dinosaurs in the Cretaceous extinction.

Kronosaurus

GROUP Sauropterygians
DATE Late Cretaceous
SIZE 30ft (9m) long
LOCATION Australia, Colombia

Named after Kronos, a Titan of Greek legend, *Kronosaurus* was among the last of the giant pliosauroids, a group of plesiosaurians with large heads and short, stocky necks. However, recent work shows that *Kronosaurus* was not as enormous as was once thought, and its estimated length has been reduced from 39ft (12m) to 30ft (9m). Like other pliosauroids, it used four large flippers to swim, but whether it used these to "row" like the oars of a boat or "fly" underwater like a sea turtle is not known. It is possible that the answer is somewhere in between.

POWERFUL JAWS
The powerful, crocodilelike jaws of *Kronosaurus* allowed it to attack other marine reptiles, but most of its diet probably consisted of large fish.

Plioplatecarpus

GROUP Lepidosaurs

DATE Late Cretaceous

SIZE 16½–20ft (5–6m) long

LOCATION North America, Europe

One of the best-known mosasaurs is *Plioplatecarpus*, a midsized predator that lived in the warm, shallow seas of North America and Europe around 80 million years ago. Its skull was long and powerful and studded with a series of thick, conical teeth like those of living crocodiles. Its body was elongated and streamlined, the hands and feet were modified into broad flippers, and the tail was deep and muscular. The jaws of this mosasaur were able to open very wide, enabling it to bite and swallow prey that was much larger than itself. This behavior is also seen in living snakes and is one of the features that marks the two groups as close relatives.

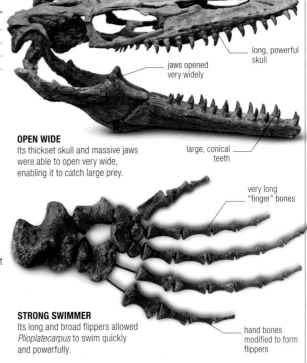

OPEN WIDE
Its thickset skull and massive jaws were able to open very wide, enabling it to catch large prey.

long, powerful skull

jaws opened very widely

large, conical teeth

very long "finger" bones

STRONG SWIMMER
Its long and broad flippers allowed *Plioplatecarpus* to swim quickly and powerfully.

hand bones modified to form flippers

Simosuchus

GROUP Crocodylomorphs
DATE Late Cretaceous
SIZE 5ft (1.5m) long
LOCATION Madagascar

Simosuchus is perhaps the strangest crocodylomorph to have ever lived. Most members of this group were powerful predators with strong, elongated skulls crammed full of sharp teeth. *Simosuchus*, however, had a short skull with a flattened face that resembled a pug dog. Not only that, but it also had a wide snout with a series of leaf-shaped teeth, perfect for shearing and chewing plants. This peculiar, herbivorous crocodile lived in Madagascar during the Late Cretaceous, about 70 million years ago. It lived alongside more characteristic crocodiles that ate meat, as well as many large theropod dinosaurs. *Simosuchus* may have avoided competition with these carnivores by evolving a plant-based diet. Other herbivorous crocodylomorphs also lived during the Cretaceous, although they were rare and none survive today. They shared ecosystems with herbivorous dinosaurs, which was a strange ecological pairing.

STRANGE CREATURE
Although the pug-nosed, plant-eating *Simosuchus* did not resemble the crocodiles of today, it was a close relative of living crocodiles, but with a comparatively bizarre lifestyle and strange body plan.

short skull with flattened face

powerful, crocodilelike tail

wide snout with leaf-shaped teeth

webbed feet for swimming

Deinosuchus

GROUP Crocodylomorphs
DATE Late Cretaceous
SIZE 39ft (12m) long
LOCATION USA, Mexico

The name *Deinosuchus* means "terror crocodile"—and for good reason. Along with *Sarcosuchus*, *Deinosuchus* was one of the largest crocodylomorphs to ever live, weighing up to 10 tons (9 tonnes). However, it lived much more recently than *Sarcosuchus* and is a member of the alligatorid family that also includes the modern alligators. It was one of the most ferocious predators of the North American coastal regions, and in some areas overlapped with tyrannosaurids such as *Daspletosaurus* (see p.301). In these ecosystems, it was *Deinosuchus*, not the tyrannosaurids, that was the largest and most powerful predator. Its anatomy and overall body plan was very similar to that of living crocodiles, so it is easy to imagine it as a giant version of living species. It probably hunted in a similar way to modern crocodiles—by lurking around the water's edge and preying on fish, marine reptiles, and the occasional land animal.

TERROR CROC
Deinosuchus was as long as *Tyrannosaurus*. In a similar way to modern crocodiles, it killed its prey by inflicting great wounds and dragging it under the water and drowning it.

eyes on top of head

long, powerful jaws armed with large teeth

bony, platelike scales

very short legs

Pteranodon

long, toothless jaws

eye socket

huge skull crest tapers to point

knuckle joint

elongated finger formed frame for wing membrane

shoulder joint

elbow joint

CRESTED SKULL
The crest on the back of *Pteranodon*'s skull may have been used to attract mates or even possibly used as a rudder for maneuvering during flight.

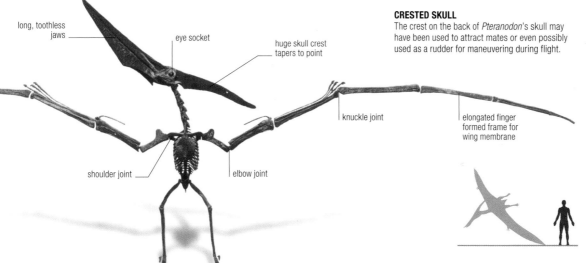

Pteranodon

GROUP Pterosaurs
DATE Late Cretaceous
SIZE 6ft (1.8m) long
LOCATION USA

Pteranodon soared across the shallow seas of North America during the Late Cretaceous. It is likely that it flew and hunted in the same way as an albatross. Vast flocks probably glided over the ocean while looking for fish in the surface waters. It is certain that they ate fish because fish bones have been found in the fossilized stomach of one specimen. The skull of *Pteranodon* was also well adapted for fishing, with long, toothless jaws and a streamlined skull for diving into the water.

Quetzalcoatlus

GROUP Pterosaurs
DATE Late Cretaceous
SIZE 39ft (12m) wingspan
LOCATION USA

The colossal Late Cretaceous pterosaur *Quetzalcoatlus* was one of the largest flying animals of all time. Its wingspan was larger than many small planes, and it was about the same height as a giraffe. Despite its monstrous size, however, *Quetzalcoatlus* weighed no more than 550lb (250kg) thanks to a complex system of air sacs inside most of its bones. For a long time, scientists thought that most pterosaurs ate fish and spent much of their time gliding over the sea, only venturing over land to catch small mammals and lizards. However, it is now believed that *Quetzalcoatlus* and its azhdarchid relatives may have spent most of their time flying over land and targeting large vertebrates as prey. It is likely, therefore, that *Quetzalcoatlus* fueled its enormous metabolic needs by stalking and feeding on dinosaurs.

Quetzalcoatlus was **named after** the ancient Aztec **feathered serpent god** Quetzalcoatl, who was the patron god of Aztec priests.

MONSTER PTEROSAUR
Quetzalcoatlus is the most fantastic of all of the flying pterosaurs. This monstrous creature, named after an Aztec god, was larger than many small planes. While most other pterosaurs ate fish, *Quetzalcoatlus* was a fierce predator that hunted dinosaurs and other vertebrates.

Ornithocheirus

GROUP Pterosaurs
DATE Early Cretaceous
SIZE 16½ft (5m) wingspan
LOCATION Europe

In the Early Cretaceous, about 40 million years before the gigantic *Quetzalcoatlus* stalked dinosaurs across the North American plains (see p.293), another giant pterosaur ruled supreme in Europe. This large beast, *Ornithocheirus*, has long caused headaches for paleontologists because it is known only from fragmentary fossils and often confused with its South American relative *Tropeognathus*. However, the bones suggest that its wingspan may have been around 17ft (5m), about half the size of the more familiar *Quetzalcoatlus*. Another close relative of *Ornithocheirus*, a similar-sized pterosaur called *Anhanguera*, is known from exceptionally well-preserved fossils that have even allowed scientists to examine details of the pterosaur's powerful brain. *Anhanguera* clearly had sharp senses and a strong sense of balance, all of which are necessary for flight, a complex and dangerous activity.

More than **10 species** have been assigned to the genus *Ornithocheirus,* but most of these are based on **fragmentary scraps** of fossils that are **difficult to study.**

SKY GIANTS

Ornithocheirus is one of the most mysterious and majestic pterosaurs, but it has also been a source of frustration for paleontologists for many years. Although it is only known from small pieces of fossil, scientists used to think that *Ornithocheirus* was one of the largest flying animals to ever live, with a wingspan approaching 33ft (10m)—the size of a small airplane. Recent analysis suggests that the wingspan was substantially smaller, and the estimate has been downgraded.

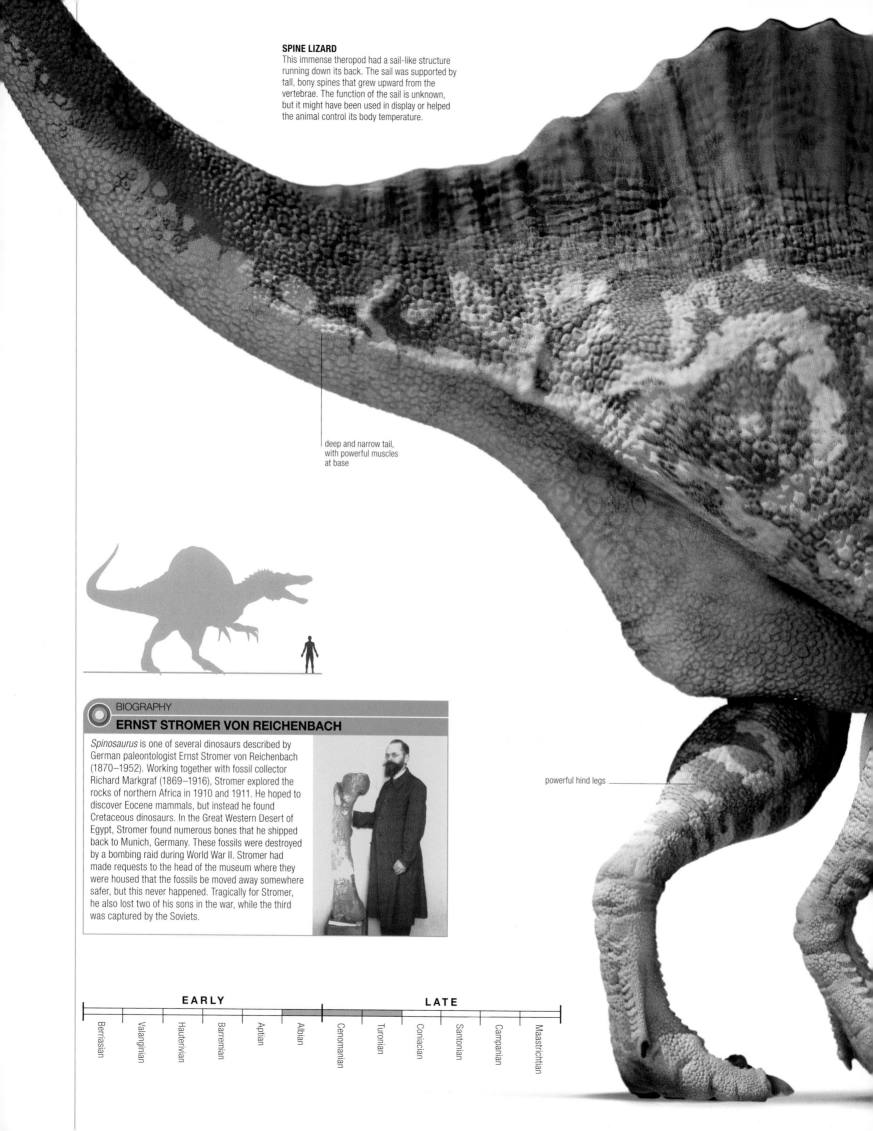

SPINE LIZARD
This immense theropod had a sail-like structure running down its back. The sail was supported by tall, bony spines that grew upward from the vertebrae. The function of the sail is unknown, but it might have been used in display or helped the animal control its body temperature.

deep and narrow tail, with powerful muscles at base

powerful hind legs

BIOGRAPHY

ERNST STROMER VON REICHENBACH

Spinosaurus is one of several dinosaurs described by German paleontologist Ernst Stromer von Reichenbach (1870–1952). Working together with fossil collector Richard Markgraf (1869–1916), Stromer explored the rocks of northern Africa in 1910 and 1911. He hoped to discover Eocene mammals, but instead he found Cretaceous dinosaurs. In the Great Western Desert of Egypt, Stromer found numerous bones that he shipped back to Munich, Germany. These fossils were destroyed by a bombing raid during World War II. Stromer had made requests to the head of the museum where they were housed that the fossils be moved away somewhere safer, but this never happened. Tragically for Stromer, he also lost two of his sons in the war, while the third was captured by the Soviets.

	EARLY							LATE			
Berriasian	Valanginian	Hauterivian	Barremian	Aptian	Albian	Cenomanian	Turonian	Coniacian	Santonian	Campanian	Maastrichtian

Spinosaurus

GROUP Theropods
DATE Early to Late Cretaceous
SIZE 49ft (15m) long
LOCATION Morocco, Libya, Egypt

Spinosaurus is one of the most famous theropods and is also thought to have been the largest. Unfortunately, the first and best specimen of *Spinosaurus* was destroyed by an Allied bombing raid on Germany during World War II. This specimen included a lower jaw and well-preserved vertebrae with tall spines. Since that original discovery by Ernst Stromer (see panel, left), only a few articulated remains have been unearthed. However, a huge number of partial remains show that *Spinosaurus* was a relatively common animal in the Cretaceous Period.

In many ways, *Spinosaurus* was typical of a large theropod. However, in two areas—its skull and vertebrae—it was atypical. Other theropods did not have the spines on the vertebrae, thought to support a vertical sail. The snout was long, like that of a crocodile, with the nostril openings placed well back from the snout tip and closer to the eyes than in most theropods. The tip of the upper jaw was expanded and rounded relative to the rest of the snout, and the teeth in this area radiated outward like the spokes of a wheel. The teeth were rounded in cross-section (most other theropods had oval teeth), and the keels of the teeth were not serrated. All of these features suggest that *Spinosaurus* plunged its jaws into the water and caught fish. However, it was large and powerful enough to also prey on small- and medium-sized dinosaurs on land.

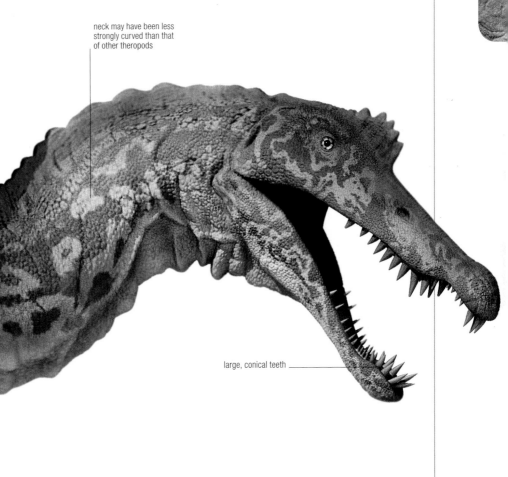

vertical sail supported by spines

neck may have been less strongly curved than that of other theropods

large, conical teeth

stout, muscular arms with three-fingered hands

three long, forward-facing clawed toes

At 49 feet long and **more than 12 tons in weight,** *Spinosaurus* is estimated to be one of the **largest theropods yet discovered.**

Suchomimus

GROUP Theropods
DATE Early Cretaceous
SIZE 33ft (10m) long
LOCATION Niger

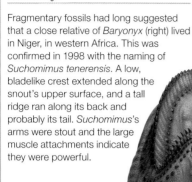

Fragmentary fossils had long suggested that a close relative of *Baryonyx* (right) lived in Niger, in western Africa. This was confirmed in 1998 with the naming of *Suchomimus tenerensis*. A low, bladelike crest extended along the snout's upper surface, and a tall ridge ran along its back and probably its tail. *Suchomimus*'s arms were stout and the large muscle attachments indicate they were powerful.

CROCODILE MIMIC
Suchomimus means "crocodile mimic" so named because it had a very long, narrow, crocodilelike snout.

Baryonyx

GROUP Theropods
DATE Early Cretaceous
SIZE 30ft (9m) long
LOCATION British Isles, Spain, Portugal

Baryonyx walkeri was discovered in 1983 by an amateur paleontologist, and it proved to be one of Europe's most interesting dinosaur fossils.

Crocodilelike jaws and teeth were combined with a fairly typical theropod skeleton. *Baryonyx* is a spinosaurid and belongs to a family named after *Spinosaurus* (see p.296). *Spinosaurus* is known for its crest of tall spines running along its backbone. When first discovered, *Baryonyx* was thought to possess only short spines on the tops of its vertebrae. However, a new specimen has shown that these spines were longer, although still not as long as those of *Spinosaurus*. It seems that *Baryonyx*, and perhaps all spinosaurids, were specialized fish-eating dinosaurs that also ate other prey, including small dinosaurs.

FIRST DISCOVERY
The original *Baryonyx* specimen was found laying on its side. It died by the edge of a pool of water, and its remains were later buried by mud.

ANATOMY
LONG SKULL

Baryonyx had a long, crocodilelike skull. Its nostrils were positioned farther back from the snout tip, and its upper jaw was curved. The skull shape suggests it was a fish-eating theropod that could plunge its long jaws into water.

narrow skull

tall sail along back

slender, curved neck

Irritator

GROUP Theropods
DATE Early Cretaceous
SIZE 26ft (8m) long
LOCATION Brazil

A close relative of *Spinosaurus* (see p.296), *Irritator* was named in 1996 when a near-complete skull was found in the Cretaceous rocks of Brazil. The skull had been modified by its finder to look like a pterosaur skull. As a result, the scientists who studied the specimen were initially misled about its identity, and the name they chose reflects their irritation at having been deceived. As with *Spinosaurus*, a bony crest was present on the upper surface of *Irritator*'s snout. Unlike the crest of *Spinosaurus*, *Irritator*'s extended over the eye sockets. The nostril openings were well away from the tip of the snout, and the teeth had conical crowns. Because no skeleton of *Irritator* has been found to date, little is known of the biology or behavior of this animal. Like other spinosaurids, it might have caught fish but could also have eaten carrion and land animals. Evidence for this feeding behavior comes from the discovery of a spinosaurid tooth embedded within a pterosaur neck bone.

powerful calf muscles

three-fingered hands

short, raised first toe

HEAVY CLAW
Baryonyx had a massive curved thumb claw—its name means "heavy claw"—and its upper arm bone (humerus) had attachment sites for huge muscles.

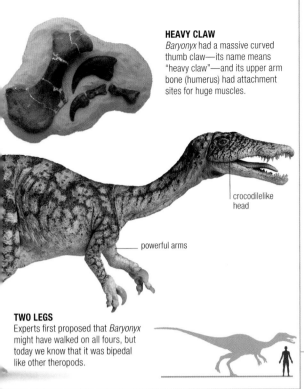

crocodilelike head

powerful arms

TWO LEGS
Experts first proposed that *Baryonyx* might have walked on all fours, but today we know that it was bipedal like other theropods.

TALL SAIL
It is assumed that, like *Spinosaurus*, *Irritator* had a tall sail along its back. However, this remains speculative because *Irritator* is only known from its skull.

long, narrow tail

Scientists who studied *Irritator* were **initially misled** about its identity. The name they chose reflects their **irritation** at having been deceived.

Carcharodontosaurus

GROUP Theropods
DATE Late Cretaceous
SIZE 39ft (12m) long
LOCATION North Africa

Carcharodontosaurus was a giant African allosauroid theropod named in 1931. Its serrated teeth resemble those of *Carcharodon*, the great white shark, hence the dinosaur's name. *Carcharodontosaurus* had a narrow head that was taller at the back than at the front of the snout. A recently discovered partial skull of *Carcharodontosaurus* was more than 5¼ft (1.6m) long. Bony ridges on the top of the skull overhung the eyes, and the bones on the sides of its skull had a distinctive wrinkled texture. Using its massive jaws and long teeth, *Carcharodontosaurus* may have preyed on sauropods and other dinosaurs.

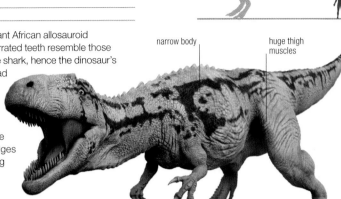

narrow body

huge thigh muscles

MASSIVE SKULL
With its huge frame and sharp teeth, *Carcharodontosaurus* was certainly an accomplished predator.

Giganotosaurus

GROUP Theropods
DATE Late Cretaceous
SIZE 39ft (12m) long
LOCATION Argentina

Giganotosaurus (meaning "giant southern lizard") was similar in size to the largest known individuals of *Tyrannosaurus* (see p.302). Its skull and skeleton strongly resembled *Carcharodontosaurus* (above). The bones above and in front of the eye of *Giganotosaurus* had low, hornlike projections. *Giganotosaurus* lived alongside sauropods like *Limaysaurus*, *Andesaurus*, and *Argentinosaurus* (see p.312), and may have preyed on them.

POWERFUL NECK
Remains suggest that the neck of *Giganotosaurus* was stout and powerful, and supported a large head.

weak arms

DISTINCTIVE FACE
Shelflike bony ridges featured above the eye socket of *Giganotosaurus*. Its lower jaw was squared-off with a distinct chinlike bony lump.

Acrocanthosaurus

GROUP Theropods
DATE Early Cretaceous
SIZE 39ft (12m) long
LOCATION USA

Acrocanthosaurus was a giant North American allosauroid, closely related to the African *Carcharodontosaurus* (above). It had a wide, muscular ridge of tall spines running along its back that would have stiffened if *Acrocanthosaurus* had latched onto a large prey animal. This stiffening might have helped *Acrocanthosaurus* to anchor its own body weight as it pulled and tore at its prey. Its powerful arms had a limited range of movement, but its fingers were equipped with massive curved claws that could withstand a lot of bending. This suggests that its hands might have grabbed prey, but it used its jaws for killing.

deep snout bones

LETHAL JAWS
The jaws of *Acrocanthosaurus* were its primary killing weapon. Its skull was vaguely triangular, the eye sockets were narrow, and a large bony boss was located over each eye.

stout lower jaw

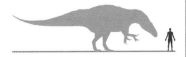

Aucasaurus

GROUP Theropods
DATE Late Cretaceous
SIZE 20ft (6m) long
LOCATION Argentina

Aucasaurus was closely related to *Carnotaurus* (see opposite) and they are united in a group called the Carnotaurini. Its short, deep-snouted skull was not as short or deep as that of *Carnotaurus*. Instead of horns, *Aucasaurus* had low swellings over the top of each eye. *Aucasaurus*'s small arms were also similar to those of its horned relative but were proportionally longer, and the bones lacked some of the unusual proportions and extra bony processes present in *Carnotaurus*. The hand of *Aucasaurus* was unusual: four metacarpal bones were present, but the first and fourth lacked fingers. The second and third were attached to short fingers, but these lacked claws. *Aucasaurus* is known from findings in the Rio Colorado Formation—Late Cretaceous rocks in Argentina that have yielded many dinosaurs, including the theropods *Alvarezsaurus* and *Velocisaurus* and the sauropod *Neuquensaurus*. Numerous sauropod eggs are also known from this deposit.

Known from a **beautifully preserved, near-complete skeleton** that is only **missing the end half of the tail**, *Aucasaurus* was named in 2002.

LIKELY PREDATOR
Little is known about the behavior of *Aucasaurus*. However, like most large theropods, it was almost certainly a predator of other dinosaurs. Perhaps it preyed on smaller theropods and on ornithischians.

Carnotaurus

GROUP Theropods
DATE Late Cretaceous
SIZE 30ft (9m) long
LOCATION Argentina

Carnotaurus, which means "meat-eating bull," was named in 1985 from a well-preserved partial skeleton. Its skull was short and deep for a theropod, and prominent thick, blunt-tipped horns projected from the skull roof above the eyes, which may have been used in display or in fighting. The puny arms, which seemed to lack an obvious function, may have been used in display. Skin impressions preserved with the skeleton show that large, keeled scutes were arranged in rows on the neck and body. Until recently, *Carnotaurus* was the best-known member of the Abelisauridae family. However, the Madagascan *Majungasaurus* is now known from even more complete remains.

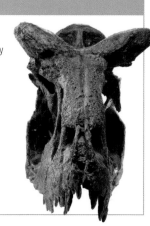

PECULIAR ARMS
The upper arm bones of *Carnotaurus* were long and straight, but the lower arms and hands were very short. The shoulder joint was very mobile, so it may have been able to move its arms more freely than most theropods.

skin was covered in small armor plates

shallow, weak lower jaw

tiny four-fingered hands

feet have yet to be discovered

ANATOMY
BLUNT HORNS

Carnotaurus is famous for possessing blunt horns over its eyes. In life, these were covered by horny sheaths, so their shape in the live animal might have been different from their shape in the fossil: they may have been longer or more pointed, for example. It remains unknown what the horns were for. Some paleontologists suggest that the horns could have been used to intimidate members of other theropod species. Others think that they were used in disputes between rival males.

Santanaraptor

GROUP Theropods
DATE Early Cretaceous
SIZE 5ft (1.5m) long
LOCATION Brazil

The single specimen of this small, poorly known theropod consists of bones from the pelvis, hind limbs, and tail. These provide little information on *Santanaraptor*'s overall appearance, but it was probably a small, generalized, fast-running coelurosaur. *Santanaraptor* is presumed to be similar to animals such as *Dilong* and *Guanlong* (see p.250) and had slim hind limbs. Fragments of preserved muscle and skin tissue were discovered, but unfortunately, there was no trace of the skin's external covering.

Daspletosaurus

GROUP Theropods
DATE Late Cretaceous
SIZE 30ft (9m) long
LOCATION North America

Large and robust, *Daspletosaurus* was a close relative of *Tyrannosaurus* but was geologically older. Its skull was proportionally larger and longer than that of most tyrannosaurids, although not as large, deep, or wide as that of *Tyrannosaurus*. Unlike *Tyrannosaurus*, it had a short, triangular hornlet above and in front of the eyes; a snout with a bumpy, gnarled upper surface; and broad cheeks. A blunt, hornlike lump projected sideways from beneath the eye socket. These features were typical of large tyrannosaurids.

STRONG JAW
The lower jaw of *Daspletosaurus* was powerfully built. The teeth were thick and robust, with long roots that were well embedded in the sturdy jaw bones.

stout, deep jaw bone

Albertosaurus

GROUP Theropods
DATE Late Cretaceous
SIZE 30ft (9m) long
LOCATION Canada

Albertosaurus was closely related to another North American tyrannosaurid, *Gorgosaurus*. In the past, it was thought that the two were similar enough to be included in the same genus. However, they are currently regarded as distinct because they differ in various skull details. *Albertosaurus* also seems to have had more slender hind limbs and proportionally smaller forelimbs than *Gorgosaurus*, but both were similar in size overall. There is an *Albertosaurus* bone bed that contains the remains of numerous juvenile and adult individuals, a discovery that suggests that *Albertosaurus* was a social animal.

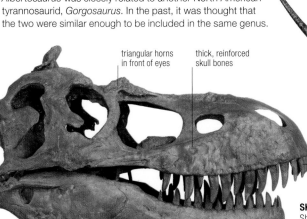

triangular horns in front of eyes

thick, reinforced skull bones

short teeth at jaw tip

SKULL ANATOMY
Shared anatomical features in the skulls of *Albertosaurus* and *Gorgosaurus* indicate that the two tyrannosaurids fed in the same way—by slicing up large prey.

long, slender hind limbs

short, two-fingered arms

strong jaw with deep-rooted teeth

LIGHTWEIGHT RUNNER
Albertosaurus was lightly built for a tyrannosaurid. Some experts suggest that it was good at pursuing and capturing fast-running prey like hadrosaurids.

The strength of *Tyrannosaurus's* **bite** is estimated to be greater than that of **any other animal**.

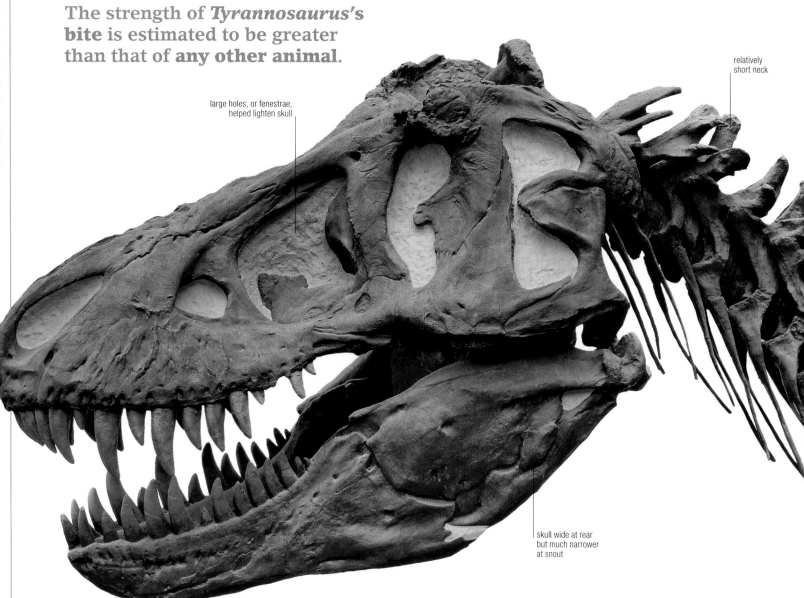

large holes, or fenestrae, helped lighten skull

relatively short neck

skull wide at rear but much narrower at snout

CLAWED FINGERS
Tyrannosaurus had very short arms and hands with two prominent fingers and one vestigial (reduced) finger. The two larger fingers bore sharp, curved claws, although their function was probably very limited.

MASSIVE MOUTH
The massive mouth of *Tyrannosaurus* contained up to 58 serrated teeth. They were different sizes, but the biggest were over 12in (30cm) long, including the root anchoring them in the jaws. Those at the front of the jaws were more closely packed in than those toward the back.

BODY CAVITY
Although it had a very heavily built body and a wide body cavity, some of *Tyrannosaurus*'s back vertebrae had holes that reduced their weight.

LIZARD HIPS
Tyrannosaurus was a saurischian, or "lizard-hipped" dinosaur, with a hipbone arrangement similar to that of most modern reptiles.

sharp claws

Tyrannosaurus

GROUP Theropods
DATE Late Cretaceous
SIZE 39ft (12m) long
LOCATION North America

The genus *Tyrannosaurus* contains the most famous dinosaur, *Tyrannosaurus rex*. Originally discovered in the Late Cretaceous Hell Creek Formation of Montana, it was widespread throughout western North America. *Tyrannosaurus* was one of the largest ever terrestrial carnivores. A fierce, agile predator with a highly developed sense of smell, keen hearing, and a large brain, *Tyrannosaurus* also had an extremely powerful bite force; worn tooth tips and bone fragments preserved in its fossilized feces indicate that it routinely crushed and swallowed bones. Fossils of *Tyrannosaurus rex* are not generally found in the kind of environments that preserve soft tissues such as feathers. However, the discovery in China that fossilized skeletons of its close relatives, *Yutyrannus* and *Dilong*, were covered in feathers has led many scientists to believe that *Tyrannosaurus rex* also had feathers.

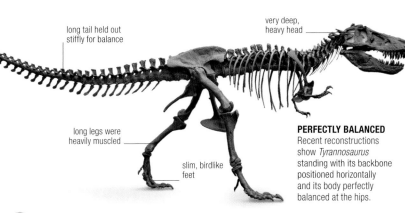

long tail held out stiffly for balance

very deep, heavy head

long legs were heavily muscled

slim, birdlike feet

PERFECTLY BALANCED
Recent reconstructions show *Tyrannosaurus* standing with its backbone positioned horizontally and its body perfectly balanced at the hips.

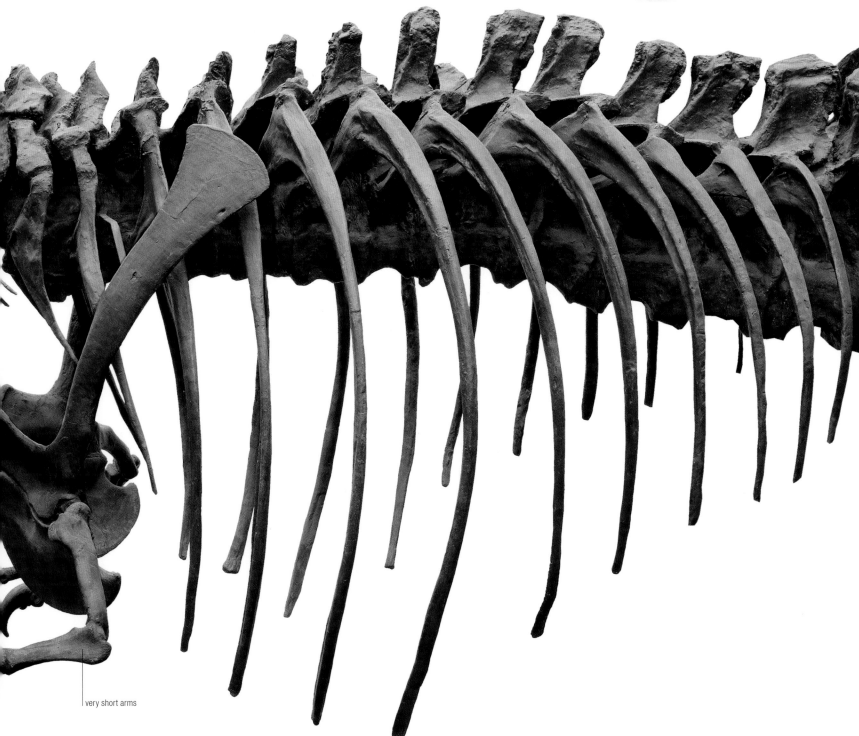

very short arms

BULKY SKULL
Tyrannosaurus's snout and lower jaws were very deep, and the back of its skull was particularly broad, especially across the cheek region. The eye sockets faced forward more than the eyes of any other tyrannosaurid, showing that *Tyrannosaurus* would have had acute binocular vision.

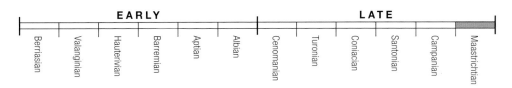

EARLY | LATE

Berriasian | Valanginian | Hauterivian | Barremian | Aptian | Albian | Cenomanian | Turonian | Coniacian | Santonian | Campanian | Maastrichtian

Tarbosaurus

GROUP Theropods
DATE Late Cretaceous
SIZE 39ft (12m) long
LOCATION Mongolia, China

Tarbosaurus was a large Asian tyrannosaurid and a close relative of North America's *Tyrannosaurus* (see p.302). Indeed, several experts have proposed that they should be regarded as different species of the same genus. However, the two differ in many details: for example, *Tarbosaurus* had a different skull and snout and slightly more teeth than *Tyrannosaurus*. These differences might have evolved because the two predators relied on different prey. *Tyrannosaurus* lived alongside giant horned dinosaurs, while *Tarbosaurus* presumably hunted sauropods, hadrosaurs, and ankylosaurs. One *Tarbosaurus* fossil seems to preserve a throat sac beneath its lower jaw. This might have been used as an inflatable display structure during the breeding season.

POWERFUL PREDATOR

Tarbosaurus may have used its powerful jaws and teeth to attack the flanks or thighs of larger prey, including sauropods like the *Nemegtosaurus* shown here.

The **two-fingered forelimbs** of *Tarbosaurus* were even smaller than those of its cousin *Tyrannosaurus*.

fenestrae to lighten skull

stout, serrated teeth

WEAKER SKULL

Tarbosaurus had a narrower, less strongly built skull than the similar *Tyrannosaurus,* but the bones along the top of its snout locked together in the same way.

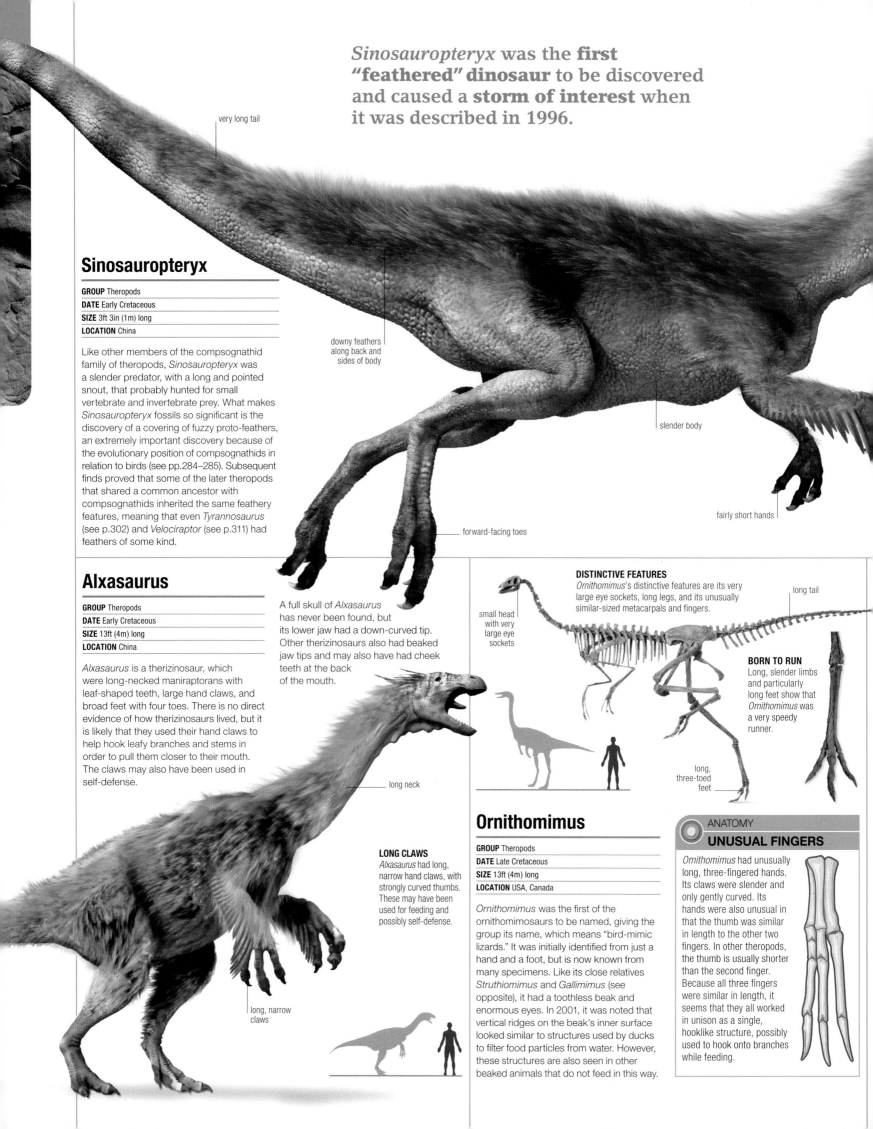

Sinosauropteryx was the **first "feathered" dinosaur** to be discovered and caused a **storm of interest** when it was described in 1996.

very long tail

Sinosauropteryx

GROUP	Theropods
DATE	Early Cretaceous
SIZE	3ft 3in (1m) long
LOCATION	China

Like other members of the compsognathid family of theropods, *Sinosauropteryx* was a slender predator, with a long and pointed snout, that probably hunted for small vertebrate and invertebrate prey. What makes *Sinosauropteryx* fossils so significant is the discovery of a covering of fuzzy proto-feathers, an extremely important discovery because of the evolutionary position of compsognathids in relation to birds (see pp.284–285). Subsequent finds proved that some of the later theropods that shared a common ancestor with compsognathids inherited the same feathery features, meaning that even *Tyrannosaurus* (see p.302) and *Velociraptor* (see p.311) had feathers of some kind.

downy feathers along back and sides of body

slender body

fairly short hands

forward-facing toes

Alxasaurus

GROUP	Theropods
DATE	Early Cretaceous
SIZE	13ft (4m) long
LOCATION	China

Alxasaurus is a therizinosaur, which were long-necked maniraptorans with leaf-shaped teeth, large hand claws, and broad feet with four toes. There is no direct evidence of how therizinosaurs lived, but it is likely that they used their hand claws to help hook leafy branches and stems in order to pull them closer to their mouth. The claws may also have been used in self-defense.

A full skull of *Alxasaurus* has never been found, but its lower jaw had a down-curved tip. Other therizinosaurs also had beaked jaw tips and may also have had cheek teeth at the back of the mouth.

long neck

LONG CLAWS
Alxasaurus had long, narrow hand claws, with strongly curved thumbs. These may have been used for feeding and possibly self-defense.

long, narrow claws

DISTINCTIVE FEATURES
Ornithomimus's distinctive features are its very large eye sockets, long legs, and its unusually similar-sized metacarpals and fingers.

small head with very large eye sockets

long tail

BORN TO RUN
Long, slender limbs and particularly long feet show that *Ornithomimus* was a very speedy runner.

long, three-toed feet

Ornithomimus

GROUP	Theropods
DATE	Late Cretaceous
SIZE	13ft (4m) long
LOCATION	USA, Canada

Ornithomimus was the first of the ornithomimosaurs to be named, giving the group its name, which means "bird-mimic lizards." It was initially identified from just a hand and a foot, but is now known from many specimens. Like its close relatives *Struthiomimus* and *Gallimimus* (see opposite), it had a toothless beak and enormous eyes. In 2001, it was noted that vertical ridges on the beak's inner surface looked similar to structures used by ducks to filter food particles from water. However, these structures are also seen in other beaked animals that do not feed in this way.

UNUSUAL FINGERS

Ornithomimus had unusually long, three-fingered hands. Its claws were slender and only gently curved. Its hands were also unusual in that the thumb was similar in length to the other two fingers. In other theropods, the thumb is usually shorter than the second finger. Because all three fingers were similar in length, it seems that they all worked in unison as a single, hooklike structure, possibly used to hook onto branches while feeding.

thin, pointed head and snout

Struthiomimus

GROUP Theropods
DATE Late Cretaceous
SIZE 14¾ft (4.5m) long
LOCATION Canada

Struthiomimus is one of several closely related ornithomimosaurs. The first *Struthiomimus* specimen was discovered in Alberta, Canada, and consisted of only pelvic and hind limb fragments. However, a far superior specimen, missing only the end of the tail and the skull roof, was discovered later on. *Struthiomimus* would have looked similar to *Ornithomimus*—its closest relative—but had a longer body and tail and shorter hind limbs. Its hands and hand claws were particularly long, and its thumbs did not oppose its fingers, reducing its grasping ability.

long, thin hand claws

very long tail

long, powerful legs

OSTRICHLIKE
Struthiomimus means "ostrich mimic," so called due to their similar leg structures, which were both designed for speed.

GROUNDBREAKING DISCOVERY
Sinosauropteryx was the first "feathered" dinosaur to be discovered and caused a storm of interest when it was described in 1996. Its proto-feathers seem to closely resemble the down feathers found on modern birds.

Gallimimus

GROUP Theropods
DATE Late Cretaceous
SIZE 20ft (6m) long
LOCATION Mongolia

Gallimimus is one of the biggest and best-known ornithomimosaurs. It was originally thought to have a snout that curved upward at its tip, but recent evidence suggests that it actually had a broad and blunt snout tip. Its lower jaw was deeper and shorter than that of other ornithomimosaurs. *Gallimimus* had proportionately shorter arms, smaller hands, and shorter hand claws than other members of this group, which suggests that it used its forelimbs differently to other ornithomimosaurs. It might have raked the ground to uncover food.

slender, flexible neck

long tail

long, powerful legs

fairly short, grasping claws

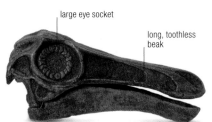

large eye socket

long, toothless beak

BLUNT BEAK
Gallimimus had a long, toothless beak that is noticeably blunt at the end, but it is not clear exactly what it ate. It had large eyes, but did not possess binocular vision.

CHICKEN MIMIC
The anatomy of the neck in *Gallimimus* reminded its describers of the neck of a chicken, which explains why its name means "chicken mimic." Its slender legs and long tail suggest it was a fast runner.

Chirostenotes

GROUP Theropods
DATE Late Cretaceous
SIZE 13ft (4m) long
LOCATION USA, Canada

Chirostenotes was a large North American oviraptorosaur with a long skull and deep, rounded crest located over the top of its head. Its lower jaw was long and shallow, with an upturned, shovel-shaped tip. Two toothlike prongs projected from the middle of the palate, but there were no true teeth. The claw on its second finger was straight in contrast to the curved finger shown on other oviraptorosaurs.

powerful, curved claw

UNIQUE TECHNIQUE
Chirostenotes had two curved claws and one straight one on each hand. It might have used these to probe under rocks, impaling small animals.

Ajancingenia

GROUP Theropods
DATE Late Cretaceous
SIZE 6½ft (2m) long
LOCATION Mongolia

Ajancingenia was a toothless oviraptorosaur with a short, rounded skull. Compared to those of its close relatives, its arms were particularly short and its hands were stout and strong, while its first finger was much larger than the other two. Also, its tail was deeper than that of other oviraptorosaurs. These unusual features all suggest that *Ajancingenia* was doing something quite different to its relatives, but its habits and lifestyle remain a mystery. However, like all oviraptorosaurs, *Ajancingenia* was feathered and birdlike.

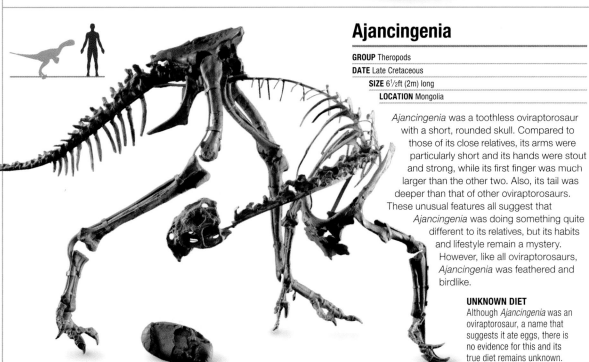

UNKNOWN DIET
Although *Ajancingenia* was an oviraptorosaur, a name that suggests it ate eggs, there is no evidence for this and its true diet remains unknown.

OVIRAPTOR EGGS

Discovered in the Gobi Desert, Mongolia, these fossilized dinosaur eggs were once thought to belong to *Protoceratops* (see p.330). The presence of an *Oviraptor* fossil found next to the nest was assumed to be evidence of its egg-stealing behavior. However, further analysis has revealed that these were, in fact, *Oviraptor* eggs.

Caudipteryx

GROUP Theropods

DATE Early Cretaceous

SIZE 3ft 3in (1m) long

LOCATION China

Despite being a theropod dinosaur, which were typically carnivores, *Caudipteryx* used its large beak to eat plants and seeds, although it may have taken small animal and insect prey as well. Unlike other theropods, *Caudipteryx* did not have a bony crest on the top of its head. A number of complete fossils of this dinosaur have been found, giving us a good idea of how it looked. The abundance of remains also suggest that *Caudipteryx* was a common animal.

EXTENSIVE PLUMES
Fossil feathers show *Caudipteryx* bore large plumes on the arms and a large tail fan, but it was not a flying dinosaur.

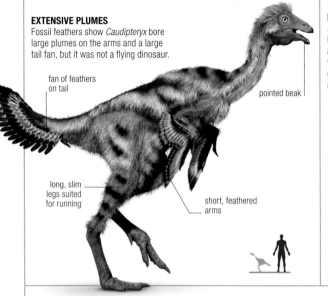

fan of feathers on tail

pointed beak

long, slim legs suited for running

short, feathered arms

Dromaeosaurus

GROUP Theropods

DATE Late Cretaceous

SIZE 6½ft (2m) long

LOCATION North America and Europe

Dromaeosaurus was the first dromaeosaurid to be described. Ironically, however, it remains one of the most poorly known members of the group, and only a partial skull and a few bones from the hand and foot have been described. The skull was deep and broad for a dromaeosaurid, and the teeth at the tip of the upper jaw were wide. The lower jaw was also deep and robust compared to the far more shallow lower jaws of dromaeosaurids such as *Velociraptor* (opposite). These features suggest that *Dromaeosaurus* had a more powerful bite than other dromaeosaurid species.

LONG FEATHERS
Dromaeosaurus would have been feathered, and particularly long feathers would have grown from its hands, arms, and tail.

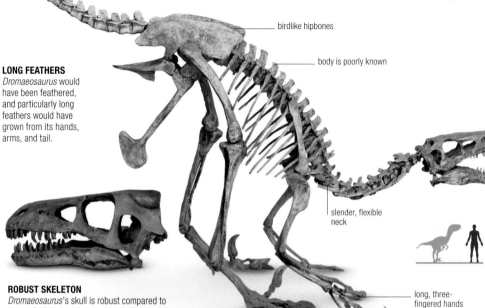

birdlike hipbones

body is poorly known

slender, flexible neck

long, three-fingered hands

ROBUST SKELETON
Dromaeosaurus's skull is robust compared to those of other dromaeosaurids, so the rest of the skeleton may also have been strongly built.

Troodon

GROUP Theropods

DATE Late Cretaceous

SIZE 9¾ft (3m) long

LOCATION North America

Troodon was named for just a single tooth (the name means "wounding tooth"), but we now know that skull and skeletal material originally named *Stenonychosaurus* belong to *Troodon*. *Troodon*'s teeth were coarsely serrated, so much so that some paleontologists have suggested that it might have been able to shred leaves. It was probably mostly predatory, however, and could have preyed on animals ranging from small lizards and mammals to medium-sized ornithischians. In some places, large numbers of *Troodon* teeth are preserved alongside the bones of baby hadrosaurs. It might be that *Troodon* stayed close to hadrosaur colonies during the nesting season, grabbing unguarded young when it could. *Troodon* had large eyes and probably had well-developed binocular vision. In proportion to its size, *Troodon* had a relatively large brain—proportionally one of the biggest of any dinosaur—which suggests that it was a highly intelligent animal and among the smartest dinosaurs. The bowl-shaped nests, eggs, and even some *Troodon* embryos have been discovered, as have preserved adults sitting on top of nests. A typical *Troodon* nest contained up to 24 eggs.

SICKLE-SHAPED CLAW
Deinonychus's most famous feature was the enlarged claw on the second toe, which was probably used to disembowel prey in a series of raking kicks.

long, backward-pointing hipbones

Deinonychus

GROUP Theropods

DATE Early Cretaceous

SIZE 9¾ft (3m) long

LOCATION USA

When named in 1969, *Deinonychus* was used to promote the idea that dinosaurs were not slow and cumbersome animals destined for extinction, but successful, often agile, and perhaps even warm-blooded. Its long fingers were equipped with three large curved claws. Like other maniraptorans, *Deinonychus* was almost certainly feathered, and long feathers called rectrices must have grown from the upper surfaces of its arms and hands.

BIOGRAPHY

JOHN OSTROM

Dinosaurs were considered unsuccessful until Ostrom argued in the 1960s that they were exciting and successful, with complex social lives. He described *Deinonychus*, showed that birds almost certainly descended from *Deinonychus*-type theropods, and studied the biology of hadrosaurs and horned dinosaurs. Many of his ideas and discoveries initiated a new phase in dinosaur research.

Velociraptor

GROUP Theropods

DATE Late Cretaceous

SIZE 6½ft (2m) long

LOCATION Mongolia

Velociraptor was discovered in the Gobi Desert during the 1920s and has become one of the most familiar dromaeosaurids. Its snout was long and narrow, with a concave upper border. Like other dromaeosaurids, *Velociraptor* had long hands, an enlarged claw on its second toe, and a fairly stiff but lightly built tail. A second species of *Velociraptor*, *V. osmolskae*, was named in 2008. It differed from the other species, *V. mongoliensis*, in minor details of skull and tooth anatomy.

FEATHERED DINOSAUR
Quill knobs found on arm bones confirm that, like other dromaeosaurids, *Velociraptor* had feathers.

enlarged claw on second toe

LUCKY FINDS
One spectacularly complete *Velociraptor* specimen is preserved locked in combat with a *Protoceratops*. Thanks to this and other specimens, we now know *Velociraptor*'s anatomy in detail.

SCIENCE

MALE PARENTING

Small theropods, including *Troodon*, sat on egg-filled nests. *Troodon* folded its legs under its body and may have used its feathered arms to cover the eggs. Like female birds, female dinosaurs possessed a special type of bone, known as medullary bone, used in the production of eggshell. This is absent in the brooding dinosaur specimens, suggesting that they might be males. This supports the idea that both male and female *Troodons* contributed to nest care, much like modern ostriches.

OPPORTUNISTIC PREDATOR
Troodon was probably an opportunistic predator that ate a variety of animal prey. Its long, slim legs show that it was a fast runner. Like other small maniraptorans, it would have had a feathery coat.

Microraptor

GROUP Theropods
DATE Early Cretaceous
SIZE 4ft (1.2m) long
LOCATION China

This small, feathered dinosaur from China is related to the more famous *Velociraptor* (see p.311) and other dromaeosaurs. However, unlike those speedy runners, *Microraptor* was more at home in the trees, gliding from branch to branch. It spent most of its time hunting small prey, such as lizards and early mammals, and used its aerobatic skills to avoid predators. *Microraptor* did not fly, but glided like modern "flying" squirrels. Its feathered arms acted as glide surfaces, and its legs may have helped steer the animal through the air. *Microraptor* was not a direct ancestor of birds, but its lifestyle suggests a form of locomotion that may have been the forerunner of powered flight in birds.

DINOSAUR FEATHERS
This fossilized *Microraptor* clearly shows the long arms and associated wing feathers. The long feathers on the hind legs are also visible, on the right below the stiff tail.

Argentinosaurus

GROUP Sauropodomorphs
DATE Late Cretaceous
SIZE 98ft (30m) long
LOCATION Argentina

One of the largest-known sauropods, *Argentinosaurus* was probably a primitive member of the group. Its broad vertebrae had small, peg-and-socket articulations that were located just above the spinal cord opening. Peg-and-socket structures were common in saurischians, and probably kept the animals' backbones rigid. These spinal features were absent in later titanosaurs (the lithostrotians), but it is unknown why they evolved more flexible backbones. *Argentinosaurus*'s immense ribs were hollow, cylindrical tubes.

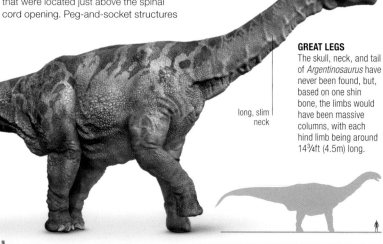

long, slim neck

GREAT LEGS
The skull, neck, and tail of *Argentinosaurus* have never been found, but, based on one shin bone, the limbs would have been massive columns, with each hind limb being around 14¾ft (4.5m) long.

Amargasaurus

GROUP Sauropodomorphs
DATE Early Cretaceous
SIZE 33ft (10m) long
LOCATION Argentina

Amargasaurus was an unusual-looking sauropod. It was relatively small and short-necked, with pairs of long spikes that projected from the top of its 12 neck vertebrae. The function of these spikes is as yet unknown. *Amargasaurus* may have fed on ground-level vegetation, while taller sauropods foraged for foliage higher up.

MYSTERIOUS SPIKES
Amargasaurus's spines may have supported skin sails or formed an array of horn-covered spikes.

FAMILY LIKENESS
This *Nemegtosaurus* has been modeled on *Rapetosaurus*, a close relative from Madagascar. All titanosaurs had wide bodies, and their flexible necks allowed them to feed high up in the trees.

Nemegtosaurus

GROUP Sauropodomorphs
DATE Late Cretaceous
SIZE 36ft (11m) long
LOCATION Mongolia

Only the skull of the Mongolian sauropod *Nemegtosaurus* has ever been found. At first glance, it resembles that of the Jurassic diplodocid *Dicraeosaurus*, and as a result *Nemegtosaurus* was originally thought to be a late-surviving diplodocid. More recent work, however, has shown that *Nemegtosaurus* was actually a

titanosaur, and therefore more closely related to animals such as *Saltasaurus* (see p.314). While some titanosaurs had stout, spoon-shaped teeth and a short skull, *Nemegtosaurus* had pencil-shaped teeth and a long snout. Some experts have reconstructed the dinosaur's head showing a rounded, bony hump on the top of its skull. If such a hump had existed, *Nemegtosaurus*'s head would have looked very much like a brachiosaur's. However, it was more likely to have been low and subtle—the most recent studies show that the back of the skull was very tall compared to the snout, and that the whole skull was long and boxy. It is possible that *Opisthocoelicaudia*, a Mongolian titanosaur, could be a *Nemegtosaurus* specimen.

Whereas most **titanosaurs** have been found **without their heads,** only the skull of *Nemegtosaurus* has ever been found.

Saltasaurus

GROUP Sauropodomorphs
DATE Late Cretaceous
SIZE 39ft (12m) long
LOCATION Argentina

Saltasaurus is one of the best known of the titanosaurs. These were a group of sauropods traditionally thought to have been mostly restricted to the Southern Hemisphere but now known to have been more widespread. Unlike other sauropods, some titanosaurs were armored, and *Saltasaurus* was one of the first titanosaurs found that demonstrated this. The upper surface and sides of its body were covered with large, oval armor plates, some of which may have been tipped with spikes. Thousands of smaller, rounded bones covered the skin between the large plates. As was the case in most other titanosaurs, *Saltasaurus* had very broad hips and its body would have been wide and rounded. Its limbs were stout and it had a flexible tail.

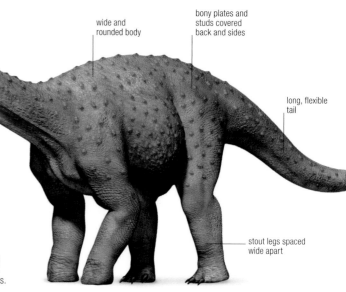

wide and rounded body

bony plates and studs covered back and sides

neck shorter than tail

long, flexible tail

stout legs spaced wide apart

ARMOR-PLATED
Unlike most other sauropods, *Saltasaurus* was covered with bony plates and studs. It is thought that its armor covering might have helped to protect it from attacks by large theropods.

Minmi

GROUP Ornithischians
DATE Early Cretaceous
SIZE 9¾ft (3m) long
LOCATION Australia

Minmi is unique in having strange extra bones, called paravertebrae, along its back. These might have helped provide increased support for its back muscles.

Small, rounded armor plates covered its body, including its belly. One specimen includes a skull that is broad and deep, with a short, narrow snout and large eye sockets. Another *Minmi* specimen provides direct evidence of its diet: preserved gut contents show that it ate fruit.

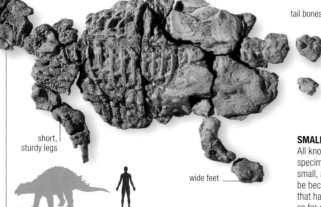

tail bones

short, sturdy legs

wide feet

SMALL SPECIMENS
All known *Minmi* specimens are relatively small, although this might be because all of those that have been discovered so far are juveniles.

tail spikes would have been deadly weapons

long shoulder spines

flattened, triangular spikes

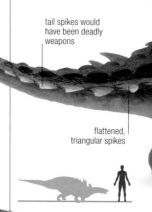

Gastonia

GROUP Ornithischians
DATE Early Cretaceous
SIZE 16½ft (5m) long
LOCATION USA

Gastonia is one of the world's best-known ankylosaurs, and most parts of its skeleton have been discovered. *Gastonia*'s skull was shallow and broad with a wide, square-tipped beak. As in nearly all

IMPRESSIVE DEFENSE
Flattened, triangular spikes projected from the side of *Gastonia*'s body and tail, and long spines pointed upward from the shoulder region. Smaller, oval plates were arranged across the animal's back.

ankylosaurs, its teeth were small and leaf-shaped. The bones forming the roof of the skull were thickened and domed, and a special joint around the bones that housed the brain may have provided a shock-absorbing function. Some experts speculate that these dinosaurs butted heads together when fighting.

continuous armor covering on back and tail

long, conical spikes on neck and shoulders

Sauropelta

GROUP Ornithischians
DATE Early Cretaceous
SIZE 16½ft (5m) long
LOCATION USA

Sauropelta was a large, long-tailed, North American nodosaurid. Thanks to a well-preserved specimen that includes a near-complete skull, its anatomy is reasonably well known. *Sauropelta* had more teeth in its lower jaw than most other nodosaurids. The back of the skull was much broader than the snout, and the top of the skull was flat. The upper surfaces of its body and tail were covered in oval bone plates that formed a continuous armor covering. The name *Sauropelta* means "shield lizard." Long, conical spikes projected upward and sideways from its neck and shoulders. Its relatively long tail was made up of more than 40 vertebrae.

FORMIDABLE SPIKES
It had been thought that only a single row of spikes was present on each side of *Sauropelta*'s neck, but new work has shown that there were two rows on each side. These spikes would have been formidable weapons.

Ankylosaurus

GROUP Ornithischians
DATE Late Cretaceous
SIZE 23ft (7m) long
LOCATION North America

The largest of the ankylosaurids, *Ankylosaurus* was a giant, club-tailed animal with large, triangular horns at the back of its skull. Its snout was short and broad. Small, leaf-shaped teeth lined the sides of its jaws, and at the front, its toothless beak was broad and deep. The sides of the snout appeared to bulge outward, and the nostrils were directed sideways. *Ankylosaurus* was very similar to one of its closest relatives, *Euoplocephalus* (see pp.316–317).

head spike

nostril

teeth

BULGING SKULL
The bulging shape of *Ankylosaurus*'s snout was created by complicated air passages running through the skull. These were also present in several other ankylosaurids, but their function is unknown.

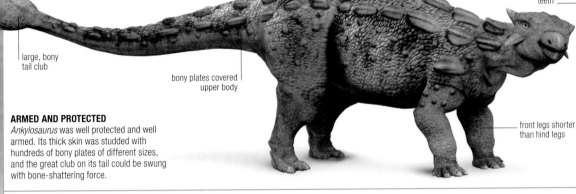

large, bony tail club

bony plates covered upper body

front legs shorter than hind legs

ARMED AND PROTECTED
Ankylosaurus was well protected and well armed. Its thick skin was studded with hundreds of bony plates of different sizes, and the great club on its tail could be swung with bone-shattering force.

Edmontonia

GROUP Ornithischians
DATE Late Cretaceous
SIZE 23ft (7m) long
LOCATION North America

Edmontonia was one of the largest and most widely distributed nodosaurids. One excellent specimen has allowed a good understanding of its anatomy and appearance. Bands of armor plates covered the upper surface of the neck and shoulder region and smaller plates covered the rest of the back and tail. Several long spikes projected from each shoulder. The first two spikes pointed diagonally forward, and the two farther back pointed sideways. *Edmontonia*'s skull had a long, low snout and its eye sockets were placed far back.

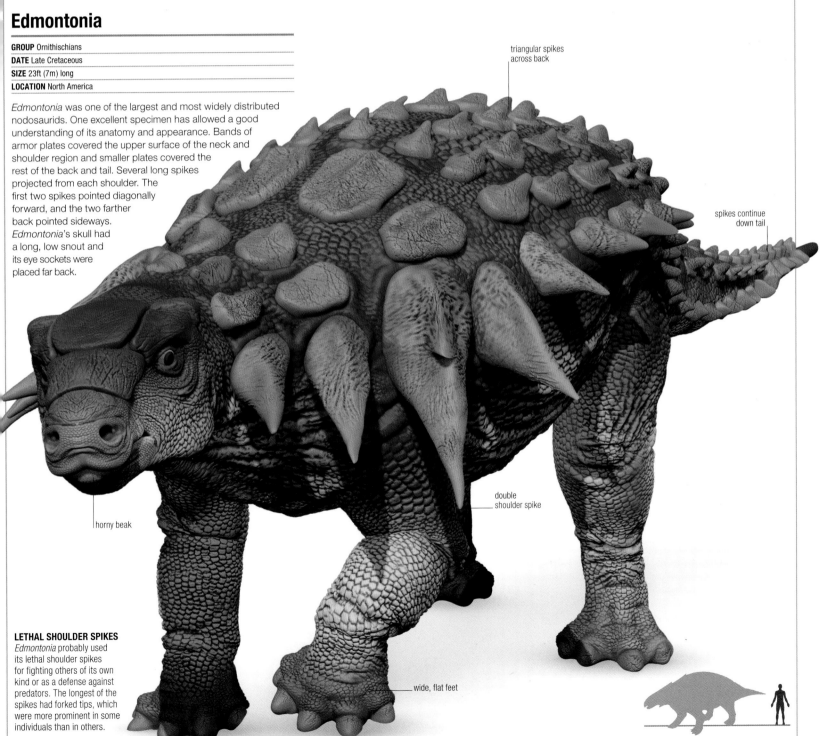

triangular spikes across back

spikes continue down tail

horny beak

double shoulder spike

wide, flat feet

LETHAL SHOULDER SPIKES
Edmontonia probably used its lethal shoulder spikes for fighting others of its own kind or as a defense against predators. The longest of the spikes had forked tips, which were more prominent in some individuals than in others.

mouth surrounded
by a horny beak

long, convoluted nasal
passages inside the skull
suggest *Euoplocephalus*
had a strong sense of smell

three toes on
hind feet

each toe tipped
with a blunt hoof

Euoplocephalus

GROUP	Ornithischians
DATE	Late Cretaceous
SIZE	20ft (6m) long
LOCATION	North America

Euoplocephalus, which means "well-armored head," is one of the largest and best known of the ankylosaurids. Specimens have been found with almost all their thick plates of studded armor in place. *Euoplocephalus* was closely related to *Ankylosaurus* (see p.315), and shared many of the same characteristics. The low-slung body and hips of *Euoplocephalus* were so broad that it must have been almost round in cross-section. The limbs were short and stocky, and only three

toes, tipped with blunt hooves, were present on each hind foot—other ankylosaurs had four toes. The vertebrae at the end of the tail were fused to form a stiffened, rodlike structure that helped to support the large, rounded club at the tip of the tail. The club was formed of four bony plates, held off the ground, and probably used in defense against predators. The base of the tail was almost certainly flexible enough to allow sideways movement, and the tail would have been quite muscular. *Euoplocephalus* was a herbivore, probably grazing on low vegetation and possibly even digging up roots and tubers. Remains are usually found singly, but the discovery of 22 young *Euoplocephalus* has raised the possibility that they may have lived in herds.

Mobile **bony shutters in its eyelids** protected *Euoplocephalus's* eyes.

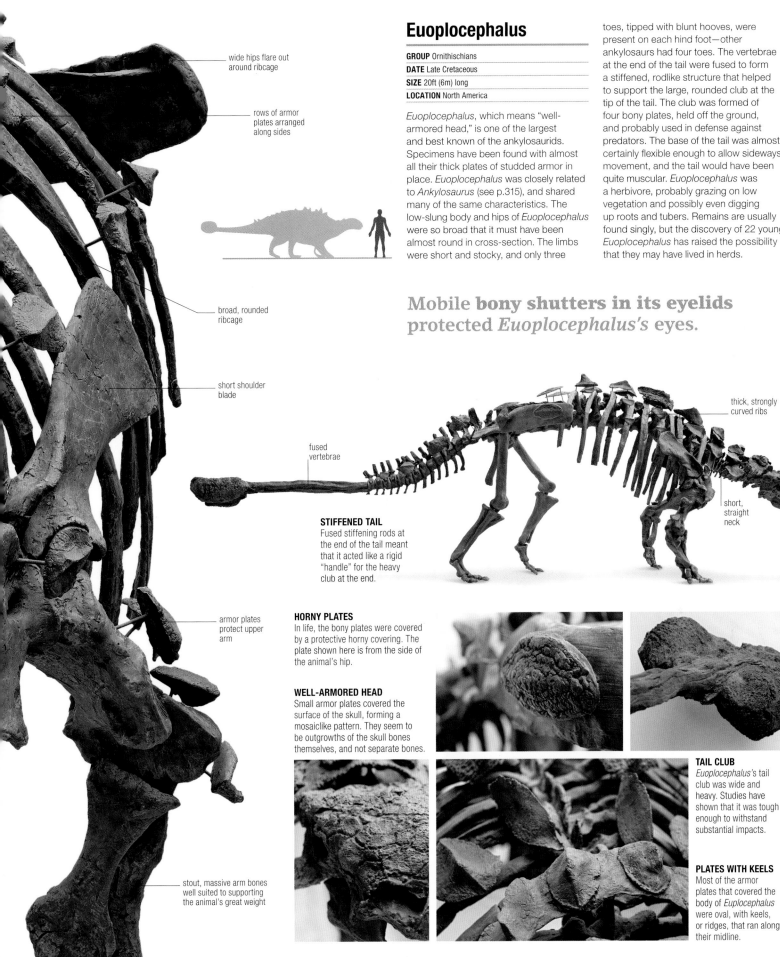

wide hips flare out around ribcage

rows of armor plates arranged along sides

broad, rounded ribcage

short shoulder blade

armor plates protect upper arm

stout, massive arm bones well suited to supporting the animal's great weight

thick, strongly curved ribs

short, straight neck

fused vertebrae

STIFFENED TAIL
Fused stiffening rods at the end of the tail meant that it acted like a rigid "handle" for the heavy club at the end.

HORNY PLATES
In life, the bony plates were covered by a protective horny covering. The plate shown here is from the side of the animal's hip.

WELL-ARMORED HEAD
Small armor plates covered the surface of the skull, forming a mosaiclike pattern. They seem to be outgrowths of the skull bones themselves, and not separate bones.

TAIL CLUB
Euoplocephalus's tail club was wide and heavy. Studies have shown that it was tough enough to withstand substantial impacts.

PLATES WITH KEELS
Most of the armor plates that covered the body of *Euoplocephalus* were oval, with keels, or ridges, that ran along their midline.

	EARLY						LATE					
Berriasian	Valanginian	Hauterivian	Barremian	Aptian	Albian	Cenomanian	Turonian	Coniacian	Santonian	Campanian	Maastrichtian

Iguanodon

GROUP	Ornithischians
DATE	Early Cretaceous
SIZE	33ft (10m) long
LOCATION	Belgium, Germany, France, Spain, England

One of the most famous ornithopods, *Iguanodon* is best known for the many near-complete skeletons found in a Belgian coal mine. These were originally reconstructed standing upright in a kangaroo-style posture. However, it is now thought that *Iguanodon* walked mainly on all fours, with its body and tail held parallel to the ground. Its arms were long and robust and well adapted for bearing its weight.

fingers joined together

finger able to grip

hand bone

HAND FOSSIL
In *Iguanodon*'s hands, the middle three fingers were joined together, the fifth finger could curl to grasp food, and the thumb was armed with a vicious spike.

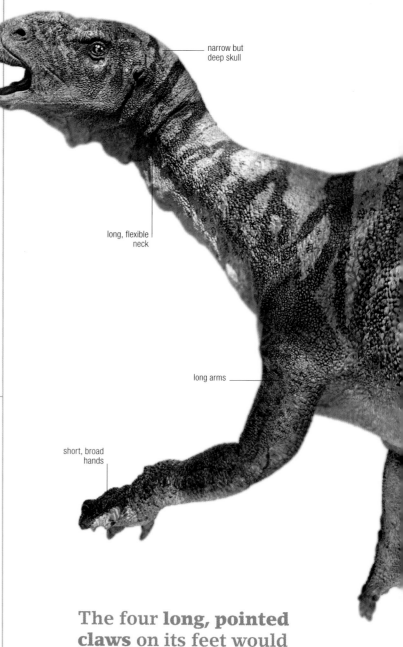

narrow but deep skull

long, flexible neck

long arms

short, broad hands

The four long, pointed claws on its feet would have given *Tenontosaurus* a **dangerous kick.**

Ouranosaurus

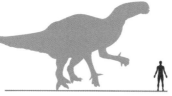

ducklike bill

GROUP	Ornithischians
DATE	Early Cretaceous
SIZE	23ft (7m) long
LOCATION	Niger

Ouranosaurus was discovered in the desert in Niger during the 1960s, and has become one of the most famous ornithopods. This is due to the remarkably tall bony spines that grew upward from the vertebrae on its back. In life, these would have been embedded in a sail formed of muscle and skin. The function of this sail is unknown, but it may have been used in display or in controlling body temperature. Many other unrelated dinosaurs, such as the giant theropod *Spinosaurus* (see pp.296–297), had similar sails. Small rounded horns in front of its eyes made *Ouranosaurus* the only known horned ornithopod.

spines longest just behind shoulder

DUCKLIKE BILL
Ouranosaurus was unusual in having a broad, ducklike bill, which resembled the mouth of hadrosaurs. Because of this, some experts have argued that they were closely related.

Leaellynasaura

GROUP	Ornithischians
DATE	Early Cretaceous
SIZE	3ft 3in (1m) long
LOCATION	Australia

Leaellynasaura is one of several small ornithopods known from a famous fossil site in Victoria, Australia, called Dinosaur Cove. When *Leaellynasaura* was alive, southern Australia was within the Antarctic Circle, and although the polar regions were less cold than they are today, it would have been continually dark for several months a year. An internal cast of *Leaellynasaura*'s braincase shows that it had large optic lobes—the parts of the brain associated with eyesight. So it probably had large eyes and good sight for seeing in the dark.

narrow tail

long, slender legs

very large eyes

STRIKING EYES
A small, bipedal herbivore, *Leaellynasaura*'s most striking feature was its very large eyes, used for seeing in the dark.

Hypsilophodon

GROUP	Ornithischians
DATE	Early Cretaceous
SIZE	6½ft (2m) long
LOCATION	England, Spain

One of the best-known small ornithopods, *Hypsilophodon* is known from several near-complete skeletons. At one time, the misconception that it had grasping hands and a first toe that was directed backward led some to think that it was a tree-climber. In fact, its rigid tail and long hind limbs and feet show that it was a fast-running ground animal. Leaf-shaped teeth show that, like other small ornithopods, it browsed on low-growing plants. Pointed teeth were present at the front of the upper jaw, and the beaked jaw tips were pointed.

slim tail held out straight

VERTEBRAE FOSSIL
Hypsilophodon fossils, like this spine section, were first discovered on the Isle of Wight, England.

SPECIALIZED TENDONS

Tenontosaurus takes its name from the specialized tendons present in its back, hips, and tail—"tenon" is the Greek word for tendon. These enabled it to raise its tail when walking on all fours.

specialized tendons in back, hips, and tail helped to suspend long tail

extremely long, deep tail

long, powerful hind legs

clawed feet

Tenontosaurus

GROUP Ornithischians
DATE Early Cretaceous
SIZE 23ft (7m) long
LOCATION USA

Tenontosaurus was a large and particularly long-tailed iguanodontian. Its long forelimbs and short, broad hands suggest that it walked on all fours, although it was probably able to rear on its back legs when feeding or fighting. Its skull was deep, and its nostril openings were long. Two species are known. In one, the front of the upper jaw was toothless, but in the other, teeth were present in the same region. The outer rim of the beak at the tip of the lower jaw was serrated. *Tenontosaurus* is best known for having been discovered alongside remains of the theropod *Deinonychus* (see p.311). This probably shows that the theropod preyed on *Tenontosaurus*. Groups of juvenile specimens have been discovered at two different locations, which suggests that young animals remained in groups after hatching.

HEAD STRUCTURE

Hypsilophodon's head was tall and short, with large eyes. Behind a bony beak, its mouth had cheek pouches that it used when chewing food.

very large eyes

pointed beak

short, weak arms

five-fingered hands

four-toed feet with claws

Muttaburrasaurus

GROUP Ornithischians
DATE Early Cretaceous
SIZE 23ft (7m) long
LOCATION Australia

Muttaburrasaurus was a large iguanodonlike ornithopod with a deep bony bump on the top surface of its snout. The skull bones beneath its eye sockets were thick and strong. Some experts have suggested that these bones might show

that *Muttaburrasaurus* was adapted for biting and chewing especially tough plants. It was thought to be related to *Iguanodon* (see opposite)—a spike-shaped bone was thought to be a thumb spike like that of *Iguanodon*. However, more recent studies have shown that *Muttaburrasaurus* was far more primitive.

deep, bony lump on snout

relatively long neck

GIANT NOSE

Muttaburrasaurus might have used its giant nose lump to make loud, resonating calls. The size and shape of its nose differed between individuals, probably depending on sex or species variations.

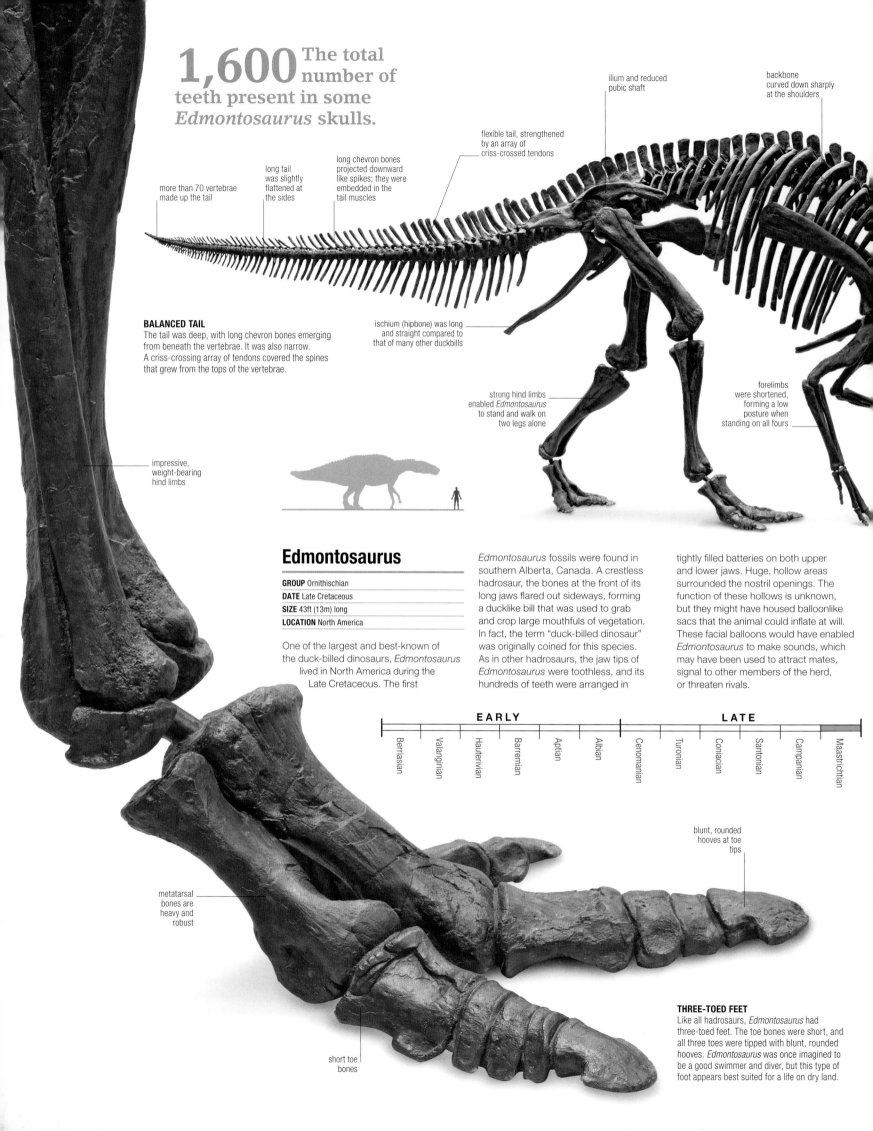

1,600 The total number of teeth present in some *Edmontosaurus* skulls.

more than 70 vertebrae made up the tail

long tail was slightly flattened at the sides

long chevron bones projected downward like spikes; they were embedded in the tail muscles

flexible tail, strengthened by an array of criss-crossed tendons

ilium and reduced pubic shaft

backbone curved down sharply at the shoulders

BALANCED TAIL
The tail was deep, with long chevron bones emerging from beneath the vertebrae. It was also narrow. A criss-crossing array of tendons covered the spines that grew from the tops of the vertebrae.

ischium (hipbone) was long and straight compared to that of many other duckbills

strong hind limbs enabled *Edmontosaurus* to stand and walk on two legs alone

forelimbs were shortened, forming a low posture when standing on all fours

impressive, weight-bearing hind limbs

Edmontosaurus

GROUP Ornithischian

DATE Late Cretaceous

SIZE 43ft (13m) long

LOCATION North America

One of the largest and best-known of the duck-billed dinosaurs, *Edmontosaurus* lived in North America during the Late Cretaceous. The first

Edmontosaurus fossils were found in southern Alberta, Canada. A crestless hadrosaur, the bones at the front of its long jaws flared out sideways, forming a ducklike bill that was used to grab and crop large mouthfuls of vegetation. In fact, the term "duck-billed dinosaur" was originally coined for this species. As in other hadrosaurs, the jaw tips of *Edmontosaurus* were toothless, and its hundreds of teeth were arranged in

tightly filled batteries on both upper and lower jaws. Huge, hollow areas surrounded the nostril openings. The function of these hollows is unknown, but they might have housed balloonlike sacs that the animal could inflate at will. These facial balloons would have enabled *Edmontosaurus* to make sounds, which may have been used to attract mates, signal to other members of the herd, or threaten rivals.

EARLY						LATE					
Berriasian	Valanginian	Hauterivian	Barremian	Aptian	Albian	Cenomanian	Turonian	Coniacian	Santonian	Campanian	Maastrichtian

metatarsal bones are heavy and robust

blunt, rounded hooves at toe tips

short toe bones

THREE-TOED FEET
Like all hadrosaurs, *Edmontosaurus* had three-toed feet. The toe bones were short, and all three toes were tipped with blunt, rounded hooves. *Edmontosaurus* was once imagined to be a good swimmer and diver, but this type of foot appears best suited for a life on dry land.

large head with no bony cranial crest

hollows on the sides of the jaws show that duckbills had cheeks

DUCK-BILLED SKULL
Viewed from the front, the skull really does look "duck-billed," thanks to its expanded jaw tips. The leading edge of the lower jaw is wide and scooplike. A distinctive pitted bone texture shows where beak tissue covered the jaws; some hadrosaur specimens have beak tissue still preserved in place. The large eye sockets indicate that *Edmontosaurus* had a wide field of vision. The snout is dominated by the hollows surrounding the nostrils.

unique skull form, resembling that of a modern duck

eye sockets positioned to allow a wide field of view

large eye sockets faced partly forward but also sideways

part of nostril opening; most of nostrils hidden from view

NECK VERTEBRAE
Edmontosaurus had 13 neck vertebrae, which formed a gently curved neck. Ball-and-socket-type articulations show that the neck was quite flexible.

GRINDING TEETH
The teeth were packed in vertical rows at the backs of the jaws. Only the teeth in the top row form the grinding surface.

pitted surface where beak tissue covered jaw tips

extra, beaklike bone at tip of lower jaw

PROJECTIONS ON SPINE
The bony spines on the tops of the vertebrae were short in *Edmontosaurus*. The ribcage was deep and narrow.

cervical vertebra

SLIM HANDS
The four-fingered hand was slim, and at least three of the fingers were tipped with blunt hooves. There was no thumb.

Brachylophosaurus

GROUP Ornithischians
DATE Late Cretaceous
SIZE 30ft (9m) long
LOCATION North America

Brachylophosaurus, whose name means "short-crested lizard," had a flattened, sheetlike crest that grew backward from the snout, overhanging the back of the skull. An excellent *Brachylophosaurus canadensis* specimen was found in Alberta, Canada, after which it is named. Some individuals—presumably the males—were more heavily built than others, with deeper, lower jaws, stouter skulls, and wider crests that extended farther along the skull. In 2000, an exceptionally complete specimen was discovered in Montana. It is covered in large amounts of preserved skin and promises to provide much information on the appearance of this dinosaur.

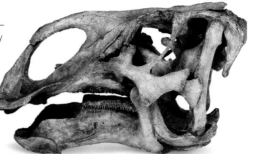

SKULL FEATURES
Brachylophosaurus had a particularly deep snout and a rectangular skull. Its nostril openings were enormous. Its jaw tips were stout and wide.

Maiasaura

GROUP Ornithischians
DATE Late Cretaceous
SIZE 30ft (9m) long
LOCATION USA

Maiasaura became world-famous thanks to the discovery of nests, eggshell fragments, and the remains of juveniles, all found alongside the skeletons of adults. As suggested by the generic name, which means "good mother lizard," the remains indicate that *Maiasaura* formed nesting colonies, with parents constructing crater-shaped nests in which the hatchlings stayed for an extended period, being fed and looked after by their parents. It is possible that these behavioral traits were true of all hadrosaurids. *Maiasaura*'s skull possessed an expanded bill and a solid crest that extended across the top of the skull above the eyes. Several views have been proposed on the lineage of *Maiasaura*, but it shares some features with *Brachylophosaurus*, and the two appear to be close relatives.

COVERED NESTS
The rounded eggs of *Maiasaura* were covered by vegetation and sediment. When they hatched, the babies may have dug their own way out, or they may have been helped by their parents.

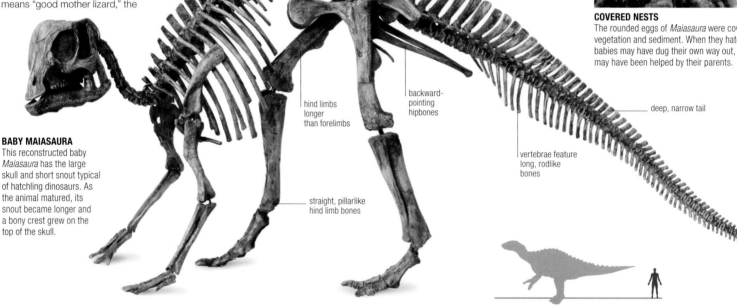

BABY MAIASAURA
This reconstructed baby *Maiasaura* has the large skull and short snout typical of hatchling dinosaurs. As the animal matured, its snout became longer and a bony crest grew on the top of the skull.

straight, pillarlike hind limb bones

hind limbs longer than forelimbs

backward-pointing hipbones

vertebrae feature long, rodlike bones

deep, narrow tail

Parasaurolophus

GROUP Ornithischians
DATE Late Cretaceous
SIZE 30ft (9m) long
LOCATION North America

This most remarkable hadrosaurid is famous for the tubular crest on the back of its skull. As was the case in all lambeosaurines, *Parasaurolophus*'s crest was hollow and contained complex internal passages. The chambers within the crest may have been used in making deep, resonating calls. *Parasaurolophus* was a particularly heavily built hadrosaurid with shorter, stouter limbs than most other kinds. Its particularly large shoulder and hip girdles show that it had large, powerful muscles. These features suggest to some experts that it was an inhabitant of deep forests, where it pushed its way through undergrowth. Several species of *Parasaurolophus* are known, and they differ in the length and shape of the crest.

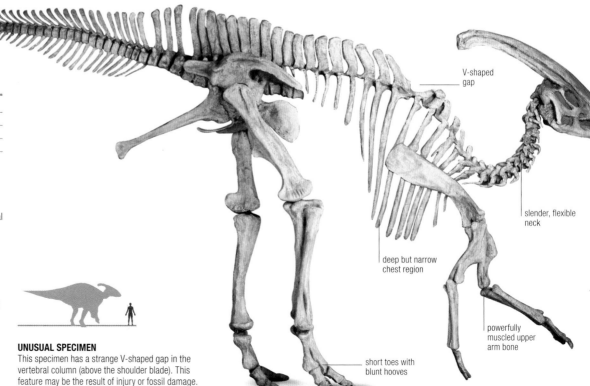

UNUSUAL SPECIMEN
This specimen has a strange V-shaped gap in the vertebral column (above the shoulder blade). This feature may be the result of injury or fossil damage.

V-shaped gap

slender, flexible neck

deep but narrow chest region

powerfully muscled upper arm bone

short toes with blunt hooves

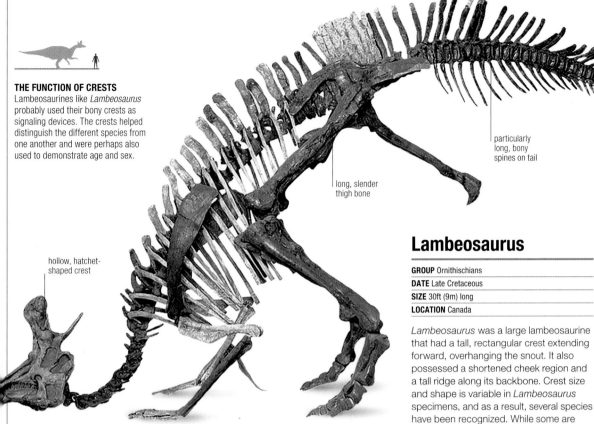

THE FUNCTION OF CRESTS
Lambeosaurines like *Lambeosaurus* probably used their bony crests as signaling devices. The crests helped distinguish the different species from one another and were perhaps also used to demonstrate age and sex.

short, bony spines on tail tip

particularly long, bony spines on tail

tail held well up off ground

long, slender thigh bone

hollow, hatchet-shaped crest

Lambeosaurus

GROUP Ornithischians
DATE Late Cretaceous
SIZE 30ft (9m) long
LOCATION Canada

Lambeosaurus was a large lambeosaurine that had a tall, rectangular crest extending forward, overhanging the snout. It also possessed a shortened cheek region and a tall ridge along its backbone. Crest size and shape is variable in *Lambeosaurus* specimens, and as a result, several species have been recognized. While some are probably distinct species, others might be males, females, or juveniles of one species. In *L. lambei*, the crest had a rectangular front portion that pointed upward and a spike at the back that was directed upward and backward. *L. clavinitialis* had a much shorter spike at the back and might be the female of *L. lambei*. In *L. magnicristatus*, there was no posterior spike, and the crest blade was large. Skin impressions show that small, nonoverlapping nodulelike scales covered *Lambeosaurus*, and it seems to have lacked the large conical tubercles present on the underside of *Corythosaurus* (see pp.324–325).

Gryposaurus

GROUP Ornithischians
DATE Late Cretaceous
SIZE 30ft (9m) long
LOCATION North America

One of the "hook-nosed" hadrosaurines, *Gryposaurus* and its relatives had forelimbs that were about two-thirds the length of the hind limbs, unlike most hadrosaurids, which had forelimbs that were about half as long as the hind limbs. Why these hadrosaurids had such long arms is unknown, but it is likely that this adaptation allowed them to feed from higher up in the vegetation than the other herbivores that shared their environment. The best known *Gryposaurus* species is *G. notabilis*; but a second species, *G. latidens*, was named in 1992; and a third, *G. monumentensis*, in 2007. A very similar hadrosaurid—*Kritosaurus incurvimanus*—is regarded by some experts as another species of *Gryposaurus*. It has even been suggested that *K. incurvimanus* and *G. notabilis* might be the female and male of the same species.

HOOK-NOSED
The bones that formed the upper surface of *Gryposaurus*'s snout arched upward and a large depression surrounded the nostril. This enlarged nasal region may have been brightly colored and used in display or to shove against rivals in disputes.

Skin impressions show that *Gryposaurus* had **frills running along the backbone** that were made of small, **triangular segments of skin.**

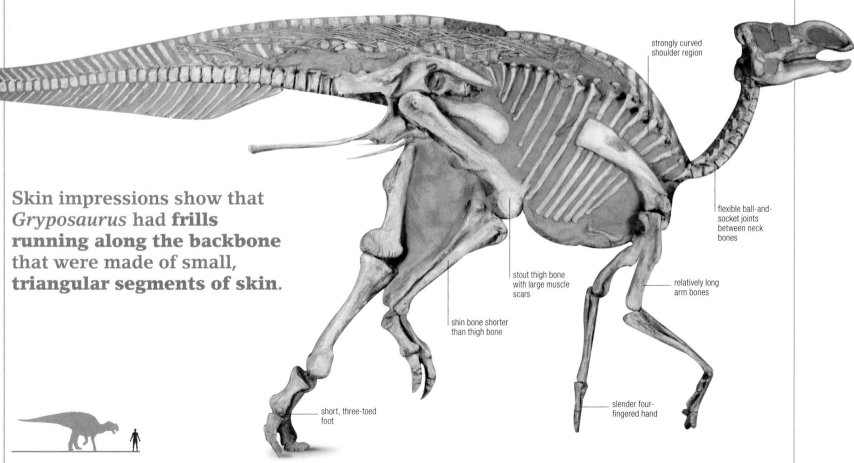

strongly curved shoulder region

flexible ball-and-socket joints between neck bones

relatively long arm bones

stout thigh bone with large muscle scars

shin bone shorter than thigh bone

slender four-fingered hand

short, three-toed foot

Corythosaurus

GROUP Ornithischians
DATE Late Cretaceous
SIZE 30ft (9m) long
LOCATION Canada

The name *Corythosaurus* means "helmet lizard." It was known for its hollow, platelike crest, and was most closely related to *Velafrons* from Mexico, *Nipponosaurus* from Russia, and *Hypacrosaurus* from the USA. Together, these hadrosaurids are known as the fan-crested lambeosaurines. *Corythosaurus* was a large dinosaur with tall spines in a ridge along its back. Its snout was shallow and delicate compared to that of many other hadrosaurids, suggesting that it may have been a more selective feeder that browsed for the juiciest fruits and youngest leaves. Much is known about *Corythosaurus*, because several complete specimens have been found. Some even have preserved skin impressions, which show that *Corythosaurus* had lines of conical tubercles along its underside and a continuous skin frill along the top of its backbone.

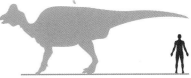

FORAGERS
Although it may have wandered through swampy areas, *Corythosaurus* probably spent most of its time foraging for leaves in woodland habitats.

 ANATOMY
CREST FOR CALLS

The crest of *Corythosaurus* was not solid bone—it contained tubes that were connected to the nostrils. All lambeosaurine species had different crests. It was once thought that the crests were air tanks, but this theory was dismissed when it was realized the animals were not aquatic. It has also been proposed that the crests were lined with nasal tissue, to help provide a sensitive sense of smell. But the most popular theory is that they were used to make loud, resonating calls.

Corythosaurus's helmetlike crest

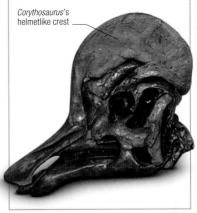

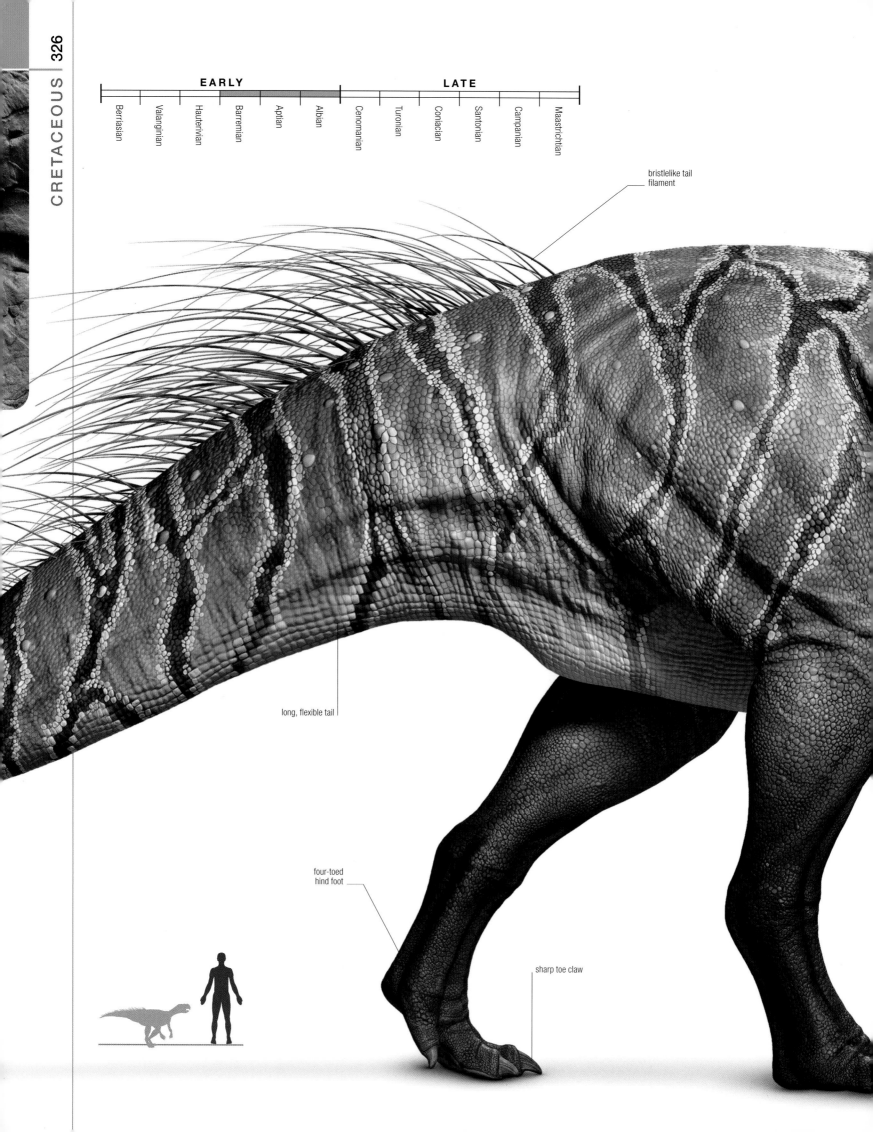

EARLY

Berriasian
Valanginian
Hauterivian
Barremian
Aptian
Albian

LATE

Cenomanian
Turonian
Coniacian
Santonian
Campanian
Maastrichtian

bristlelike tail filament

long, flexible tail

four-toed hind foot

sharp toe claw

Psittacosaurus

GROUP Ornithischians
DATE Early Cretaceous
SIZE 6¹⁄₂ft (2m) long
LOCATION Asia

Psittacosaurus is one of the earliest members of Ceratopsia, the horned dinosaur group. It is also one of the best-known Mesozoic dinosaurs owing to the number of specimens that have been discovered. This has enabled experts to identify over 10 species according to skull shape, although not all experts agree with the classifications. Unlike later ceratopsians, *Psittacosaurus* was probably bipedal. It had four-fingered hands and long hind limbs. Its short, deep skull had a narrow, toothless beak, which gave *Psittacosaurus* its name, meaning "parrot lizard." Hornlike bony growths projected sideways from the cheeks, and differed in size and shape between the species. Young *Psittacosaurus* specimens have also been discovered. In one case, dozens of young were preserved along with the remains of an adult. One *Psittacosaurus* specimen was found with numerous long filaments growing from the upper surface of its tail. Nothing like this has been seen in any other specimens, and the function of these bizarre structures remains a mystery.

covering of
scaly skin

short, stout
neck

flaring horn on
the cheek

neck vertebra

jumbled skull
bones

hind limb folded
up beneath
its body

skin impressions
around the skeleton

bristles embedded
within skin of the tail

tail vertebra

short, stout
arm bone

PERFECT PRESERVATION
This spectacular *Psittacosaurus* fossil includes virtually the entire skeleton, as well as skin impressions and tail bristles. Bristlelike structures like this were present elsewhere in ornithischians, so perhaps they were widespread among these dinosaurs. The specimen is preserved lying on its back, and its skull bones are jumbled.

gently arched back

massive shoulder blades

massive, wide chest cavity

very sturdy front legs

SOLID SKULL

Triceratops, which means "three-horned face," was one of the largest ceratopsians. In addition to being very distinctive, with its neck frill and horns, its skull was also very solidly built. This has resulted in *Triceratops* skulls being more readily fossilized than most dinosaur skulls. More than 50 *Triceratops* skulls have been found.

Pachycephalosaurus

GROUP Ornithischians
DATE Late Cretaceous
SIZE 16½ft (5m) long
LOCATION North America

Pachycephalosaurus was the biggest of the pachycephalosaurs. It is best known from a single 24in (60cm) long domed skull. Its name means "thick-headed lizard." It had large eyes, which suggests it had good vision, and small teeth, indicating it was either a herbivore or an omnivore. Two other North American pachycephalosaurs lived at the same time as *Pachycephalosaurus*: *Stygimoloch* and *Dracorex*. Both had smaller domes and larger horns than *Pachycephalosaurus*. However, some experts think that all these animals simply represent different growth stages of the same species.

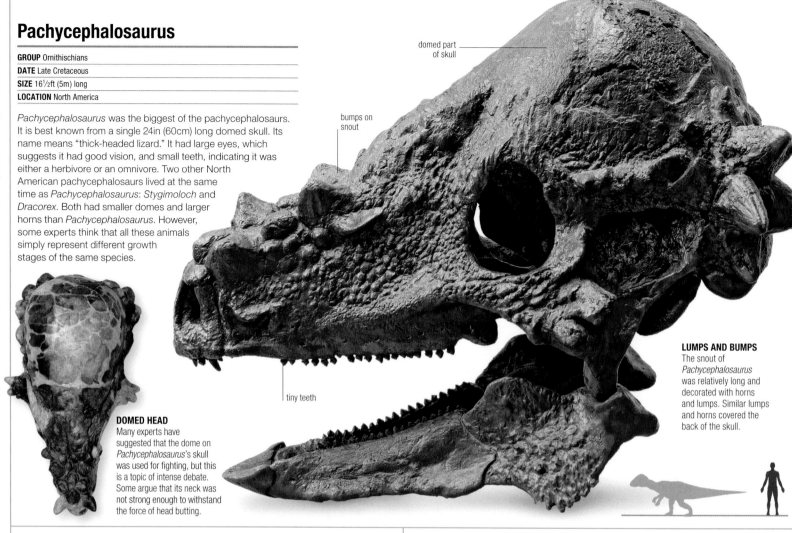

domed part of skull

bumps on snout

tiny teeth

LUMPS AND BUMPS
The snout of *Pachycephalosaurus* was relatively long and decorated with horns and lumps. Similar lumps and horns covered the back of the skull.

DOMED HEAD
Many experts have suggested that the dome on *Pachycephalosaurus*'s skull was used for fighting, but this is a topic of intense debate. Some argue that its neck was not strong enough to withstand the force of head butting.

Stegoceras

GROUP Ornithischians
DATE Late Cretaceous
SIZE 6½ft (2m) long
LOCATION North America

Stegoceras is one of the best known of the pachycephalosaurs. A prominent bony shelf decorated with bony knobs and spikes projected from the back of its domed skull roof—*Stegoceras* means "roofed horn." *Stegoceras* had a short face and, while its snout was narrow, its cheeks flared out. Its small, coarsely serrated teeth were most likely used in chewing and shredding leaves. Two species previously thought to belong to *Stegoceras* have recently been reclassified to separate genera. *Colepiocephale* lacked the bony shelf, as did *Hanssuesia*, which also had wider, flatter edges to its skull dome.

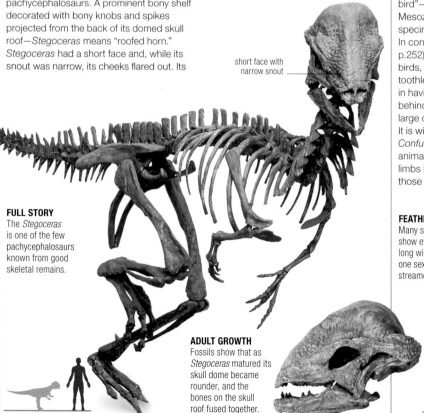

short face with narrow snout

FULL STORY
The *Stegoceras* is one of the few pachycephalosaurs known from good skeletal remains.

ADULT GROWTH
Fossils show that as *Stegoceras* matured its skull dome became rounder, and the bones on the skull roof fused together.

Confuciusornis

GROUP Theropods
DATE Early Cretaceous
SIZE 12in (30cm) long
LOCATION China

Confuciusornis—the "Confucius bird"—is one of the best-known Mesozoic birds and hundreds of specimens have been discovered. In contrast to *Archaeopteryx* (see p.252) and other more primitive birds, *Confuciusornis* was entirely toothless. It was also unusual in having a robust, bony bar right behind the eye, and exceptionally large curved thumb claws. It is widely thought that *Confuciusornis* preyed on aquatic animals. Its bill shape and hind limbs have been compared to those of modern kingfishers.

FEATHERY FRIENDS
Many specimens of *Confuciusornis* show evidence of feathers. The bird had long wing feathers, and possibly only one sex exhibited the two long, streamerlike tail feathers shown here.

long tail feathers

Iberomesornis

GROUP Theropods
DATE Early Cretaceous
SIZE 6in (15cm) long
LOCATION Spain

Originally described from a single, headless skeleton, *Iberomesornis* was a small, finch-sized Cretaceous bird. It is one of the most primitive members known of the entirely Cretaceous bird group known as the enantiornithines, or "opposite birds." They were closer to modern birds than they were to primitive species in that they had shortened tails and large chest bones. However, like the much older *Archaeopteryx* (see p.252), they had teeth.

MODERN FEATURES
The wing bones of *Iberomesornis* show that it was capable of flight. Its curved claws suggest that it perched in trees like modern birds.

curved foot claws and back-turned toe

relatively short tail

Gansus

GROUP Theropods
DATE Early Cretaceous
SIZE 8in (20cm) long
LOCATION China

Gansus was an amphibious diving bird with large feet. Fossilized skin impressions show that its feet were webbed all the way to the tips of the toes, so *Gansus* was probably similar to modern loons and grebes in its diving ability. Unfortunately, its skull remains unknown so its feeding behavior is still a mystery. *Gansus* does not have any close relatives. However, its features suggest that it was an early member of Ornithurae, the group that includes later Cretaceous forms, as well as all modern birds.

DIVING BIRD
This fossil, discovered along with a number of others in China in 2006, clearly shows *Gansus*'s long wing bones, indicating that it was a strong flier. Prior to this discovery, only one specimen was known.

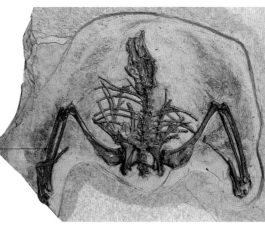

long wing bones

Hesperornis

GROUP Theropods
DATE Late Cretaceous
SIZE 3ft 3in (1m) long
LOCATION North America

Hesperornis (meaning "western bird") was a large, flightless seabird with tiny wings, massive feet, and a toothed bill. First named in the 1870s, it is the best-known member of a group known as the hesperornithines, most of which were flightless. The wings on *Hesperornis* were so small that the hands and even lower arms were absent, and only a small, rod-shaped upper arm bone remained. The legs were placed far back on the body, as they are in modern birds that swim underwater. The legs were long and could close up tightly when the bird was pulling its leg back toward its body while diving. Skin impressions preserved with one specimen show that the toes were not webbed paddles but had large, fleshy lobes projecting from their sides.

WATER CREATURE
The flightless *Hesperornis* probably fed while in the water. It had small, conical teeth in its long, pointed bill, which was well-suited for catching fish.

long toes aid movement through water

legs suited to swimming

very small wings

long, slender neck

Ichthyornis

GROUP Theropods
DATE Late Cretaceous
SIZE 12in (30cm) long
LOCATION USA

Ichthyornis (the "fish bird") is one of the most famous of fossil birds. When originally described, in 1872, it was one of only a handful of Mesozoic birds that seemed to bridge the gap between *Archaeopteryx* and modern birds. Today, many more Mesozoic birds are known. Studies have shown that *Ichthyornis* is more closely related to modern birds than were more archaic groups like the enantiornithines. However, unlike modern birds, *Ichthyornis* had teeth. These were small, smooth, and strongly curved, and were well suited for grabbing small and slippery prey, such as fish.

sharp beak with teeth

webbed feet used for swimming

GULL-LIKE FEATURES
Ichthyornis probably looked much like a modern gull. Many of its fossils have been found in rocks in marine environments.

Vegavis

GROUP Theropods
DATE Late Cretaceous
SIZE 12in (30cm) long
LOCATION Antarctica

Discovered in 1992 on Vega Island in western Antarctica, *Vegavis* is a relative of *Presbyornis*, a fossil waterfowl from the Paleocene and Eocene of North America and part of the anseriform group Anatidae, which includes modern ducks, geese, and swans. The significant discovery of *Vegavis* shows that waterfowl, properly called anseriforms, were definitely alive during the Late Cretaceous. Furthermore, their closest relatives—the gallinaceous birds, or gamebirds—must have been present at this time as well, along with even earlier modern bird groups, such as the paleognaths, which include ostriches.

MISSING LINK
Vegavis would have looked like a long-legged duck. Its skull is unknown, but like other fossil anseriforms, it probably had a ducklike bill.

Vincelestes

GROUP Early mammals
DATE Early Cretaceous
SIZE 12in (30cm) long
LOCATION Argentina

Compared to many Mesozoic mammals, *Vincelestes* is known from excellent fossils. Of nine specimens, six include skulls. These show that *Vincelestes* had large, stout canine teeth and fewer premolars and molars. It also had a short, deep snout. These features suggest that it was a predator, perhaps preying on smaller mammals, as well as on reptiles and large insects. *Vincelestes* was large compared to most Mesozoic mammals, and its length was increased by a very long tail.

Repenomamus

GROUP Early mammals
DATE Early Cretaceous
SIZE 3ft 3in (1m) long
LOCATION China

Repenomamus, meaning "reptilelike mammal," is one of the most famous of Mesozoic mammals. Whereas most mammals from this time were the size of mice and shrews, *Repenomamus* was a giant in comparison. One species, *R. giganticus*, had a skull that was 6½in (16cm) long and a total body length of about 3ft 3in (1m) long. Its jaws were stoutly constructed, and it was definitely a predator of smaller animals, including baby dinosaurs.

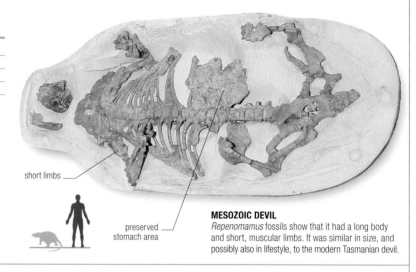

short limbs

preserved stomach area

MESOZOIC DEVIL
Repenomamus fossils show that it had a long body and short, muscular limbs. It was similar in size, and possibly also in lifestyle, to the modern Tasmanian devil.

Volaticotherium

GROUP Early mammals
DATE Middle Jurassic to Early Cretaceous
SIZE 8in (20cm) long
LOCATION China

It was once thought that Mesozoic mammals were all tiny, scurrying, shrewlike animals. However, new discoveries have shown more diversity than was once thought. *Volaticotherium* is especially surprising because it was clearly a glider. Large skin membranes stretched between its body and long limbs, and the shapes of its finger and toe bones and its large claws show that *Volaticotherium* was a good climber. It was probably an insectivore.

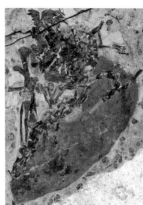

GLIDING MAMMAL
The skin membranes preserved in this *Volaticotherium* fossil suggest that it would have looked like a living flying squirrel.

WING MEMBRANES

Flying squirrels climb trees in pursuit of insects, then spread their limbs to open up the membranes between them and glide from one tree to another. Like flying squirrels, *Volaticotherium* presumably had muscles within the membranes. These would have helped it control its gliding and allowed it to fold away the membranes.

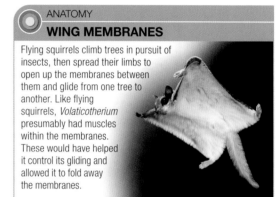

Nemegtbaatar

GROUP Multituberculate mammals
DATE Late Cretaceous
SIZE 4in (10cm) long
LOCATION Mongolia

Nemegtbaatar is a member of an important group of Cretaceous mammals called the multituberculates, or "multis" for short. Most multis were similar in size to modern mice and rats, but some could grow as big as beavers. They seem to have been mostly herbivorous and probably lived like rodents. However, they were not closely related to rodents, nor to any other group of placental mammal. *Nemegtbaatar* is one of many multis named since 1970. Its front incisors were large and protruding and it had no canine teeth. Instead, the cheek teeth were separated from the incisors by a short gap. *Nemegtbaatar* had a deep skull and a particularly wide snout, and blood vessels passed through small holes in its snout bones. This extra blood flow may have supplied a gland or a patch of sensitive skin on the top of its skull, the function of which is unknown.

wide snout

short, deep skull

body covered in hair

large, protruding incisors

MOLELIKE SKULL
Nemegtbaatar belonged to a Late Cretaceous Asian group of multis called Djadochtatherioidea. These mostly had short, deep skulls that superficially resembled those of modern voles.

clawed toes

Teinolophos

GROUP Monotreme mammals
DATE Early Cretaceous
SIZE 4in (10cm) long
LOCATION Australia

Teinolophos is a poorly understood Mesozoic mammal currently known only from a few partial lower jaw bones. Several jaw features demonstrate that *Teinolophos* was a monotreme—sometimes called egg-laying mammals. The platypus and the echidnas (or spiny anteaters) are the only living monotremes and are found only in Australasia. *Teinolophos* was originally thought to be an early monotreme, only distantly related to the platypus and echidnas. However, work published in 2008 showed that it possessed several features unique to platypuses but not echidnas.

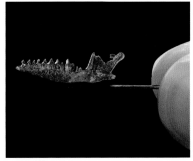

DISTINCTIVE JAW
Teinolophos's jaw shares many features of the living platypus, including a particularly large canal inside the bone of the lower jaw.

LIVING RELATIVE

DUCK-BILLED PLATYPUS

Teinolophos was similar in many ways to its modern relative, the duck-billed platypus of Australia. The platypus is an aquatic predator that forages along the bottoms of streams and pools for crustaceans and other prey. It finds these by using specialized sensory cells embedded in the rubbery skin of its soft, sensitive bill.

Sinodelphys

GROUP Eutherian mammals
DATE Early Cretaceous
SIZE 6in (15cm) long
LOCATION China

Sinodelphys was a small animal that would have resembled an extant opossum. It had a slender snout and dainty jaws, and its hand and foot bones show that it was a good climber. Discovered in the Yixian Formation in Jehol, China, its fossils have hair preserved around their bodies and limbs. For many years, *Sinodelphys* was thought to be one of the oldest and earliest known members of Metatheria—the group that includes marsupials and all their fossil relatives—and its similarity in age to the primitive eutherian *Eomaia* (see below), an indication that the split between placental and marsupial mammals occurred in Asia during the Early Cretaceous. The discovery of *Ambolestes* (see below) has caused scientists to reassess, and they now believe that *Sinodelphys* is more closely related to placental mammals than marsupials.

small, slender body

feet suited to climbing

FLEXIBLE ANKLES
Sinodelphys's ankle bones show that it was able to rotate its feet backward when climbing down tree branches head first.

thick coat of fur

clawed toes

FUR COAT
Eomaia is known from a single, remarkable fossil that shows rarely preserved details, including ears and a thick coat of fur.

relatively long tail

Eomaia

GROUP Eutherian mammals
DATE Early Cretaceous
SIZE 8in (20cm) long
LOCATION China

Mouse-sized *Eomaia* is currently one of the oldest and most primitive members of Eutheria, the mammal group that includes modern placental mammals and their fossil relatives. Its name means "dawn mother," reflecting its crucial position within our own family tree. *Eomaia* is from the Yixian Formation of Jehol, China. Like other fossils from this site, the outline of *Eomaia*'s body has been preserved. Its bones are surrounded by a thick coat of fur, and its long tail is covered in short hairs. The hands and feet of *Eomaia* were similar to those of modern climbing mammals, such as opossums and dormice, so it is thought that *Eomaia* clambered about in bushes and trees. Tall, sharp points, or cusps, on its teeth suggest that it was a predator of insects and other small animals.

Ambolestes

GROUP Eutherian mammals
DATE Early Cretaceous
SIZE 9¾in (25cm)
LOCATION China

Ambolestes is one of the oldest and earliest known members of Eutheria. It was small, about the size of a modern squirrel, and an adept climber. Its teeth indicate that it was probably insectivorous. Anatomical details of *Ambolestes* have allowed scientists to revise the evolutionary tree of early eutherian mammals. Its addition shows that *Sinodelphys* (see above) may be more closely related to placental mammals than marsupial mammals and that placental and marsupial mammals split during the Early Cretaceous in North America, not in Asia as previously proposed.

WELL PRESERVED
The fossil discovered in China is remarkably complete, with an intact hyoid at the base of the skull. Hair is also preserved around the body and limbs.

Zalambdalestes

GROUP Eutherian mammals
DATE Late Cretaceous
SIZE 8in (20cm) long
LOCATION Mongolia

A rat-sized mammal from the Late Cretaceous of Mongolia, *Zalambdalestes* is best known for its long, narrow snout and long, slender hind limbs. Its incisors at the tips of its jaws were long and grew continuously throughout its life—in the same way as a modern rodent's front teeth. A gap separated the front teeth from those farther back. The tall, pointed teeth suggest *Zalambdalestes* had a diet of insects and perhaps seeds. With its long hind legs and shorter front legs, it is thought that this animal probably hopped like a jerboa.

RODENTLIKE APPEARANCE
Zalambdalestes's narrow, upturned snout and long tail gave it a rodentlike appearance, though it was not a rodent relative.

PALEOGENE

342 Plants

346 Invertebrates

350 Vertebrates

Paleogene

Following the major extinctions at the end of the Cretaceous, new forms of life emerged. Mammals, previously small and insignificant by comparison with dinosaurs, were able to exploit many of the niches left vacant. Earth, after an interval of greenhouse conditions, began its long period of cooling that continued into the Quaternary ice age. Grasses, better able to cope with cooler conditions, evolved and spread.

CRETACEOUS–PALEOGENE
A thin band of rocks containi the Cretaceous–Paleogene b California, USA, (above) and It has been suggested that th collision of a meteorite with E

Oceans and continents

During the Paleogene, Earth's geography began to take on a more familiar shape, as Gondwana continued to break up. New oceans widened the gaps between continents, and ocean currents became established. South America and Africa spread further apart, expanding the South Atlantic Ocean. The North Atlantic Ocean also widened, and the ancestor of the powerful Atlantic current, the Gulf Stream, began to strengthen. The Rocky Mountains were forming in western North America, and the Himalayas and Tibetan Plateau were being folded and uplifted as India continued to plow northward into the Asia plate. Antarctica remained attached to the southern tip of South America until the Late Eocene, when it finally separated and the Antarctic circumpolar current became established. In the mid-Paleogene, Australia had begun to separate from Antarctica and as it moved northward, the Southern Ocean opened. As Africa began to move in the same direction, the Tethys Ocean began to contract, forcing up the Alps. As continents separated, individual land masses, such as

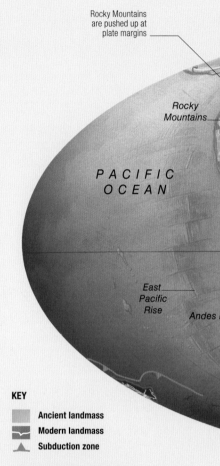

Rocky Mountains are pushed up at plate margins

Rocky Mountains

PACIFIC OCEAN

East Pacific Rise

Andes

South America, Australia, and New Zealand, became isolated from each other, allowing their faunas and floras to develop independently. Large seaways in western Russia, such as the Turgai Strait that extended north of the Caspian Sea to the Arctic and separated Europe from Asia, also restricted some animal populations to certain areas. Marine life, devastated by the mass extinction at the end of the Cretaceous, slowly recovered and became more abundant by the end of the Paleocene. By the mid-Paleogene, mammals had arrived in the oceans, in the form of whales.

ROCKY MOUNTAINS
New mountain ranges emerged as a result of collisions between lithospheric plates. The North American Rockies were the result of one such mountain-building episode.

KEY

▬	Ancient landmass
◠	Modern landmass
▲	Subduction zone

MID-PALEOGENE WORLD MAP
North and South America were now separate, the Himalayas and Andes were being pushed up, and the Alps were formed by the closure of the Tethys Ocean.

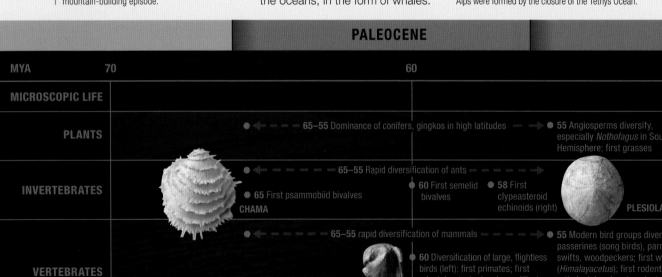

PALEOCENE

MYA	70	60
MICROSCOPIC LIFE		
PLANTS	● ◄ ─ ─ ─ 65–55 Dominance of conifers, gingkos in high latitudes ─ ─ ► ●	55 Angiosperms diversify, especially *Nothofagus* in Sou Hemisphere; first grasses
INVERTEBRATES	● ◄ ─ ─ ─ 65–55 Rapid diversification of ants ═ ═ ═ ═ ► ● 65 First psammobiid bivalves CHAMA	● 60 First semelid bivalves ● 58 First clypeasteroid echinoids (right) PLESIOLA
VERTEBRATES	● ◄ ─ ─ ═ 65–55 rapid diversification of mammals ═ ═ ─ ─ ► PACHYDYPTES (ARM BONE)	● 60 Diversification of large, flightless birds (left); first primates; first edentates (anteaters/armadillos); first carnivorous mammals; first lipotyphlan mammals (hedgehogs, shrews, moles); first owls ● 55 Modern bird groups divers passerines (song birds), parr swifts, woodpeckers; first wh (*Himalayacetus*); first rodent artiodactyl and perissodacty including horses; first lagom (rabbits); first armadillos; first (elephants); first sirenians (d

SUBTROPICAL SEAS
Even in the High Arctic, sea surface temperatures were subtropical during the Paleocene/Eocene Thermal Maximum (PETM), a short period when there was extreme, widespread climate warming.

Climate

During the 40 million years of the Paleogene, Earth experienced dramatic climate change, with a number of warm and cold cycles lasting roughly 10 million years. After a drop in global temperatures at the end of the Cretaceous, they increased again in the early Paleogene, then soared dramatically about 55.4 million years ago. During a short interval of between just 10,000 and 30,000 years, sea-surface temperatures in the tropics rose by about 9°F (5°C), while in high latitudes they increased by about 16°F (9°C). Higher temperatures were accompanied by higher rainfall, with the result that much of the world was covered by tropical rain forests. These conditions persisted for about 170,000 years, until temperatures quickly dropped back to earlier levels, then began to rise again for a few million years during the later part of the Early Eocene epoch. After this a period of cooling began that was to lead to the onset of major glaciations in the Northern Hemisphere some 50 million years later. This decline was slow and gradual, until at the Eocene–Oligocene boundary, the mean annual temperature fell substantially, by more than 14°F (8°C) in just 400,000 years. Mean annual temperatures dropped from about 68°F (20°C) to just 54°F (12°C). Falls in ocean temperatures were less, perhaps just 3 to 5°F (2–3°C). As the climate became cooler, it also became drier.

ANTARCTIC ICE SHEET
By the end of the Paleogene, much of Antarctica was covered by an ice sheet. The temperature drop at the end of the Eocene epoch marked Earth's transition from a greenhouse to an icehouse world.

India ploughs northward into the Asia plate, creating the Himalayas

Turgai Strait

EUROPE

ASIA

Arabia

RICA

Himalayas

India

INDIAN OCEAN

Australia

NTARCTICA

Australia moves northward

CARBON DIOXIDE LEVELS
Carbon dioxide levels gradually decreased through the Paleogene, while temperatures rose. This sudden global warming may have been a result of the release of methane from deep in the oceans.

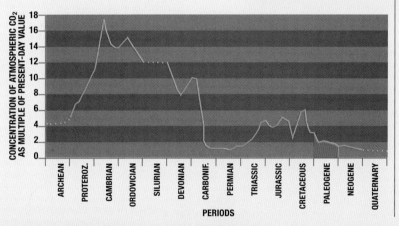

CONCENTRATION OF ATMOSPHERIC CO_2 AS MULTIPLE OF PRESENT-DAY VALUE

PERIODS: ARCHEAN, PROTEROZ., CAMBRIAN, ORDOVICIAN, SILURIAN, DEVONIAN, CARBONIF., PERMIAN, TRIASSIC, JURASSIC, CRETACEOUS, PALEOGENE, NEOGENE, QUATERNARY

OLIGOCENE

40 | 30 | 20 | MYA

MMULITES

40 Peak diversity of planktic foraminifers (left)

35 Increase in diversity of cold-tolerant ostracods and foraminifers

MICROSCOPIC LIFE

35 Beginning of expansion of grassland ecosystems

30 First eucalypts in Australia

PLANTS

40 Appearance of modern moths and butterflies

35 Major extinctions in gastropods

VELATES

30 First balanids (barnacles)

INVERTEBRATES

First embrithopod (extinct, noceroslike) mammals

40 Major diversification of perissodactyl mammals

35 Major extinctions in reptiles and amphibians; toothed and baleen whales diversify; many modern mammal groups appear: glyptodons, ground sloths, dogs; first peccaries; first eagles and hawks

30 First pigs; first cats

25 First deer

VERTEBRATES

PALEOGENE PLANTS

The mass extinction at the end of the Cretaceous had a less dramatic impact on plant life than on other organisms. Most groups of plants survived and continued to diversify. Later changes in climate, as well as the evolution of new animals, had more profound consequences for the world of plants.

The rapid evolution of different land animals during the Paleogene (see pp.346, 350) contributed to the development of ecosystems that were much more similar to those of today than those that had existed in the Cretaceous.

Tropical forests

An important new feature of Paleogene vegetation was the development of closed forests composed of large angiosperm trees. There is little evidence for these kinds of communities in the Cretaceous. By the time of the Eocene, very warm climates with abundant rainfall led to the extensive development of

FLOWERING PLANTS
The legume family, illustrated here by the genus *Robinia*, is one of the many flowering plants that made its first appearance during the Paleogene.

rain forests, not only in equatorial regions but also as far north as southern England. The Eocene London Clay flora of southern England contains the fruits and seeds of mangroves and many other plants that are

MODERN TROPICAL RAIN FORESTS
Tropical rain forests were not widespread until the Paleogene. They included many groups of plants that are characteristic of tropical vegetation today.

characteristic of tropical Southeast Asia today. Eocene floras of a similar age from other parts of the world often contain a high proportion of large leaves, a characteristic of tropical vegetation today. At the height of the warm Eocene climate, trees also grew very close to the North Pole, well above the latitudes at which forests now grow.

Coevolution with animals

During the Eocene, there is increased evidence of coevolution between angiosperms and the new groups of vertebrates and insects that flourished after the Cretaceous–Paleogene boundary. Important insect pollinators, such as bees and moths, became more common in the Paleogene, and some fossil flowers from the Eocene show specializations for insect pollination, such as the presence of oil glands and bilaterally symmetrical flowers. The average size of angiosperm fruits and seeds also increased in Paleogene floras, reflecting an increased reliance on mammals and birds for dispersal.

GROUPS OVERVIEW

Most of the important lineages of living angiosperms, as well as other groups of plants, had probably begun to differentiate by the Paleogene. Most of these Paleogene plants are similar to plants of today. By the Paleogene, there were also early indications of ecological associations among different groups of living plants that have persisted up to the present.

FERNS

The development of modern kinds of forests provided new opportunities in understory habitats. These were quickly exploited by the explosive evolution of several groups of ferns. Many different groups of ferns, as well as some lycopods, also flourished as epiphytes in the canopy habitats of these new forests.

CONIFERS

Swamp cypresses and their relatives were important in Paleogene floras. The dawn redwood and other conifers flourished in the cool climates close to the North Pole. Fossil plants preserved in ancient mountain lakes also show that the differentiation of modern pines and their relatives was well underway during the Paleogene.

MONOCOT ANGIOSPERMS

The fossil record of monocots increases dramatically during the Paleogene. Many different groups of living monocots appear for the first time during the Eocene. These include grasses, sedges, and pondweeds. Especially common are groups of monocots characteristic of aquatic and low-lying habitats.

EUDICOT ANGIOSPERMS

The Paleogene saw the continuing diversification of groups of eudicots that were established in the Cretaceous, as well as the first appearance of many others. The diversification of many groups of woody angiosperms, as well as the modernization of their pollination and dispersal biology, is a particular feature of the Paleogene.

Lygodium

GROUP Ferns
DATE Cretaceous (possibly Triassic) to present
SIZE Sterile leaflet up to 4½in (11cm) long
LOCATION Europe, North America, South America, China, Australia

Lygodium contains about 40 species of ferns, which are characterized by having leaves that continue growing to undetermined lengths, with slender twisting axes that allow them to climb. These climbing leaves are unique among ferns, giving, at first sight, the wrong impression—that of a climbing leafy stem. The pinnules are either untoothed, toothed, or regularly lobed, and there is a thin vein running down the middle of each lobe, which splits into secondary veins. Its fertile leaflets are also unusual in having very narrow leaf segments with their spore cases (sporophores) on conelike extensions. It is the distinctive sterile and fertile foliage that permits the recognition of *Lygodium* in the fossil record. Leaves in the fossil record take *Lygodium* back to the Late Cretaceous, while spores similar to those of *Lygodium* are known from the Triassic onward. Today *Lygodium* grows in tropical and subtropical regions on all continents. Several species can be problematic invasive weeds.

fertile pinnule produced spores

UNEQUAL LOBES
Lygodium skottsbergii has lobed sterile leaflets, which usually have three, as here, or five lobes of unequal length.

three-lobed leaflet

LYGODIUM SKOTTSBERGII
This fossil from the Paleogene of Chile in South America shows the characteristic mass of fertile branched leaf axes with fruiting cones at the tips.

Osmunda

GROUP Ferns
DATE Permian to present
SIZE Fertile leaf up to 6½ft (2m) long
LOCATION Worldwide

The Osmundaceae appeared in the Late Permian of both the Northern and Southern Hemispheres. Evolution of the family was rapid during the Late Paleozoic and Early Mesozoic, giving it the longest fossil record of any of the ferns. There are more than 150 extinct species, and a number of living genera, including *Osmunda*. Much of the geological history of the Osmundaceae is based upon mineralized trunks. These fossil stems are very similar to those of living members of the Osmundaceae. Some fossil leaves have been assigned to living genera when they are nearly identical to the foliage of living ones, and there are even Triassic specimens from Antarctica that are indistinguishable from the living species *Osmunda claytonensis*. The fossil record also shows that the family had produced different fertile and sterile fronds by about the Jurassic.

vascular strand

hard root mantle layer

OSMUNDA SP.
Mineralized sections of *Osmunda* trunks, such as this specimen, show the stems to have had many vascular strands embedded in softer tissue surrounded by a mantle of leaf bases and roots.

Metasequoia

GROUP Conifers
DATE Cretaceous to present
SIZE Up to 130ft (40m) tall
LOCATION North temperate regions

The earliest records of this genus are from the Cretaceous, but *Metasequoia* became one of the most abundant conifers in the Paleogene and early Neogene Northern Hemisphere, growing as far north as Arctic Canada. Large petrified trunks and stumps of *Metasequoia* have been found in North America. *Metasequoia* shoots can be easily recognized by the stems that branch simultaneously on either side, the two ranks of paired leaves flattened into one plane, and by the structure of the cones, if they are still attached to the shoots. *Metasequoia* has, like other conifers, both male pollen-producing cones and female seed cones. Differentiating *Metasequoia* species can be a problem because many are based on just a few fossil specimens. Now four or five fossil species are reliably distinguished by shoot and cone structures or by their seedlings. There are no known fossils of *Metasequoia* in the late Neogene, and it was thought that they had died out until living plants were found in China (see panel, below).

paired leaves on opposite sides of stem

spiked leaf

METASEQUOIA OCCIDENTALIS
This is the most common and best-known fossil foliage species of *Metasequoia* and is the most similar species to the living *Metasequoia glyptostroboides* (see panel, left).

LIVING RELATIVE
DAWN REDWOOD

Metasequoia glyptostroboides, like the ginkgo, is a living fossil. Amazingly, it was not discovered until 1944, when it was found growing in west central China—3 years after it had been named as a fossil. *Metasequoia* is a fast-growing tree that is now commonly grown in temperate regions around the world. It can reach over 130ft (40m) in height and nearly 6½ft (2m) in diameter.

opposing, paired leaves

seed cone

Picea

GROUP Conifers
DATE Paleogene to present
SIZE Up to 295ft (90m) tall
LOCATION North America, Europe, Asia, Japan

There are about 35 species of spruce living in the northern temperate and boreal regions today. They are tall evergreen trees with broadly spirelike crowns and shoots that can be identified by the woody peglike structures at the bases of the leaves. *Picea* bear cylindrical, pendant seed cones on leafy branchlets on their upper branches. Today the species is found in montane and subalpine forests in North America, Europe, and Asia.

PICEA SP.
The scales of cones, such as this one, are thin and fan shaped. Their winged seeds mature and are shed in the fall.

Pinus aristata is a slow-growing tree, 16½–49ft (5–15m) tall, and often contorted with strips of living bark separated by sun-bleached wood. It grows in the mountains of Colorado, New Mexico, and Arizona, where some can live up to nearly 2,500 years old.

Macginitea

GROUP Angiosperms
DATE Eocene
SIZE Leaves up to 14in (35cm) long
LOCATION W. North America

Macginitea is an extinct genus of plane tree that has been reconstructed by putting together different *Platanus*-like parts of plants that were repeatedly found next to each other at various localities in western North America. The leaves are large, with five to seven palmately arranged lobes. The trees bore stalked globular masses of male flowers or female flowers. They were adapted for early colonization in open disturbed areas, especially close to water.

Podocarpus

GROUP Conifers
DATE Cretaceous to present
SIZE Up to 148ft (45m)
LOCATION North America, Australia, Asia, South America, Africa

These conifers have needle-shaped leaves that are flattened into one plane. Individual leaves are narrowed at the base and have rounded tips. Fossil *Podocarpus* can be distinguished from the similar shoots of *Taxodium*, *Taxites,* and *Sequoia* by anatomical details of their leaf surfaces. There are reports of *Podocarpus* pollen grains from the Cretaceous of North America, but by the Paleogene, the genus was restricted to the Mississippi Basin area. *Podocarpus* leafy shoots are known from the Paleogene of Australasia, South America, and the southern part of North America. The probable origin of the genus was in Australasia, with subsequent migration to South Africa and across the Pacific from west to east. There are about 100 living species of *Podocarpus* in the warm temperate and subtropical regions of the Southern Hemisphere.

PODOCARPUS INOPINATUS
This shoot from the Paleogene of Chile shows the spirally developed flattened leaves arranged into one plane, so that all are oriented toward the light.

rounded leaf tip

narrowed base of leaf

needle-shaped leaf

paired leaf arrangement

Platanus

GROUP Angiosperms
DATE Cretaceous to present
SIZE Trees up to 138ft (42m) tall
LOCATION Europe, Asia, North America

Most of the information on the early fossil history of the Platanaceae family comes from fossil leaves. The earliest come from the Cretaceous of Europe and North America, but they are far more common in the Paleogene and Neogene of Europe, Asia, and North America. Many of the leaves have leaf venation patterns and flower structures that exclude them from *Platanus* and refer them to other genera, such as *Macginitea* (see below, left). Leaves of *Platanus* and similar plants are distinctive in being large, stalked, and palmate with five or more lobes. The leaf margins are smooth or have serrations near the tips. There is a midrib in each lobe, either coming from a common point above the stalk or from extra branches a short distance above it. There are numerous secondary and tertiary veins, although the latter are hard to see. *Platanus* fruits have hairs to help them disperse on the wind. Similar plants, such as *Macginitea*, had hairless fruits.

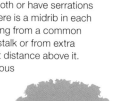

midrib of lobe

slightly serrated edge

secondary vein

"PLATANUS"
This five-lobed leaf has clear midribs and secondary veins. The marks around the tips of the lobes on this specimen show where the overlying rock was removed to uncover the leaf.

There are six species of large, spreading deciduous plane trees that are found in Southern Europe, western Asia to India, North America, and Mexico. They all have maplelike dissected leaves as shown here, and wind-pollinated flowers in spherical balls on long, pendulous stalks. After pollination, the fruits hang on the trees through the winter. The London plane, *Platanus* x *hispanica*, is a hybrid between two *Platanus* species: the European oriental plane and the American sycamore. This hybrid grows even in compacted soil and, being pollution resistant, is ideal for towns.

Florissantia

GROUP Angiosperms
DATE Paleogene
SIZE Flower 1–2¼in (2.5–5.5cm) in diameter
LOCATION North America

There are three known fossil species of these flowers and fruits. The flowers are radially symmetrical and borne on long, slender stalks (pedicels). They have five petal-like organs (sepals), fused to at least halfway along their length, and a prominent radiating network of veins. Only one species has petals, and these are free and smaller than the sepals. The projecting ovary has the bases of five pollen-producing stamens fused around it. After pollination and fertilization, the ovaries swelled to form five-lobed nutlike fruits similar to those of linden trees, but they were still attached to the remains of the sepals and stamens, suggesting that the whole flower was wind dispersed.

petal-like sepals

radiating veins

FLORISSANTIA QUILCHENENSIS
The stalks of *Florissantia* species, such as this one, suggest that the flowers hung down, and dense, hairy areas at the base of the sepals may have produced nectar to attract insects or birds.

Banksia

GROUP Angiosperms
DATE Paleogene to present
SIZE Up to 98ft (30m) tall
LOCATION Australia

There are many species of fossil *Banksia* based on the leaves and the fruits from the Paleogene onward, in Australia. Records from outside of Australia are very debatable. Fossil leaves suggest that *Banksia* was well adapted to limiting water loss because it can be seen that the breathing pores (stomata) are confined to the lower surfaces of the leaves and in depressions lined with surface hairs. In living *Banksia*, only a few of the flowers in each flower cluster ever develop and expand into the distinctive two-valved fruits. The valves of the fruits open in order to liberate the seeds for dispersal.

BANKSIA ARCHAEOCARPA
Fossil fruiting bodies, like the one illustrated here, show the positions of the spirally arranged former flowers and, in this case, three mature fruits.

former flower

open valve

mature fruit

BANKSIA COCCINEA
Commonly known as scarlet banksia, this shrub is found in southwestern Australia, where it grows to about 26ft (8m) tall, producing spiked flowering heads that are about 3¼in (8cm) in height and width.

toothed leaf margin

BANKSIAEPHYLLUM SP.
These fossilized leaves, usually assigned to the genus *Banksiaephyllum*, show imprints of the midrib, secondary veins, and a network of tertiary veins.

Nypa

GROUP Angiosperms
DATE Late Cretaceous to present
SIZE Fruit 3¼–4½in (8–12cm) long
LOCATION Northern Hemisphere

The earliest palm fossils are leaves, stems, and pollen from the Late Cretaceous. Their remains were widespread and abundant in the Paleogene, and fruits of the mangrove palm, *Nypa*, are common in European deposits. The size and number of fossil *Nypa* suggest that conditions in northwest Europe at that time were similar to present-day brackish mangroves in India and Southeast Asia.

mesocarp

NYPA BURTINI
This *Nypa* fruit eroded partially while drifting in the sea. The outer coat (epicarp) remains only at the top, exposing the inner coat (mesocarp) below.

Cyclocarya

GROUP Angiosperms
DATE Paleogene to present
SIZE Up to 98ft (30m) tall
LOCATION Europe, North America, Asia

Cyclocarya is a genus in the walnut family with only one living species. It is a deciduous tree growing to about 98ft (30m) tall with pinnately divided leaves and male and female catkins. Its distinctive fruits are surrounded by disklike wings that help them disperse in the walnut family. Fossils of the walnut family are known from the Late Cretaceous of North America and Europe, but the first genus of living plants to appear is *Cyclocarya* in the Paleogene—a genus that then quickly spread from North America to Europe and Asia. The fossil assemblages where *Cyclocarya* has occurred have also included such plants as *Glyptostrobus*, *Metasequoia*, and *Liquidambar*—all of which suggest that these plants were living in a warmer temperate to subtropical climate. Today *Cyclocarya* is found only in China.

Sabalites

GROUP Angiosperms
DATE Cretaceous to present
SIZE Leaf about 3¼–6½ft (1–2m) long
LOCATION North America, Mexico, Europe

The fossil record of palms is both diverse and cosmopolitan; leaves are among its most distinctive elements. The costapalmate leaf is an easily recognizable form in which leaf segments radiate in a fan shape from the edges of the main petiole. The sabal palm was one of a variety of palms with divided leaves found in deposits in North America, Mexico, and Europe. Like the living cabbage palm, *Sabal palmetto*, some Late Cretaceous species grew into slender trees up to 46ft (14m) in height. Palms are found in tropical and subtropical areas today; they are good indicators of warm climates.

divided leaf

petiole

SABALITES SP.
In this fossil, the stalk, or petiole, can be seen below the base of the sabal leaf. The fine lines indicate the leaf divisions.

PALEOGENE
INVERTEBRATES

Following recovery from the Late Cretaceous extinction, the fifth such event, marine invertebrate faunas similar to those of today occupied the available ecospace. A shell-littered shoreline of the Paleogene would look much like today's. Although it was warm, the first stages in global cooling had begun.

Establishment of coral reefs

One factor in the diversity of marine invertebrates during the Paleogene is the evolution of coral reefs. Scleractinian corals came into existence during the Triassic and flourished through the Mesozoic, but there were no true coral reefs until the Paleogene. Living coral reefs are of three kinds: fringing reefs, barrier reefs, and atolls. Fringing reefs develop along shorelines, especially those of volcanic islands. If the island subsides, a circular or horseshoe-shaped atoll will form, and it has been shown by deep-drilling through Pacific atolls that coral reefs began in the Paleogene. Scleractinian corals have two advantages as frame-builders. First, they can cement themselves to the substratum, unlike the Paleozoic corals that had no facility for attachment. Second, they have symbiotic algae in their tissues, which provide them with oxygen and carbohydrates, enabling the corals to grow rapidly. As a result, they have been able to build up substantial frameworks inhabited by many other living creatures, which form the basis for the most complex and productive ecosystems of all time.

WORM WOOD
This fossilized log contains the preserved remains of the shipworm *Teredo*. It still exists today and lives in groups in either floating or submerged timber.

CORAL REEF
This living Pacific coral reef contains colonies of many different species living side by side. Reefs provide support for other animal and plant life.

The Danian

Much of Denmark is underlain by Upper Cretaceous Chalk. On the shore of the island of Sjaelland, cliff sections reveal a gray, iridium-rich layer with scattered fish bones, marking the Cretaceous mass-extinction event. Above this, chalky sediment continues well into the Paleogene. This is known as the Danian. The last ammonites and belemnites perished in the catastrophe. Nautiloids persisted, and in the Danian, they take the place of the ammonites.

SPIRED SHELL
Turritella is a large, high-spired gastropod, typical of the Paleogene, and ornamented with ribs.

GROUPS OVERVIEW

Gastropods continued to evolve and spread during the Paleogene, with new forms appearing and adapting. Bivalves also appear to have diversified. Many echinoids emerged from the Late Cretaceous relatively unscathed. New forms of crustaceans appeared and quickly became established as highly effective predators.

GASTROPODS
During the Paleogene, gastropods continued to diversify and colonize new habitats. At this time, large—often highly ornamented—gastropods appeared, along with the pteropods, which are small gastropods with straight shells, adapted for a floating life in the plankton. Some freshwater Paleogene limestones may consist entirely of gastropods.

BIVALVES
Many of the common bivalve genera living today first appeared in the Paleogene. During this period, bivalves appear to have evolved many different life habits. These include: byssal attachment (in which the animal is secured to a hard surface by a threadlike structure), bacterial feeding, and both deep and shallow burrowing.

ECHINOIDS
Most echinoid groups were not greatly affected by the late Cretaceous mass extinction, and both surface-living genera, such as *Echinocorys*, and deeper-burrowing genera, such as *Linthia*, continued as an important component of the Paleogene fauna.

CRUSTACEANS
Lobsters and crabs are the largest living crustaceans, and they became highly abundant and diverse in the Paleogene. Their grasping pincers (unknown in Paleozoic crustaceans) were a major factor in their success.

ROTULARIA BOGNORIENSIS
This gathering of *Rotularia* were probably swept together by currents on the seabed.

Rotularia

GROUP Worms
DATE Middle Jurassic to Paleogene
SIZE Up to ¾in (2cm) across
LOCATION Worldwide

Because of the soft nature of their bodies, worms are rarely fossilized, and the most we know about many shell-less worms are the burrows they made into the sediments. There are some notable exceptions, namely the serpulid worms. These marine polychaete worms secreted tubes composed of calcium carbonate around themselves to protect their soft bodies. Serpulid worms are therefore preserved as fossils in the same way as other shelled invertebrates. Unlike most surviving serpulid worms—which live cemented to the seabed or to the shells of other invertebrates—*Rotularia*, although attached in its early stages of growth, was free-living. The tubular shell it lived in was at first coiled in a low spiral, but at maturity the last part of the shell straightened out and broke free from the coiled part.

Aturia

GROUP Cephalopods
DATE Paleogene to Early Neogene
SIZE Up to 6in (15cm) across
LOCATION Worldwide

Aturia was a nautiloid, similar in appearance to modern species belonging to the genus *Nautilus*. It had a very tightly coiled shell with heavily overlapping whorls and a very small umbilicus at the center of the coil. This type of shell is said to be extremely involute. One of *Aturia*'s most distinctive features is its complex suture line with sharply angled folds (lobes). These V-shaped lobes face away from the shell's opening, or aperture. In cross-section, the whorls are somewhat compressed, with flattened sides, but the venter (outer edge) is rounded. The shell is smooth, and ornamented only by very fine growth lines. *Aturia* is likely to have lived in fairly open water and probably fed on small fish and small crustaceans.

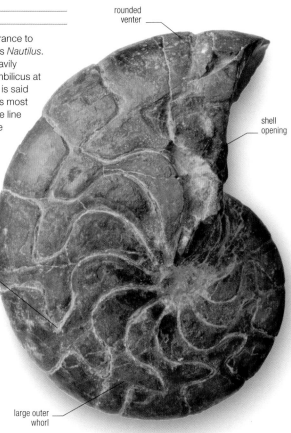

rounded venter

shell opening

sharp, backward-facing angular fold (lobe)

large outer whorl

ATURIA PRAEZIGZAC
Clearly visible on this internal mold, the suture lines of *Aturia* are quite complex for a nautiloid. These superficially resemble the suture lines of some of the earlier ammonoids.

Xenophora

GROUP Gastropods
DATE Paleogene to present
SIZE Up to 1½in (4cm) long
LOCATION Worldwide

Xenophora is a genus of marine gastropods that still exist in the world's seas today. Among other distinctive features, it had a conical shell with a fairly low cone (spire) and very little overlap between successive whorls. There are strong growth markings on the shell, most clearly visible on its base. Also on the underside, the central umbilicus is narrow and steep-sided. Unlike many gastropods, which have nodes on their shells, *Xenophora*'s bumpy outline was the result of shell fragments or rock particles embedded in its shell. The animal picked up pieces of shells and held them in the mantle margin against the outside of its soft body until other shell material was secreted around the foreign objects to fix them into place.

XENOPHORA CRISPA
This gastropod's apex (spire) contains one shell that was incorporated at an early stage in its life.

overhanging shell

recessed lower lip

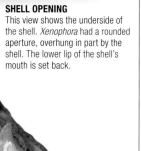

SHELL OPENING
This view shows the underside of the shell. *Xenophora* had a rounded aperture, overhung in part by the shell. The lower lip of the shell's mouth is set back.

debris fixed onto shell

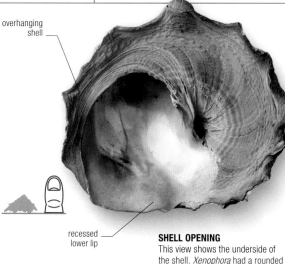

peripheral lobes

Athleta

GROUP Gastropods
DATE Paleogene to present
SIZE 2½–4in (6.5–10cm) long
LOCATION Worldwide

Gastropods of the genus *Athleta* were predatory carnivores. *Athleta* fossils have a distinctive shape, with the tip of the shell forming a small, steep cone. As the animal matured, the whorls of its shell broadened. The most prominent parts of the shell's ornament are the spiny nodes on the shoulder of the outermost whorl at the widest part of the shell. From these nodes, ribs run up to the whorl suture, where the whorls join, and down the side of the outermost whorl. On the lower part of the outermost whorl, there are strong spiral ribs, but because of the heavy overlap between whorls, these are not visible on the earlier parts of the shell. The mouth (aperture) of the shell is long, with a notch at the side for the inhalant siphon, a tubular structure used by the animal to take in water.

steep cone

spiny node

shallow ribs

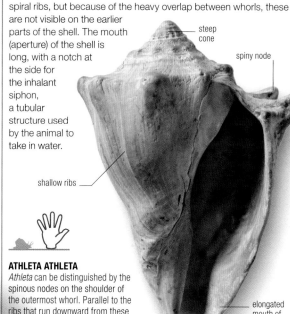

elongated mouth of the shell

ATHLETA ATHLETA
Athleta can be distinguished by the spinous nodes on the shoulder of the outermost whorl. Parallel to the ribs that run downward from these nodes are fine growth increments.

Clavilithes

GROUP Gastropods

DATE Paleogene to Neogene

SIZE 5in (13cm) long

LOCATION Europe, Africa, North America, South America

Clavilithes was a stout gastropod with a large shell that had flat-sided whorls that barely overlapped each other. The whorl suture, where one whorl joined the next, was distinctly stepped, and the outside of the shell was covered with growth lines. The shell opening was large and oval in shape. Comparisons with modern, related forms suggest that *Clavilithes* was carnivorous.

CLAVILITHES MACROSPIRA
The shell of *Clavilithes* is covered with growth lines, which are deflected backward in the mid-whorl position.

INSIDE A SHELL

The soft body parts of a mature gastropod would only have occupied the newest two whorls of the shell. The pillarlike structure in the center is called the columella. It is formed by fusion of the wall of each whorl as it coils around the axis.

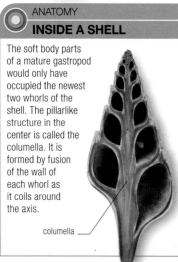

columella

Venericor

GROUP Bivalves

DATE Paleogene

SIZE 3/4–3 1/4in (2–8cm) long

LOCATION North America, Europe

Venericor was a filter-feeding bivalve. Its shell was a rounded triangular shape with forward-facing, curved beaks. The outside of the thick shell had strong radial ribs that became broader and flatter toward the ventral margin. *Venericor* had two teeth and sockets in each valve. Behind the teeth, a long, curved depression housed the ligament. The interior margins of the valves had small crenulations (notches).

beak

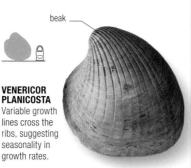

VENERICOR PLANICOSTA
Variable growth lines cross the ribs, suggesting seasonality in growth rates.

Chama

GROUP Bivalves

DATE Paleogene to present

SIZE 1 1/2in (4cm) long

LOCATION Europe, North America

Chama is a bivalve with a strongly asymmetrical shell. The animal is attached to the seabed by its larger, more convex left valve throughout its life. The right valve is smaller and much flatter than the left, although still convex. Both valves had well-developed beaks. The valves grew spirally, along a horizontal plane. The surface of each valve shows concentric frills and radially arranged, flattened spines; these two features combine to give the shell a scaly appearance. Internally, there are two muscle scars, where the adductor muscles, which open the shell, attach in life. The pallial line, where the mantle attached, is smooth. On the hinge-line there is a single, rather blunt tooth in each valve, and above the hinge-plate there is a curved groove that housed the ligament.

wide frill

spines

CHAMA CALCARTA
The spines of *Chama* might have attracted the growth of algae and small animals, which camouflaged it.

Crassatella

GROUP Bivalves

DATE Late Cretaceous to Early Neogene

SIZE 1 1/2in (3.5cm) long

LOCATION Europe, North America

Crassatella was a bivalve with very strong concentric ribs parallel to the shell growth. The fossil shows a strong shoulder on each valve running from the beak to the ventral margin. Internally, it has deep, equally sized adductor muscle impressions, with a distinct pallial line running between them. The adductor muscles were responsible for closing the shell, and the pallial line marks where the animal's mantle would have been attached. The hinge has two teeth below the beak in each of its valves.

beak

CRASSATELLA LAMELLOSA
Crassatella fossils are made up of two halves (valves) that are roughly symmetrical. On the outside of each valve, there are very clear growth ribs.

strongly marked ribs

LOVENIA SP.
In the center of the upper surface of the test is the apical disk, made up of four small, circular pores through which eggs and sperm were released.

genital pore

pore where tube foot protruded

PROTECTIVE SPINES
There is a fine covering of bumps (tubercles) on both sides of the test. All the tubercles were sites of articulation for protective spines.

upper surface of the test

ambulacral area

site of spine attachment

anus

fine tubercles

large tubercles

crescent-shaped mouth

Lovenia

GROUP	Echinoderms
DATE	Late Paleocene to present
SIZE	Up to 1½in (3.5cm) long
LOCATION	Indian Ocean, Pacific Ocean

Lovenia was a flattened, heart-shaped, burrowing echinoid that lived buried in the sand of the seabed, usually in inshore waters. Some species of the genus, called heart urchins, still exist today. *Lovenia* fossils consist of the hard "test" or endoskeleton of the animal. Like all echinoderms, *Lovenia's* body was split into five different sections, which appears as a star-shape on the upper side of the fossilized test. This star-shape comprised the five ambulacral areas, which had large pores around their margins, each of which would have been the site of a tube foot. One of the ambulacral areas formed a deep furrow that continued down to the oral surface, where the mouth was situated. There were several fascioles—smooth areas of the test surface that carried tracts containing small, hairlike extensions (cilia).

LIVING RELATIVE
HEART URCHIN

The heart urchin shown to the right (*Lovenia cordiformis*) is a relation of the fossil *Lovenia*. It is well adapted to life buried in the sediment and relies upon the water currents that its small, hairlike extensions generate to bring food particles and oxygenated water down to its body. The earliest urchins had their mouth located in the center of their underside and the anal opening positioned in the middle of their upper side; such forms are still living (see *Temnocidaris*, p.283). During the course of evolution, the anal opening migrated backward, while the mouth moved forward, as in the fossil *Lovenia*.

Palaeocarpillius

GROUP	Arthropods
DATE	Paleogene
SIZE	Up to about 2¼in (6cm) long
LOCATION	Europe, Egypt, Somalia, India, Zanzibar, Java, Mariana Islands

The domed carapace (shell) of the fossil of this early crab is oval in outline and smooth, with spiny margins at the front and sides. The orbits, which housed the stalked eyes, are well developed. The pincers on the anterior limbs (chelipeds) are clearly visible; the right one is markedly larger and more robust than the one on the left. The movable finger on the right claw had a serrated inner edge, which contrasted with the comparatively slender left pincer. Although crabs began to diversify greatly during the Cretaceous, their numbers increased spectacularly during the Cenozoic Era. This highly successful group is more numerous today than it has been at any time during its history.

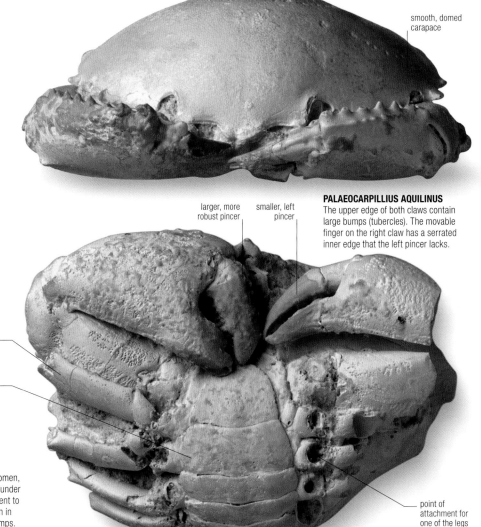

smooth, domed carapace

PALAEOCARPILLIUS AQUILINUS
The upper edge of both claws contain large bumps (tubercles). The movable finger on the right claw has a serrated inner edge that the left pincer lacks.

larger, more robust pincer

smaller, left pincer

jointed leg

abdomen segment

point of attachment for one of the legs

Crabs are a highly successful group that are **more numerous** today than they have been at **any time in their history.**

CURLED ABDOMEN
The segments of the abdomen, which were curled tightly under the carapace, are equivalent to the outstretched abdomen in modern lobsters and shrimps.

PALEOGENE
VERTEBRATES

Following the demise of the dinosaurs at the end of the Cretaceous, mammals and birds began to radiate into diverse forms on land, underwater, and in the sky. Early in the Paleogene, fish and reptiles had taken on more recognizable forms and did not change very much.

At the start of the Paleocene epoch, 66 million years ago, mammals were small and insectivorous, with little limb or teeth specialization. They soon began to change in order to exploit the ecological niches left vacant by the dinosaurs—growing in size, evolving new modes of locomotion, and diversifying their diets to include plants and other vertebrates. The mammalian radiation continued during the Oligocene.

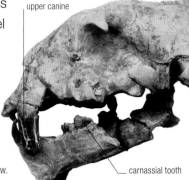

FOSSIL BAT
Fossil bats from the Messel oil shale pit in Germany are astonishingly well preserved. The wings and skin of this specimen are clearly seen.

Taking to the air

Bats are sparsely represented in the fossil record because their light bones and delicate wing membranes do not fossilize well. For this reason, the precise time in evolutionary history that bats took to the air is not known. However, a few excellent bat fossils that have been found at two locations dating from the Eocene epoch—an oil shale pit in Germany and the Green River Formation in Wyoming—show that flying bats existed at least 52.5 million years ago. Their arm and greatly elongated finger bones support a stretchy skin membrane that, powered by muscles, enables flapping flight. Bats probably evolved from a nocturnal insectivore that glided from tree to tree. The first fruit-eating bats did not appear until the Oligocene fossil record.

Carnivores and carnassials

At the beginning of the Paleogene, mammals had unspecialized teeth that could cope easily with their diet of insects. During the Paleocene epoch, however, two major groups of mammals, the carnivores and the creodonts, evolved teeth suited to a diet of meat. Carnassial teeth—the defining feature of modern-day carnivores—are supremely adapted to shearing flesh from bone in a scissorlike action, while the large, pointed canines deliver the killing bite. Carnivore ancestors, the miacids, were small, weasel-sized mammals. Dogs appeared during the middle Eocene and true cats in the early Oligocene, in parallel with a catlike group called nimravids (also known as false saber-tooths).

upper canine

SLICING MOLARS
In primitive carnivores (shown here) and their modern descendants, the carnassial teeth are the last upper premolar and the first lower molar. Creodont carnassial teeth were farther back in the jaw.

carnassial tooth

GROUPS OVERVIEW

By 50 million years ago, only 15 million years after the extinction at the end of the Cretaceous, primitive representatives of most mammalian orders had appeared. Early hoofed mammals (ungulates) were preyed upon by carnivores and creodonts, while whales took to the oceans and bats to the skies. The first primates, rabbits, and rodents also appeared.

CREODONTS
Creodonts were the dominant predators throughout the Eocene and Oligocene. They occupied many of the same niches as carnivores, which eventually replaced them. There were two families, *Oxyaenidae* and the more widespread *Hyaenodontidae*, which included *Megistotherium*—one of the largest land predators to have ever lived.

PERISSODACTYLS
The first odd-toed ungulates appeared at the end of the Paleocene and radiated during the Eocene into primitive horses, rhinos, tapirs, and brontotheres. The first species browsed leaves and were relatively small and unspecialized. They gradually increased in size and further diversified throughout the remainder of the Paleogene.

ARTIODACTYLS
The first even-toed ungulates also appeared in the early Eocene and were small and long-legged. They soon radiated into numerous different forms, including the precursors of pigs, camels, and ruminants such as giraffes and cattle. All had crescent-shaped cheek teeth and a special double-pulley bone in the ankle characteristic of artiodactyls.

CETACEANS
Throughout the Eocene, a branch of hoofed mammals gradually returned to the water as primitive whales. Early species still had well-developed limbs and could walk as well as swim, probably entering the water to hunt. Toward the end of the epoch, the first fully aquatic whales appeared. They had small hind limbs that were essentially useless.

Diplomystus

GROUP Actinopterygians

DATE Early Eocene to Miocene

SIZE 26in (65cm) long

LOCATION USA, Lebanon, Syria, South America, Africa

A distant relative of herrings and sardines, *Diplomystus* was a widespread freshwater fish. Many of the best specimens have been found in the Green River deposits of Wyoming. It had a single dorsal fin, pairs of small pectoral and pelvic fins, and an anal fin that stretched back to the narrow base of the tail. The tail itself was deeply forked, with the upper and lower parts of the fin being of equal size. A number of smaller fish, such as *Knightia* (see below), have been found preserved in the stomachs of *Diplomystus* fossils. *Knightia* was once thought to be a relative, as well as prey, of *Diplomystus*. However, this is now known to be incorrect. *Diplomystus* can be distinguished from *Knightia* by two rows of modified scales. The first runs along the midline of the back, from the skull to the dorsal fin; the other row runs along the belly.

UPTURNED MOUTH
Diplomystus's mouth faced upward, with the jaw set at an angle to the backbone rather than being parallel to it. This suggests that it hunted smaller fish swimming just below the surface.

Knightia

GROUP Actinopterygians

DATE Middle Eocene

SIZE 10in (25cm) long

LOCATION USA

Knightia is thought to have been a freshwater relative of the modern herring. Every year, huge numbers of *Knightia* fossils are unearthed at quarries in the Green River region (see panel), and they are a common sight in rock shops around the world. Most *Knightia* specimens are small but very well preserved. The sheer number of specimens suggests that *Knightia* lived in large schools in the Green River. *Knightia*'s body size and abundance tells us that the fish was probably a secondary feeder, filtering out ostracods, diatoms, and other microscopic plankton in the lake water. Many *Knightia* skeletons have been found in the stomachs of larger fish.

OFFICIAL FOSSIL
In life, *Knightia* swam in large schools. It is so common in the Green River Shale that it has been designated as the official fossil of the state of Wyoming.

KEY SITES
GREEN RIVER

Knightia is the most abundant fossil in the Green River Formation, an extensive series of lake deposits in Wyoming, Colorado, and Utah. Shales more than 9,800ft (3km) thick were deposited there between 53 and 46 million years ago. The Green River Shale is world famous for revealing hundreds of beautifully preserved skeletons with many fine details.

Sciaenurus

GROUP Actinopterygians

DATE Early Eocene

SIZE 8in (20cm) long

LOCATION England

The oldest known representative of the porgy family, *Sciaenurus* was described by the scientist Louis Agassiz in 1845. The head had large eyes and a wide mouth. The body was about four times longer than it was deep. Many *Sciaenurus* specimens have been found in the London Clay deposits, which were formed by a subtropical sea that once covered the southeast of England.

pointed teeth

ACTIVE HUNTER
The lower jaw had a single row of sharp teeth, which suggests that *Sciaenurus* was an active hunter of other fish.

Mioplosus

GROUP Actinopterygians

DATE Eocene

SIZE 10in (25cm) long

LOCATION USA

A member of the perch family, *Mioplosus* was a freshwater lake fish. Its fossils are found singly, never in groups, and the fish is thought to have been a solitary hunter. It was also capable of attacking fish up to half its own size using its many sharp teeth. A large tail with symmetrical forks extended from a narrow base suggests that *Mioplosus* was a powerful swimmer. It had two dorsal fins. The first one lay just behind the pectoral fins, and the rear dorsal fin was positioned above the anal fin.

CONSUMING PREY
As shown here, *Mioplosus* fossils are sometimes found in the process of consuming prey.

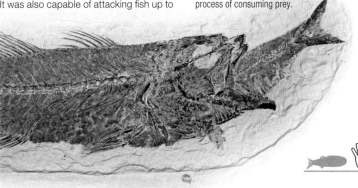

Mene

GROUP Actinopterygians

DATE Early Eocene to present

SIZE 10in (25cm) long

LOCATION Worldwide

Two species of *Mene*—*M. rhombea* and
M. oblonga—have been found in significant
numbers at Monte Bolca in northern
Italy. This site, nicknamed "the fishbowl,"
has yielded a number of fish fossils with
a remarkably high degree of preservation.
It is even possible to tell what color
many of the specimens would have been
in life. There is one surviving representative
of the *Mene* genus: the moonfish, *Mene
maculata*, from the Indian and Pacific
Oceans. This living species is typical of the
genus, with its large eyes and protruding,
upturned mouth. It is a marine fish with a
very deep body, due mainly to downward
expansion of its lower half. At the same
time, the body is also highly compressed
from side to side. *Mene*'s shape suggests
that its body was rather stiff and capable
of only limited movement. It is likely that
most of the propulsive force was
generated by rapid movement of its tail.

DIFFERENT FINS

Mene had a small dorsal fin, a long anal fin, and
small pectoral fins high on its side. The two pelvic
fins were reduced to long spines that trailed in the
water and may have extended beyond the tail.

Anguilla is the only recognized surviving genus of the freshwater eel family Anguillidae. There are 19 species, found mainly in Europe, Asia, and North America. Eels spend their adult lives in fresh water but return to the deep ocean to spawn. Their larvae (young) eat plankton as they drift on ocean currents toward the coasts, where they then enter river systems. Eels navigate along the river bottom, avoiding the stronger currents in the middle of the river.

Anguilla

GROUP Actinopterygians
DATE Middle Eocene to present
SIZE 3ft 3in (1m) long
LOCATION Worldwide

Anguilla is a living genus better known by its common name: eel. Eels are easily recognized by their distinctive, elongated bodies. The earliest *Anguilla* fossils are from Monte Bolca in Italy. Unlike today's forms (see panel, left), these early specimens were marine in origin. Only later did the genus begin to exploit the brackish and freshwater habitats that it occupies today.

UNIQUE FORM
Eels are fish like no others. The head is small, but the mouth can gape widely. The long, muscular body is the main source of power, and the fins are small and ineffective.

Heliobatis

GROUP Chondrichthyans
DATE Middle Eocene
SIZE 3ft 3in (1m) long
LOCATION USA

Heliobatis may have been an ancient relative of the stingray. Its fossils are common in the Green River Shale of Wyoming (see p.351). When the long tail spine is included, *Heliobatis* was about 3ft 3in (1m) long, a similar size to many living stingrays. However, it was not as large as today's freshwater stingray of Southeast Asia, which is up to 16½ft (5m) long and weighs 1,323lb (600kg). Most modern stingrays live in seawater, but a few species inhabit river and lake habitats. The Green River deposits that contained *Heliobatis* specimens formed on the bottom of freshwater lakes.

ROUNDED OUTLINE
Most specimens of *Heliobatis* have a rounded outline formed by their large pectoral fins.

Titanoboa

GROUP Squamates
DATE Middle Paleocene
SIZE 43ft (13m) long
LOCATION Colombia

Following the sudden demise of the dinosaurs, the snake *Titanoboa* was the largest land hunter of the Paleocene. It was also the largest snake ever known, being 30 percent larger than today's longest species. Its back would have been 3ft 3in (1m) off the ground. *Titanoboa's* enormous vertebrae were found in coal deposits in Colombia, along with the remains of the crocodiles and turtles preyed upon by the giant snake.

huge vertebra

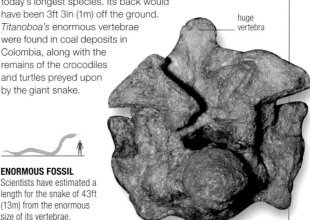

ENORMOUS FOSSIL
Scientists have estimated a length for the snake of 43ft (13m) from the enormous size of its vertebrae.

Puppigerus

GROUP Turtles
DATE Early to Middle Eocene
SIZE 35in (90cm) long
LOCATION USA, England, Belgium, Uzbekistan

Puppigerus was a sea turtle that lived in subtropical seas between the Early and Middle Eocene. Fossils found in the USA, England, and Belgium all show the same species, *P. camperi*. Until 2005, this was thought to be the only fossil species in the genus. However, in that year, the discovery of *P. nessovi*, a new species from Uzbekistan, was announced. *Puppigerus* is an extinct genus in the Cheloniidae family. All living sea turtles belong to this family—except the leatherback. The cheloniids appeared in the Cretaceous, but *Puppigerus* most resembled modern forms. For example, the shell was completely ossified into solid bone. In addition, the pygal, the rearmost plate in the upper shell, or carapace, also lacked the notch seen in earlier cheloniids.

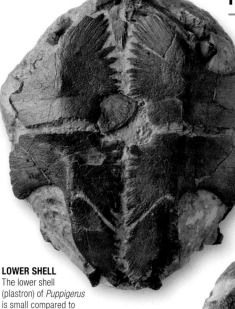

LOWER SHELL
The lower shell (plastron) of *Puppigerus* is small compared to that of land tortoises, allowing plenty of room for the paddle-shaped feet.

HUGE EYES
Puppigerus's eyes point sideways, not upward like those of primitive sea turtles.

Primapus

GROUP Theropods
DATE Early Eocene
SIZE 6in (15cm) long
LOCATION England

Primapus is known from a single species whose specimens have been found in the London Clay deposits of southeast England. *Primapus* was an apodiform, a member of the order Apodiformes, whose living members include swifts and hummingbirds. It is uncertain whether *Primapus* is an early member of the swift family, Apodidae, or belonged to a now-extinct sister family called Aegialornithidae.

ARM BONE
The apodiform skeleton points to *Primapus* being an agile flyer.

Gastornis

GROUP Theropods
DATE Early to Middle Eocene
SIZE Over 6½ft (2m) tall
LOCATION North America, Europe

Gastornis was a huge, flightless bird. It is known from fossil fragments found in Europe, as well as complete skeletons—previously called *Diatryma* (see panel)—discovered in Early and Middle Eocene rocks in the western United States that provide evidence of a densely forested landscape. Large *Gastornis* specimens are more than 6½ft (2m) tall and have a huge skull bearing a robust beak with a hooked tip. This giant bird also had stout legs with massive clawed feet, but the wings were tiny and vestigial because it was too heavy to fly. *Gastornis* is currently thought to have been an obligate herbivore—fruits and plants were probably its favored foods—but it may have supplemented its diet by scavenging on dead animals. The powerful, hooked beak was strong enough to crack open a coconut, and would have been ideal for crushing nuts and processing other plant materials. *Gastornis* was well suited to feeding in the Eocene forests, making use of their high-density food supplies. However, as the global climate warmed, these forests gradually turned into open, grassy plains. A transition in climate is one possible explanation for the extinction of these massive ground birds.

Gastornis was one of the **biggest birds ever to evolve**, and was **considerably larger** than Early Eocene mammalian carnivores.

FOREST DWELLERS
During the Eocene, these giant birds stalked through heavily forested areas of Europe and North America, foraging for fruits and other plant foods in the undergrowth.

MISTAKEN IDENTITY

The first complete skeletons of *Gastornis* were found in Wyoming in the 1870s, but were given the name *Diatryma*. Although a similar large bird—*Gastornis*—had been described in Europe in 1855, the European fossils were incomplete and incorrectly reconstructed, so few noticed the similarities. Recently, better fossils of *Gastornis* have been found; they show that the European bird was almost identical to North America's *Diatryma*. Many scientists now think that the two names apply to the same bird, which means that the first name, *Gastornis*, is used to describe both the European and North American specimens.

Presbyornis

GROUP Theropods
DATE Late Paleocene to Middle Eocene
SIZE 3ft 3in (1m) tall
LOCATION North America, South America, Europe

Presbyornis has been found in large numbers in the Green River Shale of Wyoming and in Eocene deposits representing shallow freshwater lakes. Eggs and nests have also been found in the same rocks, and *Presbyornis* probably lived in large flocks along the lake shores. It would have waded in the shallows and used its beak to filter out food from the water, as many ducks do today. It was one of the most successful species of its time, living for 20 million years.

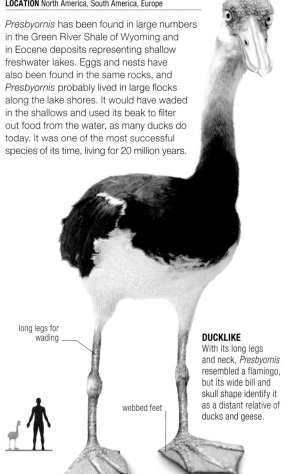

long legs for wading

webbed feet

DUCKLIKE
With its long legs and neck, *Presbyornis* resembled a flamingo, but its wide bill and skull shape identify it as a distant relative of ducks and geese.

Leptictis

GROUP Eutherian mammals
DATE Middle Eocene to Late Oligocene
SIZE 10in (25cm) long
LOCATION USA

Leptictis was an insectivorous mammal similar to modern hedgehogs, moles, and shrews. *Leptictis* would have stalked insects, amphibians, and lizards. It had a long snout filled with small teeth. These included simple, V-shaped molars like those of some modern insectivores. Along the top of the skull was a pair of long ridges, where strong jaw muscles would have been attached.

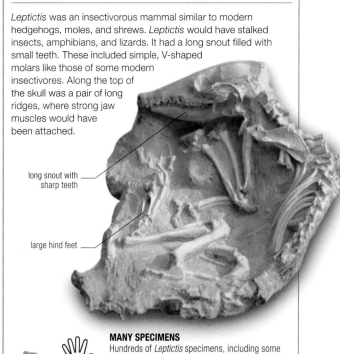

long snout with sharp teeth

large hind feet

MANY SPECIMENS
Hundreds of *Leptictis* specimens, including some complete skeletons, have been found in rocks from the Middle Eocene to the Late Oligocene.

Plesiadapis

GROUP Placental mammals
DATE Middle Paleocene to Early Eocene
SIZE 7in (18cm) long
LOCATION North America, Europe, Asia

Plesiadapis looked like a ground squirrel in size and overall body plan, but it was a relative of the primates and an adept climber. Even so, it had many rodentlike features: a pair of protruding incisors, a toothless gap between the front and back teeth, and eyes on the side of its head for spotting predators. *Plesiadapis* was so abundant that its species are used as index fossils for dating Paleocene deposits.

PRIMATE FEATURES
Although it resembled modern rodents, the teeth, skull, and ankle bones of *Plesiadapis* show that it was a close relative of the primates.

Ectoconus

GROUP Placental mammals
DATE Early Paleogene
SIZE 5ft (1.5m)
LOCATION North America

Ectoconus is a periptychid condylarth, a group of archaic mammals thought to be closely related to modern ungulate (hoofed) mammals. The periptychid condylarths were among the first large herbivorous mammals to evolve after the end-Cretaceous mass extinction that killed off the dinosaurs. *Ectoconus* lived in the Early Paleogene, around 500,000 years after the asteroid impact. It was a heavily built animal, about the size of a sheep and weighing approximately 220½lb (100kg). It had big teeth for chewing tough plants and a remarkably small brain for its body size. Fossils of *Ectoconus* are known from North America, and it is one of the most completely known species from the Early Paleogene.

Uintatherium

GROUP Placental mammals
DATE Eocene
SIZE 12½ft (3.8m) long
LOCATION North America, Asia

Uintatherium was a giant horned mammal from the Middle Eocene. Most of its fossils have been found in Utah and Wyoming, but it was widespread across North America and Asia. It had huge upper canine tusks that protruded downward and were protected by a bony flange on the lower jaw. The snout had a series of blunt, hornlike growths;

like modern rhinoceroses, both sexes had horns. As is true of horned mammals today, they probably used their horns for display and species recognition. Large upper canines are also used in display of some small deer today. *Uintatherium* has no living descendants, and where it fits in the family tree of hoofed animals is a controversial subject.

Uintatheres once lived **across North America** and Asia but vanished about **40 million years ago.**

HEAVY HERBIVORE
The size of a modern rhinoceros, *Uintatherium* was a heavy-limbed herbivore with massive bones and a bulky, barrel-shaped body. It was probably covered in a thick hide.

small cheek teeth

blunt horns

UPPER SKULL
At 3ft 3in (1m) long, *Uintatherium*'s skull was similar to the size of a modern white rhino's. In the Middle Eocene, that made *Uintatherium* one of the largest mammals in the world.

coarse, thick hide

sturdy, pillarlike legs

broad feet

huge canine tusks

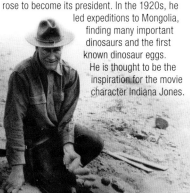

Andrewsarchus

GROUP Placental mammals
DATE Middle Eocene
SIZE 12¼ft (3.7m) long
LOCATION Mongolia

Andrewsarchus was a giant predator that lived between about 48 and 37 million years ago. Most of what we know about this mammal comes from a single skull found in Mongolia. The skull is huge—more than 3ft 3in (1m) long—and it is estimated that the animal weighed about 550lb (250kg). That makes *Andrewsarchus* the largest predatory land mammal that has ever lived. The pointed teeth are often worn and blunt. It has been suggested that *Andrewsarchus* not only hunted large prey but was also a scavenger. Its enormous jaws would have crushed the bones of animal carcasses. However, it is more closely related to modern artiodactyls (even-toed hoofed mammals) than to carnivores.

FORMIDABLE PREDATOR
The huge skull of this formidable predator suggests that it would have been bigger than the largest bear species.

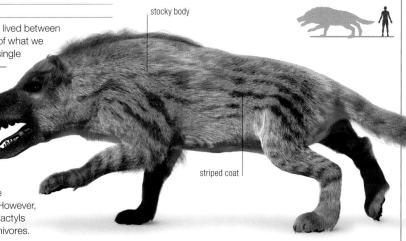

stocky body

striped coat

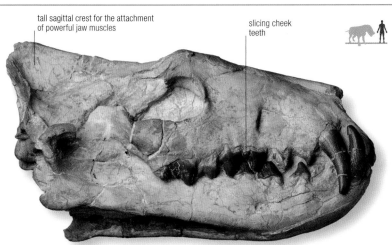

tall sagittal crest for the attachment of powerful jaw muscles

slicing cheek teeth

Hyaenodon

GROUP Placental mammals
DATE Late Eocene to Early Miocene
SIZE 1–9¾ft (0.3–3m) long
LOCATION Europe, Asia, Africa, North America

This species was a doglike predatory mammal that lived across the Northern Hemisphere in the Oligocene and persisted into the Early Miocene in Africa, finally dying out there 15 million years ago.

LAST CREODONT
Although it looked like a modern hyena, with powerful jaws and sharp teeth for tearing flesh and breaking bones, *Hyaenodon* is not related to hyenas. It was the last surviving member of an extinct group of predators called creodonts.

There were many species of *Hyaenodon*, ranging in size from a weasel to a lion. It was one of the top predators of its time and was equipped with powerful muscles in its jaws that could crush prey animals.

Hesperocyon

GROUP Placental mammals
DATE Late Eocene to Middle Miocene
SIZE 31in (80cm) long
LOCATION North America

Hesperocyon is an extinct genus of dog that lived between the Late Eocene and Middle Miocene in North America. It is the earliest known member of the dog family, Canidae, and all modern species of dog, fox, and wolf that have evolved over the past 30 million years are believed to be closely related to this mammal. However, *Hesperocyon* looked very different to modern dogs. It had a long, flexible tail, like that of a coati, but relatively short limbs and a delicate skull capable of eating only small prey, such as birds and rodents. *Hesperocyon*'s teeth suggest that it was an omnivore and probably foraged on the ground and in bushes for fruits and other plant items to supplement its diet of meat.

RACCOONLIKE
Although it is related to dogs and foxes, the small skeleton of *Hesperocyon* resembles that of a modern raccoon, with a long tail and short legs.

Icaronycteris

GROUP Placental mammals
DATE Middle Eocene
SIZE 5½in (14cm) long
LOCATION North America, Europe, Asia

Icaronycteris is one the earliest types of bats. It is known from a complete skeleton from the Middle Eocene shales of Wyoming's Green River formation (see p.351). Like other insect-eating bats, *Icaronycteris* was small and a skillful flyer. It may have used an echolocation system to catch prey, but it was much more primitive than any living bat. It had a long tail that was not connected to the hind legs, and its first finger bore a claw and was not fused to the wing membrane as in modern bats. An even more primitive bat, *Onychonycteris*, that was recently found in the same deposits lacks the inner ear features of modern bats indicative of echolocation.

free tail

wing formed by skin membrane

finger bones

clawed first finger

extended finger bone

clawed second finger

HAND WING
As in modern bats, the wing membranes of *Icaronycteris* were supported by extended finger bones. The name of the bat order, Chiroptera, means "hand wing."

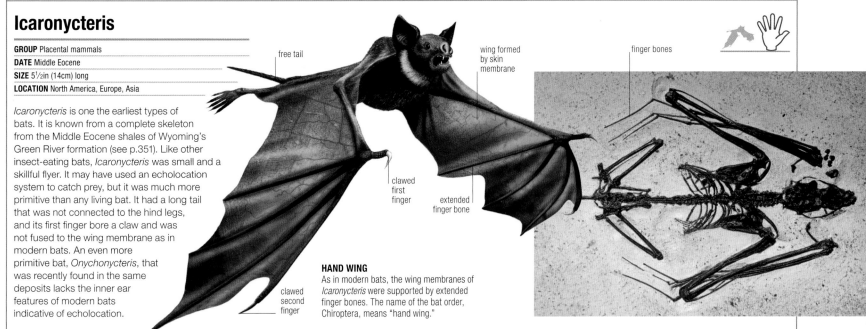

Eurotamandua

GROUP Placental mammals
DATE Middle Eocene
SIZE 3ft 3in (1m) long
LOCATION Germany

One the earliest known anteaters, *Eurotamandua* is known from a single complete skeleton fossil. It was thought to be related to tamanduas, tree-climbing anteaters from Central and South America (see panel, below). However, recent research has shown that *Eurotamandua* lacks the specialized vertebrae of anteaters and their relatives, the sloths and armadillos. It may be related to modern pangolins, which are found in Africa and Asia, but it does not have the scales characteristic of this group. *Eurotamandua* was toothless and had a long tongue for licking up ants and termites. Large claws on its forefeet were used to rip open insect nests. Like modern tamanduas, it had a long, flexible tail for gripping branches.

SCIENCE
TAMANDUA

Eurotamandua was named because it resembled the tree-climbing anteaters or tamanduas of Latin America. These animals have a similar appearance to the ground-dwelling anteaters, with a long, toothless snout and sticky tongue for lapping up ants and termites; long, sharp front claws for ripping their nests open; and short, bristly fur. However, they are only 3ft 3in (1m) long, weigh about 15lb (7kg), and have a long prehensile tail for climbing in trees.

COMPLETE SKELETON
This skeleton was found in the Messel quarry in western Germany. The bones were preserved in bituminous shale, a mixture of mudstone and oily tar.

Eomys

GROUP Placental mammals
DATE Late Oligocene to Middle Miocene
SIZE 10in (25cm) long
LOCATION Europe, Asia

Eomys was a small, gliding rodent from the Late Oligocene to Middle Miocene. Several nearly complete skeletons show that it had a long skin membrane between its front and back legs, similar to that of living flying squirrels. Although the Eurasian *Eomys* had evolved a gliding lifestyle, dozens of other members of the Eomyidae family appear to have been more similar to ground or tree squirrels. The eomyids became extinct 2 million years ago. They are thought to have been relatives of modern pocket gophers.

Palaeolagus

GROUP Placental mammals
DATE Late Eocene and Oligocene
SIZE 10in (25cm) long
LOCATION North America

Palaeolagus—meaning "ancient hare"—is one of the earliest known fossil rabbits. The few complete skeletons discovered show us that, in most respects, *Palaeolagus* resembled modern rabbits. However, it had shorter hind legs and its skull and teeth were also much more primitive than those of most later rabbits. At least eight species are recognized. Some of them had specializations in the ear region indicative of acute hearing. Like its modern relatives, *Palaeolagus* had numerous predators, and over time, the rabbit lineage has evolved longer and faster limbs to escape them.

primitive rabbitlike skull

SCAMPERER
The short hind legs of *Palaeolagus* probably indicate that it did not hop, but scampered more like a squirrel.

short hind legs

Protorohippus

GROUP Placental mammals
DATE Early and Middle Eocene
SIZE 15in (38cm) tall
LOCATION USA

Protorohippus (formerly known as *Hyracotherium* or *Eohippus*) is one the earliest known horses. Its fossils were unearthed in western USA. It was about the size of a beagle or terrier, with short limbs bearing four toes on the front feet and three toes on the hind feet. The name *Hyracotherium* was once applied to early American horses, but recent research shows that this name applies to a horselike European mammal, not a true horse.

primitive teeth

short snout

LOW-CROWNED MOLARS
The primitive, low-crowned molars of *Protorohippus* were very like the teeth of early tapiroids—the ancestors of modern tapirs.

Subhyracodon

GROUP Placental mammals
DATE Late Eocene to Miocene
SIZE 8¼ft (2.5m) long
LOCATION North America

Subhyracodon was a hornless rhinoceros from the Late Eocene to Miocene. It has been found in rocks across North America, especially in the Big Badlands of South Dakota. *Subhyracodon* was not heavily armored like modern rhinos and avoided danger by running away on its relatively long, slender legs. The cheek teeth had low crowns suited for eating the leaves of trees and bushes, and they had the characteristic π-shaped pattern of crests that is typical of living rhinos.

At one time, many different species of *Subhyracodon* were named, although they have been reduced to just three valid species today. *Subhyracodon* is closely related to *Diceratherium*. This was a larger animal, and it had paired bony ridges on its nose, which are thought to have supported short horns.

attachments for powerful shoulder muscles

elongated vertebrae

SLENDER LIMBS
Subhyracodon had slenderer limbs than living rhinos and was a little smaller—although it was still about the size of a modern cow. This specimen is the skeleton of a fetus found in Wyoming.

SMALL SKULL
The snout and braincase were smaller than those of later horses, and there was a smaller gap between the front and back teeth.

ANATOMY
HOOF EVOLUTION

Protorohippus had feet adapted for walking on rough terrain, with short hooves on each of its four fingers and three toes. As horses evolved, their middle digits enlarged and their side toes became smaller, leaving three fingers on each forefoot and three toes on each hindfoot. In modern horses, the side toes are reduced to tiny splints and all the weight is borne on the middle digit, making them very effective runners.

FOREFOOT OF PROTOROHIPPUS

Mesohippus

GROUP Placental mammals
DATE Late Eocene to Late Oligocene
SIZE 23½in (60cm) tall
LOCATION North America

An extinct horse about the size of a labrador dog, *Mesohippus* was found in the Big Badlands fossil beds of South Dakota, as well as in Canada and Mexico. Like modern horses, *Mesohippus* had a long snout with a gap between its front teeth and cheek teeth. Its teeth also show that *Mesohippus* was a specialized leaf-eater. During the Early Oligocene, there were more than a dozen *Mesohippus* species. By the end of that epoch, *Mesohippus* lived alongside its later relative *Miohippus* in several places.

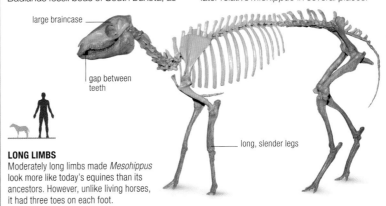

large braincase

gap between teeth

long, slender legs

LONG LIMBS
Moderately long limbs made *Mesohippus* look more like today's equines than its ancestors. However, unlike living horses, it had three toes on each foot.

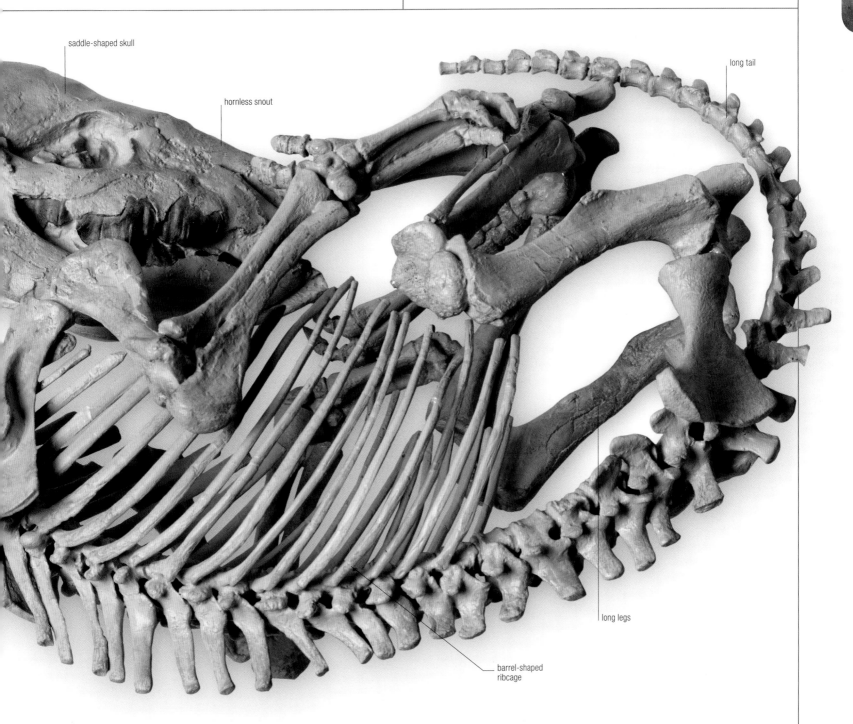

saddle-shaped skull

hornless snout

long tail

long legs

barrel-shaped ribcage

Megacerops

GROUP Placental mammals

DATE Late Eocene

SIZE 9¾ft (3m) long

LOCATION North America

This animal was the last and largest member of the brontotheres—meaning "thunder beasts." *Megacerops* was the size of a white rhinoceros and lived on the Great Plains of North America. It had a forked horn on its nose, which may have been used for butting heads (rutting) during sexual displays.

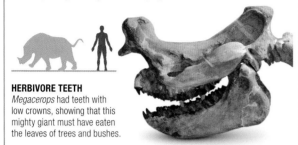

HERBIVORE TEETH
Megacerops had teeth with low crowns, showing that this mighty giant must have eaten the leaves of trees and bushes.

Mesoreodon

GROUP Placental mammals

DATE Late Oligocene

SIZE 4¼ft (1.3m) long

LOCATION USA

Mesoreodon was an oreodont, an extinct sheep-sized, even-toed hoofed mammal (artiodactyl). Like other oreodonts, it had sharp canine teeth, presumably used for display and defense. Oreodonts were not very fast runners and probably lived in herds for protection. Several complete *Mesoreodon* skeletons have been found, one of which had preserved, ossified (bony) vocal cords that show that it could make loud sounds like modern howler monkeys. These hooting calls may have warned the herd of attacks or frightened off predators.

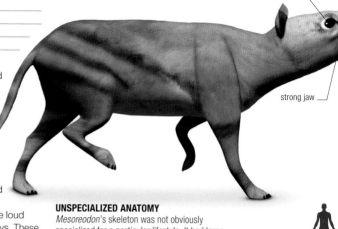

large eyes

strong jaw

UNSPECIALIZED ANATOMY
Mesoreodon's skeleton was not obviously specialized for a particular lifestyle. It had large eyes, a robust jaw, and cheek teeth with crescent-shaped ridges for eating leaves.

Leptomeryx

GROUP Placental mammals

DATE Late Eocene to Late Oligocene

SIZE 3ft 3in (1m) long

LOCATION North America

Leptomeryx was a hornless ruminant (cud-chewing hoofed mammal) from the Late Eocene and Oligocene. Its fossils have been found in many parts of North America, and it is one of the most abundant fossils in the Big Badlands fossil beds of South Dakota. *Leptomeryx* was about the size of a mouse deer or chevrotain. It had relatively slender limbs bearing two hoofed toes on each foot—it was an artiodactyl, or an even-toed hoofed mammal. Although it is only distantly related to living chevrotains, *Leptomeryx* had many similar features, including a delicate stature; the absence of any antlers or horns; and, in the case of the males, enlarged upper canines that protruded as small tusks. Recent research has revealed that *Leptomeryx* was among the most primitive of ruminants, and therefore, a distant relative of all ruminants, including deer, cattle, and camels.

During the Early Oligocene, *Leptomeryx* was **one of the most common land mammals** in North America.

SIMILAR SKELETONS
Six species of *Leptomeryx* have been described. Their skeletons are too similar to tell them apart, so each species is identified by the shape of its teeth.

Darwinius

GROUP Placental mammals
DATE Middle Eocene
SIZE 35in (90cm) long
LOCATION Germany

This squirrel-sized primate, nicknamed "Ida," is remarkable for being the most complete primate fossil ever found—even its final meal of leaves and fruit is preserved in its stomach. *Darwinius*'s skeleton shows it to have been an agile but generalized climber, and some scientists think that this animal, while still technically a prosimian, is close to the origins of anthropoid primates.

REMARKABLE FOSSIL
First discovered in 1983, this fossil remained hidden in a private collection until it was revealed to the world as *Darwinius masillae* in 2009.

Eosimias

— grasping hands

GROUP Placental mammals
DATE Middle Eocene
SIZE 2in (5cm) long
LOCATION China

Eosimias is an extinct primate genus that is one of the oldest known basal anthropoids, the group that contains monkeys and apes. It is known from the Middle Eocene beds in Shanxi, China. *Eosimias* was a tiny primate—some were only about the size of a human thumb. Its grasping hands and long tail gave it the appearance of a tiny marmoset, but it is unrelated. Like marmosets, *Eosimias* probably caught insects to supplement its fruit diet. The presence of this animal and other species of primitive anthropoids in the Eocene of Asia suggests that Asia was the place of origin for this group rather than Africa as previously assumed.

GROOMING CLAW
Eosimias fossils show that its second toe possessed a grooming claw similar to those seen in certain extant primates. It would have used this specialized claw like a comb for personal grooming.

Ambulocetus

GROUP Placental mammals
DATE Early Eocene
SIZE 9¾ft (3m) long
LOCATION Pakistan

An ancestor of modern whales, *Ambulocetus* retained strong legs and could walk as well as swim. It may have hunted like crocodiles, lurking in the shallows and then lunging forward onto land to catch prey that wandered near the water. We know *Ambulocetus* was a relative of modern whales because of the shape of its skull and teeth. The similarity of its jaw and middle ear to that of modern whales suggests that, like them, it was specialized for underwater hearing and might have lacked the external ear that picks up sounds on land. *Ambulocetus* is one of many transitional fossils that show how whales evolved from land mammals.

ABLE SWIMMER
Ambulocetus had webbed feet for paddling. It probably swam with an up-and-down flexing motion of the spine and tail, like otters do today.

Moeritherium

GROUP Placental mammals
DATE Late Eocene to Early Oligocene
SIZE 9¾ft (3m) long
LOCATION Africa

Moeritherium was one of the earliest fossil relatives of the Proboscidae, the family that includes the elephants and mammoths. Its short proboscis made it look similar to a tapir. However, it also had short tusks like other members of the Proboscidae, and many other features of the skull and skeleton that show it is a proboscidean, not a relative of tapirs. The semiaquatic nature of *Moeritherium* has lead scientists to speculate that the earliest proboscideans were semiaquatic and that their lineage became fully terrestrial only later.

GRINDING TEETH
Moeritherium's teeth were adapted for eating leaves and soft aquatic vegetation.

GENOGENE

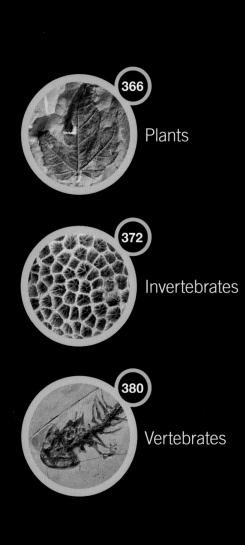

Neogene

During the Neogene Period, the Earth was plunged into an ice age. More water circulating in the atmosphere, drawn from an Atlantic Ocean, resulted in high snowfall on Antarctica, and the ice sheet increased to its present size. As Earth cooled and became drier, grasslands spread at the expense of forests. Many large and small herbivorous mammals and their predators had to adapt to a demanding life on the open plains.

MOUNTAIN RANGES
Continued collision between Earth's lithospheric plates threw up great mo ranges, such as the Andes and the Himalayas, as well as the Tibetan Pla

Oceans and continents

Neogene, meaning "newborn," is a period when many modern animals evolved, including *Australopithecus,* one of our early ancestors. Major features of today's world geography were also taking shape. The Indian plate continued to plow beneath Asia, pushing the Himalayas even higher, and closing the great Tethys Ocean. Due to the northward movement of the African, Arabian, and Australian plates, Spain collided with France, forming the Pyrenees; Italy with France and Switzerland to form the Alps; Arabia with Iran to form the Zagros mountains; and, more recently, Australia with the Asian plate to form Indonesia. In the process of mountain-building, the continental crust was compressed horizontally as rocks were forced together by converging plates. The area occupied by the continents slightly decreased and the size of ocean basins increased, causing sea levels, on average, to fall. These changes may have all contributed to the cooling of the Earth, which transformed its ecosystems. As forests gave way to grasslands in mid-latitudes, hoofed mammals, including horses, developed teeth suitable for grazing, and long legs to gallop away from predators. The newly formed Panama Isthmus also affected the world's climate, by altering ocean current circulation (see p.24). It linked the American continents, allowing fauna and flora that were once separated to migrate freely. Cats, snakes, tapirs, wolves, and deer moved southward, while opossums, armadillos, hummingbirds, and sloths traveled northward.

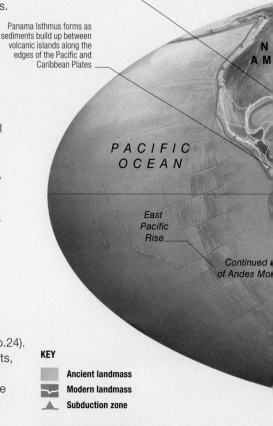

Area that is now Florida is still partly covered by sea

Panama Isthmus forms as sediments build up between volcanic islands along the edges of the Pacific and Caribbean Plates

N
A M

PACIFIC OCEAN

East Pacific Rise

Continued of Andes Mo

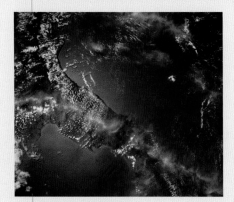

PANAMA LAND BRIDGE
About 3 million years ago, the Panama Isthmus bridged the gap between the American continents, changing the routes of currents in the Pacific and Atlantic Oceans.

KEY

▬	Ancient landmass
∿	Modern landmass
▲	Subduction zone

MID-NEOGENE WORLD MAP
Familiar mountain chains have appeared, including the Himalayas, Alps, and Pyrenees. The Panama Isthmus links North and South America.

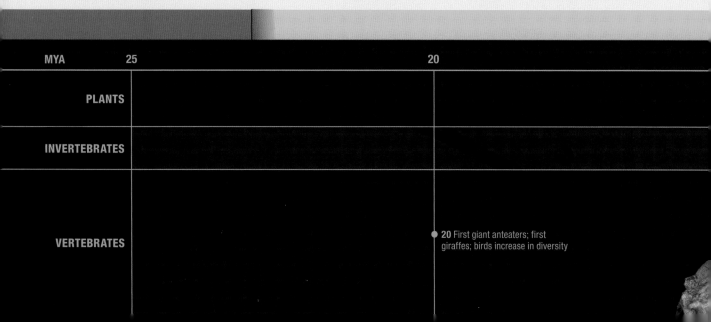

MYA	25		20	
PLANTS				
INVERTEBRATES				
VERTEBRATES			● **20** First giant anteaters; first giraffes; birds increase in diversity	

SAVANNA VEGETATION
As savannalike plants evolved in arid low-latitude areas, artiodactyls, including the giraffe, adapted to browse on and digest the tough vegetation in their grassland habitats.

Climate

During the early Neogene, the Antarctic Circumpolar Current became firmly established, isolating Antarctica as a cold region. The change from warm to cooler water is reflected in a change from carbonates to silica-rich sediments. By 12 to 14 million years ago, global temperatures had fallen and more ice built up in eastern Antarctica. While ocean temperatures dropped abruptly at the poles, they remained relatively warm in low latitudes, at about 72–75°F (22–24°C). Dry, savanna-type vegetation developed in low latitudes about 15 million years ago. By the end of the Miocene, the ice cap expanded greatly, becoming even larger than it is today. Two periods of glacial expansion, lasting about 15,000 years, occurred 5.2 and 4.8 million years ago, separated by a relatively warm phase. Yet, at the North Pole, ice had not started to form all year round. The colder conditions lowered sea levels by trapping water in the polar regions. Although the climate warmed for a short period in the early Pliocene, it then began to cool again, with large ice sheets forming in the Northern Hemisphere. This changing climate required new strategies for survival, such as herd behavior and seasonal migrations in hoofed mammals, and burrowing and hibernation in early rodents.

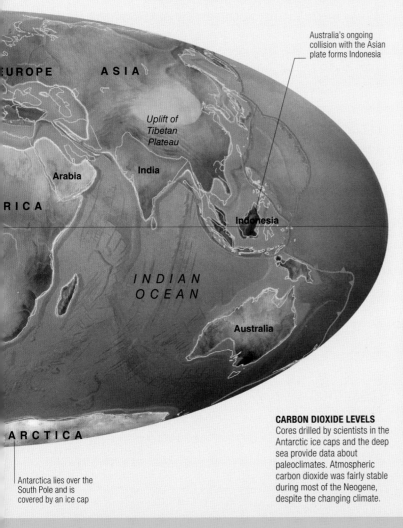

Australia's ongoing collision with the Asian plate forms Indonesia

EUROPE ASIA

Uplift of Tibetan Plateau

Arabia India

RICA

Indonesia

INDIAN OCEAN

Australia

ARCTICA

Antarctica lies over the South Pole and is covered by an ice cap

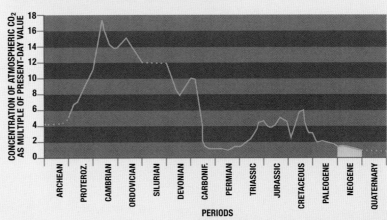

STRAIT OF GIBRALTAR
The Mediterranean Basin was isolated 5.7 million years ago, when sea levels lowered. It partially dried out, leaving a huge salt pan, until 5 million years ago, when the Gibraltar Strait (above, left) opened.

CARBON DIOXIDE LEVELS
Cores drilled by scientists in the Antarctic ice caps and the deep sea provide data about paleoclimates. Atmospheric carbon dioxide was fairly stable during most of the Neogene, despite the changing climate.

CONCENTRATION OF ATMOSPHERIC CO$_2$ AS MULTIPLE OF PRESENT-DAY VALUE

ARCHEAN · PROTEROZ. · CAMBRIAN · ORDOVICIAN · SILURIAN · DEVONIAN · CARBONIF. · PERMIAN · TRIASSIC · JURASSIC · CRETACEOUS · PALEOGENE · NEOGENE · QUATERNARY

PERIODS

PLIOCENE

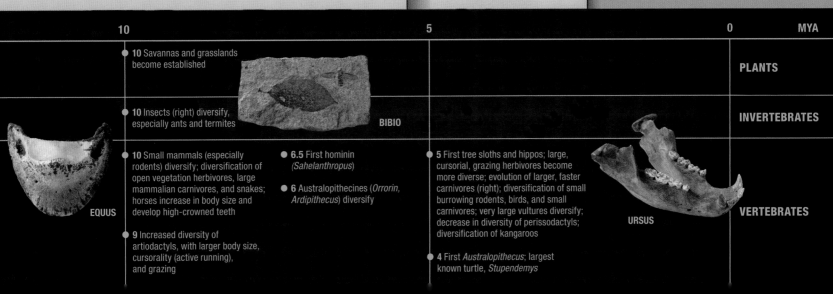

10   5   0   MYA

10 Savannas and grasslands become established — **PLANTS**

10 Insects (right) diversify, especially ants and termites — BIBIO — **INVERTEBRATES**

EQUUS

10 Small mammals (especially rodents) diversify; diversification of open vegetation herbivores, large mammalian carnivores, and snakes; horses increase in body size and develop high-crowned teeth

9 Increased diversity of artiodactyls, with larger body size, cursorality (active running), and grazing

6.5 First hominin (*Sahelanthropus*)

6 Australopithecines (*Orrorin, Ardipithecus*) diversify

5 First tree sloths and hippos; large, cursorial, grazing herbivores become more diverse; evolution of larger, faster carnivores (right); diversification of small burrowing rodents, birds, and small carnivores; very large vultures diversify; decrease in diversity of perissodactyls; diversification of kangaroos

4 First *Australopithecus*; largest known turtle, *Stupendemys*

URSUS — **VERTEBRATES**

NEOGENE
PLANTS

The dramatic climatic cooling that began around the Paleogene–Neogene boundary, together with significant mountain building in Asia and the Americas, had a major impact on global vegetation. Increased separation of the continents also led to distinctive vegetation in different parts of the world.

The Neogene marks the debut of a truly modern world, with widely separated continents and a steep gradient of temperature from cold high-latitude areas to hot equatorial regions. As climates and geography became more similar to those of today, Neogene vegetation also became increasingly modern in appearance.

Seasonal forests

The cooler climates of the Neogene saw warmth-loving plants become increasingly restricted to the tropics and subtropics. In the Northern Hemisphere, to the north of the subtropics, areas with more strongly seasonal climates became dominated by trees such as oaks, maples, and beeches. A new flora of spring ephemerals became adapted to life below the canopy of these deciduous trees. Farther north still, boreal forests developed, dominated by conifers. During the Neogene, temperate forests were widespread across the Northern Hemisphere. Their broadly similar

NEW KINDS OF HABITATS
Drier climates, perhaps combined with increased pressure from grazing animals and an increased frequency of natural fires, created new habitats during the Neogene.

composition probably reflects the relative ease of migration across the Bering Strait and northern Atlantic during the warmth of the Paleogene. Temperate forests developed much less extensively in the Southern Hemisphere, where there is much less land at the equivalent latitudes. Temperate forests in the Southern Hemisphere, for example in New Zealand, Chile, and Tasmania, remained dominated by southern beeches during the Neogene, as they had since the Late Cretaceous.

MAPLE FOREST
Modern kinds of temperate forest, dominated by familiar trees such as maples and oaks, became widespread for the first time as climates cooled during the Neogene.

Grasslands

In addition to global cooling, a key feature of the Neogene was drier climates in many parts of the world. In Australia, this drying trend was especially marked, but it also occurred in Asia, and North and South America. It was accentuated by the development of rain shadows linked to the uplift of the Andes, Rocky Mountains, and Himalayas. Grasses were the key group of angiosperms that exploited these new dry habitats effectively.

GROUPS OVERVIEW

Many new kinds of flowering plants, especially groups that are characteristic of open habitats, increased in diversity and became more prominent during the Neogene. The changing climatic conditions resulted in further diversification at species level in almost all groups of land plants during the period.

FERNS
Ferns have a relatively sparse fossil record during the Neogene, especially compared to their significance in many Mesozoic and earlier floras. However, the diversity of living ferns reveals that this group continued to diversify through the Cenozoic. Many living ferns show all the signs of groups that are still in a very active state of evolution.

CONIFERS
Changes in global climate during the Neogene resulted in the regional extinction of many conifer groups. Once widespread forms, such as redwoods, became restricted to a much smaller area. The exceptionally well-preserved Miocene fossil flora from Clarkia, Idaho, contains many conifers restricted to eastern Asia today.

MONOCOT ANGIOSPERMS
While most of the major groups of monocots were in place during the Paleogene, changing climates during the Neogene facilitated further diversification. In addition to the great Neogene explosion of grasses, the diversification of many groups of monocots in Mediterranean regions probably also dates from this time.

EUDICOT ANGIOSPERMS
The rapid diversification of eudicots continued through the Neogene. Many groups of herbaceous eudicots appear in the fossil record for the first time. These groups account for a very substantial proportion of the total diversity of living angiosperms, indicating that much of the diversity of living angiosperms is of relatively recent origin.

Taxodium

GROUP Conifers
DATE Cretaceous to present
SIZE Up to 130ft (40m) tall
LOCATION High northern latitudes

The early Tertiary northern flora that grew throughout what is now the Arctic belt began in the Late Cretaceous. As the climate cooled, the flora migrated south. Plant communities in these warmer, southern areas were dominated by *Taxodium* broad-leaved trees and shrubs. In areas where the flora could not migrate southward because of mountain ranges, the plants became extinct. The Cretaceous *Parataxodium* was the ancestor to *Taxodium*, *Sequoiadendron*, *Sequoia* (see p.276), and probably *Metasequoia* (see p.343). Living *Taxodium* are easy to recognize, but fossils can be hard to identify. Its shoots have two flat rows of needlelike leaves, but they are similar to those of *Metasequoia*. Its male and female cones are more distinctive, but far less common.

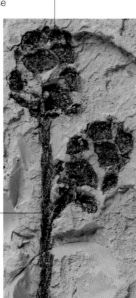

cone axis — shoot —

— pollen cone

seed cone

TAXODIUM DUBIUM
This conifer produces small male cones in large numbers off the sides of hanging axes.

SEED CONES
Large, ball-shaped seed cones occur on the ends of the leafy shoots.

The three living species of *Taxodium* are confined to southern North America, Mexico, and Guatemala. The swamp cypress thrives around the Gulf of Mexico, where it grows to nearly 130ft (40m) tall. In wet areas it has a buttressed trunk and aerating roots. Its leaves turn brown and are shed in late fall.

Taxites

GROUP Conifers
DATE Cretaceous to present
SIZE Shoot up to 4½in (12cm) long
LOCATION High northern latitudes

Taxites is a generic name that has sometimes been given to leafy shoots that are in fact the smaller "short shoots" of *Taxodium* (see left) that were naturally shed (abscized) from the trees. This happened either seasonally or annually. The linear leaves were spirally attached to the shoot. Because of their abscission, shoots are often found in large numbers, with hundreds on a single bedding-plane surface.

TAXITES LANGSDORFI
The linear, flattened leaves of *Taxites langsdorfi* are spirally attached to the axis; they are twisted and appear to be spread in two ranks.

Glyptostrobus

GROUP Conifers
DATE Cretaceous to present
SIZE Up to 115ft (35m) tall
LOCATION High latitudes in the Northern Hemisphere

The first records of *Glyptostrobus* are in the Late Cretaceous of Japan, North America, and Spitzbergen. There are many more Paleogene and Neogene records of shoots and their cones around the Northern Hemisphere and up to the present high-Arctic latitudes. The leaves were spirally arranged on the shoots, forming three ranks along them. Today, the only living species, the Chinese swamp cypress, grows in subtropical southeastern China and Vietnam.

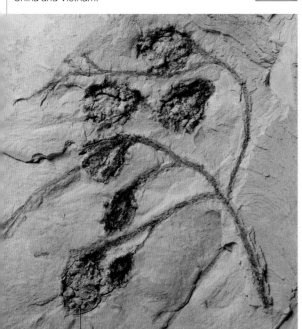

seed-bearing cone

GLYPTOSTROBUS SP.
The seed-bearing cones of this conifer were about ½–¾in (1.5–2cm) long, and were borne on the ends of the leafy shoots.

secondary vein

prominent midrib

MAGNOLIA LONGIPETIOLATA
Fossil leaves like these have often been assigned to living genera on the basis of superficial similarity to those of the living plants.

Magnolia

GROUP Angiosperms
DATE Cretaceous to present
SIZE Up to 145ft (45m) tall
LOCATION Temperate Northern Hemisphere

The characters that allow fossilized leaves of this genus to be assigned to the living genus *Magnolia* are their elongated shapes with curved edges, gradually tapering tips, stout stalks, obvious midribs, parallel secondary veins that depart at less than 45 degrees, and a meshwork of tertiary veins. Leaf size varies considerably on a single plant, so it is not a good feature for identifying a species. Fossil remains of *Magnolia* are known from the Paleogene and Neogene. During these periods, *Magnolia* grew in slightly humid environments, either as understory shrubs and trees or in openings along stream banks, and extended into higher ground on the fringes of more open habitats. Today, *Magnolia* lives naturally in the open forest in temperate and warm-temperate areas of eastern Asia.

Magnolia includes about 70 species of trees and shrubs with simple, leathery leaves and large, bisexual flowers. The showy, white or pink petal-like structures surround numerous stamens and a central mass of free carpels. *Magnolia* is deciduous and in many species the flowers emerge before the leaves.

Tectocarya

GROUP Angiosperms

DATE Middle Paleogene to Neogene

SIZE Fruit ³⁄₄in (2cm) long

LOCATION C. Europe

Tectocarya was a member of the dogwood family. Their fruits had a fleshy outer layer that attracted animals and an inner, thickened, woody wall that was resistant to digestion. Fossils were formed when the partly digested seeds that had passed through animals were swept away and covered by sediments. In time, minerals infiltrated the seeds, sometimes preserving traces of the cell walls.

— fossilized seed

TECTOCARYA RHENANA
The woody outer walls of these mineralized, nutlike seeds have water-worn surface walls.

Alnus

GROUP Angiosperms

DATE Cretaceous to present

130ft (39m) tall

LOCATION Northern Hemisphere, South America

Alnus, or alders, first appeared in the Cretaceous. The leaves are mostly elliptical or ovate and broadest in the middle or upper half. Some living alders have heart-shaped leaves. The trees have small catkins of male and female flowers. After pollination, the female catkins turn into overwintering, woody, conelike structures that open the following year to release the seeds.

— toothed margin

ALNUS CECROPIIFOLIA
Fossilized alder leaves, such as the one shown here, can be difficult to identify without associated flowers or their conelike fruits.

Podocarpium

GROUP Angiosperms

DATE Paleogene to Neogene

SIZE Seed pod about 1¹⁄₄in (3cm) long

LOCATION Europe, North America, China

Podocarpium was a member of the legume family. There are around 14,000 living species of legumes, which include peas, beans, clover, peanuts, and lupins, and many of them are economically important. This family may have had its origins in the late Cretaceous in the southern continent, Gondwanaland, but by the Eocene, fossil leaves, flowers, seedpods, and pollen grains reveal that the three major groups of legumes had evolved and spread into the northern continents.

— leaflet   fruit

PODOCARPIUM SP.
The fruits of *Podocarpium* (above) were like small pea pods on a stalk. The compound leaves (left) had one terminal and several pairs of lateral leaflets.

Hymenaea

GROUP Angiosperms

DATE Neogene to present

SIZE Tree up to 80ft (25m) tall

LOCATION Tropical Central and South America, E. Africa

Amber-preserved specimens of *Hymenaea* were formed when flowers and leaves fell from large, evergreen, leguminous trees to become entombed in resin oozing from the trunk and branches of the same plants. The resin completely encased the plant tissue before hardening and preserving cell walls, chloroplasts, xylem, and even nuclei and some cell membranes. *Hymenaea* had an extensive range in the Neogene of Central and South America and Africa. Today there are 13 tropical species in the Caribbean and from southern Mexico to Brazil, and one on the east coast of Africa and its islands.

— preserved flower head

HYMENAEA SP.
The flower shown here is one of several species of *Hymenaea* preserved in Neogene Dominican amber.

Betula

GROUP Angiosperms

DATE End of the Cretaceous to present

80ft (24m) tall

LOCATION Northern temperate regions in Asia, Europe, North America

Betula, or birch, leaves are ovate to rounded or triangular in shape, with toothed margins. They are stalked and have a main vein with branching secondary veins. There are about 40 living species of birch, and all are comparatively short-lived deciduous trees with both male and female catkins on the same tree. Some species grow to nearly 98ft (30m) tall, and because they shed many of their small branches, they are usually slender trees with few large branches. They all grow in open areas with poor soils and are pioneer trees on disturbed or abandoned land. Today birches are found in Europe, Asia, and North America.

secondary vein

BETULA ISLANDICA
These fossil leaves are preserved as impressions but still show their *Betula* characters quite clearly.

toothed margin

Many elm species hybridize with each other, especially when grown together. Artificial propagation can produce varieties of such hybrids that show desirable traits. *Ulmus minor* 'Dicksonii' is slow growing, with leaves that turn a bright golden yellow in the fall.

Ulmus

GROUP Angiosperms
DATE Paleogene to present
120ft (36m) tall
LOCATION Northern temperate regions

Ulmus, or elm, leaves are elongated, broadest in the middle, with toothed margins and an asymmetric rounded base; some species have an asymmetric apex as well. The leaves have a prominent midrib with equally prominent, straight, regularly spaced, and sometimes branched lateral veins that extend to the leaf margin, where they usually end in the teeth. These are all characteristics of elm leaves, but the problem, as in many other fossils, is identifying the exact species. There is a natural variation in the leaves on living elms, which should be taken into account when studying fossils. There are about 45 elm species living today in Europe, Asia, North Africa, and North America. All are large, deciduous trees with some species reaching 120ft (36m) in height. Small clusters of bisexual flowers are produced in the spring before the leaves open, and winged seeds are shed either later in the spring and germinate immediately or in the fall, when they overwinter before germinating.

lateral vein

prominent midrib

ULMUS SP.
Although this leaf shows many characteristics of an elm leaf, identifying the species is especially difficult if there is only one specimen to study.

Populus

GROUP Angiosperms
DATE Paleogene to present
195ft (60m) tall
LOCATION Northern temperate regions

Populus, or poplar, leaves are rounded with lobed or blunt-toothed margins, sometimes leathery, and almost maplelike in appearance. They are stalked, and the branching leaf veins are all of the same thickness without much of a prominent midrib. Poplars today are fast-growing,

deciduous trees that bear their flowers in catkins. The flowers emerge in spring before the leaves, enabling the wind to carry the pollen to the female catkins relatively unhindered. Individual trees have either the male pollen-producing catkins or the female catkins. There are about 30 species spread across the northern temperate regions of Europe, Asia, and North America. The American west coast black cottonwood is the largest today, growing up to 195ft (60m) in height. A commonly planted tree in Europe and North America is the Lombardy poplar, which grows to 98ft (30m) tall and has an elegant, spirelike shape.

This is a young plant of the vigorous hybrid cross between American cottonwood and the European black poplar. The particular variety shown here (*Populus* x *canadensis* 'Serotina Aurea') is commonly called the golden poplar because its leaves turn golden yellow in fall.

branching veins

midrib

network of tiny veins

rounded shape

POPULUS LATIOR
These fossils reveal the fine network that the smallest leaf veins form between the major veins of a poplar leaf.

Cedrelospermum

GROUP Angiosperms
DATE Eocene to present
SIZE 130ft (40m) tall
LOCATION North America, Asia, Europe

Cedrelospermum is an extinct member of the elm family, which today contains some 45 species of trees and shrubs that are common in northern temperate latitudes. The leaves and fruits are frequent fossils in the Early Eocene through Miocene of Europe and North America. The fruits are distinctive, with one or two flattened wings of fibrous tissue.

Fagus

GROUP Angiosperms
DATE Late Cretaceous to present
145ft (45m) tall
LOCATION Northern temperate regions

Fagus comprises 10 large, deciduous beech trees that are found throughout the Northern Hemisphere in Europe, Turkey, Japan, China, and eastern North America. The trees produce male flowers in hanging bunches and female ones in small clusters, with each flower producing two seeds. Leaves of *Fagus* are characteristic of many Neogene floras and are also known from the Paleogene. There are two basic forms of leaves: elongated and slender with 12–16 pairs of lateral veins in North America, and shorter and broader

with 6–9 pairs of lateral veins in Europe. Intermediates occur in other parts of the world. The earliest assemblages of fossils showed all the variations, but the forms gradually became separated by distribution during the Neogene.

FAGUS GUSSONII
This leaf of *Fagus gussonii* was discovered in Greece, where the plant material has been compressed in layers of sediments.

lateral vein

Nothofagus

GROUP Angiosperms
DATE Late Cretaceous to present
SIZE Up to 145ft (45m) tall
LOCATION Australia, Papua New Guinea, New Zealand, South America, Antarctica

Nothofagus is known from fossil leaves, wood, seed cupules, and pollen. Its leaves are symmetrical and elliptical in outline, with broadly rounded tips and toothed margins. *Nothofagus* was the most significant tree in the temperate zone of the Southern Hemisphere throughout the Paleogene and Neogene. It is known as fossils from Antarctica, Australasia, and South America, and first appears in the Late Cretaceous. It invaded the earlier conifer forests and became the dominant tree. In the Neogene, its distribution was more limited in Australia, as the continent drifted north and became drier. *Nothofagus* became extinct in Antarctica, along with nearly all the vegetation, as ice sheets expanded to cover most of the continent.

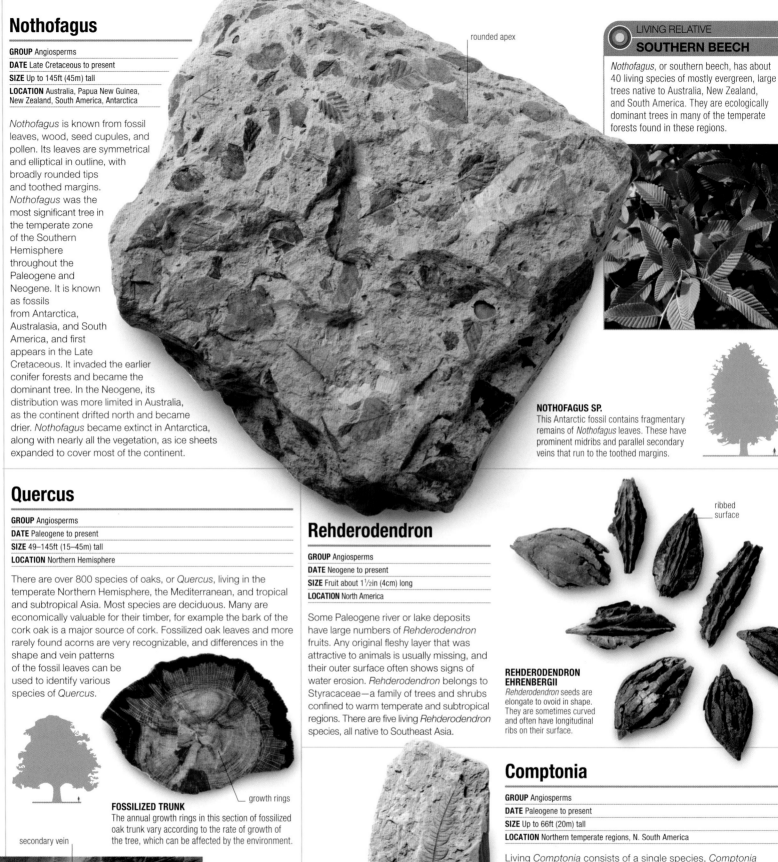

rounded apex

NOTHOFAGUS SP.
This Antarctic fossil contains fragmentary remains of *Nothofagus* leaves. These have prominent midribs and parallel secondary veins that run to the toothed margins.

Quercus

GROUP Angiosperms
DATE Paleogene to present
SIZE 49–145ft (15–45m) tall
LOCATION Northern Hemisphere

There are over 800 species of oaks, or *Quercus*, living in the temperate Northern Hemisphere, the Mediterranean, and tropical and subtropical Asia. Most species are deciduous. Many are economically valuable for their timber, for example the bark of the cork oak is a major source of cork. Fossilized oak leaves and more rarely found acorns are very recognizable, and differences in the shape and vein patterns of the fossil leaves can be used to identify various species of *Quercus*.

FOSSILIZED TRUNK
The annual growth rings in this section of fossilized oak trunk vary according to the rate of growth of the tree, which can be affected by the environment.

growth rings

secondary vein

lobed margin

QUERCUS FURUHJELMI
Oak leaves are elliptical to roughly oval, with distinctly lobed margins and stalks. They have a central vein and secondary veins that end in the leaf lobes.

Rehderodendron

GROUP Angiosperms
DATE Neogene to present
SIZE Fruit about 1½in (4cm) long
LOCATION North America

Some Paleogene river or lake deposits have large numbers of *Rehderodendron* fruits. Any original fleshy layer that was attractive to animals is usually missing, and their outer surface often shows signs of water erosion. *Rehderodendron* belongs to Styracaceae—a family of trees and shrubs confined to warm temperate and subtropical regions. There are five living *Rehderodendron* species, all native to Southeast Asia.

ribbed surface

REHDERODENDRON EHRENBERGII
Rehderodendron seeds are elongate to ovoid in shape. They are sometimes curved and often have longitudinal ribs on their surface.

lobed margin

long, oval leaf

Comptonia

GROUP Angiosperms
DATE Paleogene to present
SIZE Up to 66ft (20m) tall
LOCATION Northern temperate regions, N. South America

Living *Comptonia* consists of a single species, *Comptonia peregrina*, which grows in eastern North America. *Comptonia* is closely related to *Myrica*, a genus that consists of about 35 to 50 species of medium-height trees—up to 66ft (20m) tall—and aromatic shrubs that grow up to 3ft 3in (1m) tall. These plants can tolerate drier soil, but some species prefer acid, boggy ground. *Myrica* leaves are simple, but the leaves of *Comptonia* are distinctively and deeply toothed. *Comptonia* is first recorded from the Eocene; it then spread to most temperate northern regions by the Neogene.

COMPTONIA DIFFORMIS
This is an impression of a leaf on fine-grained sediment. The leaf is long and oval in outline with a distinctly lobed margin.

Acer

GROUP Angiosperms
DATE Paleogene to present
SIZE 30–98ft (9–30m) tall
LOCATION North America, Europe, Asia

Living *Acers*, commonly known as maples, are medium to large trees. They have distinctive stalked leaves with toothed margins. Although most *Acer* leaves have five lobes, some fossils have just three. The extent of the incisions between the lobes, the size of the marginal teeth, and the size of the leaf vary between species. Sometimes, as in the living Japanese maple, the lobes are virtually separate and are themselves lobed. The fruits of maple trees are produced in pairs and, being winged, can travel some distance from the parent tree by spinning in the wind. Fossils of these winged fruits, very much like those of sycamores, are also known.

ACER TRILOBATUM
Maples have easily recognized, lobed leaves, with stalks and toothed margins.

five-lobed leaf

toothed margin

DISTINCTIVE LEAVES
The fossil above shows a maple leaf with the usual five-lobed arrangement. On the left is a species with only three lobes, but it is still a distinctive maple leaf shape.

LIVING RELATIVES
MAPLES

There are over 200 species of maples living in North America, Europe, and Asia today. The European sycamore and the North American sugar maple are the largest, growing to over 98ft (30m) tall. Sugar maples yield up to 5¼ pints (2.5 liters) of syrup a year. Many maple leaves turn red or yellow in the fall, as shown here by this red maple.

Sapindus

GROUP Angiosperms
DATE Neogene to present
SIZE 98ft (30m) tall
LOCATION Worldwide tropics and subtropics

There are about 12 living species of this genus, living in warm temperate to tropical regions around the world. The plants have clusters of creamy white flowers that form fruits called soap nuts. These contain natural saponins and have been used for washing by various people, such as Native Americans, for thousands of years.

narrow leaflet

SAPINDUS FALSIFOLIUS
This is a pinnately divided *Sapindus* leaf, with the leaflets arranged alternately on the shoot.

Chaneya

GROUP Angiosperms
DATE Paleogene to Neogene
SIZE Up to 66ft (20m) high
LOCATION China, Europe, North America

Until recently called *Porana* and thought to be a climbing vine in the bindweed family, the plants that produced these small five-petaled flowers are now believed to be more closely related to Rutaceae. Today, this family includes *Citrus*, an economically important genus from which oranges, lemons, and grapefruits are cultivated.

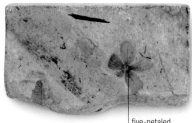

five-petaled flower

CHANEYA SP.
Although the *Chaneya* flower's five free petals can be clearly seen, the overall habit of these herbs, shrubs, or small trees remains elusive.

Typha

GROUP Angiosperms
DATE Paleogene to present
SIZE Erect stem up to 9¾ft (3m) long
LOCATION Worldwide

Typha, or bullrush, is a genus of about 11 species of monocotyledonous plants that grow in wetlands around the world, but primarily in the Northern Hemisphere. *Typha* fossils may have originated in the Late Cretaceous, but are definitely known from the Paleogene of the USA and the Early Neogene of Europe. By the Quaternary, *Typha* was widespread and common around the world.

female flowers

TYPHA SP.
Bullrush flowers are packed into dense cylindrical spikes. Fossil fruits and pollen of *Typha* are common, but flower spikes, such as this, are rare.

Phragmites

GROUP Angiosperms
DATE Paleogene to present
SIZE Up to 9¾ft (3m) tall in the flowering season
LOCATION Worldwide

Fossil leaves that can be assigned to the genus of living monocotyledononous grasses *Phragmites* have large numbers of parallel veins of the same thickness. This is unlike most dicotyledonous angiosperm leaves, with a series of main, secondary, and tertiary veins. *Phragmites* leaves are commonly found in freshwater deposits from the Paleogene onward. Its remains are also commonly found in Quaternary peats, where a layer of *Phragmites* indicates an increase in the water table and a regrowth of the reed beds.

vein

PHRAGMITES ALASKANA
This fossil leaf of *Phragmites alaskana* shows its characteristic parallel veins.

LIVING RELATIVE
COMMON REED

The common reed, *Phragmites australis*, is the only species of this genus of perennial grasses. It grows into extensive reed beds in wetlands in tropical and temperate regions, spreading by as much as 16½ft (5m) a year by means of its rooting horizontal runners.

Palmoxylon

GROUP Angiosperms
DATE Neogene
SIZE Up to 98ft (30m) tall
LOCATION Warm temperate to subtropical areas in Northern Hemisphere

Palms are monocotyledonous angiosperms. Their trunk structure differs from that of normal conifers and angiosperm trees, in that the "wood" has no growth rings and is made up of a central core of vascular bundles. *Palmoxylon* from the Late Cretaceous of Texas is one of the earliest known monocotyledons.

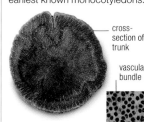

cross-section of trunk

vascular bundle

PALMOXYLON SP.
This slab of a fossilized palm trunk shows the detail of its internal anatomy, which consists of many individual vascular bundles.

NEOGENE INVERTEBRATES

This was a time of continued global cooling. The first ice sheets in Antarctica were beginning to spread about 25 million years ago, although this was followed by a warming event. By 14 million years ago, however, permanent ice sheets draped the Antarctic continent, and invertebrates adapted accordingly.

Continuing evolution

Neogene invertebrate faunas were similar to those of the Paleogene, with bivalves, gastropods, bryozoans, and echinoids as the main preservable components. The genera and species changed through time, and the original divisions of the Tertiary—Eocene, Oligocene, Miocene, and Pliocene—were based upon the ratios of the bivalves and gastropod faunas of these times to genera still living today. In Neogene sediments there are also crab, lobster, and fish remains. Neogene sediments are widespread: notable examples are the "Crag deposits" exposed in low cliffs on the coast of East Anglia, Britain, replete with well-preserved fossils, which extend into the Quaternary. The unique Coralline Crag, exposed above the shoreline, is a shallow-water bank consisting almost entirely of the remains of bryozoans of many genera and species with attendant mollusks, echinoids, and giant barnacles. Similar banks of living species can be found off south Australia today.

BIBIO MACULATUS
Known as the March fly, *Bibio* lived in grasslands. The delicate veins on the wings can be seen in this specimen.

COLEOPLEURUS PAUCITUBERCULATUS
This echinoid from the Neogene is preserved without its spines. *Coleopleurus* lived on hard, rocky substrates in shallow water.

Adaptations to cooling seas

During the Neogene, many echinoids that lived in cold Antarctic waters, where growth and development were slow, have developed "marsupiae" or brood pouches. This is an adaptation for incubating fertilized eggs, which develop directly without the free-swimming larval stages. The marsupiae form deep depressions on the surface of the echinoids. In the Paleogene and Neogene of Australia, marsupiate forms were abundant—Australia was originally situated closer to the South Pole and gradually moved northward. Marsupiate echinoids became progressively fewer during the Neogene and there are none in the warm Australian seas now.

MODERN FORM
This Neogene crustacean, *Archaeogeryon peruvianus*, was similar to crabs of the present day. It probably hunted in deep waters.

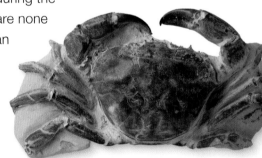

GROUPS OVERVIEW

Some invertebrate species that were prevalent during the Neogene are still living today, or they have some very similar modern-day equivalents. This is the case for bryozoans, bivalves, echinoids, and insects, which all continued to evolve and adapt to the changing habitats brought about by the gradual drop in global temperature throughout this period.

BRYOZOANS
The genus *Cupuladria*, still in existence today, is known from as far back as the early Neogene. Because the temperature limits of living colonies are well known, and assuming no change in habitat, *Cupuladria* can be used to gauge the temperature of the Neogene sediments in which its remains occur.

BIVALVES
Many of the species now found in cool–temperate waters, such as those found around the British coasts, are the same as, or very closely related to, those of the Neogene. There is a progressive increase in the number of cold-water forms through time.

ECHINOIDS
Both deep- and shallow-burrowing echinoids, as well as surface dwellers, continued as an abundant component of the marine fauna through the Neogene. These include the "living fossil" *Cidaris*, which has hardly changed since the Triassic.

INSECTS
Many insects are preserved in remarkable detail in amber; this was originally a sticky resin that oozed from conifers and other trees, trapping the insects for perpetuity. The most diverse faunas are from the Dominican Republic, and are probably early Neogene in age. The resin came from a now-extinct plant, *Hymenaea protera*.

Meandrina

GROUP Anthozoans
DATE Paleogene to present
SIZE Corallite 3/8–3/4in (1–2cm) across
LOCATION Worldwide

Meandrina is a colonial scleractinian coral that is often found in reefs. It is formed by budding from the initial, individual corallite, the skeleton of a polyp. The top of the colony is divided into meandering "ridges" and "valleys," and the polyps, the individual members of the colony, are housed in the cup-shaped depressions of the valleys. The partitions (septa) that divide the body cavity of the corallite are fairly straight and long, and those from adjacent corallites are often in line. The axial ends of the septa form the axial plate that runs along the center of the horizontally elongated corallites. Modern species of *Meandrina* include the brain corals, so called because the meandering corallites resemble the surface of a human brain.

LONG PARTITIONS
The septa that divide each of *Meandrina*'s corallites are straight and elongated, and corallites are often arranged in lines, adding to the coral's folded appearance.

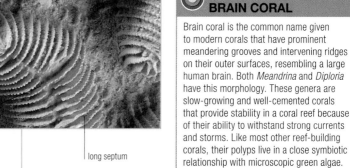

cup-shaped depression

long septum

folded surface

MEANDRINA SP.
In life, this coral would have been covered in a thin layer of colorful polyps, which collected particles of food from the water using tentacles.

hard skeleton

LIVING RELATIVE
BRAIN CORAL

Brain coral is the common name given to modern corals that have prominent meandering grooves and intervening ridges on their outer surfaces, resembling a large human brain. Both *Meandrina* and *Diploria* have this morphology. These genera are slow-growing and well-cemented corals that provide stability in a coral reef because of their ability to withstand strong currents and storms. Like most other reef-building corals, their polyps live in a close symbiotic relationship with microscopic green algae. They occur both in the Atlantic and the Indo-Pacific oceans.

Sphenotrochus

GROUP Anthozoans
DATE Eocene to present
SIZE 3/8in (1cm) high
LOCATION Worldwide

Sphenotrochus is a small, solitary, conical coral that in life was entirely covered by the polyp body. In cross-section, the skeleton (corallite) is elliptical. The thin partitions or septa that divide the inside of the corallite are fused together around the outside to form the corallite wall. There are three cycles of septa, resulting in a total of 24 septa. A distinct plate in the center of the corallite runs down its entire length, and is produced from the modified ends of the septa. *Sphenotrochus* survives to the present day. It usually lives in water around 66–900ft (20–275m) deep.

thin septum

SPHENOTROCHUS INTERMEDIUS
The septa radiate out from the center of the corallite and are fused around the outside to form the corallite wall.

corallite wall

Meandropora

GROUP Bryozoans
DATE Pliocene
SIZE Colony up to 3 1/2in (9cm) in diameter
LOCATION Europe

The bryozoan *Meandropora* formed large, rounded colonies, which were composed of radiating cylindrical clusters (fascicles) of small tubular structures called autozooecia. These clusters divided and rejoined to produce a distinctive pattern on the surface of the colony. The autozooecia were very long with thin walls, and some were partitioned with diaphragms. They housed *Meandropora*'s soft-bodied, feeding individuals (autozooids). Each autozooid possessed a feeding organ called a lophophore, which consisted of a ring of tentacles surrounding the mouth. The autozooecia may have been arranged in clusters in order to provide a focus for channeling water currents. This would have enabled the colony to remove the sediment from the areas of the feeding autozooids and prevent it from harming them.

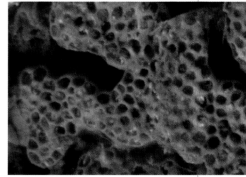

RADIATING CLUSTERS
The long, thin-walled, tubular autozooecia were arranged in clusters. This arrangement may have enabled *Meandropora* to channel water currents.

cluster of autozooecia

channel between autozooecia

MEANDROPORA SP.
The clusters of autozooecia divide and rejoin to give the colony its distinctive surface pattern. The name "bryozoan" itself means "moss animal."

Biflustra

GROUP Bryozoans

DATE Cretaceous to present

SIZE Autozooecial openings 0.1–0.2mm wide

LOCATION Worldwide

Biflustra is a cheilostome bryozoan, with a hard skeleton that is composed of calcium carbonate. It forms encrusting, flattened, or bifoliate (two-sided) colonies. Tubular chambers called autozooecia house the small, soft-bodied individuals (autozooids) of the colony. Each autozooid has a ring of tentacles around its mouth known as a lophophore. Small, beating, hairlike organs on the tentacles (cilia) create a flow of water to drive food particles toward the mouth. When the autozooids are not feeding, their tentacles can be withdrawn into the autozooecia for protection.

There can be **thousands** of tiny individuals in a bryzoan **colony.**

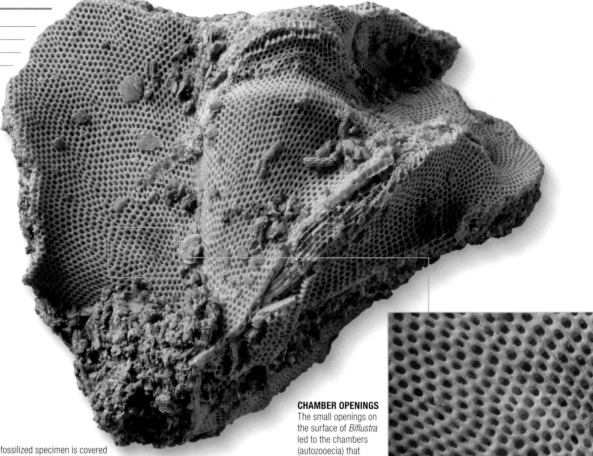

CHAMBER OPENINGS
The small openings on the surface of *Biflustra* led to the chambers (autozooecia) that housed the individual colony members.

BIFLUSTRA SP.
The surface of this fossilized specimen is covered with rectangular openings that are arranged in longitudinal, parallel rows.

Vasum

GROUP Gastropods

DATE Neogene to present

SIZE 1½–4in (4–10cm) long

LOCATION Tropics worldwide

This gastropod has a conical spire, with distinct spines that stick out around the shell and developed as part of its spiral ornament. There is one row of spines where the whorls meet each other and another below it at the shoulder of the whorl. Below the shoulder spines are five spiral ribs that have no spines. Lower down the whorl are four more rows of spirally developed spines. There are spines right back to the part of the shell that developed when the animal was a juvenile. The shell opening (aperture) is long and thin with a long anterior canal. This would have housed the animal's inhalant siphon, which brought water into the shell. Modern species of this genus predate marine worms and are found living in tropical zones worldwide.

spirally arranged spines

shoulder spines

VASUM SP.
Vasum's strong, thick, spirally ribbed shell would have offered good protection from most seafloor predators.

anterior canal

Viviparus

GROUP Gastropods

DATE Jurassic to present

SIZE ½–1½in (1.5–3.5cm) long

LOCATION Worldwide

Viviparus is a freshwater gastropod. The whorls are rounded and convex, and have well-marked growth lines. Mature examples have some thickening on the outer lip, and some have similar earlier thickenings that suggest a slowing of the growth rate, perhaps due to environmental conditions. Unlike some other freshwater gastropods, it is not air-breathing. Instead, it has a single gill, which suggests it is derived from marine, gill-bearing ancestors. Females of modern *Viviparus* species keep their fertilized eggs in the space between the body and the shell wall (the mantle cavity) and give birth to live young.

spiral ornament

VIVIPARUS SP.
A light, horny covering or operculum is attached to the foot and used to seal the shell opening.

○ LIVING RELATIVE
FRESHWATER SNAILS

Species of *Viviparus* are common and widely distributed across the globe. Some species display a burrowing habit, such as the European *V. viviparus* and the North American *V. intertextus* (shown here), which inhabits swamps. Both species have developed what is called ciliary feeding—using the leading edge of its single gill to collect food particles suspended in the water. This is similar to the filter-feeding method used among bivalves.

Terebra

GROUP Gastropods

DATE Neogene to present

SIZE 2¾–10in (7–25cm)

LOCATION Tropics worldwide

This pointed, multiwhorled gastropod is best described as "turreted." The shell is ornamented with sinuous, vertical ribs, and there is a faint spiral band about a third of the way down the side of each whorl. Modern species vary in appearance and size, with some species reaching 10in (25cm) long. Living species are all carnivores and often select a single species of invertebrate to feed on. Many bury themselves in soft sediment during the day and emerge to hunt at night. It is possible that the extinct Neogene species of *Terebra* behaved in a similar way.

TEREBRA FUSCATA
This specimen has around 15 whorls, which barely overlap each other.

faint spiral band

vertical rib

Semicassis

GROUP Gastropods

DATE Neogene to present

SIZE ¾–3¼in (2–8cm) long

LOCATION Worldwide

The whorls of this gastropod are broad with each one overlapping about three-quarters of the following whorl. Some spiral ornamentation can be seen on the upper part of the outer whorl. The area surrounding the shell opening is called the mouth border, and there is a deep notch in this where the anterior inhalant siphon would have protruded to draw in water. Modern species of the genus are carnivores that feed on echinoids by boring into their shells with a tonguelike, toothed radula. It has been assumed that extinct species fed in a similar way.

SEMICASSIS SP.
This can be identified as a mature specimen because of the thickened lip that is partly visible at the opening to the shell.

Ostrea

GROUP Bivalves

DATE Cretaceous to present

SIZE 2¾–6in (7–15cm) wide

LOCATION Worldwide

Neogene *Ostrea* were very similar to the living oysters that belong to the same genus. Like other members of the oyster group, the valves vary greatly in shape—the left valve is convex and often strongly ribbed, while the right valve is flat and lacks ribs. Unlike most bivalves, *Ostrea* incubated its eggs within the mantle cavity, the space between the body and the body wall (see p.400). When the young were finally released after 6–18 days, they already had a minute bivalved shell.

OSTREA VESPERTINA
Growth lines are visible on both valves. On the inner shell, a kidney-shaped impression has been left by the muscle that closes the valves.

strong growth lines

single adductor muscle scar

Nassarius

GROUP Gastropods

DATE Neogene

SIZE Up to 2¼in (6cm) long

LOCATION W. Europe

Nassarius had stout whorls with convex sides, each overlapping about half the height of the preceding whorl. It had strong spiral ribs and grooves that were crossed by equally strong growth-parallel ribs, which created a pattern of small rectangles. The aperture was long with a well-marked anterior notch that housed the inhalant siphon. Living members of the same family as *Nassarius* are scavengers and carrion feeders that feed on dead organisms on the seabed, so it is probable that *Nassarius* had the same feeding habits.

large lip

NASSARIUS SP.
The mouth of the shell is large with a "lip" beneath the front of the largest whorl.

anterior notch

Astarte

GROUP Bivalves

DATE Jurassic to present

SIZE ¾–2¾in (2–7cm) long

LOCATION Worldwide

Astarte is a triangular bivalve with forward-pointing beaks. It has two clear adductor muscle scars on the inside of the shell, and a well-marked pallial line. The pallial line runs between the two muscles scars and marks the attachment of the mantle (the external body wall) to the shell. The shell's exterior is marked with strong ribs parallel to its growth, and finer growth lines run between these. An external ligament, which opened the shell, was located behind the beaks, housed in a long groove. Internally, positioned under the beak in each valve, there are large, well-defined hinge teeth and sockets. These ensured accurate shell articulation. The family to which *Astarte* belongs can be traced back to the Devonian. This bivalve group evolved early and, due to its successful body-plan, it has changed very little over time.

ASTARTE MUTABILIS
All *Astarte*'s relatives are characterized by the strong growth-parallel ribs over the whole surface of the shell.

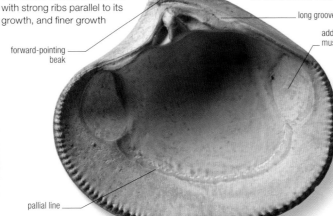

forward-pointing beak

long groove

adductor muscle scar

rib

pallial line

Amusium

GROUP Bivalves
DATE Neogene to present
SIZE 2–4¹⁄₂in (5–12cm) wide
LOCATION Worldwide

Amusium is a large, smooth, almost circular bivalve from the same family as modern scallops. The main difference is that *Amusium's* valves are virtually smooth, whereas scallop valves are ribbed. *Amusium's* thin shell has two prominent, earlike extensions on either side of the beak, with one slightly larger. Fine growth lines cover the shell surface, and faint radial striations are also visible.

smooth valve

AMUSIUM SP.
The virtually smooth valves of *Amusium* are the reason some species of *Amusium* are referred to as "moon scallops."

Anadara

GROUP Bivalves
DATE Late Cretaceous to present
SIZE ³⁄₄–2¹⁄₄in (2–6cm) long
LOCATION Worldwide

At first sight, *Anadara* looks like a modern cockle, with strong, radially arranged ribs crossed by growth lines. Inside, however, it differs greatly from modern cockles. Beneath the forward-facing beak of each valve is a long, striated groove that, in life, houses the long ligament responsible for opening the shell. Below the ligament groove is a long, straight hinge-line, with many, almost vertical teeth and sockets that help the shell stay together when shut. There are two scars, where the muscles that controlled the movement of the valves would have been attached. An inconspicuous line between the scars is where the mantle was attached.

ANADARA RUSTICA
With its deep, radiating ribs, *Anadara* looks like a cockle, but it belongs to a group of bivalves called arcoids, which can be traced back to the Early Ordovician period.

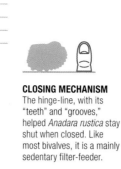

hinge-line

CLOSING MECHANISM
The hinge-line, with its "teeth" and "grooves," helped *Anadara rustica* stay shut when closed. Like most bivalves, it is a mainly sedentary filter-feeder.

ribs crossed by growth lines

Some species of *Anadara* have been used as a source of **food** by **humans** since **prehistoric times.**

Schizaster

GROUP Echinoderms
DATE Paleogene to present
SIZE 1³⁄₄–2¹⁄₂in (4.5–6.5cm) long
LOCATION Worldwide

Schizaster is a heart-shaped, burrowing sea urchin with five well-marked ambulacral areas or "petals" on its upper surface, where its tube feet protrude. The rear two petals are far shorter than the front three. The middle front ambulacral area is set in a deep groove that reaches down to the lower surface and the mouth. When viewed from the side, the shell is quite high, and there is a slight overhang at the rear above the anal opening.

smooth fascioles

SCHIZASTER SP.
Smooth fascioles are set around the petal-shaped upper ambulacra and beneath the anus. In life, hairlike cilia on these would have generated water currents.

Scutella

GROUP Echinoderms
DATE Late Paleogene to Early Neogene
SIZE 2¹⁄₄–3¹⁄₂in (6–9cm) across
LOCATION Mediterranean

Scutella was a medium-sized, flattened, burrowing sea urchin, with internal "pillars" that strengthened its shell. There were five petal-shaped ambulacral areas on its lower surface, where its many tube feet emerged. Around the edge of these areas were pairs of pores—each outer pore had a slitlike appearance. The mouth was located in the center of the animal's lower surface, and well-marked food grooves led directly to it. Where these food grooves crossed the junction between the upper and lower surfaces, they were as wide as the zones between the ambulacral areas. *Scutella's* anal opening was also positioned underneath it, about midway between the back margin of the shell and the mouth. In the center of the upper surface was the apical disk, a multiplated area with four genital pores known as gonopores. *Scutella* is one of many echinoid fossils that are sometimes referred to as "sand dollars," although the sand dollars found by beachcombers today are usually the shells of modern echinoid species.

apical disk with genital pores on upper surface

SCUTELLA SP.
With its five petal-like ambulacra and slightly domed shape, *Scutella* has an asymmetrical appearance. Like modern echinoids, it was probably covered with tiny, hairlike spines.

ambulacral "petal"

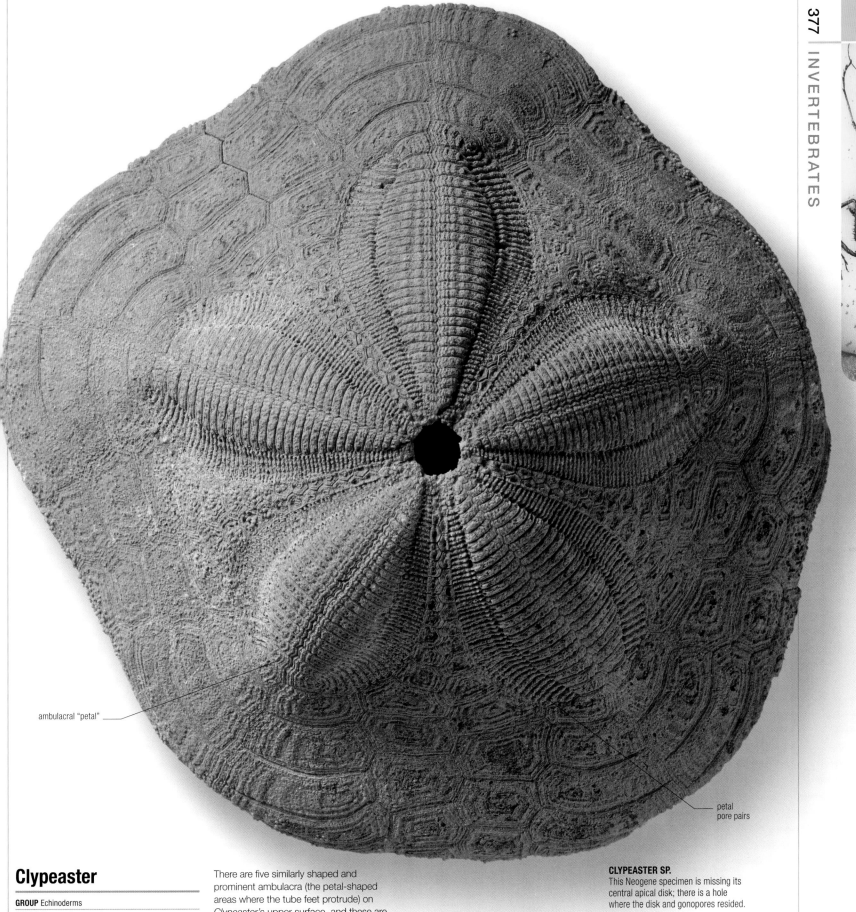

ambulacral "petal"

petal
pore pairs

CLYPEASTER SP.
This Neogene specimen is missing its
central apical disk; there is a hole
where the disk and gonopores resided.

Clypeaster

GROUP	Echinoderms
DATE	Paleogene to present
SIZE	2–6in (5–15cm) across
LOCATION	Worldwide

Clypeaster is a medium to large burrowing
echinoid with a variable shape that ranges
from a flattened disk to bell-shaped forms,
depending on the species. Sometimes,
particularly in the flattened varieties, the
shell wall is made up of two separate layers
separated by supporting pillars, which
gives the shell greater mechanical strength.

There are five similarly shaped and
prominent ambulacra (the petal-shaped
areas where the tube feet protrude) on
Clypeaster's upper surface, and these are
far wider than the interambulacral areas. On
its lower surface, the ambulacral areas are
far less obvious. The mouth is located in the
center of the lower surface, and the anus
is situated close to its edge. The disklike
structure located in the center of the upper
surface contains five genital pores. Some
species of *Clypeaster* live in burrows as
much as 6in (15cm) below the seabed. The
shells of many modern species are often
called "sea biscuits" by beachcombers.

6 inches The distance
below the seabed some
Clypeaster species dig
their burrows.

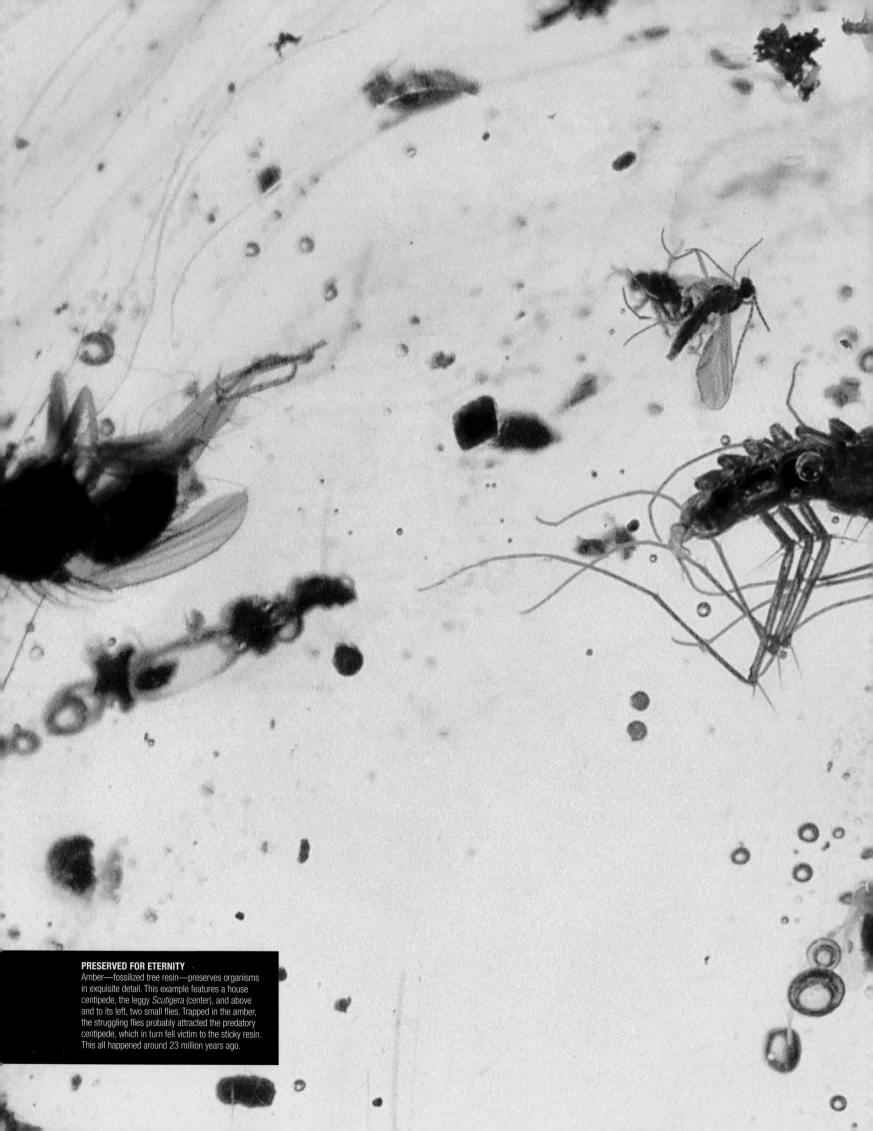

PRESERVED FOR ETERNITY
Amber—fossilized tree resin—preserves organisms in exquisite detail. This example features a house centipede, the leggy *Scutigera* (center), and above and to its left, two small flies. Trapped in the amber, the struggling flies probably attracted the predatory centipede, which in turn fell victim to the sticky resin. This all happened around 23 million years ago.

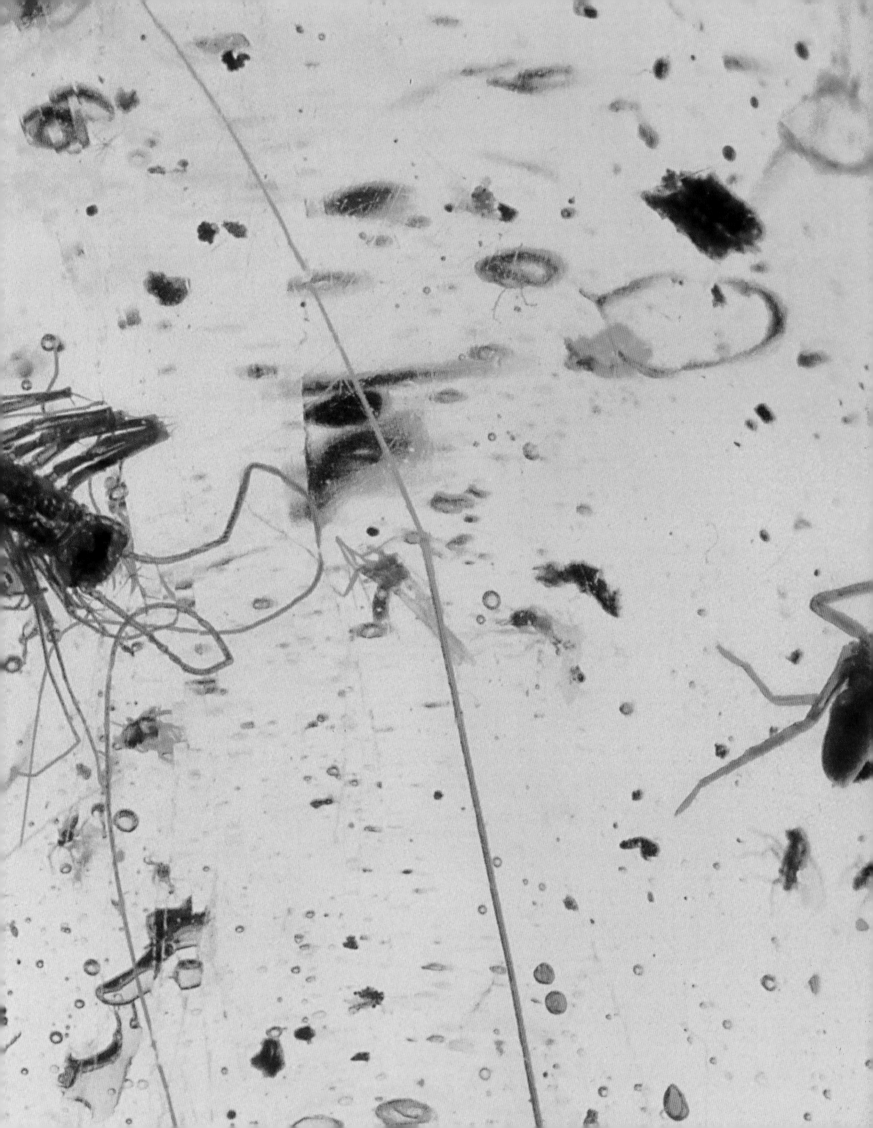

NEOGENE
VERTEBRATES

From the start of the Neogene period 23 million years ago, the world's climate became gradually cooler and drier. This caused deserts and grasslands to advance across much of the continental land surface. Predators and prey either adapted to the new conditions or became extinct.

Many mammal groups colonized new continents during the Neogene, crossing land bridges created by lowered sea levels and the collision of landmasses. Ancestral camels originated in North America but crossed to Eurasia about 7 million years ago, where they later evolved into dromedaries and Bactrian camels. Movements to South America about 3 million years ago gave rise to llamas, guanacos, and vicunas.

Grasslands and grazers

The expansion of grasslands gave rise to a new mode of feeding among herbivores—grazing. Even-toed ungulates (hoofed mammals), especially ruminants, were better suited to eating the coarse grasses low in nutritional value than odd-toed ungulates due to their more highly developed stomachs. Ruminants have four-chambered stomachs with microbes that digest cellulose. They also regurgitate and rechew tough plant material before swallowing it again. These adaptations led to a dramatic diversification of artiodactyls such as cattle, sheep, and

large premolars

high-crowned molars with many cusps

TEETH AND TOES
During the Miocene epoch, three-toed horses, such as *Hipparion*, evolved high-crowned teeth with grinding surfaces, enabling them to eat tough grasses. On each foot, the side toes became reduced and the central toe modified into a single weight-bearing hoof, allowing them to run fast across the plains.

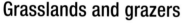

elongated foot bone

large central toe forming a hoof

antelopes during the Miocene at the expense of perissodactyls such as horses, rhinoceroses, and tapirs. The grazers began to herd together for safety while feeding and traveling on their seasonal migrations to find fresh pasture. They evolved to become bigger and faster to evade their predators. In turn, carnivores evolved strategies such as pack and pursuit hunting to help them catch their fast-moving prey. The advance of grassland in Africa also contributed to apes (hominoids) moving out of the forests and adopting a more upright posture.

Giants of the sea

Both modern lineages of whales—the mysticetes, or baleen whales, and the odontocetes, the toothed whales—existed at the start of the Neogene. Toward the end of the Miocene, representatives of several modern whale, dolphin, and porpoise families had appeared alongside some families that are now extinct. The first true seals, sea lions, and walruses also evolved in the Miocene, and the herbivorous sirenians (dugongs and manatees) diversified. All of these marine mammals were preyed upon by the gigantic shark *Carcharocles megalodon* (see opposite).

GROUPS OVERVIEW

Changing conditions during the Neogene prompted the appearance of new mammalian groups and the extinction of some older ones. The even-toed ungulates expanded at the expense of their odd-toed cousins, while carnivores took over from creodonts as dominant predators. Small mammals such as rabbits, rodents, and raccoons flourished.

PERISSODACTYLS
Horses continued to diversify during the Neogene as they exploited new grassland habitats. However, many species became extinct by the end of the Miocene around 5 million years ago, leaving only one genus—*Equus*. Many Paleogene rhinos and tapirs also died out, and the number of species continued to dwindle throughout the Neogene.

ARTIODACTYLS
Even-toed ungulates rose in prominence during the Neogene. Pigs, peccaries, and hippos were soon joined by the first giraffes; deer; and bovids, such as gazelles and goats. Bovid evolution occurred primarily in Africa and Asia, but the Bering Strait land bridge allowed the ancestors of modern-day bison and bighorn sheep to move into North America.

RODENTS
Rodents did well during the Neogene, aided by their generalist habits and rapid reproduction rates. Modern-looking squirrels had appeared by the Miocene and murids—the family that includes mice, rats, gerbils, voles, lemmings, and hamsters—underwent a dramatic radiation in the Pliocene. The land bridge from Africa to Asia allowed porcupines to colonize Eurasia.

HOMINOIDS
The apes diverged from Old World monkeys about 30 million years ago. They evolved larger brains and bodies and became gradually more terrestrial. The lesser and great apes split from one another about 20 million years ago, and species spread from Africa into Asia. The fossil genus *Sivapithecus* is considered to be the ancestor of orangutans.

Carcharocles

GROUP Chondrichthyan
DATE Early Miocene to Pliocene
SIZE 59ft (18m) long
LOCATION Europe, North America, South America, Africa, Asia

Weighing around 55 tons (50 tonnes) or more, *Carcharocles* was the largest predatory shark that ever lived. The species—*megalodon*—is named for its huge triangular teeth, measuring up to 6½in (17cm) high. Formerly, the species was assigned to the same genus as the extant great white shark (*Carcharodon carcharias*). The large teeth of *Carcharocles megalodon* have serrations running down both cutting edges, ensuring an efficient slicing action. Cut marks in bones and isolated teeth lying next to fossil carcasses show that *Carcharocles* must have preyed upon a wide

range of whales, dolphins and porpoises, seals, giant turtles, and fish. It probably ambushed its prey by attacking them at speed. Once disabled, the prey could be processed using its huge jaws and colossal bite force. There are certainly many similarities between the teeth of these two species, and in the absence of a complete specimen, *Carcharocles megalodon* is usually reconstructed using the great white as a model.

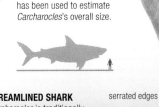

"BIG TOOTH"
This shark's species name—*megalodon*—is taken from the Greek for "big tooth." The size of its teeth has been used to estimate *Carcharocles*'s overall size.

serrated edges

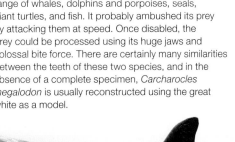
tall dorsal fin

STREAMLINED SHARK
Carcharocles is traditionally thought to have had the appearance of a scaled-up great white, with the classic streamlined body shape, pointed snout, and large fins.

huge, vertical tail fin

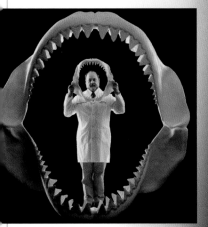

MASSIVE MOUTH
The open reconstructed jaws are the height of a man and many times larger than those of the great white.

Myliobatis

GROUP Chondrichthyan
DATE Late Cretaceous to present
SIZE 5ft (1.5m) long
LOCATION Worldwide

TOOTH PLATE
Myliobatis fossils are often identified by their tooth plates. These have one row of long teeth in the middle with rows of hexagonal teeth on each side.

Myliobatis was a common sting-ray during the Neogene. At least 11 *Myliobatis* species survive today and are found worldwide—known as the eagle rays. Many specimens of extinct species have been unearthed in Pliocene deposits in North Carolina. However, other finds in North America, Africa, and Europe show that *Myliobatis* extends back as far as the Cretaceous Period.

Gavialosuchus

GROUP Crocodylomorph
DATE Late Oligocene to Early Pliocene
SIZE 17¾ft (5.4m) long
LOCATION North America, Europe

Gavialosuchus was an extinct gharial (crocodilelike reptile) from the Late Oligocene to Early Pliocene of North America and the Early Miocene of Europe. Like living gharials, it had a very long skull with a narrow snout that it used to catch fish. *Gavialosuchus* fossils have been found in coastal deposits, which suggests that it lived in estuaries or shallow seas, where it could have fed on a variety of different types of fish. Although living gharials are restricted to India and Southeast Asia, fossils of different *Gavialosuchus* species found in Florida, Austria, and Georgia suggest that the group was once distributed in swampy and coastal habitats all over the tropical regions of the world.

narrow snout

sharp teeth for securing fish

ELONGATED SKULL
Like some modern-day crocodilians, *Gavialosuchus* had long jaws and sharp teeth.

Phalocrocorax

GROUP Aves
DATE Early Miocene to present
SIZE 17½in–39in (45–100cm) tall
LOCATION USA, France, Spain, Moldova, Bulgaria, Ukraine, Mexico, Mongolia, Australia

Modern-day members of the *Phalocrocorax* genus are the cormorants and shags. There are 36 living species of these medium-sized, fish-eating seabirds. A similar number of extinct species from all over the world have been described. *Phalocrocorax* hunted by diving and had a flexible neck and long bill for catching fish. The tip of its upper bill was slightly hooked to cling onto slippery prey. Historically, cormorants were thought to be the shoreline cousins of ravens. However, *Phalocrocorax* belongs to the Pelecaniformes, a group of water birds that includes pelicans and gannets.

finger bones

slightly hooked upper bill

long, flexible neck

CORMORANT COUSIN
Phalocrocorax had a body shape very similar to that of modern-day cormorants and shags, including the characteristic flexible neck.

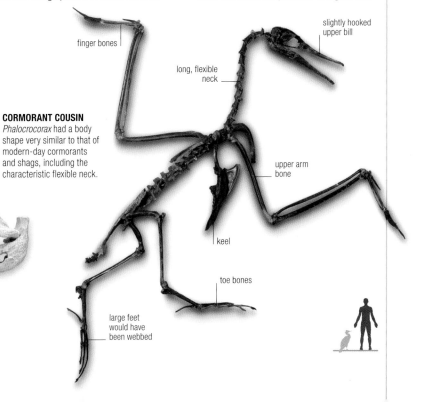

upper arm bone

keel

toe bones

large feet would have been webbed

featherless head

splayed tips help generate extra lift

broad wing

Argentavis

GROUP	Theropods
DATE	Late Miocene
SIZE	11ft (3.5m) long
LOCATION	Argentina

long, strong legs

Argentavis is by far the largest flying bird to have ever lived. It resembled a gigantic condor, with a wingspan of 26ft (8m); a large specimen would have weighed 175lb (80kg). This is more than double the wingspan and five times the weight of the Andean condor, the largest bird flying today (see panel, left). *Argentavis* had strong legs and wide feet, so it was able to walk easily. Its bill was long with a hooked tip, like that of an eagle, so it could rip open almost any carcass. *Argentavis* was almost certainly a scavenger, searching for carrion as it soared in the air, before landing and driving off other large predators from their kills. This giant of the sky would have had to cover a vast territory to find enough food to sustain itself.

MASSIVE WINGS
With its wing area of almost 75 square ft (7 square m), *Argentavis* was able to soar effortlessly in the skies above Argentina, riding the thermals and updrafts.

Thylacosmilus

GROUP	Marsupial mammals
DATE	Late Miocene to Early Pliocene
SIZE	5ft (1.5m) long
LOCATION	South America

Thylacosmilus was a large predatory marsupial that was about the size of today's jaguar. Like the saber-toothed cats such as *Smilodon* (see p.411), to which it was unrelated, it had long, saber-shaped upper canine teeth. This is a classic example of convergent evolution. However, in *Thylacosmilus*, these canines grew continuously throughout its life. When the mouth was closed, the sabers were protected by a pair of scabbardlike flanges in the lower jaw. *Thylacosmilus* was the last of the line of marsupial carnivores that diversified in isolation in South America. When North and South America became connected via the Panama land bridge, around 3 million years ago, placental carnivores moved into South America and took their place.

short lower canine

KILLER CANINES
Thylacosmilus had reduced molars; no lower incisors; and short, peglike lower canines. The jaws were dominated by its upper canines.

Thylacosmilus's **upper canines** were **even longer** than those of *Smilodon*, the saber-toothed tiger.

large, deep skull

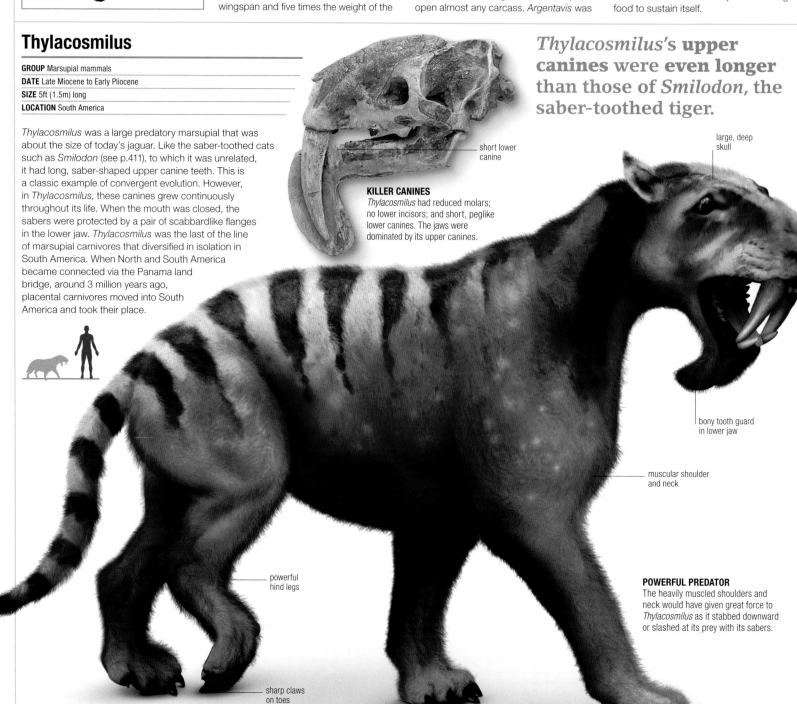

bony tooth guard in lower jaw

muscular shoulder and neck

powerful hind legs

sharp claws on toes

POWERFUL PREDATOR
The heavily muscled shoulders and neck would have given great force to *Thylacosmilus* as it stabbed downward or slashed at its prey with its sabers.

Deinogalerix

GROUP Placental mammals
DATE Late Miocene
SIZE 24in (60cm) long
LOCATION Italy

Deinogalerix was a giant, spineless, hairy hedgehog that lived on Gargano during the Late Miocene, when it was an island off the coast of Italy. Like today's hedgehogs, to which it was related, *Deinogalerix* fed on a mixture of insects, snails, and other invertebrates. However, it was huge for an insectivore—weighing as much as a large cat—and its large body size would have enabled it to hunt birds and small mammals as well. *Deinogalerix* is a classic example of how small mammals tend to grow large on islands where they do not face competition from larger predators, such as cats, dogs, and bears. However, the giant barn owl *Tyto gigantea* also lived on Gargano during the Late Miocene and may even have preyed upon *Deinogalerix*.

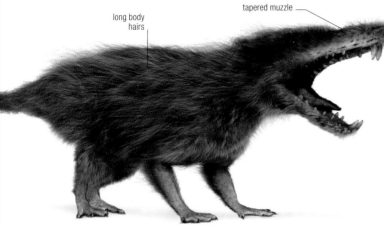

long body hairs

tapered muzzle

HAIRY HEDGEHOG
Deinogalerix means "terrible shrew," but it actually resembles a large moonrat, a primitive type of hedgehog found today in Southeast Asia, with a long tail and fur instead of spines.

Enaliarctos

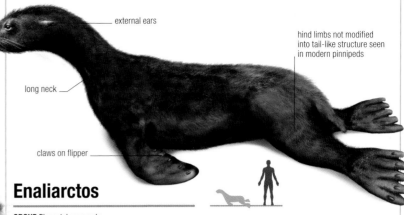

external ears

hind limbs not modified into tail-like structure seen in modern pinnipeds

long neck

claws on flipper

GROUP Placental mammals
DATE Late Oligocene to Early Miocene
SIZE 5ft (1.5m) long
LOCATION USA

Enaliarctos is one of the earliest known relatives of the pinnipeds (seals, sea lions, and walruses). Its fossils have been found in Late Oligocene to Early Miocene rocks in California and Oregon. Like later seals and sea lions, *Enaliarctos*'s limbs were modified into crude flippers, although they were not as highly specialized as those of modern pinnipeds. *Enaliarctos* used both front and hind limbs for swimming; modern sea lions use just their forelimbs for swimming, and all four flippers to move on land, while modern seals use their hind flippers for swimming and are less mobile on land. Although it had large eyes, sensitive whiskers, and good underwater hearing like its living descendents, *Enaliarctos*'s teeth were more primitive and resembled those of its bearlike ancestors. It ate fish but, unlike seals, could not eat its catch while swimming—instead, it had to drag its prey to the shore to tear it apart.

PRIMITIVE PINNIPED
Its elongated and strongly muscled hind limbs indicate that *Enaliarctos* was more active on land than modern pinnipeds.

Enaliarctos had **sensitive whiskers, large eyes, and good hearing.**

Allodesmus

GROUP Placental mammals
DATE Middle Miocene
SIZE 5ft (1.5m) long
LOCATION USA, Mexico, and Japan

Allodesmus was an early relative of sea lions and, like them, it swam with its forelimbs and could rotate its hind limbs forward to help it walk in a shuffling fashion on land. It had a long, slender skull like that of a modern leopard seal, but its teeth were blunt with bulbous crowns, more suited for catching fish and squid, which it would then have swallowed whole as modern sea lions do. Its large eyes would have helped it see while hunting underwater. *Allodesmus* showed a wide range in sizes between males and females, suggesting that the large bulls, which weighed up to 790lb (360kg), would have guarded a harem of cows.

Allodesmus displayed **sexual dimorphism, with males** being **much larger** than **females**.

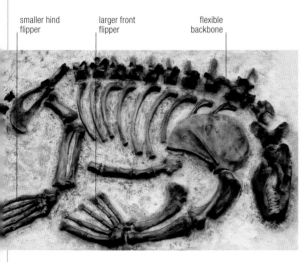

smaller hind flipper

larger front flipper

flexible backbone

⊙ KEY SITE

SHARKTOOTH HILL

Sharktooth Hill, which lies northeast of Bakersfield, California, is the preserved remains of a North Pacific near-shore oceanic ecosystem of the Middle Miocene. The sheer quantity and quality of its fossils were first documented by Swiss paleontologist Louis Agassiz in 1856. In addition to marine mollusks, the fossils of a wide variety of fishes have been found, including the teeth of 30 species of sharks (hence the site's name). Turtles, birds, and 17 species of whales and other marine mammals have also been discovered, including many *Allodesmus* fossils.

FORWARD PROPULSION
Allodesmus propelled itself through the water with its front flippers, which were larger and more powerful than its rear flippers.

Ceratogaulus

GROUP Placental mammals
DATE Middle to Late Miocene
SIZE 12in (30cm) long
LOCATION North America

Ceratogaulus was a burrowing rodent distinguished by a pair of straight horns on its nose. It was roughly the size of a modern marmot but more closely resembled a gopher, with its strong forelimbs equipped with huge claws. The role of the horns has been much debated. It has been suggested that they were used for digging, but they were not in the correct position. Both males and females had horns, so they are unlikely to have been used for display during courtship. It is now thought that the horns were used as defensive weapons when the animal was out in the open, away from the safety of its burrow.

paired horns

sturdy, gopherlike forelimb

DESIGNED FOR DIGGING
Ceratogaulus used its large claws for digging. It fed on roots, bulbs, and other underground plant parts, but it also came aboveground to forage.

Palaeocastor

GROUP Placental mammals
DATE Early Oligocene to Early Miocene
SIZE 16in (40cm) long
LOCATION North America, Asia

Palaeocastor is an early relative of living beavers, living between the Early Oligocene and the Early Miocene. Fossils have been found in western USA and Japan. Although related to beavers, *Palaeocastor* was not an aquatic tree-cutter. Instead, it was a terrestrial burrower, about the size and shape of a modern marmot.

BURROWING BEAVER
A member of the beaver group, *Palaeocastor* was a burrowing mammal that used its strong front teeth to dig into the ground.

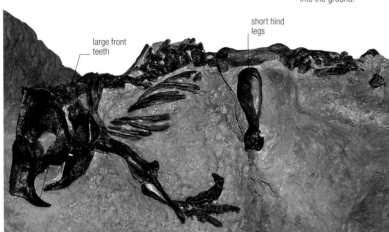

large front teeth

short hind legs

KEY FIND

DEVIL'S CORKSCREWS

Palaeocastor is most famous for making long spiral burrows that were found in Nebraska. These spirals, known as "Daemonelix," are shaped like corkscrews and went as deep as 9¾ft (3m) below the surface. At first they were a mystery, until a specimen of *Palaeocastor* was found at the bottom of one of them. The scratch marks along the sides, long-attributed to the claws of a burrower, were actually cut by the chisel-like front teeth of the beaver as it dug with its mouth.

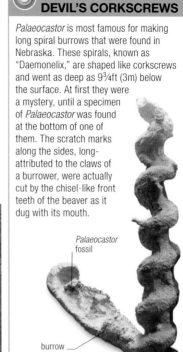

Palaeocastor fossil

burrow

Teleoceras

GROUP Placental mammals
DATE Miocene
SIZE 13ft (4m) long
LOCATION North and Central America

Teleoceras was a large rhinoceros with a single small horn on the tip of its nose. It comes from Miocene deposits of North and Central America. Although it is a true rhinoceros, its body plan is similar to that of a hippopotamus, with short, stumpy limbs, a massive barrel chest, and a skull with high-crowned cheek teeth. *Teleoceras* fossils are found in large numbers in many ancient river and pond deposits on the high plains of western North America. One such place is Ashfall Fossil Beds State Park, Nebraska (see panel, right). Here hundreds of complete *Teleoceras* skeletons are found. Some specimens have grass seeds fossilized in their throats, confirming that *Teleoceras* was a grass-eater.

Stenomylus

GROUP Placental mammals
DATE Late Oligocene to Early Miocene
SIZE 5ft (1.5m) tall at the shoulder
LOCATION North America

Stenomylus was a humpless camel that lived in North America between the Late Oligocene and the Early Miocene. Although *Stenomylus* was a true camel, it was small, reaching only 5ft (1.5m) tall at the shoulder. Its legs and body were delicate and slender, like those of a modern gazelle. Its most remarkable features were its elongated, high-crowned molars, with deep roots that reached the base of the jaw and the top of the skull. These must have been used to eat very gritty plants since they show signs of extreme wear during the life of the animal.

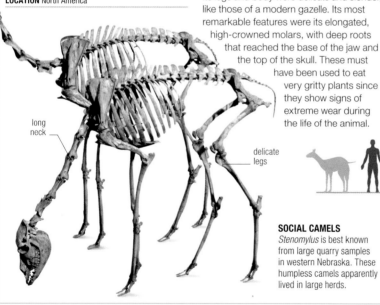

long neck

delicate legs

SOCIAL CAMELS
Stenomylus is best known from large quarry samples in western Nebraska. These humpless camels apparently lived in large herds.

Menoceras

GROUP Placental mammals
DATE Early Miocene
SIZE 5ft (1.5m) long
LOCATION North and Central America

Menoceras was a small rhinoceros that has been found in the early Miocene deposits of North and Central America, especially Nebraska and Wyoming. It was a true rhinoceros, but is more slender in its build than modern members of the group. It was also smaller and only reached the size of a large pig. Males had two bony horns positioned side by side on the tip of the nose, but the females were hornless.

hornless female

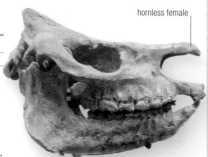

TWIN-HORNED RHINOCEROS
Menoceras is the first-known rhino to develop horns. However, only the males developed a pair of horns on the tip of their nose.

The **large number of *Teleoceras* fossils** found at the Ashfall Fossil Beds State Park has led to the site being called **"Rhino Pompeii."**

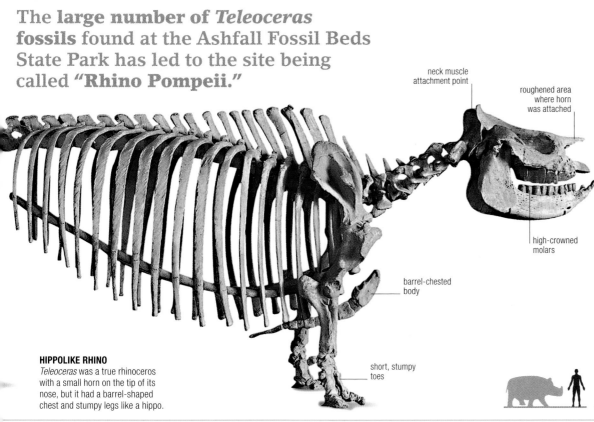

neck muscle
attachment point

roughened area
where horn
was attached

high-crowned
molars

barrel-chested
body

short, stumpy
toes

HIPPOLIKE RHINO
Teleoceras was a true rhinoceros with a small horn on the tip of its nose, but it had a barrel-shaped chest and stumpy legs like a hippo.

Teleoceras is best known from Ashfall Fossil Beds State Park, Nebraska. Here hundreds of complete skeletons of *Teleoceras* and other mammals and birds were fossilized as they died, suffocated by a volcanic ash fall that buried their waterhole 10 million years ago. Some specimens were removed, but most have been left excavated in the ground and can be viewed as they were found.

Paraceratherium

GROUP	Placental mammals
DATE	Late Oligocene to Early Miocene
SIZE	26ft (8m) long
LOCATION	Europe, Asia

Paraceratherium—also known as *Indricotherium*—was a gigantic, hornless rhinoceros from the Late Oligocene to Early Miocene of Asia. Its immense skull alone was 4¼ft (1.3m) long and had two conical upper tusks, flaring lower tusks, and retracted nasal openings suggesting a fleshy proboscis. Its immense size and long neck allowed *Paraceratherium* to feed on the tops of trees, like a giraffe does today. Despite its huge size, *Paraceratherium* still had the long toes and limbs of its ancestors, the hyracodonts, and not the stumpy toes found in similarly huge beasts, such as elephants.

IMMENSE MAMMAL
Paraceratherium was the largest land mammal that ever lived, reaching 18¼ft (5.5m) at the shoulder, 26ft (8m) in length, and weighing about 16 tons (15 tonnes).

ANATOMY
PREHENSILE LIP

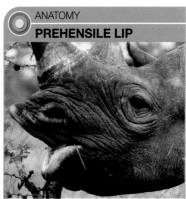

Paraceratherium, like many rhinos and tapirs, probably had a long prehensile lip that could wrap around branches and strip leaves from the trees that it fed upon. The living black rhinoceros does this with its upper lip, and so do most tapirs. Tapirs and many rhinos have the nasal bones retracted far back on the skull, leaving a large nasal notch where the complex muscles attach to manipulate their lip or proboscis.

Chalicotherium

GROUP Placental mammals
DATE Late Oligocene to Early Pliocene
SIZE 6½ft (2m) long
LOCATION Europe, Asia, Africa

Chalicotherium was a perissodactyl (odd-toed, hoofed mammal) that lived between the Late Oligocene and the Early Pliocene in Europe, Asia, and Africa. It looked somewhat like a large horse, except that it had claws on its toes rather than hooves. The purpose of these claws on a hoofed mammal was a puzzle for some time, but it is now thought that they were used to hook branches and pull them in order to feed. In addition, *Chalicotherium*'s front limbs were far longer than its back legs, and it had features on the hands and pelvic bones that suggested that it knuckle-walked, much like a gorilla does today. When it was not on the move, *Chalicotherium* sat for long periods of time on its haunches feeding on vegetation. The relationships of chalicotheres have been a mystery for many years. Recent evidence has shown that they are cousins of the tapirs, but when the claws of *Chalicotherium* were first discovered they were thought to be a type of carnivore.

Chalicotherium means **"pebble beast,"** which refers to its **pebblelike molar teeth.** At maturity, its incisors and upper canines were shed.

RE-EVOLVED CLAWS

Although *Chalicotherium* was related to tapirs and other odd-toed hoofed mammals, its hooves had evolved into claws. This is the only known instance of a hoofed mammal reverting back to the ancestral clawed toe.

Merychippus

GROUP Placental mammals
DATE Middle to Late Miocene
SIZE 3½ft (1.1m) tall
LOCATION North and Central America

During the Miocene, the world's climate became drier and forests thinned out into grasslands. *Merychippus* was a primitive horse that lived in large herds on these new plains. It had evolved from its forest-dwelling ancestors to become a long-legged, fast-running grazer, similar to today's equines. *Merychippus* was the first equine to have a single-hoofed middle toe on each foot. Although two other toes flanked the middle one, in most species these were only long enough to touch the ground while it was galloping. *Merychippus* is closely related to *Pliohippus* (right) and today's *Equus* species.

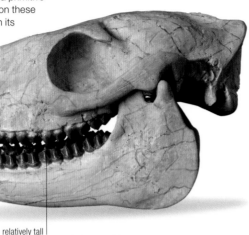

relatively tall molars

GROUND DOWN
The teeth of *Merychippus* were taller than those of its leaf-eating ancestors, so they could grind gritty grasses without being worn down too much.

Pliohippus

GROUP Placental mammals
DATE Middle Miocene to Early Pliocene
SIZE 4ft (1.2m) tall
LOCATION North and Central America

Pliohippus was an early, one-toed horse that still had two vestigial side toes on each hoof. It was about the size of a zebra and slightly smaller than most species of modern horses. For many years, it was thought that *Pliohippus* was a close ancestor of *Equus*, but *Pliohippus* has deep pockets on the facial region, and its teeth were strongly curved, while those of *Equus* are straight. Scientists currently think that *Equus* is more closely related to *Dinohippus*, and *Pliohippus* is part of a lineage of horses that included *Calippus* and *Astrohippus*.

long muzzle

one-toed feet

ONE-TOED HORSE
Pliohippus was one of the earliest horses to walk on a one-toed foot. Although modern horses also have one-toed feet, they did not evolve from *Pliohippus*.

Deinotherium

GROUP Placental mammals
DATE Middle Miocene to Early Pleistocene
SIZE 14¾ft (4.5m) tall
LOCATION Europe, Africa, Asia

Deinotherium was a relative of the mastodonts and modern elephants, with tusks that curved down and back from the front of the lower jaw. The purpose of its tusks is still controversial, but it is likely they were used for hooking onto branches to drag them down and make it easier to reach the leaves. *Deinotherium* was slightly larger than today's African elephant, weighing about 15.5 tons (14 tonnes), which makes it the third-largest known land mammal that ever lived.

TERRIBLE BEAST
The name *Deinotherium* means "terrible beast." The scientists who discovered it gave it this name because they were so impressed by its huge size, strange appearance, and unusual tusks.

tusk growing from the lower jaw

deeply retracted nasal bones

SHORT TRUNK
Deinotherium's huge skull was almost 3ft 3in (1m) long. The nasal bones are very deep, suggesting that the trunk was wider and shorter than that of today's elephants.

Gomphotherium

GROUP Placental mammals
DATE Early Miocene to Late Pliocene
SIZE 9¾ft (3m) tall
LOCATION North America, Europe, Asia, Africa

This four-tusked proboscidean first evolved in Africa. During the Early Miocene, it became the first mastodont to escape its homeland, migrating to Eurasia and North America, where it flourished. It was about the size of a small elephant, with a long, flat skull; a large pair of long upper tusks that extended straight out; and a smaller pair of shovel-shaped tusks that grew out of the lower jaw. These lower tusks were probably used to scrape up vegetation and strip bark and leaves off trees.

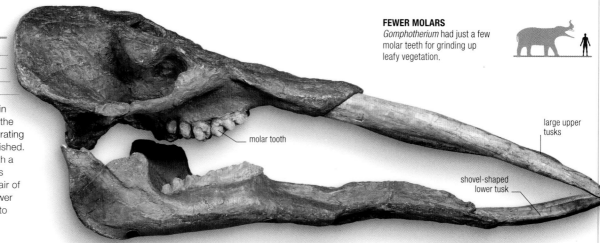

FEWER MOLARS
Gomphotherium had just a few molar teeth for grinding up leafy vegetation.

large upper tusks

molar tooth

shovel-shaped lower tusk

Orycteropus

GROUP Placental mammals
DATE Early Miocene to present
SIZE 5ft (1.5m) long
LOCATION Africa, Europe

The sole living *Orycteropus* species is the aardvark. Not only is it the only species in the genus, it is also the only survivor in the entire mammal order Tubulidentata. The aardvark lives in Sub-Saharan Africa, where, for protection from predators, it sleeps during the day and comes out at night to feed on insects. It uses its sharp front claws to tear open termite nests, then laps up the insects with its extremely long tongue. Fossil relatives of the aardvark include *Orycteropus gaudryi*, which has been found on the Greek island of Samos.

GREEK AARDVARK
Although aardvarks are found only in Africa today, they were much more widespread in the past. This incomplete skull is from Samos.

simple peg-like tooth

Amphicyon

GROUP Placental mammals
DATE Oligocene to Middle Miocene
SIZE 6½ft (2m) long
LOCATION North America, Europe, Asia

With its robust limbs and skull, long tail, and wolflike teeth, *Amphicyon* looked like a large dog with the build of a bear, which is why it is also known as "bear dog." Being the largest predator in the Middle Miocene of North America, it was able to kill and eat almost any animal and also drive off other predators to scavenge their kills. *Amphicyon* became extinct in North America but persisted a little longer in Europe before being driven to extinction by true bears.

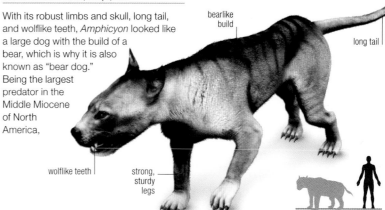

bearlike build

long tail

wolflike teeth

strong, sturdy legs

Proconsul

GROUP Placental mammals
DATE Early Miocene
SIZE 26in (65cm) long
LOCATION Kenya

Proconsul was the size of a modern baboon or gibbon and was the first fossil anthropoid (monkeylike primate) to be found in Africa. It was positioned close to where the Old World monkeys and apes split apart on the primate family tree. Like Old World monkeys, *Proconsul* had thin tooth enamel, a narrow chest, short forelimbs, and a light build. This tells us that it was primarily a tree-dweller, living on soft fruits. However, *Proconsul* also shared some similarities with apes, having a large brain, apelike elbows, and no tail.

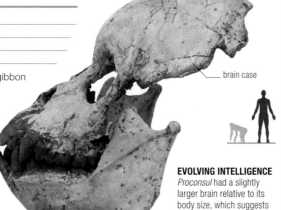

brain case

EVOLVING INTELLIGENCE
Proconsul had a slightly larger brain relative to its body size, which suggests evolving intelligence.

TOUGH TEETH
Sivapithecus had large canine teeth and heavy molars, which suggests that its diet included tough food, such as seeds and savanna grasses.

large canine tooth

Sivapithecus

GROUP Placental mammals
DATE Middle to Late Miocene
SIZE 5ft (1.5m) long
LOCATION Asia

Sivapithecus is a fossil ape closely related to the ancestors of orangutans. It was the size of an orangutan, but it was built like a chimpanzee, and spent most of its time on the ground, rather than in the trees like orangutans. Another extinct ape genus, *Ramapithecus*, was based on a partial lower jaw that looked like an early hominid (member of the human family). This became the basis for arguing that hominids arose 12 million years ago in Asia, but recent specimens have shown that *Ramapithecus* was a small species of *Sivapithecus*.

Dryopithecus

GROUP Placental mammals
DATE Late Miocene
SIZE 24in (60cm) long
LOCATION Africa, Europe, Asia

Dryopithecus was an ape the size of a monkey with a build like a chimpanzee. Its body proportions, and the shape of the limbs and wrists, show that it could walk on all fours like a chimpanzee. The limbs also suggest that it spent most of its life in trees, swinging like a gibbon. *Dryopithecus* had molars with thin enamel, suggesting it ate soft fruit, and the same pattern of cusps as the great apes and hominids.

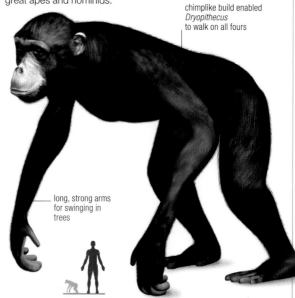

chimplike build enabled *Dryopithecus* to walk on all fours

long, strong arms for swinging in trees

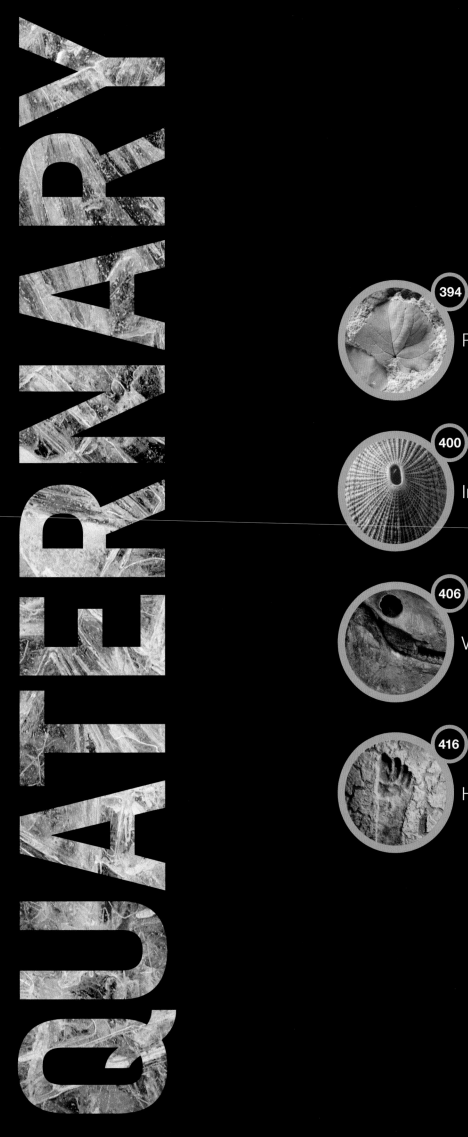

QUATERNARY

Quaternary

The Quaternary Period is part of a continuing ice age, during which intense periods of glaciation lasting nearly 100,000 years have been alternating with warmer interglacials 20,000 to 30,000 years long. Major changes in mammal fauna occurred during the last glacial and interglacial periods, with the extinction of the Pleistocene megafauna. Our own species may have played a significant role in their demise.

PO RIVER DELTA
Quaternary was the name given to one of four subdivisions of Earth's rocks made by early geologists, and typified by these alluvial deposits in Italy's Po River Valley.

Oceans and continents

Earth's oceans and continents have been more influenced by climate during the Quaternary than in any other period. While India continues to nudge into Asia, Australia into Indonesia, and Africa and Arabia into Europe and Asia, with associated and often catastrophic earthquakes, the oceans have risen and fallen like global soufflé. As the ice sheets advanced with a pincer movement from north and south, much of the water that otherwise would have returned to the oceans became locked in the frozen continents and the sea levels fell substantially. In Europe, the Rhine and Thames rivers converged in a huge estuary that emptied into the North Sea off the northern coast of England, and the English channel did not exist. Rising oceans during the last 10,000 years of the Pleistocene brought dramatic changes to the geographies of these and many other parts of the world. During this period, humans, who had previously lived as hunter-gatherers, began to domesticate animals and establish permanent farming settlements. Our influence on the landscape has increased and we have come to dominate the environments in which we live, at the expense of many other species. Yet the world geography we see today could be transient. Human activities seem to be causing global warming on a scale that threatens to flood of terrestrial areas. No one can predict whether the Quaternary's natural cycles of glacial advance and retreat will continue and they do, whether the next glaciation could overcome human impact on the Earth's temperatures. If it does, the size of the continents will increase, as once again there is a return to a windy, frozen world.

ENGLISH CHANNEL
The English Channel did not separate England and France 25,000 years ago. With much of Earth's water frozen in glacial ice, sea levels were as much as 425ft (130m) lower than they are today.

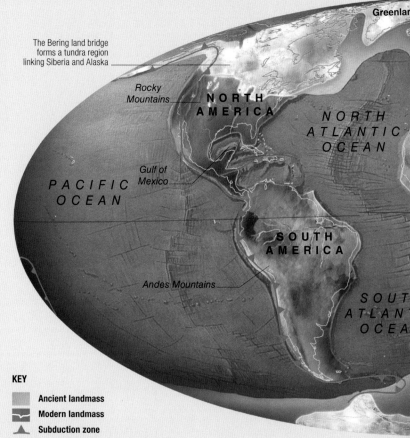

The Bering land bridge forms a tundra region linking Siberia and Alaska

Greenlan

Rocky Mountains

NORTH AMERICA

NORTH ATLANTIC OCEAN

PACIFIC OCEAN

Gulf of Mexico

SOUTH AMERICA

Andes Mountains

SOUT ATLAN OCEA

KEY

- Ancient landmass
- Modern landmass
- Subduction zone

QUATERNARY WORLD MAP
With climatic fluctuations, the flooding of land masses alternated with the drying out of some shallow seafloors. At intervals, land bridges formed.

PLIOCENE

PLANTS

● **2** Conifers at high latitudes become more diverse

● **2.35** First *Homo*

● **2** Extinction of australopithecines

● **1.7** Evolution of *Homo erectus*; continued expansion of North American carnivores into South America; expansion of South American opossums, sloths, armadillos, and glyptodonts into North America

VERTEBRATES

HOMO HABILIS

DIDELPHIS

FRESH WATER SUPPLY
Locked in the ice of the Perito Moreno Glacier, in Argentine Patagonia, is the world's third largest reserve of fresh water. Unlike many glaciers in the current interglacial period, it is not in retreat.

Climate

Earth's climate has fluctuated dramatically over the last 1.8 million years, and particularly during the last 600,000, while the planet has been locked in a series of cold glacial and warmer interglacial periods. Temperature, precipitation, and carbon dioxide levels have all varied cyclically. About 80 percent of Quaternary climate change has been due to orbital forcing (see p.23) that has caused cyclical variations in the solar input to the Earth. From about 1.3 million years ago there were 30,000-year pulses of colder periods. Longer cold periods of nearly 100,000 years began about 900,000 years ago. During the cold phases ice sheets expanded enormously, especially in the last half a million years. With much of the world's water frozen and low evaporation, the climate was very dry, even in ice-free areas. Ice sheets covering Europe, Asia, and North America during the last glacial advance reached their maximum between about 15,000 to 20,000 years ago, and only started to retreat about 12,000 years ago. To cope with the cold, giant mammals developed thick protective pelts of fur, or hair. Bison, woolly mammoths, and deer were hunted by saber-toothed tigers, cave bears, and by our ancestors—large-brained hominins that had begun to spread across the globe. Human impact on the climate and environment has been so extensive that some scientists have proposed a new epoch in Earth's history—the Anthropocene.

The Mediterranean Sea is linked to the Atlantic by the Gibraltar Strait

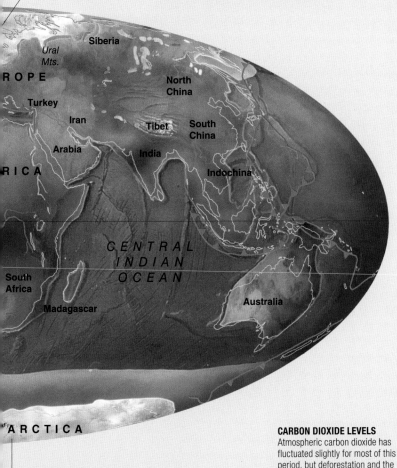

Siberia
Ural Mts.
North China
ROPE
Turkey
Iran
Tibet
South China
Arabia
India
RICA
Indochina
CENTRAL INDIAN OCEAN
South Africa
Madagascar
Australia
ARCTICA

Large, permanent ice sheets cover Antarctica, advancing and retreating at intervals

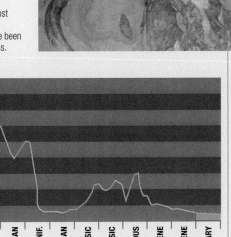

FOSSILIZED MAMMOTH SKELETON
Many large mammals in Europe, including most mammoths, became extinct at the end of the Pleistocene. These large herbivores could have been exterminated by human hunting or habitat loss.

CARBON DIOXIDE LEVELS
Atmospheric carbon dioxide has fluctuated slightly for most of this period, but deforestation and the use of fossil fuels have increased levels by about 35 percent since the Industrial Revolution began.

CONCENTRATION OF ATMOSPHERIC CO_2 AS MULTIPLE OF PRESENT-DAY VALUE

18
16
14
12
10
8
6
4
2
0

ARCHEAN | PROTEROZ. | CAMBRIAN | ORDOVICIAN | SILURIAN | DEVONIAN | CARBONIF. | PERMIAN | TRIASSIC | JURASSIC | CRETACEOUS | PALEOGENE | NEOGENE | QUATERNARY

PERIODS

EISTOCENE

HOLOCENE

1   0   MYA

● **0.9** Great expansion of steppe floras   **PLANTS**

● **0.43** Evolution of *Homo neanderthalensis*; giant flightless birds diversify (*Moa*, New Zealand; *Aepyornis*, Madagascar; *Genyornis*, Australia)

● **0.3** Evolution of *Homo sapiens*

● **0.05** Arrival of *Homo sapiens* in Australia

● **0.7** Evolution of *Homo heidelbergensis*

● **0.1** Major reduction in mammalian diversity in North America, especially mammoths, camels, perissodactyls, gomphotheres, saber-toothed tigers, horses, sloths

● **0.04** Extinction of *Homo neanderthalensis*

● **1.2** Evolution of *Homo antecessor*

● **0.02** Humans reach America

● **0.01** Megafaunal extinctions in Europe (e.g. mammoths)

VERTEBRATES

● **0.01** Expansion of megafauna in Australia

● **0.0002** About 80% of small Australian mammals extinct

● **0.05** Megafaunal extinctions in Australia

● **0.00007** Extinction in 1936 of last member of the family Thylacinidae (Tasmanian tiger)

HOMO ANTECESSOR

HOMO HEIDELBERGENSIS

HOMO NEANDERTHALENSIS

QUATERNARY PLANTS

During the Quaternary, the geographic distribution of plants was shaped by dramatic changes in the global climate. The Quaternary marks the final phase in the evolution of land plants before the great modification of terrestrial environments caused by human activity over the past few thousand years.

The Quaternary was a time of rapid climatic change that had profound effects on global vegetation, and as the climate changed, it had a major impact on where new plants could establish themselves. The cumulative effect over hundreds, or thousands, of years is seen as "migration," with plants staying in step with the climatic conditions in which they grow best.

Regional extinction

Successive expansion and contraction of ice sheets and glaciers in the Northern Hemisphere during the Quaternary resulted in successive southerly and northerly migrations by different species. One effect of these migrations was significant regional extinction, especially in Europe, where east–west running mountain ranges partially blocked migration routes to the south. The strong similarities between the rich temperate forests of China and those of eastern North America largely reflect the elimination of many species from Europe during the Late Neogene and Quaternary. Fossil evidence shows that

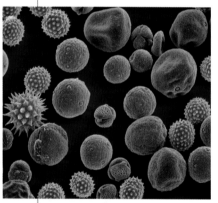

POLLEN GRAINS
By studying ancient pollen and spore assemblages it has been possible to reconstruct the main vegetational changes that occurred during the Quaternary.

TUNDRA
Distinctive tundra vegetation, often dominated by dwarf birches and willows, first became widespread during the successive expansions of ice sheets and glaciers that took place through the Quaternary.

many of these plants, such as the tulip tree, flourished in Europe until relatively recently. There were also Late Neogene and Quaternary regional extinctions in western North America, most likely due to the increasingly dry climate.

Human landscapes

Many of our most familiar landscapes have been modified to a greater or lesser extent by human action. This trend is destined to continue, with the loss of much of the great diversity that has accumulated over more than 450 million years of land plant evolution.

HUMAN IMPACT
In the last few thousand years, and especially over the past few hundred years, human populations have exerted ever greater influence over the ecosystems that have grown up on land.

GROUPS OVERVIEW

There are probably more land plant species living today than at any time in the past. This extraordinary variety has arisen as group upon group has undergone bursts of diversification in response to changing environmental conditions. The fossil record provides ample evidence of extinction, but over time new species have arisen faster than they have disappeared.

CLUBMOSSES
Living clubmosses include forms very similar to ancient lycopods from the Early Devonian, alongside other plants that are the product of more recent bursts of diversification. The diversity of clubmosses is a wonderful mixture of ancient and modern forms that reflects more than 400 million years of plant evolution.

FERNS
Ferns are second only to angiosperms in the number of living species. Their success has been based mainly on exploiting the forested habitats that angiosperm trees created. However, among the great diversity of modern ferns there are also more archaic forms that have persisted largely unchanged since the Paleozoic and Mesozoic.

CONIFERS
With fewer than a thousand living species, conifers are nevertheless very important in the vegetation of many parts of the world. They have proved especially successful in the Northern Hemisphere, where they dominate the relatively uniform, cool temperate forests that extend across much of Asia, Europe, and northern North America.

ANGIOSPERMS
Despite their recent origin compared to the much longer evolutionary history of most other groups of land plants, angiosperms dominate the vegetation over most of Earth's surface. They exhibit bewildering diversity in almost every aspect of their structure and biology. There are about 300,000 species of living angiosperms.

Chara

GROUP Charales
DATE Silurian to present
SIZE Up to 3ft 3in (1m) long
LOCATION Worldwide fresh and brackish water

These fresh- or brackish water plants possess both algal and higher plant characteristics, which makes them an isolated group between the green algae and the bryophytes. Fossilized whole plants are extremely scarce, but the distinctive female reproductive organ (oogonium) is more common; it can become heavily calcified and is therefore more resilient. When found as fossils, these oogonia are termed gyrogonites. The plants are bushy and up to 3ft 3in (1m) long. Individual shoots are only 1/16in (1–2mm) wide and organized in short lengths called internodes with whorls of branches arising at the joints (nodes). The Charales has an extensive fossil record, first appearing in the Late Silurian and diversifying during the Devonian and the Carboniferous. The genus *Chara* first appeared during the Early Paleogene.

preserved stem

CHARA STEMS
Chara gyrogonites can be recovered from all kinds of sediments, but the shoots need to have been infiltrated and mineralized to survive. They may be preserved silicified or calcified in limestone, as in this rare specimen from China.

Marchantia

GROUP Bryophytes
DATE Quaternary to present
SIZE Thallus up to 4in (10cm) long
LOCATION Worldwide, except in arid regions

Marchantia are liverworts, which make up a major part of the bryophyte group. Some resemble leafy mosses, while others have a flattened body called a thallus. Spores similar to those of liverworts are in evidence from the Ordovician onward, with the earliest fossils appearing in the Devonian. By the Late Triassic there were forms similar to *Marchantia* and many more are known from the Cretaceous and the Tertiary.

M. POLYMORPHA
This specimen shows impressions of thalli preserved in postglacial calcrete from Eskatorp, Sweden. The faint forking lines mark the centers of the thallus.

LIVING RELATIVE
COMMON LIVERWORT

The common liverwort (above) has dull green mats of branching thalli that can grow up to 4in (10cm) long and up to 1/2in (1.4cm) wide. It has separate male and female plants. The green, umbrellalike structures that form on the ends of the stalks bear the reproductive organs: female underneath, or male on top. The spore-bearing part of the life cycle is minute, developing from the fertilized egg on the underside of the umbrellalike structures, where it remains living parasitically.

Sphagnum

GROUP Bryophytes
DATE Jurassic to present
SIZE Up to 12in (30cm) long
LOCATION Cool and wet regions worldwide

Commonly called peat moss, *Sphagnum* has between 150 and 300 species. It occurs mainly in the Northern Hemisphere, as far north as Svalbard, Norway, but also in the cool and wet regions of the Southern Hemisphere. Its leaves have a lattice of small, green photosynthetic cells, and larger, dead cells that have a capacity to hold large quantities of water. *Sphagnum* can build up great thicknesses of peat through its ability to increase the acidity of its surroundings, which prevents decay by bacterial and fungal activity. When rainfall is high, *Sphagnum* bogs can grow into domes above the level of the water table. *Sphagnum*-like plants are known from the Russian Permian, but *Sphagnum* itself probably arose in the Jurassic.

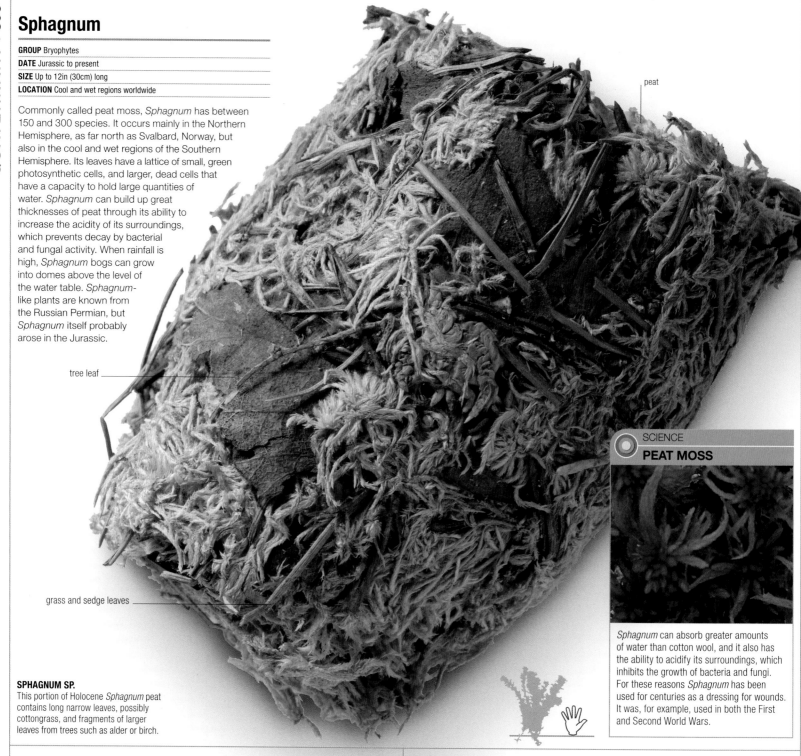

peat

tree leaf

grass and sedge leaves

SPHAGNUM SP.
This portion of Holocene *Sphagnum* peat contains long narrow leaves, possibly cottongrass, and fragments of larger leaves from trees such as alder or birch.

SCIENCE
PEAT MOSS

Sphagnum can absorb greater amounts of water than cotton wool, and it also has the ability to acidify its surroundings, which inhibits the growth of bacteria and fungi. For these reasons *Sphagnum* has been used for centuries as a dressing for wounds. It was, for example, used in both the First and Second World Wars.

Amblystegium

GROUP Bryophytes
DATE Neogene to present
SIZE Mat up to 8in (20cm) across
LOCATION Americas, Europe, Asia, Australasia, Pacific

This creeping moss has tapering leaves, ⅛–¼in (3–6mm) long, that end in fine points. It grows and divides irregularly and horizontally to form a mat of interlocking filaments. There are about 15 species and all prefer moist, temperate climates. The earliest record of fossil *Amblystegium* is in the Early Neogene of southern Germany, and there are a number of records of it in the Quaternary.

AMBLYSTEGIUM SP.
This fossil shows a moss mat that has been infiltrated by calcium carbonate in mineral-rich rivers. It then crystallized, preserving the plant in three dimensions.

interlocking filaments

LIVING RELATIVE
FIR CLUBMOSS

Huperzia selago grows in northern grassland and rocky areas, and northern and more southerly mountainous regions above the tree line in Europe, Russia, Asia, Japan, USA, and Canada. This distribution was a result of its migration away from advancing woodland as the climate warmed after the last ice age.

Huperzia

GROUP Lycophytes
DATE Cretaceous to present
SIZE Shoot up to 24in (60cm) long
LOCATION Worldwide, except for arid areas

Huperzia is a simple clubmoss, and perhaps one of the most primitive vascular plants living today. It has small, spirally arranged leaves and instead of cones, like the other clubmosses, *Huperzia* has zones of spore-producing structures (sporangia) in the leaf axils. Nearly 400 species (about 85 percent of all clubmosses) exist in temperate and tropical regions worldwide. *Huperzia* represents a lineage of living plants that stretches back to Devonian genera such as *Asteroxylon* (see p.116). The living species of *Huperzia* probably evolved, diversified, and spread during the Paleogene and Neogene.

Andreaea

GROUP Bryophytes
DATE Quaternary to present
SIZE Up to ⅜in (1cm) high
LOCATION Arctic regions, mountains throughout the Northern Hemisphere

Andreaea is one of two genera of mosses that are commonly found on granite rock faces in northern mountainous regions and the Arctic. Their hairlike, rooting rhizoids can penetrate small cracks in the rock and the plants form small, wind-resistant cushions. There are up to 100 species of *Andreaea*. Fossil mosses are known from the Carboniferous onward, but there are no undisputed records of *Andreaea* before the Quaternary.

ANDREAEA ROTHII
Andreaea is protected against exposure to ultraviolet light by specialized compounds in its cell walls. The stalked red spore capsules release their spores through four longitudinal slits.

Marattia

GROUP Ferns
DATE Carboniferous to present
SIZE Frond up to 16½ft (5m) long
LOCATION Tropics worldwide

The group that includes *Marattia* is the most primitive of the living ferns, having large leaves (fronds) that are enclosed when young by two basal scales (stipules). Their sporangia are large and borne in clusters on the undersides of the fronds. The group can be traced back to numerous Carboniferous and Permian fossil stems and foliage. The best known of these is the tree fern called *Psaronius* (see p.141). The earliest fossils referred to as *Marattia* come from the Jurassic of Yorkshire, in northern England, where there are also specimens of another similar living genus, *Angiopteris*. The earliest of these ferns probably grew near streams, and they spread into other wetland and humid habitats as they underwent evolutionary changes and the habitats changed. The cooling of the climate during the Late Neogene and Quaternary caused their extinction in the temperate regions of the world, limiting the remaining 26 species to the tropics.

MARATTIA SALICINA
The king fern, with fronds up to 16½ft (5m) long, is native to New Zealand and the islands of the South Pacific. The starchy stem was once a traditional food for the Maori.

Molinia

GROUP Grasses
DATE Quaternary to present
SIZE Flowering head up to 8¼ft (2.5m) tall
LOCATION Europe, Iran, Siberia, E. Canada, N.E. USA

Molinia caerulea, purple moor-grass, is a densely tufted perennial grass that forms large sward in places that are at least seasonally wet. It is found on heaths, moors, bogs, fens, lake-shores, mountain grasslands, and cliffs. Its purple flower heads give it its common name. This grass grows best in the wettest areas and can dominate large parts of bogs and moors, where it contributes to the growth of peat. Sections through peat from both the Quaternary and today's bogs and moors reveal layers that are densely packed with leaves of this species.

MOLINIA CAERULEA
This compact, tufted grass is found in damp places, where it can cover large areas. Its stiff leaf-blades taper to form sharp points.

yellow pollen cones

leaf

EPHEDRA MAJOR
All species of *Ephedra* are woody shrubs, with photosynthetic stems, and leaves that are reduced to brown scales.

Ephedra

GROUP Gnetales
DATE Triassic to present
SIZE Up to 6½ft (2m) tall
LOCATION Arid and semiarid regions of Asia, Europe, N. Africa, W. North America, South America

The 35–45 living species of *Ephedra* are woody shrubs. It belongs to the same group of gymnosperms that includes *Gnetum* and *Welwitschia* (see p.275). *Ephedra* has both seed-producing and pollen-producing cones. Its pollen is distinctive, with variable thickenings producing longitudinal ridges and grooves. Similar pollen first appeared in the Permian, but more reliable pollen is known from the Cretaceous.

Eriophorum

GROUP Angiosperms
DATE Quaternary to present
SIZE Flower heads up to 28in (70cm) across
LOCATION Europe, Asia, North America, Arctic

Cottongrasses (*Eriophorum*) are actually sedges, and not grasses. *Eriophorum* is a genus with about 25 species. It is found in acid bog habitats throughout the temperate Northern Hemisphere, in Europe, Asia, North America, and particularly in the higher latitude Arctic tundra. They are herbaceous and perennial, spreading by underground stems. Their bisexual flowers are on stalks and produce masses of seeds that are covered in fluffy cotton, which allows the wind to carry them. The sedge family is known from the Early Neogene, where flower parts of *Carex* have been found in Nebraska, but remains of *Eriophorum* are only found in Quaternary interglacial and postglacial peat. Layers of *Eriophorum* leaves and rhizomes occur in such peat, representing a time when the plants dominated the bog's surface.

ERIOPHORUM VAGINATUM
The hare's-tail cottongrass can form tussocks that are difficult to walk over. It has only one cottony flower head.

SCIENCE
BOGS AND FENS

Swards of cottongrass grow on acid bogs and fens where there is shallow, standing water. In the foreground of this photograph is common cottongrass (*Eriophorum angustifolium*), which has several cottony heads on each flower, and leaves that are V-shaped in section. The bog's acidity prevents the plants from rotting after death so an accumulation of cottongrass remains builds up. Water table or rainfall changes may favor the growth of other plants, such as *Sphagnum* (see p.396), which in turn dominate the bog's surface.

Davidia

GROUP Angiosperms
DATE Paleogene to present
SIZE Tree up to 59ft (18m) tall
LOCATION North America, Russia, China

Fossil leaves that are similar to those of living *Davidia* show that it was prominent in the Early Paleogene vegetation of western North America, and suggests that the family had its origins in the Cretaceous. Fossil fruits matured from flower heads that appear to have been borne between two large bracts, as in the living genus. *Davidia* spread to eastern Russia and China during the Paleogene and Neogene, and narrowly survived extinction during the Quaternary ice ages. There is only one living species in this genus.

LIVING RELATIVE
DOVE TREE

Davidia involucrata has large, green leaves with pink stalks. Its flowers are reddish and clustered into tight heads that hang on the ends of drooping stalks, enclosed by two enormous, petal-like, pure white bracts. The bracts give it another common name, the "pocket handkerchief tree."

pure white bracts

Corylus

GROUP Angiosperms
DATE Late Paleogene to present
SIZE Up to 79ft (24m) tall
LOCATION North America, Europe, Asia

There are about 15 species of *Corylus*, or hazel, from northern temperate regions. All are bushes, except the Turkish and Chinese hazels, which are trees. They are deciduous and have male flowers in catkins and female flowers enclosed within buds. *Corylus avellana* was an early colonizer after the retreat of the Quaternary glaciers.

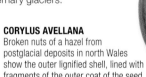

seed coat

CORYLUS AVELLANA
Broken nuts of a hazel from postglacial deposits in north Wales show the outer lignified shell, lined with fragments of the outer coat of the seed.

Trapa

GROUP Angiosperms
DATE Neogene to present
SIZE Fruit up to 1¹⁄₂in (4cm) long
LOCATION Europe, Asia, Africa

The pollen and ornately shaped, nutlike fruits of *Trapa* have been found across the Northern Hemisphere, from Europe to Africa, India, and China. Evidence of the genus stretches back from the present to the Early Neogene. Each fruit has a single, very large, starchy seed that contains toxins. However, these toxins are destroyed during cooking, and *Trapa* seeds are known to have been used as a food source for thousands of years by early people in present-day Europe and China, with the oldest recorded use being by the nomadic Maglemosian people. These people lived in northern Europe around 10,000 to 8,000 years ago.

spine

TRAPA NATANS
Trapa natans has four sharp barbed spines on its fruits. Its generic name is derived from the Latin word for "thistle."

LIVING RELATIVE
WATER CHESTNUT

The water chestnut (*Trapa natans*) is an aquatic plant that grows naturally in slow-moving water in warm temperate regions of Europe, Asia, and Africa. It is used as a food source in China and India. It has also been introduced into parts of North America and Australia, where it has become an invasive weed.

Tilia

GROUP Angiosperms
DATE Paleogene to present
SIZE Up to 120ft (36m) tall, leaf up to 5in (12.5cm) long
LOCATION Temperate regions of Northern Hemisphere

Tilia (linden trees) are large, deciduous trees that are now spread across the temperate regions of the Northern Hemisphere, except in northwest America. Their fossils may be leaves, flower parts, wood, or pollen grains. The fossil leaves shown here are heart-shaped impressions in the surface of the rock, with no original plant material remaining. Some leaves landed the right way up, which resulted in their veins becoming embedded into the rock. Those that landed upside down show the smoother upper surface on the rock. The five veins that depart from the base of each leaf branch into secondary veins; a further transverse network of smaller veins is visible between them.

LIVING RELATIVE
COMMON LINDEN

serrated margin

flower cluster

Tilia x *europaea* is a cross between the large-leaved linden from southern continental Europe and the small-leaved linden from northern Europe. It grows larger than the two parents, but looks untidy because of the growth of small shoots around its base and on its larger branches.

LINDEN POLLEN
Linden pollen found in Quaternary peat reflects climatic and other factors. After the last ice age, it gradually increased, but then decreased when humans started to fell lowland forests.

upper side of leaf

TILIA SP.
These impressions of linden leaves on calcrete are from the postglacial deposits at Benestad, in southern Sweden. Differences in appearance reflect the orientation of the leaves.

QUATERNARY
INVERTEBRATES

The start of the Quaternary is now taken as 2.6 million years ago. Temperatures dropped gradually, and many large mammals became extinct during the coldest time, but in general terms the extinctions on land and in the sea were considerably less severe than at the end of the Cretaceous.

Dominance of invertebrate groups

In the oceans, the same kinds of animals prevailed as in the Neogene, although their distribution changed as the icesheets advanced and retreated. The remains of both marine and land-living animals have been used to understand the climatic changes that took place, especially from cores drilled through Cenozoic and Quaternary marine sediments. Foraminiferids (fossilized plankton) are sensitive indicators of climatic change, as successive cold- and warm-water groupings change through time. The foraminiferids themselves preserve a record of stable oxygen isotopes, which can be used to measure the temperatures that existed when these microfossils were alive. Cold-water bivalves are different from those of warmer seas and are used in a

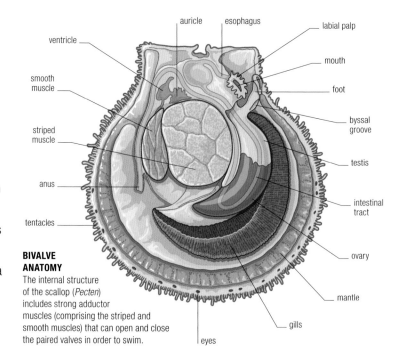

BIVALVE ANATOMY
The internal structure of the scallop (*Pecten*) includes strong adductor muscles (comprising the striped and smooth muscles) that can open and close the paired valves in order to swim.

Labels: ventricle, auricle, esophagus, labial palp, mouth, foot, byssal groove, testis, intestinal tract, ovary, mantle, gills, eyes, tentacles, anus, striped muscle, smooth muscle

LIVING MUSSELS
These mussels (*Mytilus*) and limpets (*Patella*) are cold-water mollusks, similar to those that existed during the Quaternary. They live on the shoreline in rock pools, where they feed on algae.

similar way. On land, the remains of beetles in peat and soil are used to understand past climatic changes. Many living species of beetles, the same as those of the Quaternary, have restricted temperature preferences. Successive groupings of warm- and cold-loving beetles in a peat core can be used to tell relative temperatures through time. This data all points to a dramatic temperature rise about 11,500 years ago.

GROUPS OVERVIEW

Invertebrate groups, both marine and land-dwelling, remained largely the same in the Quaternary as they were in the Neogene. However, the distribution of these groups was greatly affected by the changing climatic conditions. Gastropod, bivalve, and echinoid fossils all play a role in our understanding of how and where temperatures changed.

GASTROPODS
Cold-adapted gastropods, along with other specialized marine invertebrates, were first recognized as long ago as 1780, in Otto Fabricius' great work *Fauna groenlandica* (Greenland fauna). Several species of wholly Arctic whelks are known, for example, and the specific temperature preferences of many such gastropods are now known in detail.

BIVALVES
Arctic bivalves, fairly distinct from those of temperate conditions, are well known from the Quaternary. The mussel *Yoldia arctica*, for example, gave its name to the Yoldia Sea, a great area of cold brackish water that existed in what is now the Baltic region. The Yoldia Sea was formed by glacial meltwater from the retreating Scandinavian ice sheet.

ECHINOIDS
These do not appear to have been adversely affected by the climatic changes of the Quaternary, the cold-adapted types tracking the edge of the ice sheets as these advanced and retreated. Echinoids survived to become an integral component of today's fauna, at various depths in the sea.

ARTHROPODS
The crustaceans (amphipods particularly) have temperature-restricted occurrence, with many forms confined to polar regions. These, however, have only a limited preservation potential, so are not of great use in tracking climate change.

Heterocyathus

GROUP Anthozoans
DATE Late Neogene to present
SIZE Up to ³⁄₈in (1cm) long
LOCATION Indian and Pacific Oceans

Heterocyathus is a small tabular coral with a flat base and top and roughly rounded sides. The center of the coral has a series of slender, vertical pillars arranged across from the ends of the dividing walls, which look like rings of small dots. Like all corals, *Heterocyathus* starts life as a planktonic larva. This eventually settles on the sea bed, usually attaching itself to a small shell, which it soon outgrows. Modern forms sometimes live in symbiosis with sipunculid worms (peanut worms), and it is likely that fossil forms did, too. Living species can be found at depths of 33–1,800ft (10–550m).

HETEROCYATHUS SP.
Radiating out from the center of these coral fossils are numerous vertical walls, or septa, which become thicker toward the edge.

vertical septum

central, vertical pillars

Magellania

GROUP Brachiopods
DATE Neogene to present
SIZE ³⁄₄–1¹⁄₄in (2–3cm) long
LOCATION Australia, South America, Antarctica

This is a typically shaped, weakly ornamented terebratulid brachiopod shell. Its pedicle opening lies close beneath the beak of the pedicle valve; in this example, the opening has been slightly enlarged toward the beak by a small breakage. There is a well-marked triangular area beneath the pedicle opening that closes off its front side. The brachial valve houses the loop-shaped lophophore support, which, together with the fine perforations over the surface of the shell (punctae), are characteristic features of terebratulid brachiopods. The ornamentation consists of fine growth lines.

pedicle opening

fine growth lines

MAGELLANIA SP.
This specimen has a single, well-marked growth increment, possibly caused by a pause in growth. The growth lines are crossed by faint radial ribs that strengthen the front edge of the shell.

ANATOMY
PEDICLES

Most brachiopods, such as *Magellania*, have a fleshy stalk called the pedicle, which emerges through an opening in the larger pedicle valve and anchors them permanently. In some inarticulate brachiopods, such as *Lingula*, the pedicle is larger and emerges between the two valves to act as a movable anchor rather than a permanent fixture. In yet other Paleozoic brachiopods, the pedicle died away in the adult, its opening grew over, and the brachiopod was anchored by the shell's weight.

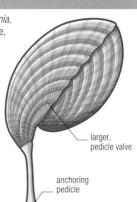

larger, pedicle valve

anchoring pedicle

Hippoporidra

GROUP Bryozoans
DATE Paleogene to present
SIZE Can be over 4in (10cm) across
LOCATION Europe, Afric, North America

Colonies of the bryozoan *Hippoporidra* encrusted the surface of gastropod shells in many thick layers, sometimes forming large spiral extensions. On the colony surface, many circular features can be seen. These are the openings to structures called autozooecia, which contained the soft-bodied feeding animals (autozooids). The autozooids had a ring of tentacles surrounding the mouth, called a lophophore. When the zooid was not feeding, the lophophore could be withdrawn into the autozooecia for protection. The frontal walls of the autozooecia are perforated with pores. Raised areas (monticules) can be observed on the colony surface, where the autozooecia are larger. Avicularia, which housed defensive, nonfeeding zooids, can be found between the monticules.

circular opening

raised area (monticule)

HIPPOPORIDRA EDAX
This fossil suggests a symbiotic relationship between the bryozoan and the hermit crab, although the crab has not been preserved here.

Hippoporidra colonies have been found to **live in symbiosis** with hermit crabs.

Capulus

GROUP Gastropods
DATE Paleogene to present
SIZE Up to 1in (2.5cm) across
LOCATION Atlantic and Mediterranean regions

Capulus is a cap-shaped gastropod with a tight, asymmetrical initial spiral. Modern *Capulus* are often found with bivalves, particularly *Pecten* (see p.404), *Chlamys*, and *Modiolus*. They attach themselves by their large foot to the open edge of the bivalves' shells, close to the inhalant current that sucks in nutrient-rich water. In this position, *Capulus* can use its long, extendable proboscis to collect food particles that the bivalve is drawing in. It may also remove food particles that are trapped on the edges of the bivalve's gills. This semiparasitic mode of life does not appear to affect the bivalves adversely.

fine pitting caused by boring organism

growth ridges

CAPULUS SP.
This fossil shell looks like a tiny elf's cap, with a wide opening, a brim, and a tip swept back in one direction.

Neptunea

GROUP Gastropods
DATE Neogene to present
SIZE Up to 3¹⁄₄in (8cm) long
LOCATION North Atlantic Ocean, North Africa, Mediterranean Sea

Neptunea's shell has a tall spire, coiled so that each whorl overlaps the preceding one for about a third of its height. Fine spiral ribs are crossed by variably strong growth lines, which may represent seasonal variations in growth. *Neptunea* is essentially a marine scavenger, feeding on dead fish, crustaceans, and mollusks.

fine, spiral ribs

growth lines

NEPTUNEA CONTRARIA
Unusually, this species has a sinistral shell, coiled so that the aperture is on the left side of the upright shell.

D-shaped opening

Lymnaea

GROUP Gastropods
DATE Paleogene to present
SIZE Up to 2¼in (6cm) long
LOCATION Worldwide

Lymnaea is a genus of freshwater snail that includes the great pond snail (*Lymnaea stagnalis*), common across much of the Northern Hemisphere. The shells are pointed, with the whorls forming a long spire. *Lymnaea* snails have flattened feelers and curtainlike edges to the foot. All freshwater snails and land snails are hermaphroditic—individuals produce both eggs and sperm. Unlike marine gastropods, which have gills, *Lymnaea* snails breathe air and must surface to take it in.

LYMNAEA SP.
As in this specimen, the outside of the shell lacks ribs or any other noticeable structure. However, faint and closely spaced growth lines are often visible.

opening toward right of shell

Littorina

GROUP Gastropods
DATE Neogene to present
SIZE Up to 2in (5cm) long
LOCATION NE Atlantic Ocean

Living species of *Littorina* include winkles and periwinkles. These mollusks graze algae and other microorganisms from the surface of rocks in the intertidal zone. At low tide, they may be found out of the water, stuck to rocks. When covered by the tide, they release their grip and search for food. Like all marine gastropods, *Littorina* breathe by removing oxygen from the water with their gills. When exposed by the tide, they secrete a mucous plug to keep water in. They also have a "lid" (operculum) on their shells, with which they can seal the opening.

spiral ribs

LITTORINA RUDIS
The surface of this shell displays typical spiral ribs, which are often crossed by faint growth lines.

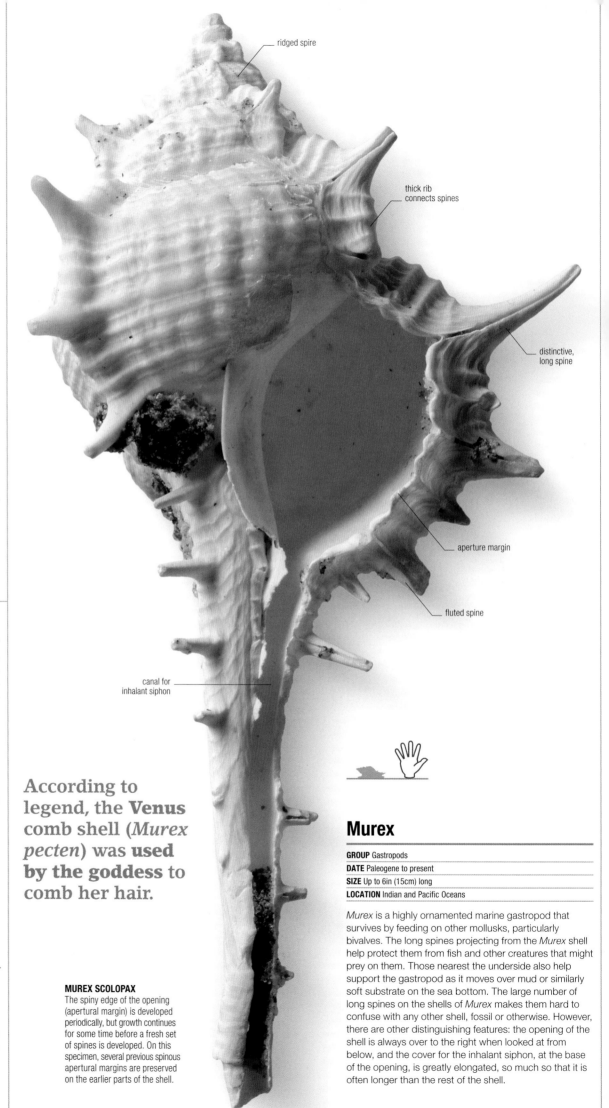

ridged spire

thick rib connects spines

distinctive, long spine

aperture margin

fluted spine

canal for inhalant siphon

MUREX SCOLOPAX
The spiny edge of the opening (apertural margin) is developed periodically, but growth continues for some time before a fresh set of spines is developed. On this specimen, several previous spinous apertural margins are preserved on the earlier parts of the shell.

According to legend, the **Venus comb shell** (*Murex pecten*) was **used by the goddess** to comb her hair.

Murex

GROUP Gastropods
DATE Paleogene to present
SIZE Up to 6in (15cm) long
LOCATION Indian and Pacific Oceans

Murex is a highly ornamented marine gastropod that survives by feeding on other mollusks, particularly bivalves. The long spines projecting from the *Murex* shell help protect them from fish and other creatures that might prey on them. Those nearest the underside also help support the gastropod as it moves over mud or similarly soft substrate on the sea bottom. The large number of long spines on the shells of *Murex* makes them hard to confuse with any other shell, fossil or otherwise. However, there are other distinguishing features: the opening of the shell is always over to the right when looked at from below, and the cover for the inhalant siphon, at the base of the opening, is greatly elongated, so much so that it is often longer than the rest of the shell.

Fusinus

GROUP Gastropods

DATE Paleogene to present

SIZE Up to 6in (15cm) long

LOCATION Widely distributed in warmer seas

This genus of gastropod mollusks contains a large number of species, many of which are still alive today. *Fusinus* shells vary in shape among species, but all have certain features in common, such as a long, pointed spire and a slender canal extending down from the shell's opening. In life, this acts as a cover for the long inhalant siphon, a tube that is used for sucking in water to pass over the gills. Living *Fusinus* species are known collectively as spindle shells. They are large, carnivorous sea snails from tropical and subtropical seas that feed on smaller gastropods.

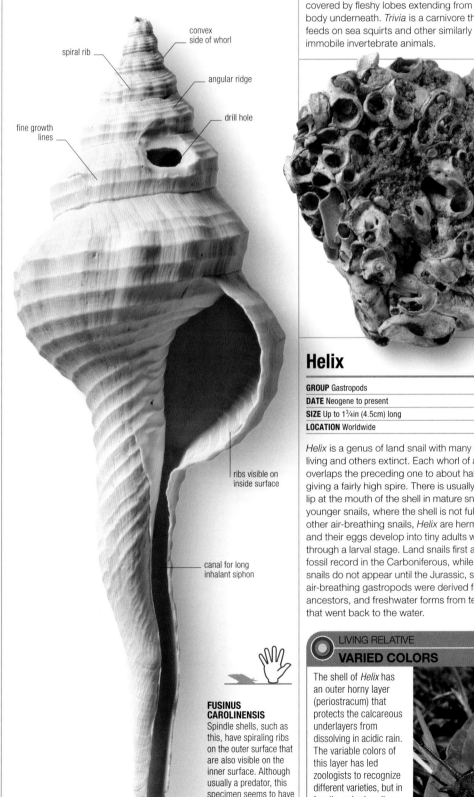

convex side of whorl

spiral rib

angular ridge

drill hole

fine growth lines

ribs visible on inside surface

canal for long inhalant siphon

FUSINUS CAROLINENSIS
Spindle shells, such as this, have spiraling ribs on the outer surface that are also visible on the inner surface. Although usually a predator, this specimen seems to have been prey itself— there is a clear drill hole in one of the whorls.

Trivia

GROUP Gastropods

DATE Paleogene to present

SIZE Up to 1¼in (3cm) long

LOCATION Worldwide

Trivia is a gastropod similar to the modern cowrie shell. Looked at from the outside, *Trivia*'s spiral shape is invisible. A cross-section through the shell, however, shows it to be spiraled. As it grew, each whorl completely enveloped the preceding one, hiding it from view. Extending out from its long, narrow opening (aperture) are slender ridges. In life, its shell is at least partly covered by fleshy lobes extending from the body underneath. *Trivia* is a carnivore that feeds on sea squirts and other similarly immobile invertebrate animals.

transverse rib

TRIVIA AVELLANA
This fossil shows how the narrow opening runs the full length of the shell. Both its inner and outer edges are strongly ridged.

narrow aperture

twisting tube

Helix

GROUP Gastropods

DATE Neogene to present

SIZE Up to 1¾in (4.5cm) long

LOCATION Worldwide

Helix is a genus of land snail with many species, some living and others extinct. Each whorl of a *Helix* shell overlaps the preceding one to about half of its height, giving a fairly high spire. There is usually a pronounced lip at the mouth of the shell in mature snails, absent in younger snails, where the shell is not fully grown. Like other air-breathing snails, *Helix* are hermaphrodites and their eggs develop into tiny adults without going through a larval stage. Land snails first appear in the fossil record in the Carboniferous, while freshwater snails do not appear until the Jurassic, suggesting that air-breathing gastropods were derived from marine ancestors, and freshwater forms from terrestrial snails that went back to the water.

LIVING RELATIVE
VARIED COLORS

The shell of *Helix* has an outer horny layer (periostracum) that protects the calcareous underlayers from dissolving in acidic rain. The variable colors of this layer has led zoologists to recognize different varieties, but in fossils, color banding is seldom preserved.

Vermetus

GROUP Gastropods

DATE Neogene to present

SIZE Tube up to ¼in (6mm) across

LOCATION Worldwide

Vermetus is a most unusual gastropod that looks more like a worm tube. The first part of its shell is coiled as in most gastropods, but then the spiral opens out to form an irregular, twisted tube. The tubular shell has flattened sides, which in places are slightly concave. It also has fairly thick walls bearing finely spaced growth plates and thicker ribs parallel to the direction of shell growth. *Vermetus* often lives in clusters of several individuals.

VERMETUS INTORTUS
The holes bored into the tubes shown here were probably created by a predatory gastropod such as *Natica*.

pronounced lip

HELIX ASPERSA
This shell displays the large, rounded whorls and wide, rounded opening typical of *Helix*. The lipped mouth indicates that this was a mature snail.

rounded whorl

Land snails are **hermaphrodites** and have a variety of **complex mating rituals.**

Clinocardium

GROUP Bivalves
DATE Neogene to present
SIZE Up to 1¹⁄₂in (4cm) long
LOCATION North Pacific and North Atlantic coastal regions

C. INTERRUPTUM
This specimen shows three distinct growth ridges, suggesting that its growth was interrupted seasonally.

Clinocardium is a small, cocklelike bivalve that still exists today. The shell has clear ribs radiating out from the beak to the rounded edges. These ribs are crossed by concentric growth increments, which are strong on some parts of the shell but very weak on others. Inside there are two teeth and sockets below the beak in each valve. There are also clear, equal-sized scars where the adductor muscles were attached.

Corbicula

broad beak

growth-parallel ribs

GROUP Bivalves
DATE Cretaceous to present
SIZE Up to 1in (2.5cm) long
LOCATION Worldwide

Corbicula is a small bivalve with a rounded triangular shape and a broad beak on each valve (half) of its shell. Inside, each valve has clear scars where the adductor muscles attached, and the hinge-line has prominent teeth. The central tooth below the beak of one valve fits in a central socket in the other. Outside the central teeth, there are lateral teeth and sockets that run parallel to the edge of the shell. The elastic ligament that opened the valves sat outside, behind the beaks.

CORBICULA FLUMINALIS
This shell was found with fresh-water fossils, so unlike some early forms, it probably was not marine.

Pecten

GROUP Bivalves
DATE Paleogene to present
SIZE Up to 4in (10cm) along the hinge-line
LOCATION Worldwide

Both valves of a *Pecten* shell are ornamented with ribs that radiate out from the hinge end. The tops of the ribs are flat and the spaces between them narrower than the ribs themselves. The hinge-line of each valve is fairly straight. Inside each valve, the most obvious feature is the very large adductor muscle scar. As one might expect for a bivalve that is able to swim by opening and closing its shell, the adductor muscle in a live animal is large and powerful. Most people know members of the genus *Pecten* as scallops.

flattened rib

PECTEN MAXIMUS
The valves of *Pecten* differ in shape; the valve above, which lay on the seabed, is convex, while the one below is almost flat.

hinge-line

LIVING RELATIVE

SCALLOP

Very few bivalves are free-swimming, and *Pecten*'s ability to clap its valves together enables it to avoid many predators. It needs to sense their approach and has developed sophisticated eyes on the edges of its mantle. It also has chemoreceptors, enabling the bivalve to taste and smell the water, and statocysts to help determine its orientation.

Cerastoderma

pointed beak

GROUP Bivalves
DATE Paleogene to present
SIZE Up to 1¹⁄₂in (4cm) long
LOCATION Europe

Cerastoderma is best known as the cockle commonly found on modern European beaches, but its history goes back to the Late Paleogene. In outline, its shell is relatively square, with strongly radiating ribs crossed by raised riblets parallel to the direction of growth. The beak of each half of the shell (valve) is pointed and runs straight across the hinge-line. There are two teeth on the underside of the beak in each valve. Inside the shell, there are two clear muscle scars where the adductor muscles (the muscles that pulled the shell closed) attached. In life, *Cerastoderma* lives in shallow burrows.

radiating rib

Cerastoderma **feeds by extending its siphons** up through the seabed **to suck in water and food.**

CERASTODERMA ANGUSTATUM
This fossil shell shows the strong, raised ribs fanning out from the beak, crossed by fine, parallel growth lines.

Venus

GROUP Bivalves
DATE Late Paleogene to present
SIZE Up to 1½in (3.5cm) across
LOCATION Europe, Africa, Indonesia, North America

The genus *Venus* includes both living and extinct bivalve species. Each shell has a beak that curves around to one side, behind which is a smooth, depressed area. On the inside of each valve (half of the shell), beneath the beak, there are three large teeth and sockets. Each valve also has two scars on the inside that are roughly equal in size: these mark the points where the adductor muscles, which closed the valves, were attached. The living species are commonly known as Venus clams. These mollusks are filter feeders that live in burrows in the mud or sand of the seabed. Like most other burrowing bivalves, they use their siphons to circulate water through the shell, using their gills to filter out food particles and for oxygen and passing the filtered water back to the seabed.

Venus **lived in burrows**; its **siphons** kept it **connected** to the **seawater**.

VENUS VERRUCOSA
This fossilized upper valve shows that the outer surface is strongly ridged with curved ribs, which are crossed by less obvious radial ribs, running from the beak to the edge of the shell.

strong concentric rib

forward-facing beak

radial rib

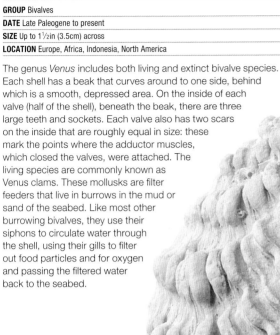

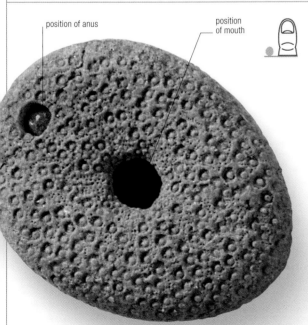

position of anus

position of mouth

Echinocyamus

GROUP Echinoderms
DATE Late Cretaceous to present
SIZE Up to ½in (1.5cm) across
LOCATION Widely distributed in Northern Hemisphere

Echinocyamus is an extremely small echinoid with a rather flattened shape. As with most sea urchin fossils, those of *Echinocyamus* lack the spines and just show the hard "test," which in life contains all of the internal organs. This is roughly egg-shaped, flattened in profile, and has a thick, rounded edge. Both the mouth and the anus are on the underside, the latter smaller than the former and positioned behind it toward the edge of the test.

ECHINOCYAMUS PUSILLUS
This lower (oral) surface has numerous spine bases, and the test has small pores in the ambulacral areas that are less conspicuous than on the upper surface.

Balanus

GROUP Crustaceans
DATE Paleogene to present
SIZE Up to ¾in (2cm) across
LOCATION Worldwide

Members of the genus *Balanus* are commonly known as barnacles; this type, both fossilized and living, has a roughly conical form. Barnacles tend to be irregular in shape, especially when crowded together, as they often are. The walls of the shell are angled inward and are composed of four to six rigid plates, cemented together.

irregular ridges

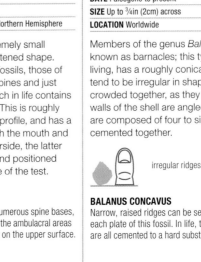

BALANUS CONCAVUS
Narrow, raised ridges can be seen on each plate of this fossil. In life, the plates are all cemented to a hard substrate.

QUATERNARY
VERTEBRATES

Throughout the Pleistocene epoch, ice caps advanced and retreated around the globe, forcing vertebrates to move and adapt. Many species succumbed to extinction after the end of the last ice age about 10,000 years ago. Since then, modern humans have had a dramatic impact on the world's fauna.

Mammals abounded beyond the edges of the ice caps during the Pleistocene. Large animals were better able to conserve body heat and survive extreme cold than small animals. In addition, several northern species, such as woolly mammoths and woolly rhinos, sported thick fur coats that kept them warm. The ranges of many animals expanded and contracted according to the changing climate and vegetation. For example, reindeer moved southward from the Arctic to southern Europe during glacial periods and, contrastingly, hippos moved northward from Africa to central Europe during interglacials.

The rise and fall of megafauna

Giant mammals were ubiquitous during the Pleistocene. Elephantlike mastodons and car-sized armadillolike glyptodonts lumbered around North America, while giant ground sloths, which were up to 20ft (6m) tall and weighed 2,200 pounds (a metric ton), inhabited both North and South America. In Eurasia, there were giant rhinoceroses and a species of giant deer, *Megaloceros giganteus*, whose antlers measured over 11½ft (3.5m) from tip to tip. Australia was home to giant marsupials, including *Diprotodon*, a hippopotamus-sized relative of the wombat and koala, and *Procoptodon*, the largest kangaroo to have ever lived. Both of these were prey to the marsupial lion *Thylacoleo*. A combination of factors is likely to

have contributed to the extinction of much of the prehistoric megafauna at the end of the last ice age. These included climate change, which caused many areas to dry out, and pressure imposed by the spread of modern humans, who hunted many of the species and also learned to use fire to clear vegetation and for cooking.

LITTLE AND LARGE
Giant beavers (*Castoroides*) inhabited North America during the Pleistocene. They were up to 8ft (2.5m) long and weighed as much as 220lb (100kg), dwarfing today's North American beaver (*Castor canadensis*), which is 39in (1m) long and weighs up to 77lb (35kg).

CASTOROIDES

NORTH AMERICAN BEAVER

GROUPS OVERVIEW

Much of the world's existing vertebrate fauna appeared during the Quaternary, but many other species became extinct. Many seabirds, for example, died out, possibly due to competition with newly evolved marine mammals. The Hominoidea included *Gigantopithecus*, which, at 9¾ft (3m) in height, was the largest ape to have ever existed.

AVES
Birds grew large during the Pleistocene too. *Aepyornis*, the ostrichlike "elephant bird" of Madagascar, weighed nearly 880lb (400kg), whereas the flightless moas of New Zealand reached a height of about 12ft (3.6m). They were preyed upon by the huge Haast's eagle, which weighed up to 33lb (15kg).

CARNIVORES
The largest terrestrial carnivore during the Pleistocene was the giant short-faced bear *Arctodus simus*, which weighed up to 2,000lb (900kg). Among the cat family, lions that were larger than today's African lions were found worldwide, and there was also the saber-toothed cat *Smilodon*. The dire wolf, a close relative of today's gray wolf, lived in North America.

EQUIDS
Horses proliferated in North America during the Pleistocene. Using the land bridge to eastern Asia, they spread to Eurasia and Africa, where nine *Equus* species survive today, including two species of wild asses, three species of zebras, the kiang, and Przewalski's horse. No wild horses remained in North America after the end of the Pleistocene mass extinction.

PROBOSCIDS
Mammoths and mastodons lived in North America and Eurasia during the Pleistocene but died out after the last ice age. Not all proboscids were large: dwarf elephants are known from several Mediterranean islands. The last woolly mammoths lived as recently as 3,600 years ago on Wrangel Island in the Arctic Ocean.

Megalania

GROUP Squamates
DATE Pleistocene
SIZE 26ft (8m) long
LOCATION Australia

Megalania was a heavily built, giant monitor lizard that resembled the Komodo dragon. It was so impressively large and powerful that it was able to eat any of the large ice-age mammals of Australia, including the rhinoceros-sized wombats known as diprotodonts and the giant kangaroos. Its great size may have been due partly to the fact that there were few large mammalian predators in Australia. Only the much smaller marsupial lion, *Thylacoleo*, would have offered *Megalania* some competition. *Megalania* vanished about 40,000 years ago, at about the time that the first humans arrived in Australia.

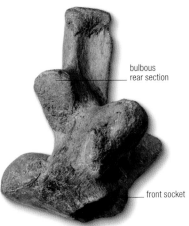

bulbous rear section

front socket

LARGE VERTEBRA
Megalania's spine was made up of 29 presacral vertebrae. Each had a strongly socketed front and a bulbous rear section.

KOMODO DRAGON

The largest living lizard is the Komodo dragon, which lives only on the Indonesian island of Komodo. This impressive creature can be as long as 9³⁄₄ft (3m) and can weigh up to 155lb (70kg). It eats carrion and hunts large prey, such as deer and pigs, by ambushing them along game trails. Even if the catch escapes its clutches, the Komodo dragon's poisonous saliva eventually kills the animal.

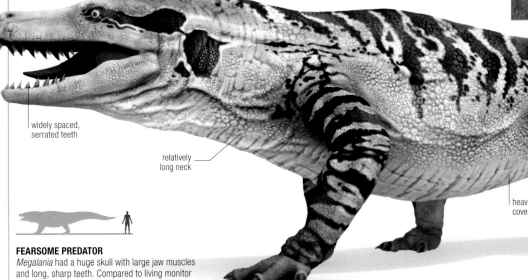

long, muscular tail

widely spaced, serrated teeth

relatively long neck

heavily built body covered in scales

four-toed feet with sharp claws

FEARSOME PREDATOR
Megalania had a huge skull with large jaw muscles and long, sharp teeth. Compared to living monitor lizards, it was bulky, weighing 1,270lb (575kg).

Trilophosuchus

GROUP Crocodylomorphs
DATE Early Miocene
SIZE 5ft (1.5m) long
LOCATION Australia

Compared with modern crocodilians, *Trilophosuchus* had a short snout, very large eyes, and three long ridges of bone running along the top of its skull between the eyes. *Trilophosuchus* fossils have been found at Riversleigh in Queensland, Australia, where fossil beds have revealed the beautifully preserved remains of the

inhabitants of a thick forest. Excavations show a rich variety of mammals, which could have provided *Trilophosuchus* with ample prey. *Trilophosuchus* was a member of the crocodilian family Mekosuchinae, which first appeared in Australia in the Eocene. Members of this family were the dominant crocodilians in much of the southwest Pacific throughout most of the Cenozoic, until saltwater crocodiles arrived from Asia. The Mekosuchinae vanished at the end of the Pleistocene, probably as a result of human hunting or the destruction of their habitat or food source.

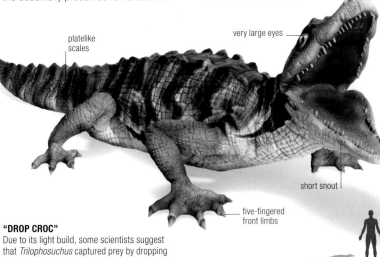

platelike scales

very large eyes

short snout

five-fingered front limbs

"DROP CROC"
Due to its light build, some scientists suggest that *Trilophosuchus* captured prey by dropping out of trees—hence the nickname "drop croc."

Aepyornis

GROUP Theropods
DATE Early Pleistocene to Holocene
SIZE 9³⁄₄ft (3m) tall
LOCATION Madagascar

A giant, flightless, ostrichlike animal, *Aepyornis* was heavily built, weighing 880lbs (400kg), making it the heaviest bird that ever lived. It fed on a wide variety of fruits, nuts, and seeds. *Aepyornis* coexisted with humans for thousands of years, but became extinct soon after Europeans arrived in Madagascar. The reason for its extinction is unclear, but remains reveal that humans butchered *Aepyornis*, so overhunting may be blamed.

ankle joint

FOOT FOSSIL
Aepyornis had relatively short but powerful legs with large, three-toed feet. This massive metatarsus was joined to the ankle at the top and has toe articulations at the bottom.

FOSSILIZED EGG

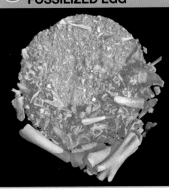

The huge fossilized eggs of *Aepyornis* are impressive by their sheer size—they have a circumference of more than 3ft 3in (1m) and a width of 13¹⁄₂in (34cm). Until recently, there was no way to see inside an intact egg without destroying its contents. However, high-resolution X-ray computed tomography (HRXCT) can now be used to photograph and digitally analyze the contents of the eggs. Images reveal nearly every bone of the skeleton as well as anatomical features that are visible only in embryos, disappearing as the bird became adult.

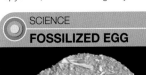

Dinornis

GROUP Theropods
DATE Pleistocene to Holocene
SIZE 11¾ft (3.6m) tall
LOCATION New Zealand

Dinornis was a moa, one of a group of giant flightless birds that lived in New Zealand from the early Pleistocene until they were wiped out by the Maori in the 16th century. It was the tallest bird that ever lived and, weighing in at 620lb (280kg), *Dinornis* resembled a gigantic, heavily built emu. Despite the relatively small size of its head and bill, it would have been able to eat almost any type of fruit, seed, or other plant food. It was covered in reddish-brown, hairlike feathers, except on its leathery feet and legs and much of its head and neck. There were at least six genera of moas living in New Zealand when humans arrived. Their extinction was caused by human hunting combined with habitat loss as the Maori slashed and burned the forests.

short bill

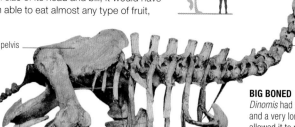

pelvis
long neck
long, heavily built legs
sternum
three-toed clawed feet

BIG BONED
Dinornis had massive bones and a very long neck that allowed it to reach a variety of vegetation. Its head and bill were small compared to other large flightless birds.

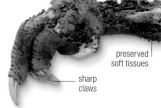

preserved soft tissues
sharp claws

MUMMIFIED FOOT
This mummified foot of *Megalapteryx*, a close moa relative of *Dinornis*, has remarkably well-preserved soft tissues, including traces of scaly skin.

Toxodon

GROUP Placental mammals
DATE Late Miocene to Early Holocene
SIZE 8¾ft (2.7m) long
LOCATION South America

Fossils of this rhinoceroslike mammal have been found in Late Miocene to Early Holocene deposits in Argentina and elsewhere in South America. It was a member of an extinct group of South American mammals known as the notoungulates. However, it was the size of a modern hippopotamus and looked remarkably like the rhinos of Eurasia and North America, making it a classic example of convergent evolution. *Toxodon* had a short proboscis, a little like that of a tapir, and a robust skeleton with short, stumpy limbs that had three toes on each foot. Its teeth were bow-shaped and suited for chewing a range of leafy vegetation. About 3.5 million years ago, the Panamanian land bridge connected North and South America, prompting the migration of many mammals between the two continents. *Toxodon* remained in South America and survived the arrival of northern mammals because no direct competitors made the journey. However, numerous remains embedded with arrowheads suggest that humans eventually hunted *Toxodon* to extinction.

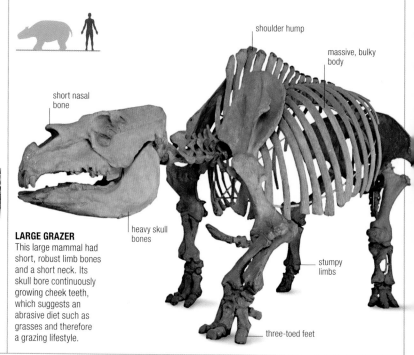

shoulder hump
massive, bulky body
short nasal bone
heavy skull bones
stumpy limbs
three-toed feet

LARGE GRAZER
This large mammal had short, robust limb bones and a short neck. Its skull bore continuously growing cheek teeth, which suggests an abrasive diet such as grasses and therefore a grazing lifestyle.

Teratornis

GROUP Theropods
DATE Pleistocene
SIZE 30in (75cm) long
LOCATION North America

This huge bird is known from the vast number of bones preserved at La Brea tar pits in California (see p.410) but is also known from Pleistocene deposits in Nevada, Arizona, and Florida. It had stout legs and feet strong enough to hold down prey as it tore its food apart. Although *Teratornis* probably ate carrion, like vultures and condors, it also had features that suggest it was good at catching fish in the water. Large numbers of these birds may have been trapped in the tar pits at La Brea not only because they were attracted to carrion, but also because they may have tried to fish in the overlying film of water or drink from the edge of the tar pools. *Teratornis* became extinct 10,000 years ago, perhaps due to the loss of carrion, as many large American mammals vanished at the end of the last ice age.

CONDORLIKE
Although *Teratornis* resembled the giant condor, it was from a separate family. It was larger than the living Andean condor and had a wingspan of up to 13ft (4m).

HOW TERATORNIS FLEW

Like most modern condors and vultures, *Teratornis* depended on soaring on thermal updrafts so that it could cover large areas in search of carcasses. Modern condors may fly several hundred miles a day in this way. Most soaring birds have long, relatively large wings and small bodies that give them a low wing loading, enabling them to soar with minimal flapping. The wing dimensions of *Teratornis* are more like those of the California condor, suggesting that it could take off from the ground by jumping up with a few simple flaps of its wings, without needing to flap into the wind from a high point, as some vultures and condors must in order to takeoff.

Glyptodon

GROUP Placental mammals
DATE Late Pliocene to Pleistocene
SIZE 8¼ft (2.5m) long
LOCATION North America, South America

Originating in South America, *Glyptodon* was a giant relative of the armadillo that migrated to southern North America when the Panamanian land bridge appeared 3.5 million years ago. It was a well-protected animal largely covered in skin-covered bones known as osteoderms, which served as body armor. Even the tail was armored, being constructed of a series of tapering bony rings made of osteoderms.

Glyptodon vanished at the end of the last ice age due to the changing climate and possible human hunting, although no direct evidence of human attacks on the genus has ever been found.

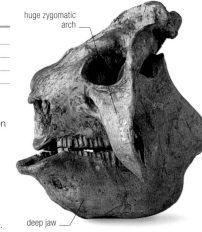

huge zygomatic arch

deep jaw

LONG IN THE TOOTH
The flat-crowned teeth grew continuously, so *Glyptodon* could eat a wide variety of tough vegetation, including gritty grasses.

WELL ARMORED
Glyptodon's shell was composed of over a thousand little tiles or osteoderms, each of which was about 1in (2.5cm) thick.

huge domed shell

short face

large, powerful claws

DOMED GIANT
Glyptodon weighed at least 1.1 ton (1 tonne) and had a huge domed shell that made this creature about the shape and size of a Volkswagen Beetle.

LIVING RELATIVE
ARMADILLO

Armadillos are placental mammals that bear live young, which are nourished in the womb by a placenta. They have a leathery armored shell and a long, tubular snout. Armadillos are xenarthrans, related to sloths and anteaters. There are 10 genera and 20 species known throughout Latin America as far north as Nebraska. Their stout legs and claws allow them to dig burrows rapidly, or dig up ants and termites (their main foods) from the nests. Considerably smaller than their extinct cousin *Glyptodon*, their average length is about 30in (75cm), although the giant armadillo can grow to twice that length.

Castoroides

GROUP Placental mammals
DATE Late Pliocene to Pleistocene
SIZE 8¼ft (2.5m)
LOCATION North America

Castoroides was a giant ice-age beaver. Its remains are known from rocks of the late Pliocene and Pleistocene in North America. Fossils have been found primarily around the Great Lakes and in the Midwest, but they have also been dug up in Alaska, Canada, and Florida. *Castoroides* was the length of a modern bear and weighed up to 220lb (100kg). Its tremendous size was not the only difference with modern beavers—it did not have their simple, chisel-like front teeth, or incisors. Instead, they were larger and broader, reaching to about 6in (15cm) long. In addition, *Castoroides*'s tail was longer and narrower than the broad, flat appendage of modern beavers, and its hind limbs were shorter.

LAKE-DWELLERS
Castoroides may not have lived like modern beavers, because its teeth were less efficient at gnawing wood for building dams and lodges.

This **truly giant beaver** had teeth about 6 inches long and **weighed as much as 220 pounds.**

DEADLY CLAWS
Megatherium had long forelimbs that were equipped with impressive claws, used for grasping branches and defending itself against predators.

Megatherium

GROUP	Placental mammals
DATE	Pliocene to Holocene
SIZE	20ft (6m) long
LOCATION	South America

A giant ground sloth, *Megatherium* weighed in at 4.4 tons (4 tonnes). It was about the size and weight of a bull elephant and was covered with long hair. It walked on the sides of its feet with its claws curled inward and could prop itself up on its back legs and massive tail. *Megatherium* vanished at the end of the last ice age, when humans arrived in South America, although there is no direct evidence of human attacks.

HIGH-CROWNED TEETH
Simple peglike teeth made of dentin, with no enamel, were very high-crowned, allowing *Megatherium* to eat a variety of leaves and grasses.

Arctodus

GROUP	Placental mammals
DATE	Pleistocene
SIZE	11¼ft (3.4m) long
LOCATION	North and Central America

This giant bear had a huge body that was covered with long hair and a relatively short skull with very powerful jaws that were capable of crushing its prey—bones and all. Although *Arctodus* was very large, it had relatively long, slender limbs for a bear and may have been a fast runner, ambushing its prey with a quick dash from cover or covering long distances to wear down its prey. It could also have been a scavenger and thief, easily able to drive other hunters away from their kills. *Arctodus* vanished at the end of the last ice age due to the changing climate and possible competition from humans, although it was so large and fierce that humans probably did not try to hunt it down.

relatively short skull

very powerful jaws

relatively long, slender limbs

sharp claws

GRACILE SKELETON
Even though it was one of the largest land predators of its time, *Arctodus* had a gracile skeleton, which suggested it was an efficient runner.

2,000 pounds The approximate **weight** of *Arctodus*, one of the **largest land carnivores** that ever lived.

Canis

GROUP	Placental mammals
DATE	Late Pleistocene
SIZE	5ft (1.5m) long
LOCATION	Canada, USA, Mexico

Canis dirus is the scientific name for the dire wolf, an extinct North American species. Its fossils have been found in rocks dating from the Late Pleistocene. It was the size of a modern gray wolf but was much heavier, weighing about 176lb (80kg). It is thought that dire wolves were scavengers, living like today's hyenas. At that time, there were no hyenas or other mammalian scavengers on the continent. Dire wolves died out at the end of the last ice age, probably because climate change and the arrival of humans in North America wiped out the wolf's food sources.

KEY SITE

LA BREA TAR PITS

Dire wolf fossils have been found at La Brea tar pits in California, the world's richest site for Pleistocene fossils. Tar from oil deposits, disguised by a layer of water, has trapped many animals over the past 38,000 years. Tens of thousands of fossils have been recovered, representing about 660 species.

COMPACT STRENGTH
About the size of a modern gray wolf, the dire wolf had a shorter, broader head, with sizable teeth for crushing bone and shorter, more powerful limbs.

Crocuta

GROUP	Placental mammals
DATE	Miocene to present
SIZE	4¼ft (1.3m) long
LOCATION	Africa, Europe, Asia

Crocuta is the genus that includes the living spotted hyena, an African scavenger and hunter that is well adapted to running long distances in pursuit of prey. The Pleistocene subspecies, *C. crocuta spelaea*, known as the "cave hyena," was considerably larger and more robust than its modern relative. It was apparently adapted for the colder climate along the edges of glaciers that covered Eurasia at that time. Although it is thought to have used caves as dens, it probably hunted and scavenged in open grasslands. First appearing in India at the beginning of the Pleistocene, the cave hyena had spread east to China, and also west to Europe and Africa by the Middle Pleistocene. Eventually, most members of Crocuta died out, leaving only the lineages that still survive in Africa today.

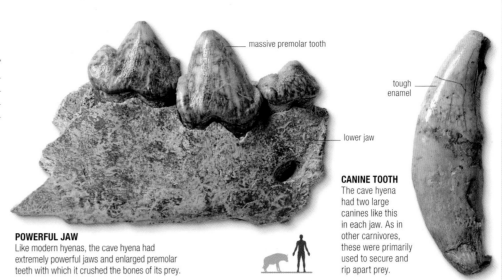

massive premolar tooth

lower jaw

tough enamel

CANINE TOOTH
The cave hyena had two large canines like this in each jaw. As in other carnivores, these were primarily used to secure and rip apart prey.

POWERFUL JAW
Like modern hyenas, the cave hyena had extremely powerful jaws and enlarged premolar teeth with which it crushed the bones of its prey.

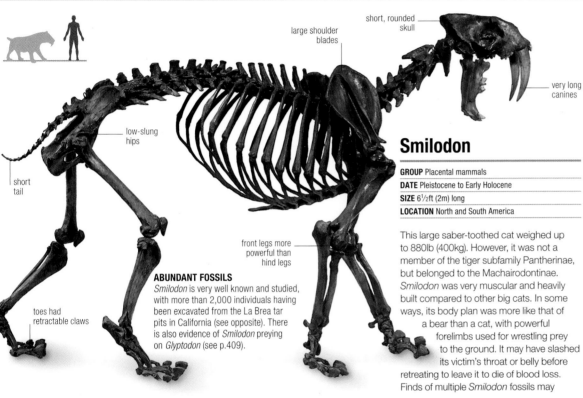

short, rounded skull

large shoulder blades

very long canines

low-slung hips

short tail

front legs more powerful than hind legs

ABUNDANT FOSSILS
Smilodon is very well known and studied, with more than 2,000 individuals having been excavated from the La Brea tar pits in California (see opposite). There is also evidence of *Smilodon* preying on *Glyptodon* (see p.409).

toes had retractable claws

Smilodon

GROUP	Placental mammals
DATE	Pleistocene to Early Holocene
SIZE	6½ft (2m) long
LOCATION	North and South America

This large saber-toothed cat weighed up to 880lb (400kg). However, it was not a member of the tiger subfamily Pantherinae, but belonged to the Machairodontinae. *Smilodon* was very muscular and heavily built compared to other big cats. In some ways, its body plan was more like that of a bear than a cat, with powerful forelimbs used for wrestling prey to the ground. It may have slashed its victim's throat or belly before retreating to leave it to die of blood loss. Finds of multiple *Smilodon* fossils may

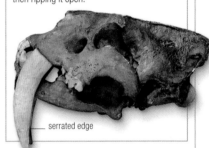

ANATOMY
SABER TEETH

Mammals exhibit a variety of specialized teeth, and the most spectacular are perhaps the long, bladelike canines. *Smilodon*'s canines were particularly impressive. They were very long, narrow recurved blades with serrated front and back edges that were used for stabbing prey in the throat or belly, then ripping it open.

serrated edge

indicate that this cat, like modern-day lions, lived and hunted in large social groups. Like many other large mammals, *Smilodon* vanished at the end of the last ice age, probably because of the changing climate, the disappearance of large herbivores, and possible competition with humans.

PREDATORY CARNIVORE
A formidable predator, *Smilodon* would probably have eaten a wide variety of large prey, including deer, horses, and mammoths.

RIDGED TOOTH
Mammoth teeth consisted of a series of enamel plates held together by a dental cement. Teeth erupted in the back of the jaw and moved forward as they wore, to be replaced by other teeth from behind.

tall, domed head

fatty lump used as energy store

trunk covered in hair

long, curved tusks

WOOLLY MAMMOTH
Mammuthus stood to about 16½ft (5m) at the shoulder and weighed 8.8 tons (8 tonnes). The species shown here, the woolly mammoth, was much like modern elephants in most features, except that it had smaller ears, was covered in thick hair, and had longer, curved tusks.

thick, flexible trunk

columnlike legs

This genus **became extinct** in the Late Pleistocene, but one **dwarf mammoth survived** on Wrangel Island near Alaska **until 3,600 years ago.**

back sloping from shoulders to hips

tusks longer than in modern elephants

large body cavity

long chin

very sturdy leg bones

HUGE TUSKS
Mammoths probably used their long, curved tusks to scrape away snow and ice from the ground when feeding, as well as for protection and in dominance rituals.

four-toed feet

hair grew up to 35in (90cm) long

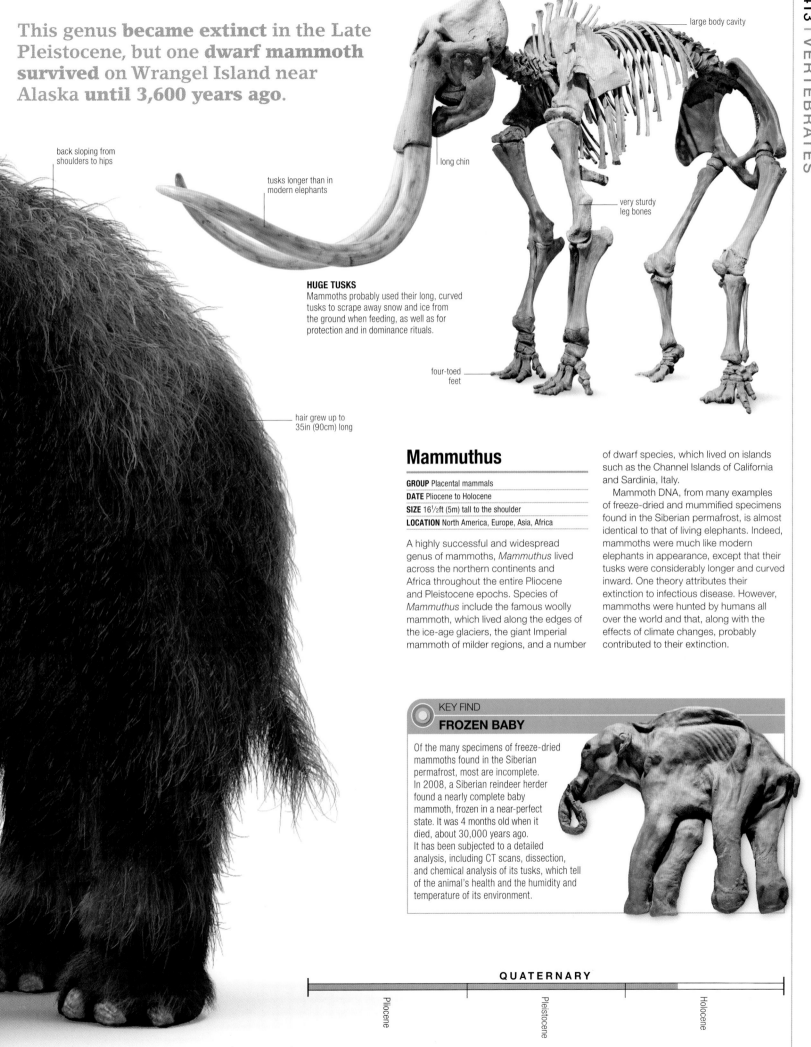

Mammuthus

GROUP Placental mammals
DATE Pliocene to Holocene
SIZE 16½ft (5m) tall to the shoulder
LOCATION North America, Europe, Asia, Africa

A highly successful and widespread genus of mammoths, *Mammuthus* lived across the northern continents and Africa throughout the entire Pliocene and Pleistocene epochs. Species of *Mammuthus* include the famous woolly mammoth, which lived along the edges of the ice-age glaciers, the giant Imperial mammoth of milder regions, and a number of dwarf species, which lived on islands such as the Channel Islands of California and Sardinia, Italy.

Mammoth DNA, from many examples of freeze-dried and mummified specimens found in the Siberian permafrost, is almost identical to that of living elephants. Indeed, mammoths were much like modern elephants in appearance, except that their tusks were considerably longer and curved inward. One theory attributes their extinction to infectious disease. However, mammoths were hunted by humans all over the world and that, along with the effects of climate changes, probably contributed to their extinction.

KEY FIND

FROZEN BABY

Of the many specimens of freeze-dried mammoths found in the Siberian permafrost, most are incomplete. In 2008, a Siberian reindeer herder found a nearly complete baby mammoth, frozen in a near-perfect state. It was 4 months old when it died, about 30,000 years ago. It has been subjected to a detailed analysis, including CT scans, dissection, and chemical analysis of its tusks, which tell of the animal's health and the humidity and temperature of its environment.

QUATERNARY

Pliocene

Pleistocene

Holocene

Mammut

GROUP Placental mammals
DATE Pliocene to Pleistocene
SIZE 9¾ft (3m) tall
LOCATION North America, Europe, Asia

Mammut is perhaps better known as the mastodon, an extinct cousin of the elephants. It weighed up to 5.5 tons (5 tonnes) and had a long, flat skull with slightly curved tusks and a back that sloped down from the hump over the shoulders. It was covered in long, shaggy hair and was well adapted to living in dense conifer forests. Preserved stomach contents show that it ate spruce and other low-fiber vegetation. It appears that *Mammut* did not vanish when humans arrived in the Americas 10,000 years ago. Small populations survived until about 6,000 years ago in places like Utah and Michigan.

VERTEBRA
Mammut fossils, like this huge vertebra, come from Pliocene and Pleistocene deposits of North America and Eurasia.

centrum

Y-shaped neural arch

cusps arranged in parallel rows

MAMMUT MOLARS
Mastodon means "breast tooth." This refers to the shape of the *Mammut's* molars, which had conical cusps shaped like breasts.

Stegomastodon

GROUP Placental mammals
DATE Pliocene to Early Pleistocene
SIZE 9¾ft (3m) tall
LOCATION North America, South America

Stegomastodon was a gomphothere—an extinct elephantlike proboscidean mammal family. It weighed over 6.6 tons (6 tonnes). It had a shortened skull and jaw compared to other gomphotheres, similar to those of mammoths and elephants, with two long, incurved upper tusks that grew up to 11½ft (3.5m) long. Its huge molars had a series of cusps that looked like three-leaf clovers when worn down. This form of tooth was well adapted for eating grass and other gritty vegetation. *Stegomastodon* originated on the grassy plains of North America. Several species crossed the Panamanian land bridge to South America.

Equus

GROUP Placental mammals
DATE Pliocene to present
SIZE 8¼ft (2.5m) tall
LOCATION Worldwide

Equus is the genus that includes today's horses, zebras, and donkeys. It first appears in the fossil record in the early Pliocene and still flourishes today in domestication, but most wild species of *Equus*—especially asses, onagers, and Przewalski's horse—are rare. In North America, *Equus* vanished with other large Pleistocene mammals about 10,000 years ago, possibly due to hunting by humans. Fossils from North America show a diverse range of species—for example, dwarf and giant forms, horses with robust legs, and others that were stilt-legged. These North American fossil species appear to be related to African zebras and Eurasian asses, suggesting that horses once had a greater geographic spread.

large eye sockets set back in skull

long, slim head

deep bottom jaw

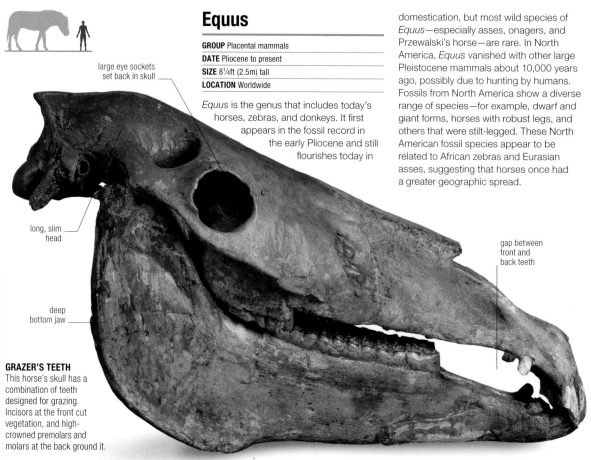

gap between front and back teeth

GRAZER'S TEETH
This horse's skull has a combination of teeth designed for grazing. Incisors at the front cut vegetation, and high-crowned premolars and molars at the back ground it.

Przewalski's horse is the species from which all domestic horses are descended. It looks like a short-legged version of a domesticated horse, with a brown coat, white belly, and dark brown mane. Originally found in the desert steppes of Mongolia, it was wiped out in the wild, surviving only in zoos, but has been reintroduced to its original habitat. There are about 1,500 alive today, both in the wild and in captivity.

Coelodonta

GROUP Placental mammals
DATE Pliocene to Pleistocene
SIZE 12¼ft (3.7m) long
LOCATION Europe, Asia

This ice-age woolly rhinoceros was about the size of a large white rhino and bore a thick coat of long, shaggy hair. It had a pair of horns that were oval in cross-section and curved backward, which it probably used to shift snow while foraging. *Coelodonta's* cheek teeth had the characteristic π-shaped pattern of crests found in all rhinos, but were also highly ornate, with many additional ridges that increased the grinding surface. Many lines of evidence show that woolly rhinos were mostly grazers, although they could probably eat almost any vegetation.

Coelodonta vanished about 10,000 years ago at the end of the last ice age, probably due to the retreat of its glacial habitat caused by climate changes. Having coexisted with humans for thousands of years, it is unlikely to have been hunted to extinction.

WOOLLY RHINO
Coelodonta lived across Eurasia during the ice ages of the Pliestocene. Several complete skeletons have been found, as well as mummified specimens.

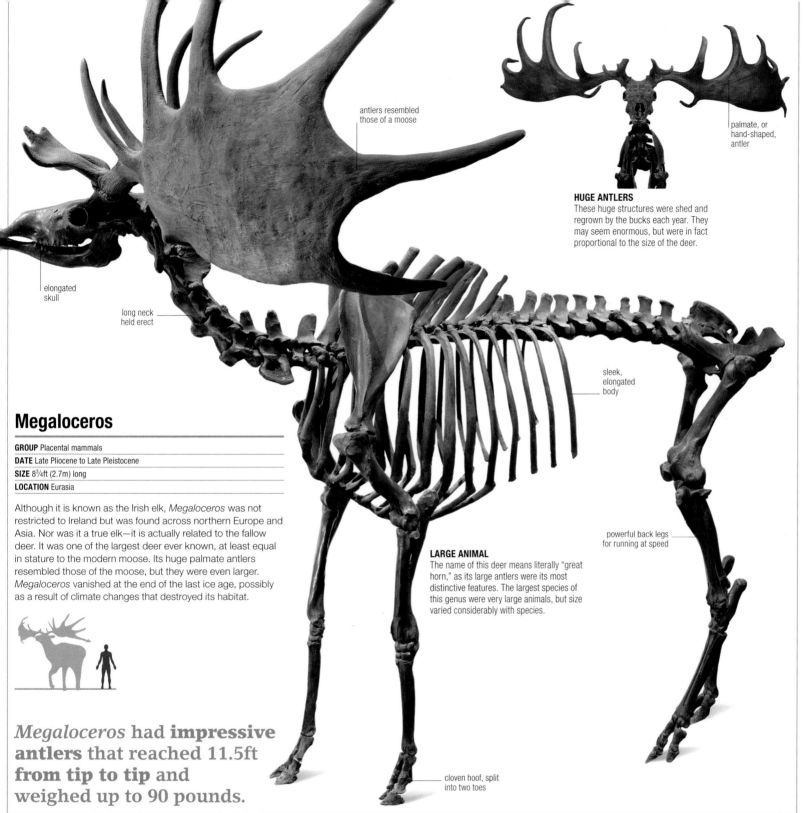

antlers resembled those of a moose

HUGE ANTLERS
These huge structures were shed and regrown by the bucks each year. They may seem enormous, but were in fact proportional to the size of the deer.

palmate, or hand-shaped, antler

elongated skull

long neck held erect

sleek, elongated body

powerful back legs for running at speed

LARGE ANIMAL
The name of this deer means literally "great horn," as its large antlers were its most distinctive features. The largest species of this genus were very large animals, but size varied considerably with species.

cloven hoof, split into two toes

Megaloceros

GROUP Placental mammals

DATE Late Pliocene to Late Pleistocene

SIZE 8¾ft (2.7m) long

LOCATION Eurasia

Although it is known as the Irish elk, *Megaloceros* was not restricted to Ireland but was found across northern Europe and Asia. Nor was it a true elk—it is actually related to the fallow deer. It was one of the largest deer ever known, at least equal in stature to the modern moose. Its huge palmate antlers resembled those of the moose, but they were even larger. *Megaloceros* vanished at the end of the last ice age, possibly as a result of climate changes that destroyed its habitat.

Megaloceros had **impressive antlers** that reached 11.5ft **from tip to tip** and weighed up to 90 pounds.

Gigantopithecus

GROUP Placental mammals

DATE Miocene to Pleistocene

SIZE 9¾ft (3m) tall

LOCATION Asia

No complete *Gigantopithecus* skeletons or skulls have been found, and most of what we know about this creature comes from its teeth and jaw bones. It was a huge ape that lived between 6–9 million and 300,000 years ago. It was built like a gorilla but was much bigger, weighing 1,190lb (540kg). It has been speculated that the Bigfoot and Yeti myths may have been based on *Gigantopithecus*. However, there is no evidence to suggest that these animals survived beyond about 300,000 years ago—many thousands of years before humans lived in the area.

LIVING RELATIVE

ORANGUTAN

Gigantopithecus's closest living relative, the orangutan is the largest tree-dwelling mammal alive, weighing up to 260lb (120kg). Orangutans have very long arms, very short legs, and are covered with a coat of long, reddish-orange hair. They are only found in the vanishing forests of Borneo and Sumatra, but fossils have been found across Southeast Asia. Orangutan means "person of the forest" in the Malay language, and indeed they look quite humanlike at times. They spend all their lives up in trees, making nests of leaves and eating mostly fruit. According to recent studies, they are the most intelligent of nonhuman primates, able to solve problems that baffle chimpanzees.

large molars

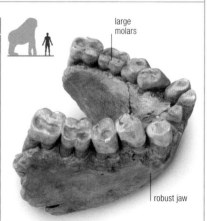

robust jaw

TEETH AND JAW BONES
Gigantopithecus's jaw bones were deep and robust, with low-crowned molars with thick enamel. Tooth-abrasion patterns suggest a diet of bamboo grass.

QUATERNARY
HUMANS

Differences between chimpanzee (*Pan* sp.) and modern human (*Homo sapiens*) DNA suggest that our lineages diverged in the Late Miocene or Early Pliocene. DNA evidence alone cannot narrow down this timeframe, but fossil primates that were probably ancestral to modern humans date from about 7–6 mya.

Many lines of evidence contribute to our understanding of human evolution. The study of fossil anatomy provides evidence of adaptive pressures faced by different species, while archaeology offers insights into their cognition and behavior. DNA studies of modern human and primate populations, and the extraction of ancient DNA from fossils, allows us to assess similarities and differences and estimate the time period over which such differences accumulated. Each type of evidence has strengths and weaknesses; to fully understand human origins, researchers must understand how it all fits together.

The human family

Fossil finds are usually highly fragmented and essentially offer limited snapshots of what was a very gradual and geographically variable process. It is not always clear whether two fossils differ because they represent different species, or variation within a species—for example, due to sex, age, or variation across space and through time. As a result, it is difficult to reconstruct ancestor/descendant relationships with certainty. Hominin evolution was not a straightforward linear process, but involved different adaptations on the part of multiple genera and species, only some of which proved successful.

Sole survivors

Given this remarkable range of adaptations, a big puzzle about human evolution is why only one species of the genus *Homo*— our own, *H. sapiens*—is still alive today. As recently as 50,000 years ago, at least three species— ourselves, Neanderthals, and the mysterious Denisovans— are known to have coexisted, interacted, and even interbred. It remains unclear why only *H. sapiens* survived, while only traces of some of our extinct relatives still live on in our DNA.

BIPEDALISM
Humans and other primates differ in our anatomies because we are adapted to different ways of moving around. Other primates are quadrupeds and/or climbers, but modern humans walk upright on two legs.

FORAMEN MAGNUM
Human heads balance over the spine, not in front of it, so the foramen magnum ("big hole") where the spinal cord connects to the brain, is positioned more centrally than in other primates.

PELVIS AND THIGH BONES
A lower, broader pelvis centers the human torso over the hips, providing stability. Thigh bones angle inward to support body weight and improve balance.

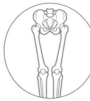

FEET
Primate's feet resemble their hands, with divergent big toes to aid climbing. In humans, all the toes are in line, and the longer heel allows a better transfer of weight while walking.

GORILLA   **HUMAN**

There is little agreement about the precise species and genera represented in the hominin fossil record, but they fall into four general groups. Each represents a burst of diversification that occurred when adaptations to the environment led to new opportunities and challenges. New species continued to appear as different environments selected for different anatomy and behavior.

BASAL HOMININS
Several fossil genera and species date to around the time of the divergence of the chimpanzee and human lineages. Although quite chimpanzeelike, anatomical changes in the pelvis and feet in particular tell us they probably walked on two legs, although they still retained some skeletal adaptations to climbing.

ARCHAIC HOMININS
These bipedal hominins also resembled chimpanzees in many ways but had slightly larger brains. Paranthropines, however, had distinctive wide faces and extremely large jaws and teeth that might relate to different diets. Some species—probably more than one, and from multiple genera—made simple stone tools.

EARLY *HOMO*
These hominins represent a significant change in human evolution; taller, heavier, and with larger brains, these species had the long legs and shorter arms characteristic of modern humans. Evidence of complex stone tools, fire, and hunting may explain their ability to spread into new territories in Eurasia, as well as Africa.

LATE *HOMO*
A number of later *Homo* species evolved across Africa and Eurasia: the result was a range of species so closely related that some are known to have interbred. All had big brains and made complex tools and objects. By about 40,000 ya, only *H. sapiens* was left to face a deteriorating climate and the last ice age.

Sahelanthropus tchadensis

DATES 7.2–6.8 mya

BRAIN SIZE 19–23 cubic in (320–380 cubic cm)

HEIGHT Unknown

FINDS Single skull, pieces of jaw and teeth

LOCATION Djurab Desert, Chad

Sahelanthropus lived around the time of the last common ancestor of humans and other apes. It is unclear whether *Sahelanthropus* lived before or after the human and ape lineages separated and to which lineage it belonged. Only a skull and teeth have been found, and while there are indications that it may have been bipedal, a lack of bones from the body means anthropologists cannot be certain. If *Sahelanthropus* was proved to be bipedal, that would place it more clearly on the human lineage.

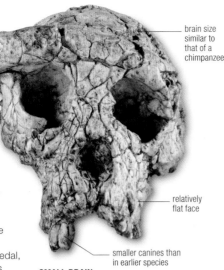

brain size similar to that of a chimpanzee

relatively flat face

smaller canines than in earlier species

SMALL BRAIN
Sahelanthropus's skull is small and relatively long and low, like those of apes. However, its face is shorter and more vertical, and its skull was balanced on top of its spine, suggesting that it may have walked on two legs.

Orrorin tugenensis

DATES 6–5.7 mya

BRAIN SIZE Unknown

HEIGHT Unknown

FINDS Jaw bone, teeth, fragmented arm and thigh bones, finger bone

LOCATION Tugen Hills, Kenya

The fragmentary remains of *Orrorin tugenensis* suggest that this species was closely related to the common ancestor of both humans and chimpanzees, and it may be the oldest known biped. Its features exhibit a mixture of apelike and humanlike traits. Like humans, *Orrorin* had teeth covered in thick enamel, although some of the teeth are more apelike in shape and size. The shape of its thigh bones suggests that the hip joints carried most of its body weight, and thus that it was bipedal. Also, marks on the bone that indicate where the hip muscles

attached are located in similar places to those of later bipedal hominins. However, features of its arm bones suggest that it may also have climbed trees.

TUGEN HILLS
At least five *Orrorin* individuals were found in the Tugen Hills of Kenya in 2000 by Brigitte Senut and Martin Pickford. *Orrorin* is Tugen for "original man."

Ardipithecus ramidus

DATES 4.5–4.3 mya

BRAIN SIZE Unknown

HEIGHT Unknown

FINDS Skull, jaw, teeth, and arm bone fragments

LOCATION Afar, Ethiopia; Tabarin, Kenya

Ardipithecus ramidus is thought to be an early member of the hominin lineage and would have looked much like a modern chimpanzee. However, several lines of evidence suggest that it was bipedal, and therefore probably an early hominin: its skull seems to have balanced squarely over the spine, and the arrangement of its toes suggests its feet were adapted for walking rather than grasping branches. Its teeth are chimpanzeelike, but with some hominin features: the canines are shaped more like those of later hominins; the front teeth are relatively small, like later hominins; and

the cheek teeth are shaped differently to chimpanzees'. *A. ramidus*'s arms show a mixture of features seen in climbing apes and in bipedal hominins.

JAW BONE
Ardipithecus ramidus's cheek teeth are similar to later australopithecine species and provide some of the strongest evidence that it was a hominin.

Paranthropus boisei

DATES 2.3–1.3 mya

BRAIN SIZE 25–33 cubic in (410–550 cubic cm)

HEIGHT 4ft–4ft 5in (1.2–1.4m)

FINDS Various skulls and fragments of jaws and teeth

LOCATION Various Rift Valley sites in Kenya and Ethiopia

Paranthropus boisei, nicknamed "nutcracker man," was the largest and most specialized of the paranthropines, with the most powerful chewing apparatus, largest cheek teeth, and thickest tooth enamel of any hominin. All the fossils we have of *P. boisei* are skull fragments or teeth, but their similarity to fossils of another paranthropine, *P. robustus*, has led anthropologists to assume that *P. boisei* was probably also rather apelike in overall body shape, despite its larger brain. At some sites, fossils of *P. boisei* are found near stone tools, but anthropologists are not convinced that the paranthropines were toolmakers: in most cases, fossils of early members of the genus *Homo* were recovered from the same sites.

brow ridge sits over wide-set eyes

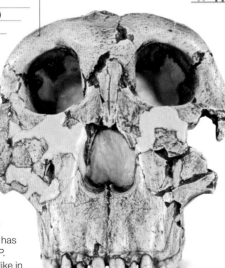

BROAD FACE
The skull of *Paranthropus boisei* was specially adapted for powerful chewing, with flared cheek bones creating a wider face. The jaw was also very strongly built, with large cheek teeth but smaller front teeth and canines.

Australopithecus afarensis

DATES 3.7–3 mya

BRAIN SIZE 23–29 cubic in (380–485 cubic cm)

HEIGHT 3ft 5in–5ft (1.05–1.51m)

FINDS Several variably complete adult, juvenile, and infant skeletons, and many other fragments

LOCATION Hadar, Ethiopia, and other sites in Ethiopia, Kenya, and Tanzania

This species is one of the best-known early hominins, with some quite complete specimens, such as "Lucy," giving us a good idea of how it looked and behaved. Surprising many, this species shows that walking upright came before brain expansion. Although *A. afarensis* was mainly bipedal, it probably habitually climbed trees. Its brain is not much bigger than that of a modern chimpanzee, but cutmarks on bones associated with the species suggest it may have used stone tools. Like other australopithecines, *A. afarensis* was smaller than modern humans, and like apes, probably matured faster and died younger. Males were a lot larger than females, suggesting that they competed for female attention.

"crest" at top of skull supported chewing muscles

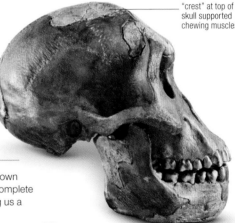

RECONSTRUCTED SKULL
This skull has been built using the fragments of Lucy's skull as a guide. These are identified in the lighter color.

long apelike arms suggest that *A. afarensis* was a good climber

cone-shaped rib cage, as in gorillas and chimpanzees

long forearms relative to upper arms, like chimpanzees

apelike wrist bones

short, wide pelvis shows adaptations to bipedalism

angled thigh bone helped Lucy to balance on its hind legs for long periods

very short thighs suggest it was not exclusively bipedal

shape and depth of notch of knee cap intermediate between ape and human

LUCY
Until the discovery of the Dikika baby in 2000, Lucy was the most complete prehistoric hominin skeleton known, with almost 40 percent of her bones recovered.

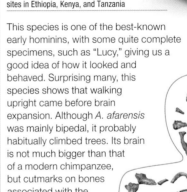

Homo habilis

DATES 2.35–1.64 mya

BRAIN SIZE 30–39 cubic in (500–650 cubic cm)

HEIGHT 3ft 3in–4ft 2in (1–1.3m)

FINDS Various skulls and fragments of cranium; fragments of lower body, including feet and hands

LOCATION Olduvai Gorge, Tanzania; East Turkana, Kenya; Sterkfontein, South Africa

Homo habilis was the first member of the genus *Homo*, to which modern humans belong, and may be ancestral to all later *Homo* species. Its body size and shape was similar to the australopithecines (see p.417), but its face looked more human: with smaller cheek teeth and chewing muscles, its jaw did not project so far forward from its face, although its brain was only slightly larger than those of australopithecines. The species was named *H. habilis*— meaning "handyman"—because some of its fossils are associated with early stone tools. While we now know that tool use was not restricted to the genus *Homo*, it is clear that *H. habilis* was routinely using stone tools to break open the bones of scavenged animal carcasses for marrow, thus enjoying a diet of more animal protein than previously—a characteristic of all later hominins. As more discoveries have been made, it has become clear that the

fossils assigned to *H. habilis* fall into two groups: one with larger brains and teeth, now known as *H. rudolfensis*, and one that remained quite similar to its earlier forebears that should perhaps be called *Australopithecus habilis*.

rounded back of skull

distinct angle between brow ridges and forehead

lower face projects forward less than in earlier species

smaller molars and premolars than earlier species

LARGER BRAINS
The use of stone tools allowed *H. habilis* to access new energy- and protein-rich foods such as buried tubers, nuts, meat, and bone marrow. This caused *H. habilis* and its descendants to develop larger brains, while huge cheek teeth for grinding grasses, seeds, and tough vegetable foods were less essential.

Homo antecessor

DATES 1.2–0.78 mya

BRAIN SIZE 61 cubic in (1,000 cubic cm)

HEIGHT 5ft 3in–6ft (1.6–1.8m)

FINDS Teeth and bone fragments from skull and body

LOCATION Atapuerca, northern Spain

Homo antecessor, the second oldest fossil hominin found in Europe, is known from just a few fossils from two sites in Atapuerca, northern Spain, although fossil footprints from Happisburgh, UK, may have been made by the same species. Like some of the Dmanisi fossils, it is similar to African *H. ergaster*, but it is unknown whether *H. antecessor* survived and evolved into later European species. It might represent instead an early small-scale and unsuccessful colonization of Europe—some of the fossils from Atapuerca show signs of having been butchered by stone tools, suggesting that they might have been short on food in these challenging environments.

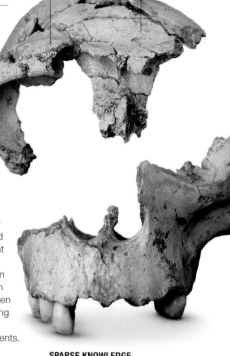

low forehead

primitive skull

SPARSE KNOWLEDGE
Most of the Atapuerca fossils are fragments of the skull, jaw, and teeth; the brain was slightly larger than *Homo ergaster*'s. Little is known of the rest of its body except that it was powerfully built.

thick "keel" of bone runs along top of skull

thick, heavy brow ridge

broad nasal bones with flat bridge to nose

wide cheek bones

large cheek teeth and quite large, "shoveled" front teeth

PEKING MAN
Asian fossils of *H. erectus*, such as this skull from Zhoukoudian in China, have a long, low brain case and generally flatter facial features than earlier hominins.

Homo erectus

DATES 1.7 mya–140,000 ya

brain size 45–79 cubic in (750–1,300 cubic cm)

HEIGHT 5ft 3in–6ft (1.6–1.8m)

FINDS Several relatively complete crania, as well as teeth and jawbones; a few limb bones

LOCATION Various sites in China and Java; African sites are contested

Homo erectus shares a number of features with *H. ergaster*, and the two species are clearly closely related. However, anthropologists disagree to what extent. Some feel that the two species are so similar they should all be considered *H. erectus*, while others prefer to separate the African forms as *H. ergaster* and the Asian specimens as *H. erectus*. It remains unclear as to whether *H. erectus* is the descendant of populations of *H. ergaster* who spread into Asia, or whether

H. erectus evolved in Asia from earlier populations of hominins that left Africa earlier than thought and then later spread back into Africa to become *H. ergaster*. Both species are much more "modern" looking than their ancestors, and the Asian fossils seem to show specialized features— such as a thicker skull and more massive jaws—that gradually disappear later in hominin evolution. Once widespread across Asia, fossils from sites in China and Indonesia suggest that these populations remained isolated without changing much until at least 300,000 and possibly less than 100,000, years ago.

Homo heidelbergensis

DATES 700,000–200,000 ya

BRAIN SIZE 67–85 cubic in (1,100–1,400 cubic cm)

HEIGHT Up to 6ft (1.8m)

FINDS Bone fragments from nearly all parts of the body, including a complete cranium and pelvis

LOCATION Across Europe and Africa

This species was probably the last common ancestor of Neanderthals and modern humans. Fossils display a mixture of features inherited from earlier hominins, as well as more modern or "derived" traits. The fossils from the African record of this time were initially assigned to a variety of species, including *H. rhodesiensis*. However, as a group, they are so similar to the European *H. heidelbergensis* that many anthropologists now consider them

to be members of the same widespread species. Individuals were tall and well muscled, with thick, heavy limb bones, suggesting that they were engaging in tough physical activity. Marks on animal bones found with fossils tell us that these African hominins were capable scavengers and possibly good hunters.

thick skull walls

EVOLVING FEATURES
H. heidelbergensis had a larger brain and a higher, broader skull. Its projecting jaw, thick skull, large brow ridges, and robust lower limbs were inherited from earlier hominins.

low, flat forehead

KEY FINDS

TOOLS

Early hominins developed "core-and-flake" technology— sharp stone flakes "knapped" from cobbles—to gain access to new foods, such as marrow from scavenged animal bones. Later *Homo* species made fine Acheulean handaxes and wooden tools such as hunting spears.

noncutting side, for grip

flaked side

tool made of quartzite

FLAKE   **FLAKED COBBLE**   **CHOPPER**   **HANDAXE**   **CLACTON SPEAR**

Homo neanderthalensis

DATES 430,000–40,000 ya

BRAIN SIZE 86 cubic in (1,412 cubic cm)

HEIGHT 5ft–5ft 6in (1.52–1.68m)

FINDS A number of entire skeletons, plus various bone fragments from more than 275 individuals

LOCATION Across Europe and southwest Asia

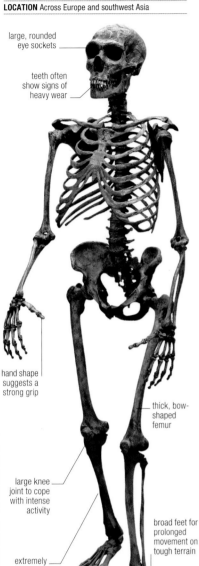

large, rounded eye sockets

teeth often show signs of heavy wear

hand shape suggests a strong grip

large knee joint to cope with intense activity

extremely robust legs

thick, bow-shaped femur

broad feet for prolonged movement on tough terrain

The best-known of all fossil hominins, the Neanderthals thrived in Europe for over 300,000 years before modern humans arrived. More than 275 individuals are known from over 70 sites in Europe and western Asia, including some virtually complete skeletons recovered from what may be intentional graves. Neanderthals were excellent toolmakers and proficient hunters, superbly adapted to their world. A physical lifestyle and harsh environment may explain their distinctive anatomy. Big brains were more heat-efficient, and large noses warmed the air before it reached the lungs. Their heavily muscled arms and bodies also particularly suited hunting. The pelvis differs slightly from modern humans: some argue that Neanderthals had longer pregnancies and more mature babies, but others believe the differences relate to their posture and lifestyle. Their relationship to *H. sapiens* is also debated: DNA analysis shows that Neanderthals interbred with both modern humans and Denisovans. Ultimately, Neanderthals died out, but some of their genes live on in us.

STOCKY BODY
Neanderthals' short, stocky bodies were adapted to cold climates and intense physical activity, as their small surface area minimized heat loss. Even today, humans in hot equatorial areas tend to be tall and thin, with longer legs than those in colder regions.

KEY FINDS
THE DENISOVANS

In 2010, DNA extracted from a tooth from Siberia revealed a genome so different from that of Neanderthals that the tooth has been assigned to a new species—the Denisovans. Little is known about them, but their DNA shows traces of interbreeding with Neanderthals and modern humans.

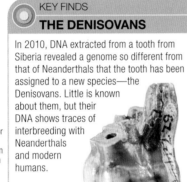

tooth provided ancient DNA

Homo floresiensis

DATES 700,000–50,000 ya

BRAIN SIZE 23–25 cubic in (380–420 cubic cm)

HEIGHT Female: 3 ft 7 in (1.1 m)

FINDS One nearly complete skull and partial skeleton, and parts of at least 11 other individuals

LOCATION Liang Bua Cave and Mata Menge on the island of Flores, Indonesia

In 2003, the smallest hominin ever found, *Homo floresiensis,* was discovered on the island of Flores in Indonesia. It is barely 39in (1m) high, and its brain is smaller than the earliest hominins. Initially dated to 12,000 years ago but later pushed back to 50,000 years ago, the discovery created significant debate. Many argued that the bones were those of modern humans suffering from a growth-stunting condition. However, the fossils differ from modern human bones in some key respects. More recently, similar finds from nearby Mata Menge dated to 700,000 years ago suggest that these fossils are probably the remains of hominins

descended from *H. erectus* groups who adapted to isolated island life by becoming smaller. Dwarf forms of many animal species are known from Flores and other islands, and *H. floresiensis* shows us that even very recently, our relatives were far more variable than we once thought.

ISLAND ISOLATION
Flores has long been isolated from Australia and Asia; over time, this may have caused its hominins to become smaller than their mainland relatives.

Homo naledi

DATES 335,000–236,000 ya

BRAIN SIZE 28–34 cubic in (465–560 cubic cm)

HEIGHT About 5ft (1.5m)

FINDS Fragmentary, disarticulated and articulated remains of almost the full skeleton from at least 18 individuals of a range of ages, including 5 skulls

LOCATION The Dinaledi and Lesedi chambers in the Rising Star cave system, Gauteng, South Africa

In 2013, while exploring the Rising Star cave system in South Africa, recreational cavers found a chamber containing many fossil hominin bones. Excavations in the Dinaledi chamber later produced the remains of at least 15 individuals, including four

prominent brow ridge similar to that of *Homo erectus*

high, round cranium with no large chewing muscle attachments

skulls that show an odd mix of primitive and more modern anatomical traits. The fossils were assigned to a new species, *Homo naledi*, and in 2017, they were dated to around 250,000 years old—about the time that *H. sapiens* was evolving in Africa. This suggests that there was much more variability among hominin species during this period than previously assumed. A further puzzle is how the fossils got to this difficult-to-access cave. No evidence has been found that they were accumulated by carnivores or scavengers or transported by a flood. One suggestion is that the remains were deliberately deposited there, perhaps in some kind of burial ritual.

MODERN CHARACTERISTICS
Alongside more primitive evolutionary traits shared with australopithecines, such as a small body and brain size, other features look rather modern, including the teeth and the shape of the skulls.

Homo sapiens

DATES About 300,000 ya–present

BRAIN SIZE 61–122 cubic in (1,000–2,000 cubic cm)

HEIGHT Up to about 6ft 1in (1.85m)

FINDS Early *Homo sapiens* fossils include complete skulls and various skull and skeleton fragments

LOCATION Across Africa, then worldwide

Early *Homo sapiens* were first found in Europe, many buried with tools, art, and jewelry. However, later finds established evidence of a long, gradual development of *H. sapiens* in Africa. Here, specimens dated to 400,000–250,000 years ago have the long, low skulls of *H. heidelbergensis* but with large faces, heavy brow ridges, a bony skull crest, and thick bones; they are variously known as *H. rhodesiensis*, *H. leakeyi*, and "archaic *H. sapiens*." Isolated finds from 250,000–125,000 years ago sporadically combine ancestral traits with modern features, such as a chin, smaller brow ridges, and a rounder skull; one from Jebel Irhoud, Morocco, looks very modern, despite dating to some 300,000 years ago. After around 125,000 years ago, fossils are recognizable as *H. sapiens*, with smaller faces, brow ridges, and teeth, and higher, more rounded braincases. *H. sapiens* lived in the Levant by 130,000–100,000 years ago but did not arrive in northern Europe until after 50,000 years ago. Neanderthals from sites nearby in Israel date to 60,000–45,000 years ago, and this period may have been when interbreeding occurred. However, other groups of *H. sapiens* may have spread east across Asia earlier, reaching China c.100,000 years ago.

LAID TO REST
Some of the early *H. sapiens* from the Levant seem to have been deliberately buried. This woman from Qafzeh, Israel, was buried with her child.

Glossary

abdomen
The belly of a vertebrate or the hind part of the body of an arthropod.

aboral surface
The body surface (for instance, in echinoids) opposite to where the mouth is located. *See also* oral surface.

acanthodians
An extinct group of jawed fish that existed from the Silurian (and possibly Ordovician) period, and died out during the Permian.

actinopterygians
A major group of bony fish that includes the great majority of living fish species. Also called ray-finned fish, their earliest known fossils date from the Silurian period. *See also* osteichthyans, teleosts.

adductor muscle
A muscle that pulls two structures together, such as the two shells of a bivalve.

aetosaurs
A group of plant-eating reptiles of the Triassic period, possibly related to crocodilians, that had backs protected by armored plates and spines.

Afrotheria
A broad group of mammals originally thought to be unrelated, but now believed to have evolved from a single ancestral group in Africa at a time when it was isolated from other continents. Afrotherians include elephants, manatees, aardvarks, and diverse smaller forms.

agnathans
Jawless fish, including living lampreys and many extinct forms.

algae (sing. alga)
Any of a wide variety of simple, mainly aquatic plantlike organisms, not all closely related, that produce their own food by photosynthesis. They include the single-celled "plants" of the plankton, as well as larger seaweeds. *See also* photosynthesis, cyanobacteria, protists.

ambulacra
The five narrower segments on the outer surface of an echinoid, through which structures of the water vascular system project. *See also* water vascular system.

amino acid
Any of about 20 different small molecules that are the "building blocks" of proteins, and thus are essential to all life.

ammonoids
An extinct group of swimming cephalopods with external, usually coiled, chambered shells. Includes ammonites and nautiloids. *See also* cephalopods, nautiloids.

amniotes
Vertebrates (tetrapods) that lay shelled eggs, or whose ancestors laid shelled eggs. They include reptiles, birds, and mammals. *See also* amphibians, tetrapods.

amniotic egg
The type of egg laid by reptiles, birds, and early mammals.

amphibians
(1) The group of vertebrates that includes present-day frogs and salamanders, and their direct ancestors. (2) More broadly, all groups of tetrapods that never evolved amniotic eggs and so had to return to water to breed. In this broader sense there are a great variety of extinct species of amphibians. *See also* lepospondyls, temnospondyls, amniotes, tetrapods.

anal fin
A small fin near the anal opening in some fish.

anapsids
Reptiles characterized by having no skull opening in the region behind the eyes.

angiosperms
The plant group that contains all flowering plants. They are distinguished from other seed plants by the fact that angiosperm seeds grow hidden and protected in a flower structure called a carpel that matures into a fruit. *See also* eudicots, eumagnoliids, monocots.

anoxic
Lacking in oxygen.

anterior
Toward the front.

anthozoans
The group of cnidarians that includes sea anemones and most living corals. *See also* cnidarians, coral.

apical disk
Part of the outer skeleton of an echinoid, usually on the upper surface and containing the anal opening.

appendage
A limblike structure, especially of an arthropod. Appendages may be modified as legs, gills, swimming organs, or in other ways.

arachnids
The group of chelicerate arthropods that includes spiders, scorpions, and relatives

Archaea
Microscopic single-celled prokaryotic organisms that were once classified with bacteria but are now treated as a separate kingdom in their own right. They are noted for flourishing in extreme environments. *See also* bacteria, prokaryotes.

archosaurs
A major group of reptiles that includes dinosaurs, pterosaurs, and crocodilians, as well as birds. They first evolved in the Triassic period. *Compare* lepidosaurs.

arthropods
A major group (phylum) of invertebrate animals with jointed legs and a hard outer skeleton. They include crustaceans, insects, chelicerates, and trilobites.

artiodactyls
Even-toed hoofed mammals, including pigs, camels, hippos, giraffes, deer, cattle, sheep, and antelopes. *See also* ruminants, perissodactyls.

asexual
Relating to reproduction not involving sex. Examples of asexual reproduction include creating new individuals by splitting or fragmenting the body, and forming specialized nonsexual structures to aid in dispersal (such as asexual spores).

australopithecine
Any member of the genus *Australopithecus* of extinct hominins. Several species are known dating from between 4.2 and 2 mya.

axil
The upper angle formed between a part of a plant, such as a leaf, and the stem that it grows from.

bacteria
Microscopic single-celled organisms that are prokaryotes. They abound in all habitats and may be free-living or parasitic. *See also* prokaryotes, Archaea.

barinophytes
An extinct subgroup of zosterophylls distinguished by the production of spores of two different sizes in a single sporangium. *See also* zosterophylls, spores, sporangium.

bedding
The laying down of sedimentary rock, or the manner in which a particular rock has been laid down.

belemnites
Extinct relatives of squids. Their long, often cigar-shaped internal shells are common fossils in Jurassic and Cretaceous rocks. *See also* cephalopods.

bennettitaleans
An extinct group of seed plants superficially resembling cycads in their leaves and their growth form but not closely related to them. *See also* seed plants.

biodiversity
The variety of different kinds of organisms found at a given time or location. Often calculated in terms of number of species, but can be measured in other ways.

bipedal
Walking on two legs.

bivalves
Mollusks, such as clams, mussels, and oysters, that have a shell made up of two hinged halves. Most species move slowly or not at all, and are filter feeders. *See also* mollusks.

blastoids
An extinct group of Paleozoic filter-feeding echinoderms. *See also* echinoderms.

bony fish
See osteichthyans.

brachial
Relating to the arms.

brachiation
A method of movement characteristic of apes such as gibbons, in which the body is swung through the trees by the arms.

brachiopods
A major group (phylum) of marine invertebrates, also known as lamp shells. Their two-part shell makes them look superficially similar to bivalve mollusks, but they are unrelated. There are some living forms, but they were far more abundant and diverse in the Paleozoic and Mesozoic.

brackish
Saltier than freshwater, but less salty than typical ocean water.

bract
A modified leaf at the base of a flower or flower head. May be small and scalelike, large and petallike, or like normal foliage.

branchial
Relating to the gills.

bryophytes
A general term covering three groups of mainly small land plants: mosses, liverworts, and hornworts. These lack specialized water-conducting (vascular) tissue. *See also* mosses, liverworts, hornworts, vascular plants.

bryozoans
Also called moss animals, a group of filter-feeding invertebrate animals consisting of colonies of small immobile individuals.

budding
In zoology, new individuals growing directly from the flesh of existing individuals.

calcareous
Consisting of or containing the mineral calcium carbonate.

calice
The upper part of a coral skeleton.

carapace
The upper shell of a turtle; also the protective outer covering of some other animals, such as crabs and lobsters.

carbon dioxide
An atmospheric gas (carbon combined with two oxygens), released by burning, volcanic activity, and when organisms respire. Conversely, plants remove it from the atmosphere during photosynthesis. *See also* respiration.

carnivore
(1) Any flesh-eating animal. (2) A member of the order of mammals that includes cats, dogs, bears, weasels, hyenas, racoons, and mongooses.

carpel
The female part of a flower, containing and protecting the unfertilized seeds, and also the seeds themselves after fertilization.

cartilage
A tough tissue formed mainly of protein that is found in the bodies of vertebrates. In

sharks and relatives the skeleton is made of cartilage; in other vertebrates cartilage occurs in joints and other locations.

cartilaginous fish
See chondrichthyans.

cast (the filling of a mold)
The material filling a mold.

Caytoniales
An extinct group of seed plants with distinctive four-parted palmate leaves and small, fleshy, fruitlike structures that contain seeds.

cellulose
A carbohydrate that is the main structural material of plants.

cephalochordates
Small marine invertebrates that are of interest in understanding how vertebrates evolved. The living species, called amphioxus or lancelets, are filter feeders that burrow in shallow seafloors, but also swim in a fishlike way. Soft-bodied, they rarely fossilize. *See also* chordates.

cephalopods
The group of advanced mollusks that includes living squids and octopuses, the mainly extinct nautiloids, and the entirely extinct ammonites and belemnites.

cephalothorax
The front part of the body of some arthropods when not obviously divided into a separate head and thorax.

cetaceans
The group of mammals that includes whales and dolphins. They evolved from artiodactyl ancestors during the Paleogene. *See also* artiodactyls.

chelicerates
A major subgroup of arthropods including arachnids (spiders and relatives) and many other forms, including the extinct eurypterids. They are characterized by joined pincers called chelicerae.

chemosynthesis
The phenomenon by which some specialized microorganisms are able to make their own food using the energy from naturally occurring chemicals such as hydrogen sulfide.

chitin
A complex substance consisting mainly of carbohydrate that forms the tough outer covering (exoskeleton) of insects and other arthropods. It is also found in fungi.

chloroplast
A structure within a plant or algal cell within which photosynthesis takes place.

chondrichthyans
The major group of fish that includes the present-day sharks and rays, plus many extinct forms. Their skeleton is made of cartilage, not bone.

chordates
A large group (phylum) of animals that includes all the vertebrates as well as some related invertebrate forms such as cephalochordates. They are named for a rodlike support of cartilage called a notochord that runs down their backs. *See also* cephalochordates.

cilia (sing. cilium)
Microscopic beating hairlike structures on the surfaces of some cells, used to aid movement in small organisms, or to create water currents.

clade
A group of species consisting of all the evolutionary descendants of a given ancestor. Mammals are a clade, for example. Reptiles, however, are not a clade unless birds are included within them, since birds are descended from dinosaurs, which were reptiles. Cladistics is the approach to classification that seeks to use only clades to classify organisms without using Linnaean ranked taxa.

cladistics
See clade.

claspers
Appendages found in some male insects and chondrichthyans that are used for grasping the female during mating.

clubmosses
See lycophytes.

cnidarians
A major group (phylum) of mainly marine invertebrates with simple bodies bearing stinging tentacles that surround a single opening (mouth). *See also* anthozoans, scyphozoans, hydrozoans, coral.

coevolution
The evolution of two or more different organisms together, so that they become increasingly adapted to one another. The mutual adaptations of flowers and the bees that pollinate them are an example.

colony
A colony of animals can consist of separate individuals working together, as in ant colonies, or animals joined by strands of living tissue, as with many corals. Individuals may be specialized for different roles, such as feeding, reproduction, and defense, in which case the colony may behave more like a single animal.

common ancestor
Any original species from which later species under discussion are all believed to have descended.

compound eye
An eye formed as a mosaic of smaller eyes, as in some arthropods.

compression fossil
A fossil where the original form has been squashed flat, destroying evidence of external structure.

condylarths
An extinct group of mammals known mainly from the Paleogene. They included plant-eating forms similar to today's hoofed mammals, although not closely related.

conifers
The group of nonflowering seed plants, mainly trees, that includes pines, firs, redwoods, cypresses, and other living and extinct species.

conodont animals
Extinct eel-like marine animals whose hard parts (small internal toothlike structures called conodonts) were known and studied long before fossils of the whole animals were discovered from the 1980s onward. Dating from the Cambrian to the Triassic, they are now thought to be chordates and possibly even primitive vertebrates. *See also* chordates.

continental shelf
A region where the seafloor is relatively shallow around the margins of a continent. At its outer edge the seafloor steepens until it reaches the depth of the deep ocean floor.

convergent evolution
The phenomenon in which unrelated organisms have evolved to look similar as a result of being adapted to similar environments or ecological niches. For example, marsupial mice in Australia look superficially similar to mice found elsewhere, but are not related.

coral
Any of various cnidarians that live fixed to the seabed or to reef structures. The true corals are members of the class Anthozoa, secrete a calcareous skeleton, and may be solitary or colonial. *See also* cnidarians.

corallite
In corals, the skeleton of an individual coral animal.

corallum
The skeleton of a colonial coral, composed of corallites. *See also* corallite.

correlation (of rocks)
The matching of rocks from different regions of the world (for instance, by comparing their fossils) so that rocks of the same age in different places can be recognized.

cranial
Relating to the brain case, or to the skull as a whole.

creodonts
An extinct group of flesh-eating mammals with a similar ecological role to modern carnivores. They were dominant predators in the Eocene and Oligocene epochs.

crinoids
Stalked filter-feeding echinoderms, also called sea lilies. Living crinoids mainly live in deep water, except for stalkless forms called feather stars. *See also* echinoderms.

crocodilians
The group that includes living crocodiles and alligators and their immediate ancestors. They and their extinct relatives (together called the crocodylomorphs) belong to the archosaur group of reptiles. *See also* archosaurs.

crocodylomorphs
See crocodilians.

crustaceans
The diverse group of mainly marine arthropods that includes crabs, lobsters, shrimps, barnacles, and many smaller forms, including members of the plankton.

cusp
The sharp apex where two curved surfaces meet, especially in the context of the surface of a tooth.

cyanobacteria
Microscopic photosynthetic relatives of bacteria; also called blue-green algae.

cycads
A group of nonangiosperm seed plants that are superficially similar to palms but are reproductively very different and produce seeds and pollen in cones.

cynodonts
A group of advanced synapsids that arose in the Late Permian. *See also* synapsids.

czekanowskialeans
An extinct group of Mesozoic seed plants with distinctive finely divided, linear leaves often borne on short shoots.

decapods
(1) The group of cephalopods that includes squid and cuttlefish. (2) The group of crustaceans that includes crabs, lobsters, and relatives. The name literally means "10 feet," referring to the number of squid tentacles and crab/lobster legs respectively.

denticle
A small toothlike projection in the skin of sharks and some other fish.

deuterostomes
Members of a major division of the animal kingdom, defined by the second opening formed in the developing embryo turning into the mouth. (The first opening becomes the anus.) They are thought to represent a single related "super-group" and include echinoderms, chordates (including vertebrates), and graptolites. *See also* protostomes.

diadectomorphs
A group of extinct tetrapods thought either to be the earliest amniotes or closely related to them. *See also* amniotes.

diapsids
A major division of reptiles whose name refers to the two openings on each side of the skull that are characteristic of the group. Two main groups within the diapsids are the archosaurs (including dinosaurs, birds, crocodilians, and pterosaurs) and the lepidosaurs (including lizards, snakes, and plesiosaurs, among others).

dicynodonts
A group of herbivorous therapsids with two tusks and a blunt beak. *See also* therapsids.

dinosaurs
A group of reptiles that evolved during the Triassic and dominated life on land in the Jurassic until their extinction (with the exception of their descendants, the birds) at the end of the Cretaceous. *See also* ornithischians, saurischians, archosaurs.

DNA
Short for deoxyribonucleic acid, a very long molecule made up of small individual units. DNA is found in the cells of all living things; the order of the small units "spells out" the genetic instructions (genes) of the individual.

dorsal
Relating to the back or upper surface of an animal. *See also* ventral.

echinoderms
A major group (phylum) of marine invertebrates that includes starfish, echinoids, crinoids, and many other extinct forms. They have calcareous protective plates under their skin, and living forms display radial symmetry.

echinoids
A group of living and extinct echinoderms. Typical echinoids, also called sea urchins, have a globular body, long spines, and a downward-facing mouth to graze algae. There are also many burrowing echinoids.

ecosystem
Any community of organisms considered together with the interactions between them and their associated physical environment.

edentates
An unusual group of eutherian mammals living mainly in South America and including armadillos, sloths, and some anteaters. ("Edentate" literally means "having no teeth.")

Ediacaran fauna
Soft-bodied invertebrates living before the Cambrian period and originally discovered fossilized in the Ediacara Hills in Australia.

embolomeres
A group of large aquatic amphibians from the Carboniferous, which may be the ancestors of amniotes. *See also* amniotes.

endemic
A species native to only one restricted region of the world. (Also used as an adjective.)

enzyme
Any of a large variety of different molecules (nearly always proteins) that promote particular chemical reaction in the body.

Eocene
The second epoch of the Paleogene period, between the Paleocene and the Oligocene. It lasted from 55.8 to 33.9 mya.

eon
The largest and highest-level unit into which geological time is divided.

epidermis
The outermost layer of skin in animals; the outermost cell layer of a plant.

epithelium
The covering of cells that forms the surface of organs and tissues in the body.

epoch
A unit of geological time that is a division of a period—for example, the Eocene epoch of the Paleogene period.

era
A unit of geological time that is a division of an eon and is itself divided into periods—for example, the Paleozoic era of the Phanerozoic eon.

eudicots
Also called eudicotyledons, this is the largest group within the flowering plants (angiosperms). It contains many familiar flowers and tree species. *See also* angiosperms.

eukaryotes
Living things whose cells have a complex structure, including a distinct nucleus that contains the genetic material and is bounded by a membrane. Eukaryotes include all animals, plants, and fungi, as well as many microscopic single-celled species. *See also* prokaryotes, protists.

eumagnoliids
A group of angiosperms thought to have branched off early in angiosperm evolution, and that retains some of the features of primitive angiosperms. The group's members include magnolias and laurels.

euphyllophytes
One of the two main subgroups of vascular plants, including all land plants having complex leaves with many veins. Euphyllophytes include all seed plants, ferns, and horsetails. The term primitive euphyllophytes, as used in this book, refers to early lineages of euphyllophytes that do not fall into any of the main subcategories within the group. *See also* lycophytes, vascular plants.

eurypterids
Extinct aquatic predatory arthropods, sometimes called sea scorpions. They occurred form the Ordovician to the Permian. *See also* chelicerates.

eutherians
Mammals in which the fetus grows to a relatively advanced stage in the mother's womb, nourished by a placenta. Includes all living mammals except marsupials and monotremes.

evaporite
A sedimentary deposit, usually made up of salts of various kinds, that has been created by the evaporation of salty water.

evolution
In its straightforward modern definition, evolution is simply any change in the average genetic makeup of a population of organisms of the same species between one generation and the next. What is often called "the theory of evolution" is based on the idea, supported by various lines of evidence, that such genetic change is not random but is largely the result of natural selection, and that the operation of such processes over time can account for the huge variety of species found on Earth. *See also* natural selection.

exhalent siphon
The projecting tube in many bivalves through which circulating water leaves the animal. *See also* inhalent siphon.

exoskeleton
A skeleton located on the outside of an animal. Exoskeletons combine support with body protection.

fasciole
A band on the outside of the test of some echinoids, with minute granules to which in life very fine spines were attached that were specialized to create water currents.

fenestra
A hole or opening in the side of the skull, or in an individual bone.

ferns
A group of nonflowering vascular plants that reproduce by spores rather than seeds. They most commonly produce their leaves from underground stems, but some are large enough to form trees. *See also* euphyllophytes.

filter feeding
Feeding by collecting and separating small food particles from the environment. When the food particles are suspended in water, it is also called suspension feeding. Collecting and sieving small particles from mud and sand is called deposit feeding.

flagellum (pl. flagella)
A microscopic beating hairlike structure borne by some cells. Flagella are similar to cilia but much longer.

form genus
A scientific name given not to a whole organism, but to a particular part of one (such as a seed) that is regularly found in the fossil record. Form genera are convenient in classifying plant fossils, which are often found in fragments that are difficult to relate to each other.

fungi
A huge and diverse kingdom of living things that includes mushrooms, molds, and numerous other species. The typical fungus "body" is a network of small threads. Structures such as mushrooms are specially produced "fruiting bodies" to disperse spores. *See also* lichens.

gastropods
The diverse group of mollusks that includes marine, freshwater, and land snails and slugs. *See also* mollusks.

genital plate
A section of the outer skeleton of an echinoid that contains small pores through which eggs or sperms are released.

genome
The complete set of genes of a given species of animal or other living thing.

Ginkgoales/ginkgos
A group of nonflowering seed plants with a single living species (*Ginkgo biloba*), but many extinct species. Designated as "gymnosperms" in traditional classification.

Gnetales
A group of living and extinct nonflowering seed plants of varying appearance. They are designated as "gymnosperms" in traditional classification.

gracile
In paleontological contexts: smaller and/or more lightly built.

graptolites
An extinct group of colonial planktonic invertebrates, typically growing in the form of long narrow colonies of small individuals, supported by a hard skeleton.

green algae
A group of relatively simple water-living plants, including some microscopic plankton and some seaweeds. Many species live in freshwater. *See also* plants, algae.

greenhouse effect
Phenomenon in which heat radiating from the Earth's surface is absorbed by certain atmospheric gases, including water vapor, carbon dioxide, and methane, causing average temperatures to be warmer than otherwise.

guard cell
A cell whose swelling and shrinkage acts to open and close stomata in plants. *See also* stomata.

gymnosperms
A general term for several groups of seed plants that are not true angiosperms. The term literally means "naked seeds" in that their seeds are not fully protected within a structure called a carpel, in contrast to angiosperms. *See also* conifers, cycads, Ginkgoales, Gnetales, progymnosperms.

Holocene
The most recent epoch of the Neogene period, lasting from 11,700 years ago to the present. It is also designated as part of the Quaternary period. *See also* Pleistocene.

hominids
The subgroup of primates that includes humans, chimpanzees, and gorillas, together with their last common ancestor and related extinct forms. Does not include orangutans, gibbons, or monkeys.

hominins
The subgroup of hominids that includes humans and their extinct ancestors and relatives, taken back to the point at which the lineage leading to humans split from that leading to chimpanzees.

hornworts
A group of small, low-growing land plants, living mainly in moist environments. They were traditionally classified with two other groups of "nonvascular" plants—mosses and liverworts—as "bryophytes." *See also* bryophytes, vascular plants.

hydrothermal vent
A fissure in a volcanically active region of the ocean floor from which superheated, chemical-laden water emerges. Found mainly at mid-ocean ridges.

hydrozoans
A group of cnidarians that includes many small colonial species with tiny polyps, some significant in coral reef formation, and also floating forms such as the Portuguese man-of-war. *See also* cnidarians, polyp.

ice age
Any episode in which the Earth's surface temperatures were much lower than today and ice cover more extensive.

ichthyosaurs
A group of marine reptiles that first appeared in the Triassic. Ichthyosaurs were predators, and most had streamlined bodies for fast pursuit, similar to present-day dolphins or tuna. They became extinct before the end of the Cretaceous.

igneous rock
Rock that originates from the solidification of magma rising from deep in the Earth.

inarticulate
Having no joints or articulations.

inhalent siphon
The projecting tube in many bivalves through which water is drawn into the animal's shell, providing it with oxygen and food particles. *See also* exhalent siphon.

internal mold
A fossil mineral impression of the inside of a now-vanished original shell or other part of an organism.

invertebrate
Any animal without a backbone, i.e. other than a vertebrate.

keystone predator
A predator whose removal would cause a drastic change to its ecosystem—for example, by allowing its former prey to proliferate.

leaflet
A single division of a leaf, in the case of a leaf that is divided into several parts. *See also* palmate, pinna, pinnate, pinnule.

lepidosaurs
The group of reptiles that includes the present-day lizards and snakes, plus rhynchocephalians, such as *Sphenodon*. *Compare* archosaurs.

lepospondyls
A diverse group of small-bodied extinct amphibians widespread in the Carboniferous and Permian periods. They included both aquatic and terrestrial forms. *See also* temnospondyls.

lichens
Composite living things made up of a fungus and an alga growing in close association. There are many types, typically growing as crusts or bushy outgrowths on rocks, tree branches, etc.

lignophytes
Vascular plants that produce specialized woody tissue. They include seed plants and some extinct relatives that are often grouped together as progymnosperms. *See also* progymnosperms.

lineage
Any given branch on an evolutionary family tree. A lineage consists of a species or group plus its forebears, traced back to an original ancestor. *See also* clade.

liverworts
A group of small, low-growing land or freshwater plants, often growing in the form of a flat sheet, but sometimes with lobelike leaves. They lack vascular tissue and were traditionally classified with two other groups of "nonvascular" plants, mosses and hornworts, as "bryophytes." *See also* bryophytes, vascular plants.

living fossil
Refers either to an individual species that has hardly changed over millions of years, or a living representative of a taxonomic group that is otherwise extinct.

lobe-fins
See sarcopterygians.

lobopods
A group of soft-bodied arthropodlike crawling invertebrate animals, superficially resembling caterpillars,

that includes the living *Peripatus*, plus many extinct forms dating back to the Cambrian and beyond.

lycophytes
A group of vascular plants that includes the lycopods and zosterophylls. Lycopods include the clubmosses of the present day, plus many extinct forms, including giant forest trees in the Carboniferous period. Lycophytes differ from all other vascular plants (euphyllophytes) in their very simple, scalelike leaves. *See also* euphyllophytes, vascular plants, zosterophylls.

mammals
Amniote vertebrates that suckle their young on milk and whose skin is typically covered in hair or fur. Mammals evolved from synapsid ancestors during the Triassic period. *See also* synapsids.

mandible
The lower jaw of a vertebrate, or (plural) the biting mouthparts of an arthropod.

mantle
In mollusks, the tissue, originally forming the upper surface of the animal, that secretes the mollusk's shell. *See also* mollusks.

marsupials
A group of mammals in which offspring are born in a relatively undeveloped state; they often continue their growth within an external pouch on the mother. Includes kangaroos, opossums, and many other living and extinct forms.

matrix
The material on which a fossil rests or in which it is embedded.

medullosans
An extinct group of seed plants with fernlike leaves and large, fleshy seeds.

Mesozoic
The second era of the Phanerozoic eon, dating from the beginning of the Triassic period to the end of the Cretaceous.

metacarpals
The bones occurring between the wrist and the fingers in the human hand; also, the equivalent bones in other species.

metamorphic rock
Any rock that has been transformed underground by natural heat or pressure into a new form. For example, marble is metamorphized limestone.

metazoa
The group that includes all multicellular animals except for sponges, which have a simpler form of body organization.

microfossil
A microscopic or very small fossil.

mid-ocean ridge
A submerged range of mountains running along any part of the deep-ocean floor, marking the place where new ocean crust is being created by molten rocks rising from deep in the Earth.

midrib
The central vein of a leaf (including its leaf-stalk).

Miocene
The first epoch of the Neogene period. It lasted from 23.03 to 5.332 mya.

mollusks
A major group (phylum) of invertebrate animals that includes the gastropods, bivalves, cephalopods, and other forms. Mollusks are soft-bodied and typically have hard shells, though some subgroups have lost their shells during evolution.

monocots
Also called monocotyledons, the group within the angiosperms that includes grasses, orchids, lilies, and palm trees, as well as many other species. They are characterized by having only one cotyledon (seed-leaf) within each seed. Most monocots have leaves with parallel veins. *See also* angiosperms.

monotremes
Egg-laying mammals, including the platypus and the echidnas (spiny anteaters). This egg-laying habit is thought to be the original mode of reproduction for mammals.

mosasaurs
A group of Cretaceous marine reptiles, thought to be close relatives of snakes.

mosses
A group of small, low-growing land or freshwater plants, often bushy or cushionlike in appearance, that lack vascular tissue. *See also* bryophytes, vascular plants.

multituberculates
An extinct group of mainly rodentlike early mammals found from the Jurassic to the Paleogene period.

mya
Million years ago.

myriapods
The group of many-legged arthropods that includes millipedes and centipedes.

natural selection
Evolutionary process by which the environment favors the fittest individuals in a given population, weeding out those that are less fit. ("Fitness" in this context means the sum total of the qualities that give an individual an increased chance of leaving descendants.) Since fitness is partly inherited, this should result in a population or species changing genetically over time and (other things being equal) becoming better adapted to their environment.

nautiloids
The group of cephalopod mollusks containing the living *Nautilus* plus many extinct forms. Their coiled external shell contains gas-filled chambers that lighten the body in the water.

Neolithic
The final phase of the stone age, occurring at different times in different parts of the world and characterized by the transition to an agricultural lifestyle and by new technologies such as pottery.

niche
Roughly, the ecological "role" that an animal or other living thing plays (for

example "small night-active insect-eater living high up in trees").

noeggerathians
A group of extinct plants that have proved difficult to classify. They have been linked to horsetails and to ferns, but recent research suggests that they were more closely allied to progymnosperms.

nothosaurs
See sauropterygians.

notochord
See chordates.

occipital
Relating to the back of the head.

Oligocene
The third and final epoch of the Paleogene period. It lasted from 33.9 to 23.03 mya.

ooze
Sediment on the deep seafloor that contains a large proportion of skeletons of dead microscopic planktonic life forms. *See also* plankton.

operculum
A term with various applications in zoology, all related to the original Latin meaning of "covering" or "lid." It can mean: the horny or calcareous disk used by many snails to shut themselves into their shells; a flap that covers the gills of bony fish and tadpoles.

oral surface
The body surface where the mouth is located. *See also* aboral surface.

ornithischians
One of the two main subgroups of dinosaurs, the name literally meaning "bird-hipped." Ornithischians include stegosaurs, ankylosaurs, ornithopods, pachycephalosaurs (dome heads), and ceratopsians (horned dinosaurs). *Compare* saurischians.

ossicles
Small bony or other hard parts collectively making up the skeletal support in some animals.

osteichthyans
The largest group of jawed fish, made up of two subgroups, the actinopterygians (ray-finned fish) and the sarcopterygians (lobe-finned fish). Also called bony fish.

ozone layer
A layer high in the atmosphere with a high concentration of ozone (a 3-atom form of oxygen). Ozone absorbs harmful ultraviolet radiation that can damage living things.

Paleocene
The first epoch of the Paleogene period and Cenozoic era. It lasted from 66 to 55.8 mya.

paleoclimatology
The study of past climates.

paleoecology
The study of the ecology and environments of past geological time periods.

palmate
Of a leaf: divided into leaflets spreading out from the base of a leaf like the fingers of a hand.

paranthropine
Any member of the genus *Paranthropus* of extinct hominins, with fossils known from between 2.7 and 1.2 mya.

parareptiles
A group of extinct amniotes considered not to be true reptiles. They include mesosaurs and several other groups.

pedicle
The flexible stalk of a brachiopod that attaches it to its substrate.

pelvic fins
A pair of fins situated on the under surface of most fish.

Pentoxylales
An extinct group of Mesozoic seed plants, known from India, Australia, and New Zealand. They have leaves similar to those of some bennettitaleans. *See also* bennettitaleans.

period
A unit of geological time that is a division of an era—for example, the Jurassic period of the Mesozoic era.

perissodactyls
Odd-toed hoofed mammals, including horses, rhinoceroses, and tapirs. *See also* artiodactyls.

permineralization
A form of fossilization where minerals occupy the spaces between the hard parts of the original organism, but do not replace the hard parts themselves. *See also* petrifaction.

petrifaction
Also called petrification, a form of fossilization where the detailed structures of an original organism are replaced by minerals, sometimes in a way that preserves very fine details. *See also* permineralization.

Phanerozoic
The most recent large time period (eon) of Earth's history. It stretches from the Cambrian period to the present.

photosynthesis
Process in plants, algae, and cyanobacteria whereby the Sun's energy, trapped by the green pigment chlorophyll, is used to build energy-containing food molecules from carbon dioxide and water.

phragmocone
The chambered part of a cephalopod shell.

phylum (pl. phyla)
The highest-level grouping in the traditional classification of the animal kingdom. Mollusks, arthropods, and chordates are examples of phyla.

phytosaurs
A group of crocodilelike aquatic reptiles that were common in the Triassic period.

pinna
In leaves: a leaflet or primary segment of a pinnate leaf. *See also* leaflet, pinnate.

pinnate
In leaves: being divided into leaflets arranged in regular rows on either side of the leaf's midrib. *See also* leaflet, midrib.

pinnule
In leaves: any of the ultimate leaflets of a bipinnately compound leaf. *See also* leaflet.

placental mammals
See eutherians.

placoderms
A group of extinct jawed fish, widespread in the Devonian period, whose skin was protected by bony "armor-plating."

placodonts
Extinct Triassic marine reptiles that fed on mollusks and other shellfish.

plankton
Any living species (plants, animals, or microorganisms) that live in open water but cannot swim strongly (or cannot swim at all) and so drift with the currents. Most plankton are small, but some larger types such as jellyfish are also included.

plate tectonics
Phenomena relating to the fact that the Earth's crust and its underlying rocks are divided into rigid sections called tectonic plates that move relative to one another.

Pleistocene
The third epoch of the Neogene period. It lasted from 1.806 to 0.0117 mya.

plesiosaurs
See sauropterygians.

pleurae (sing. pleura)
In trilobites, the lateral parts of the thoracic segments.

Pliocene
The second epoch of the Neogene period. It lasted from 5.332 to 1.806 mya.

pliosaurs
See sauropterygians.

polyp
The body form of many cnidarians, including sea anemones and coral animals. Polyps are typically tubular, attached at their base, and with a single opening (mouth) at the top surrounded by tentacles. *See also* cnidarians.

primates
The group of mammals that includes monkeys, apes, and humans, plus various less advanced forms (prosimians). Typical characteristics include grasping hands and forward-facing eyes. *See also* prosimians.

proboscidians
The group to which elephants and mammoths belong, characterized by a long, flexible, grasping trunk. *See also* Afrotheria.

progymnosperms
Spore-producing extinct plants with woody tissue similar to that of present-day seed plants. *See also* lignophytes.

prokaryotes
Living things such as bacteria whose individual cells are smaller, simpler in structure, and lack a well-defined nucleus, compared with those of eukaryotes. *See also* eukaryotes, bacteria, Archaea.

prosimians
Primates that are consider less "advanced" than the monkeys and apes. Living forms include lemurs, galagos, and tarsiers.

protists
A wide grouping of often unrelated, mainly microscopic, eukaryotic organisms, traditionally classified as a single kingdom. Protists are mainly single-celled, and include both plantlike forms (algae) and animal-like forms (protozoans). They can be defined informally as including all eukaryotes that have not been assigned to another kingdom. *See also* eukaryotes, algae, protozoans.

protozoa
Animal-like single-celled organisms, mainly microscopic, that are common in virtually all habitats and may be free-living or parasitic. The thousands of species are in many different subgroups, not all closely related. *See also* protists.

pteridophytes
A miscellaneous group of vascular plants that reproduce by spores rather than by seeds. Living pteridophytes include ferns, clubmosses, and horsetails. These plants are of diverse evolutionary relationships.

pteridosperms
A miscellaneous group of seed plants that has been used in the past to include several very different kinds of plants, such as medullosans and Caytoniales. These plants are of diverse evolutionary relationships.

pterosaurs
Flying reptiles related to the dinosaurs, with wings formed by skin stretched over their front limbs. Originating in the Triassic, they went extinct at the end of the Cretaceous.

rachis
The central stalk or midrib in a leaf or a flower shoot. *See also* midrib.

radial symmetry
A form of body symmetry analogous to the spokes of a wheel, as in echinoids and starfish.

radiation (evolutionary)
The emergence of a large number of novel evolutionary forms as the result of new adaptations.

radiometric dating
Measuring naturally occurring radioactivity to estimate the age of rock or other material.

radius (bone)
One of the two main bones of the forearm.

ratites
The group of flightless birds that includes the living ostriches, emus, rheas, and kiwis, together with related extinct forms.

rauisuchians
A group of dinosaurlike reptiles that were actually more closely related to crocodilians. Widespread in the Triassic before the rise of dinosaurs, they became extinct before the end of that period.

reducing conditions
Conditions low in oxygen, sometimes favorable to the preservation of fossil animal and vegetable material.

reptiles
Amniotes that are more closely related to lizards and crocodiles than to mammals. Reptilia plus Parareptilia make up the Sauropsida lineage of amniotes, as compared to the other main amniote lineage—the Synapsida—which lead to mammals. Within reptiles, the most successful group are the Archosauria, which includes crocodiles, dinosaurs, and birds. The reptiles first appear in the Carboniferous period.

reptilomorphs
The group of tetrapods that contains all the amniotes plus some reptilelike amphibians that are thought to be their nearest relatives.

respiration
(1) Breathing. (2) Also called cellular respiration, the biochemical processes within cells that break down food molecules to provide energy, usually by combining the molecules with oxygen.

rhizome
A creeping underground stem that helps plants spread asexually, to overwinter, and/or to store food. *See also* asexual.

rhynchosaurs
A group of plant-eating reptiles related to the archosaurs that lived only in the Triassic period. *See also* archosaurs.

rift valley
A large block of land that has dropped vertically compared with the surrounding regions, due to plate tectonic activity.

RNA
Short for ribonucleic acid. It has a similar structure to DNA and has various essential roles in cells, but is not used to store the main genetic information of an organism, except in some viruses.

robust
In paleontological contexts: larger and/or more strongly built.

rodents
A very diverse group of mammals characterized by front teeth (incisors) that are specialized for gnawing. They include mice, rats, squirrels, beavers, and porcupines, together with many other forms.

rostroconchs
A group of extinct Paleozoic mollusks, superficially similar to bivalves but not closely related. *See also* bivalves, mollusks.

rotifers
Microscopic multicelled invertebrates living in both seawater and freshwater and having a variety of lifestyles.

ruminants
A subgroup of the even-toed hoofed mammals (artiodactyls) whose members have a specialized four-chambered stomach that allows them to digest tough

plant food. Ruminants include cattle, sheep, antelopes, deer, and giraffes, but not pigs or hippopotamuses.

sarcopterygians
A mainly extinct group of fish having fleshy, muscular bases to their paired front and back fins. Also called lobe-finned fish, they are first known from the Silurian period and include the present-day lungfish and coelacanth. *See also* osteichthyans, tetrapods.

saurischians
One of the two main divisions of dinosaurs, in turn made up of two major subgroups, the theropods and the sauropodomorphs. The term "saurischian" means "lizard-hipped." *Compare* ornithischians.

sauropodomorphs
The group of dinosaurs that includes the well-known sauropods, huge herbivores with long necks and long tails that were the largest land animals to have ever lived. Other sauropodomorphs were smaller, and some early species were bipedal.

sauropterygians
A diverse group of semiaquatic and fully aquatic reptiles that existed from the Triassic until the end of the Cretaceous. Well-known members include the long-necked plesiosaurs and the shorter-necked, large-headed pliosaurs. Early forms, such as some nothosaurs, still possessed legs, but in later sauropterygians the limbs were modified as flippers.

scyphozoans
The group of cnidarians that includes jellyfish. *See also* cnidarians.

sedentary
Of animals such as worms: habitually staying in one position. *See also* sessile.

sedimentary rock
Rock formed when small particles settle out from water or are deposited by the wind, volcanic activity, etc., before later hardening to rock.

seed ferns
A general term for several groups of extinct seed plants that have fernlike leaves but belong to the seed plant group. *See also* bennettitaleans, Caytoniales, medullosans.

seed plants
Plants that reproduce by seeds as distinct from simpler spores. *See also* seed fern, gymnosperm, angiosperm, spore.

septum (pl. septa)
A partition within the body of an animal.

sessile
Of an animal: attached permanently to a surface, especially without a stalk, and not able to move around. *See also* sedentary.

sexual dimorphism
Condition in which the males and females of a species differ obviously in appearance.

sexual selection
Type of natural selection in which a feature has developed solely because of its direct advantage in furthering mating and

reproduction, even if it appears disadvantageous in other ways. A male peacock's tail is an example.

silica/silicate
The chemical compound silicon dioxide, which as the mineral quartz is the main component of sand. Silicates consist of silicon and oxygen in chemical combination with atoms of various metals.

speciation
The formation of new species.

species divergence
The increasing dissimilarity over time of two or more species derived from a common ancestor.

sphenophytes
A group of vascular plants, also known as horsetails, whose above-ground stems have a jointed appearance, with rings of small branches bearing tiny scalelike leaves. The few living species are outnumbered by the many extinct forms, some as big as medium-sized trees.

sponges
A large group (phylum) of marine invertebrates with a very simple structure that feed by creating currents through their bodies and filtering small particles from the water. They have no muscles or nerve cells. Many have skeletons built of small hard elements called spicules.

sporangium (pl. sporangia)
A structure that produces spores: found in plants other than seed plants, and also in fungi.

spore
A usually microscopic structure produced, often in large numbers, by plants (not seed plants), fungi, and many microorganisms, from which new individuals can grow. Spores are usually spread by wind or water, and may be produced by sexual or asexual reproduction.

sporophore
A spore-bearing structure.

stomata (sing. stoma)
Small openings on the surfaces of most plants, especially on the underside of their leaves, that allow gases (expecially oxygen, carbon dioxide, and water vapor) to enter or leave the plant. *See also* guard cell.

stratigraphy
The geological study of the order and relative position of strata (layers of sedimentary rock) in Earth's crust.

stromatolites
Large, hard, domelike structures, built up of thin layers, that are usually produced in shallow water by the action of many generations of cyanobacteria. Some of Earth's oldest fossils are of stromatolites.

subduction
The forcing down of oceanic crust belonging to one tectonic plate beneath another plate when two plates collide. *See also* plate tectonics.

suture
A rigid junction between hard parts of animals, such as shell or bone. In internal

molds of ammonoids, a suture line marks the junction between the internal dividing walls and the inside wall of the shell.

symbiosis
A close living relationship between two species, especially one in which both benefit.

synapsids
A major group (formerly referred to as "mammal-like reptiles") that branched off early in the evolution of amniote tetrapods, and eventually gave rise to the mammals. *See also* amniotes, teptrapods.

taxon
Any named group of organisms that forms part of a classification system.

taxonomy
The science of classifying organisms; also, a classificatory scheme resulting from the scientific study of a particular group.

teleosts
An evolutionarily advanced group of ray-finned bony fish (actinopterygians) that include the majority of living fish species.

temnospondyls
A diverse group of extinct amphibians widespread from the Carboniferous to the Triassic, and not finally becoming extinct until the Cretaceous. They included some massive crocodilelike forms. *See also* lepospondyls.

Tertiary
An old term for the geologic period following the Cretaceous that has now been replaced by Paleogene and Neogene.

tetrapods
Vertebrates with four limbs, including amphibians, reptiles, birds, and mammals.

theca (pl. thecae)
In graptolites: the organic-walled tubes housing zooids. *See also* graptolites, zooids.

therapsids
An extinct group of land vertebrates that include the direct ancestors of mammals. A subgroup of synapsids. *See also* synapsids.

theropods
A major group of bipedal dinosaurs that includes well-known top predators such as *Tyrannosaurus rex*, as well as many smaller species.

trace fossil
A fossil that preserves the activity of a living thing, rather than the living thing itself—for example, a dinosaur footprint.

trilobites
An enormously diverse group of extinct marine arthropods. They became extinct at the end of the Permian period.

type species
A species that defines the genus in which it is included, and to which the genus name originally assigned is permanently attached even if later reclassification occurs.

ulna
One of the two main bones of the forearm. *See* radius.

valve
In mollusks and brachiopods: a shell, especially a separate section of a shell when it comes in more than one part. *See also* bivalves.

vascular bundle
A longitudinal strand of tissues within the stem of a vascular plant that conducts water and food materials.

vascular plants
Plants that have specialized tissue for conducting water and nutrients between different parts of their bodies. Most land plants, except for bryophytes, are vascular plants. *See also* bryophytes, lycophytes, euphyllophytes.

ventral
Relating to the lower surface or belly of an animal. *See also* dorsal.

vertebrates
Animals with backbones, including fish, amphibians, reptiles, birds, and mammals. The backbone elements (vertebrae) are usually made of bone, but sometimes of cartilage. Vertebrates also share many other distinctive features, such as the possession of a skull (cranium) and the way the internal organs are arranged. *See also* chordates.

water vascular system
A system of hydraulically operated water-filled tubes that is characteristic of the body structure of echinoderms. It typically ends in small projecting "tube feet," although these may be used for food capture and other purposes as well as walking, depending on the type of animal. *See also* echinoderms.

zooid
An individual of a colonial animal, such as corals, graptolites, and bryozoans.

zosterophylls
An extinct group of primitive vascular plants included together with the clubmosses (lycopods) in the lycophytes. Zosterophylls are among the earliest land plants known. They had simple, usually leafless, stems. *See also* lycophytes.

Index

Page numbers in **bold** indicate the main reference, page numbers in *italics* an illustrated reference.

B

Acknowledgments

Dorling Kindersley would like to thank the following people for their help in the preparation of this book: Anushka Mody for additional design help; Regina Franke at DK Verlag and Riccie Janus for help with setting up photo shoots in Germany; Roger Jones for access to his fossil collection; Dr. Charles Wellman at Sheffield University for plant spore images; and Steve Willis and Mel Fisher for color work.

DK India would also like to thank Shreya Chauhan, Shreya Iyengar, Aashirwad Jain, Aadithyan Mohan, Priyanjali Narain, Rupa Rao, and Mark Silas for editorial assistance in preparing the second edition, and Steve Crozier for image retouching.

Many thanks go to the following for access to their fossil collections for photography: Else Marie Friis, Kamlesh Khullar, Steve McLoughlin, and the rest of the team at Naturhistoriska riskmuseet, Stockholm; Doris Von Eiff and all the staff at Senckenberg Forschungsinstitute u. Naturmuseen, Frankfurt; Kirsten Andrews-Speed, Eliza Howlett, Monica Price, Derek Siveter, Malgosia Nowak-Kemp, and Tom Kemp at Oxford University Museum of Natural History; Bob Owens, Caroline Buttler, and John Cope at Amgueddfa Cymru—National Museum Wales.

Sources for the illustrations listed below are as follows: **p.23** Oxygen isotopes diagram: Simon Lamb and David Sington, *Earth Story*, p.149; **p.24** Rain, altitude, and leaf types diagram: Simon Lamb and David Sington, *Earth Story*, p.131; **p.29** House Sparrow size map: http://evolution.berkeley.edu/evosite/evo101/IVB1aExamples.shtml; **p.38** Dinosaur locomotion: http://www.nature.com/nature/journal/v415/n6871/full/415494a.html; **p.76** Nonvertebrate and vertebrate diagram: F. Harvey Pough, Christine M. Janis, John B. Heiser, *Vertebrate Life*—seventh edition, p.25; **p.92** Dermal plate development in fish: L.B. Tarlo, *The Downtonian Ostracoderm Corvaspis kingi Woodward, with notes on the development of dermal plates in the Heterostraci*; **p.98** Tabulate corals diagram: http://faculty.cns.uni.edu/~groves/LabExercise09.pdf; **p.102** Rolled trilobite diagram: http://www.trilobites.info/enrollment.htm; **p.104** Jaw development, F. Harvey Pough, Christine M. Janis, John B. Heiser, *Vertebrate Life*—seventh edition, p.55; **p.110** Water conducting cells diagram: Wilson N. Stewart and Gar W. Rothwell, *Paleobotany and the Evolution of Plants*, p.84; **p.121** Elkinsia cross-section: Wilson N. Stewart and Gar W. Rothwell, *Paleobotany and the Evolution of Plants*; **p.123** *Heliophyllum* diagram: http://faculty.cns.uni.edu/~groves/LabExercise09.pdf; **p.156** *Falcatus*: after illustration in Fossil fishes of Bear Gulch, 2005 by Richard Lund and Eileen Grogan; **p.171** *Fenestella* artwork: http://www.kgs.ku.edu/Extension/fossils/bryozoan.html; **p.194** Ammonoid sutures diagram: http://faculty.cns.uni.edu/~groves/LabExercise09.pdf; **p.234** *Magellania*: E.N.K. Clarkson, Invertebrate Paleontology and Evolution, p.159; **p.294** *Archaeanthus*, Wilson N. Stewart and Gar W. Rothwell, *Paleobotany and the Evolution of Plants*, p.448; **p.284**

Feather evolution: http://www.nature.com/nature/journal/v420/n6913/fig_tab/nature01196_F5.html; **p.400** Scallop anatomy: http://www.fao.org/docrep/007/y5720e/y5720e07.htm; **p.401** *Magellania* pedicle: E.N.K. Clarkson, *Invertebrate Paleontology and Evolution*.

Plate Tectonic and Paleogeographic Maps on pages 54–55, 64–65, 80–81, 94–95, 108–09, 136–37, 188–89, 216–17, 270–71, 340–41, 364–65, 392–93 by C.R. Scotese, © 2007, PALEOMAP Project (www.scotese.com)

The publisher would like to thank the following for their kind permission to reproduce their photographs:

(Key: a-above; b-below/bottom; c-center; f-far; l-left; r-right; t-top)

6 Alamy Images: Louis Champion (tl). **Ardea:** Pat Morris (cra/Cambrian). **Corbis:** Michael Freeman (crb/Devonian); Layne Kennedy (cra/Archean); Kevin Schafer (br/Carboniferous). **Getty Images:** Art Wolfe (tc). **Photolibrary:** Iain Sarjeant (cra/Proterozoic). **Science Photo Library:** Hervé Conge, ISM (crb/Silurian); Paul Whitten (cr/Ordovician). **7 Alamy Images:** WaterFrame (clb/Paleogene). **Corbis:** Micha Pawlitzki / zefa (bl/Quaternary) (cla/Jurassic). **DK Images:** Natural History Museum, London (cla/Permian). **Getty Images:** Peter Chadwick / Gallo Images (tc); Ralph Lee Hopkins / National Geographic (cla/Triassic); Louie Psihoyos / Science Faction (cl/Cretaceous). **Photolibrary:** Tam C. Nguyen (clb/Neogene). **8–9 Corbis:** Michael S. Yamashita. **10–11 Alamy Images:** Louis Champion. **12 Anne Burgess:** (bl). **Corbis:** Hulton-Deutsch Collection (cb). **DK Images:** Satellite Imagemap Copyright © 1996-2003 Planetary Visions (br). **Science Photo Library:** Bernhard Edmaier (cl); Dirk Wiersma (clb). **12–13 Getty Images:** Ralph Lee Hopkins / National Geographic (tc). **13 Corbis:** Jonathan Blair (cr). **DK Images:** Natural History Museum, London (tr). **14 Corbis:** Bettmann / Mariner 10 (crb); NASA / CXC / GSFC / U. Hwang / EPA (cb). **Getty Images:** Stocktrek Images (cl). **NASA:** JPL (cr/Venus). **15 NASA:** (crb). **16 Alamy Images:** Blue Gum Pictures (b). **17 Getty Images:** Johnny Johnson / The Image Bank (tr). **Martina Menneken:** (bl). **Science Photo Library:** Michael Abbey (bc); Eye of Science (bc/archaea). **18 Corbis:** image100 (tr); David Pu'u (ca). **Bradley R. Hacker :** (br). **19 Corbis:** Roger Ressmeyer (br). **20 Corbis:** Image Plan (br). **Getty Images:** Astromujoff (ca). **20–21 Corbis:** Arctic-Images (tc). **21 Corbis:** Lloyd Cluff / Comet (cb); George D. Lepp (br/green broadbill); Jochen Schlenker / Robert Harding World Imagery (br/cockatoo); Jim Sugar / Comet (tr). **DK Images:** University Museum of Zoology, Cambridge (bl); Natural History Museum, London (lb) (bc). **22 Corbis:** Chinch Gryniewicz / Ecoscene (tr); Douglas Pearson / Flirt (b). **Paul F. Hoffman:** (ca).

23 Corbis: Tony Wharton / Frank Lane Picture Agency (tl). **Science Photo Library:** (br); British Antarctic Survey (cl); Andrew Syred (tr). **25 Corbis:** Catherine Karnow (t); Michael T. Sedam (crb). **DK Images:** Natural History Museum, London (crb/fossil mammoth tooth). **26 Corbis:** Philip Gould (tr); Elisabeth Sauer / zefa (ca). **Science Photo Library:** M.I. Walker (cra). **27 Corbis:** Michael & Patricia Fogden (br). **DK Images:** Robert L. Braun—modelmaker (crb); Natural History Museum, London (tl/bones). **Depositphotos Inc:** Ian Redding (cl). **28 Corbis:** Michael & Patricia Fogden (cla) (bc); Martin Harvey (c). **DK Images:** Oxford Scientific Films (cb). **Getty Images:** Bob Thomas / Popperfoto (ca). **Science Photo Library:** Photo Researchers (cra). **29 Corbis:** Stephen Frink (bl). **Getty Images:** Wally McNamee (br); Steve Winter / National Geographic (tr). **naturepl.com:** Dave Watts (cla). **30 Corbis:** DLILLC (crb/dolphin photo) (cra/dolphin photo); Jeffrey L. Rotman (cra/dolphin and calf). **32 Corbis:** Jonathan Blair (tr) (cra); Frans Lanting (b). **Science Photo Library:** Mark Pilkington / Geological Survey of Canada (ca). **33 Corbis:** Jonathan Blair (cb) (clb); Michael & Patricia Fogden (crb). **Science Photo Library:** George Bernard (ftr). **34 Corbis:** Jonathan Blair (cl); George H.H. Huey (cra); Vienna Report Agency (ca). **DK Images:** Natural History Museum, London (bc). **Royal Saskatchewan Museum:** (clb). **Science Photo Library:** Steve Gschmeissner (cr). **35 Corbis:** Jonathan Blair (bl); Layne Kennedy (ca). **DK Images:** Rainbow Forest Museum, Arizona (ca/petrified log); Natural History Museum, London (tl). **Derek Siveter:** (br/nymphon and haliestes). **36 DK Images:** Natural History Museum, London (bl). **Mark V. Erdmann:** (clb/living coelacanth). **38 Corbis:** Louie Psihoyos (r/main image); Visuals Unlimited (cla). **Oxford University Museum of Natural History:** (clb). **39 DK Images:** Natural History Museum, London (cl). **Getty Images:** J. Sneesby / B. Wilkins (br). **40–41 Corbis:** Louie Psihoyos. **42–45 Alamy Images:** Louis Champion. **45 Science Photo Library:** James King-Holmes (tc); Dr. Ken Macdonald (cra). **46–47 Getty Images:** Art Wolfe. **48–49 Corbis:** Layne Kennedy. **50 Martin Brasier:** (cr). **Getty Images:** Peter Hendrie / Photographer's Choice (cl); Carsten Peter / National Geographic (bl). **50–51 Corbis:** Layne Kennedy. **51 Martin Brasier:** (b/all 3 images). **naturepl.com:** Doug Perrine (t). **Science Photo Library:** B. Murton / Southampton Oceanography Centre (cl). **52–53 Photolibrary:** Iain Sarjeant (main). **54–61 Photolibrary:** Iain Sarjeant (sidebars). **54–55 Photolibrary:** Iain Sarjeant (b/background). **54–55** Plate Tectonic and Paleogeographic Map C.R. Scotese, © 2007, PALEOMAP Project (www.scotese.com).

54 Alamy Images: Randy Green (tr). **Getty Images:** O. Louis Mazzatenta / National Geographic (cl). **55 Alamy Images:** Blue Gum Pictures (ca); David Wall (cra). **Corbis:** Jonathan Blair (bl); Kazuyoshi Nomachi (tl) (b/background). **56 Photolibrary:** Iain Sarjeant (t/microscopic lettering). **58 J. Gehling, South Australian Museum:** (clb). **Photolibrary:** Iain Sarjeant (t/invertebrates lettering). **59 © Leicester City Museums Service:** (tr). **61 J. Gehling, South Australian Museum:** (ca). **62–63 Ardea:** Pat Morris (main). **64–77 Ardea:** Pat Morris (sidebars). **64–65 Ardea:** Pat Morris (b/background). **64–65** Plate Tectonic and Paleogeographic Map C.R. Scotese, © 2007, PALEOMAP Project (www.scotese.com). **64 Corbis:** Yann Arthus-Bertrand (tr); David Muench (cl). **The Natural History Museum, London:** (br). **65 Alamy Images:** All Canada Photos (cra). **Alamy Stock Photo:** ImageBroker (cla). **Getty Images:** O. Louis Mazzatenta / National Geographic (clb/Burgess Shale). **The Natural History Museum, London:** (bc). **Science Photo Library:** Jonathan A. Meyers (tl) (b/background). **66 Ardea:** Pat Morris (t/microscopic lettering). **Martin Brasier:** (cb). **67 David Siveter (University of Leicester):** (br) (b/background). **68 Ardea:** Pat Morris (t/invertebrates lettering). **John Cope:** (cra). **Getty Images:** O. Louis Mazzatenta / National Geographic (cr). **69 Science Photo Library:** Alan Sirulnikoff (cra). **70 Getty Images:** O. Louis Mazzatenta / National Geographic (cr). **The Natural History Museum, London:** (tc). **72 Alamy Images:** All Canada Photos / T. Kitchin & V. Hurst (bl). **The Natural History Museum, London:** (cla). **75 Alamy Images:** Kevin Schafer (t). **US Geological Survey:** (br). **76 Ardea:** Pat Morris (t/vertebrates lettering). **Natural Visions:** (br). **78–79 Science Photo Library:** Paul Whitten (main). **80–91 Science Photo Library:** Paul Whitten (sidebars). **80–81** Plate Tectonic and Paleogeographic Map C.R. Scotese, © 2007, PALEOMAP Project (www.scotese.com). **80–81 Science Photo Library:** Paul Whitten (b/background). **80 Corbis:** Gary Braasch (cl). **Ken McNamara:** (tr). **81 Corbis:** Stephen Frink (tl/smaller image); **Getty Images:** Richard I'Anson / Lonely Planet Images (cra). **Getty Images:** Jeff Rotman (tl). **82 DK Images:** Natural History Museum, London (cla) (b/background). **Science Photo Library:** Paul Whitten (t/invertebrates lettering) (b/background). **90 Science Photo Library:** Paul Whitten (t/vertebrates lettering). **Davide Bonadonna:** (br). **92–93 Science Photo Library:** Hervé Conge, ISM (main). **93 DK Images:** Natural History Museum, London (cra). **94–105 Science Photo Library:** Hervé Conge, ISM (sidebars). **94–95 Science Photo Library:** Hervé Conge, ISM (b/background). **94–95** Plate Tectonic and Paleogeographic Map C.R. Scotese, © 2007, PALEOMAP Project (www.scotese.com).

94 Corbis: Free Agents Limited (tr); Wilfried Krecichwost / zefa (cl).
95 Corbis: Christophe Boisvieux (ca). **DK Images:** Natural History Museum, London (clb). **Getty Images:** Medioimages / Photodisc (tl). **Science Photo Library:** Sinclair Stammers (cra) (t/plants lettering).
96 Science Photo Library: Hervé Conge, ISM (b/background).
97 Alamy Stock Photo: The Natural History Museum (tc).
98 Alamy Images: Maximilian Weinzierl (clb). **Science Photo Library:** Hervé Conge, ISM (t/invertebrates lettering).
99 Alamy Images: David Bagnall (br). **Patrick L. Colin, Coral Reef Research Foundation:** (cra). **The Natural History Museum, London:** (crb).
102 Jan Hartmann: (tc) (b/background).
104 Science Photo Library: Hervé Conge, ISM (t/vertebrates lettering).
106–107 Corbis: Michael Freeman (main).
108–133 Corbis: Michael Freeman (sidebars).
108–109 Corbis: Michael Freeman (b/background).
108–109 Plate Tectonic and Paleogeographic Map C.R. Scotese, © 2007, PALEOMAP Project (www.scotese.com).
108 DK Images: Natural History Museum, London (clb) (cb). **Getty Images:** Image Source (tr); Dr. Marli Miller / Visuals Unlimited (cl). **Ken McNamara:** (clb/fish).
109 Alamy Images: Phil Lyon / Sylvia Cordaiy Photo Library Ltd (cra). **DK Images:** Natural History Museum, London (clb). **Science Photo Library:** Sinclair Stammers (tl).
110 Corbis: Michael Freeman (t/plants lettering). **Science Photo Library:** Hervé Conge, ISM (b/background).
110–111 Corbis: Michael Freeman (b/background).
111 Science Photo Library: Dr. Jeremy Burgess (cla).
112 Courtesy of the Smithsonian Institution—Photo by Carol Hotton: (cla).
113 Science Photo Library: Sinclair Stammers (tr) (b/background).
116 Alamy Stock Photo: Philip Scalia (cra).
117 Dr. Hong-He Xu / Dr. Christopher M. Berry: (bl).
118 Corbis: Michael Freeman (t/invertebrates lettering). **From** *"The Biota of Early Terrestrial Ecosystems: The Rhynie Chert"* (www.abdn.ac.uk/rhynie/intro.htm) , © University of Aberdeen: (cra).
119 Corbis: Jeffrey L. Rotman (cr).
120 DK Images: Royal Museum of Scotland, Edinburgh (cla).
122 Corbis: Michael Freeman (t/vertebrates lettering); Stephen Frink (cb).
122–123 Corbis: Michael Freeman (b/background).
123 AAP Image : Museum Victoria (crb). **DK Images:** Courtesy of the University Museum of Zoology, Cambridge, on loan from the Geological Museum, University of Copenhagen (fcla).
133 Alamy Images: Yves Marcoux / First Light (cl).
134–135 Corbis: Kevin Schafer (main).
136–160 Corbis: Kevin Schafer (t/plants lettering))sidebars).
136–137 Plate Tectonic and Paleogeographic Map C.R. Scotese, © 2007, PALEOMAP Project (www.scotese.com).
136–137 Corbis: Kevin Schafer (b/background).
136 DK Images: Natural History Museum, London (bl) (br). **Getty Images:** Peter Essick / Aurora (cl); Travel Ink / Gallo Images (tr).
138 Corbis: Kevin Schafer (b/background). **DK Images:** Natural History Museum, London (cl).
139 Dorling Kindersley: Gary Ombler / Swedish Museum of Natural History (tr).
142 Alamy Images: blickwinkel / Ziese (cra). **Corbis:** Ashley Cooper (tl). **DK Images:** Natural History Museum, London (cb) (bl). **Valdosta State University :** Paleothyris cast obtained from Robert Carroll of McGill University's Redpath Museum; displayed on http://.fossils.valdosta.edu (cl).
143 DK Images: Natural History Museum, London (cla).
146 Alamy Images: Steffen Hauser / botanikfoto (cl) (b/background). **Corbis:** Kevin Schafer (t/invertebrates lettering).
147 PrehistoricStore.com: (cla).
149 Science Photo Library: Louise K. Broman (cr).
153 Alamy Images: The Natural History Museum (cla). **Collection of the Illinois State Museum:** (fcr).
154–155 Corbis: Kevin Schafer (b/background).
155 Corbis: Anthony Bannister / Gallo Images (cra).
156 Eileen Grogan / Richard Lund, The Bear Gulch Project: Carnegie Museum of Natural History, St. Joseph's University (bl) (tl).
157 Eileen Grogan / Richard Lund, The Bear Gulch Project: Carnegie Museum of Natural History, St. Joseph's University (tc). **DK Images:** Natural History Museum, London (cl). © **Hunterian Museum & Art Gallery, University of Glasgow:** (cr).
158 Alamy Images: John Cancalosi (bl). **Roger Jones Collection:** (tc).
160 DK Images: Royal Museum of Scotland, Edinburgh (cl). **Craig Slawson:** Geoconservation UK (cb). **Valdosta State University :** Paleothyris cast obtained from Robert Carroll of McGill University's Redpath Museum; displayed on http://.fossils.valdosta.edu (b).
161 Getty Images: Alan Marsh / First Light (cra).
162–163 DK Images: Natural History Museum, London (main).
163 DK Images: Natural History Museum, London (crb).
164–185 DK Images: Natural History Museum, London.
164 DK Images: Natural History Museum, London (br). **Getty Images:** Anthony Boccaccio (cl). **Photolibrary:** Oxford Scientific (OSF) / Konrad Wothe (tr).
164–165 DK Images: Natural History Museum, London (b/background).
165 DK Images: Natural History Museum, London (bl). **Alamy Stock Photo:** Africa Media Online (tl). **Jon Ranson, NASA Goddard Space Flight Center:** (cra).
166 Corbis: James L. Amos (cr) (t/plants lettering). **DK Images:** Natural History Museum, London (b/background). **FLPA:** Krystyna Szulecka (cl).
170 Corbis: Jonathan Blair (clb) (b/background) (t/invertebrates lettering). **DK Images:** Natural History Museum, London (cra).
174 DK Images: Natural History Museum, London (b/background).
175 Andrew Milner: (b).
176 Corbis: Jack Goldfarb/Design Pics (cla).
176–177 Alamy Images: WaterFrame (bc). **DK Images:** Natural History Museum, London (t).
177 DK Images: Natural History Museum, London (cra). **Getty Images:** Ken Lucas / Visuals Unlimited (br).
178 Corbis: Martin Schutt / DPA (cla). **DK Images:** University Museum of Zoology, Cambridge (clb). **Getty Images:** Hulton Archive (c). **Andrew Milner:** (tc).
180 DK Images: Natural History Museum, London (tl).
181 Corbis: Lester V. Bergman (ca). **Getty Images:** Ken Lucas / Visuals Unlimited (tr).
182 Corbis: W. Perry Conway (t).
183 DK Images: Natural History Museum, London (cr). **Henssen PalaeoWerkstatt , www.palaeoowerkstatt.de:** (br).
186–187 Getty Images: Ralph Lee Hopkins / National Geographic (main).
187 DK Images: Natural History Museum, London.
188–213 Getty Images: Ralph Lee Hopkins / National Geographic (sidebands).
188–190 Getty Images: Ralph Lee Hopkins / National Geographic (b/background).
188–189 Plate Tectonic and Paleogeographic Map C.R. Scotese, © 2007, PALEOMAP Project (www.scotese.com).
188 Corbis: Rose Hartman (tr). **DK Images:** Natural History Museum, London (br) (bl) (clb) (crb). **Rex by Shutterstock:** Jason Bye (clb).
189 Corbis: Fridmar Damm / zefa (cra); Martin B. Withers / Frank Lane Picture Agency (tl).
190 Corbis: George H.H. Huey (cl). **Getty Images:** Ralph Lee Hopkins / National Geographic (t/plants lettering).
191 Dr. Philippe Moisan: (tr).
192 DK Images: Natural History Museum, London (cb).
192–193 DK Images: Natural History Museum, London (c).
194 Getty Images: Ralph Lee Hopkins / National Geographic (invertebrates lettering) (b/background).
195 Dr. Hans Arene Nakram, University of Oslo: (tl).
198–199 Getty Images: Ralph Lee Hopkins / National Geographic (b/background).
199 Museum of Texas Tech University: (cla).
200 Giuseppe Buono (http://fossilspictures.wordpress.com/ and http://paleonews.wordpress.com/): (cla). **Dr. Rainer Schoch / Staatliches Museum für Naturkunde:** (tc).
201 DK Images: Royal Tyrrell Museum of Palaeontology, Alberta, Canada (br).
202 DK Images: Natural History Museum, London (bl) (c).
203 DK Images: Institute of Geology and Palaeontology, Tubingen, Germany (ca).
204 Corbis: Jonathan Blair (cra).
205 Museum of Texas Tech University: (tr).
209 Corbis: Louie Psihoyos (bc/Bob Bakker).
211 DK Images: Institute of Geology and Palaeontology, Tubingen, Germany (clb) (cla).
213 DK Images: Natural History Museum, London (cr).
214–215 DK Images: Natural History Museum, London (main).
215 DK Images: Natural History Museum, London (crb).
216–267 DK Images: Natural History Museum, London (sidebands).
216–217 Plate Tectonic and Paleogeographic Map C.R. Scotese, © 2007, PALEOMAP Project (www.scotese.com).
216 Corbis: Tom Bean (tr) (clb/flower). **DK Images:** Natural History Museum, London (bl) (bc). **Getty Images:** Jeff Foott / Discovery Channel Images (cl). **Science Faction Images:** Louie Psihoyos (tr/dinosaur tracks).
216–226 DK Images: Natural History Museum, London (b/background).
217 Corbis: Theo Allofs (cra). **DK Images:** Natural History Museum, London (clb) (bl). **Getty Images:** Fumio Tomita / Sebun Photo (tl).
218 Corbis: David Spears / Clouds Hill Imaging Ltd. (cr). **DK Images:** Natural History Museum, London (t/plants lettering). **Science Photo Library:** Dee Breger (clb).
219 Getty Images: Kevin Schafer / Photographer's Choice (bc).
221 DK Images: Natural History Museum, London (cra).
224 Corbis: Derek Hall / Frank Lane Picture Agency (cl) (b/background). **DK Images:** Natural History Museum, London (t/invertebrates lettering).
231 Corbis: Douglas P. Wilson / Frank Lane Picture Agency (bl). **Photoshot:** Ken Griffiths / NHPA (cra).
232–233 DK Images: Natural History Museum, London (b/background).
235 Corbis: Kevin Schafer (crb).
238 Photolibrary: Rich Reid / National Geographic (bl).
239 Corbis: Naturfoto Honal (cla).
241 Science Photo Library: D.J.M. Donne (tr).
243 DK Images: Robert L. Braun—modelmaker (crb).
249 DK Images: Staatliches Museum für Naturkunde, Stuttgart (tc); American Museum of Natural History (tr).
250 Corbis: Bettmann (cr).
252 DK Images: Natural History Museum, London (br).
253 Corbis: Louie Psihoyos (bl); University of the Witwatersrand / EPA (crb).
254 The Natural History Museum, London: (crb).
255 DK Images: Natural History Museum, London (bl). © **Frank Luerweg, Uni Bonn:** (cra).
258 DK Images: Carnegie Museum of Natural History, Pittsburgh (cla).
262 DK Images: Leicester Museum (clb). **Martin Williams:** (cla).
262–263 DK Images: Natural History Museum, London (bc).
263 Martin Williams: (br).
266 Visuals Unlimited, Inc.: Ken Lucas (cra).
267 Visuals Unlimited, Inc.: Albert Copley (bl).
268–269 Getty Images: Louie Psihoyos / Science Faction (main). **Science Photo Library:** Andrew Syred (tr).
270–273 Getty Images: Louie Psihoyos / Science Faction (b/background).
270–337 Getty Images: Louie Psihoyos / Science Faction.
270–271 Plate Tectonic and Paleogeographic Map C.R. Scotese, © 2007, PALEOMAP Project (www.scotese.com).
270 DK Images: Natural History Museum, London (clb) (br). **Getty Images:** Don Klumpp / Photographer's Choice (ftr); Panoramic Images (cl).
271 DK Images: Natural History Museum, London (clb) (bc) (bl). **Getty Images:** Travel Ink / Gallo Images (tl); Eric Van Den Brulle (cl).
272 Getty Images: Louie Psihoyos / Science Faction (t/plants lettering). © **Sharon Milito, Paleotrails Project:** (crb).
274 Alamy Images: Scenics & Science (bl). **Photolibrary:** Phototake Science (cla). **Dorling Kindersley:** Gary Ombler /

Swedish Museum of Natural History (crb).
275 DK Images: Royal Museum of Scotland, Edinburgh (cb); Natural History Museum, London (b/araucaria). **Barry Thomas:** (crb/*Araucaria*).
276 DK Images: Natural History Museum, London (tr). **Ardea:** John Mason. **DK Images:** Natural History Museum, London.
277 DK Images: Natural History Museum, London (cl). **Getty Images:** Jonathan Blair / National Geographic (cl). **DK Images:** Royal Tyrrell Museum of Palaeontology, Alberta, Canada (b) (b/background).
278 Getty Images: Louie Psihoyos / Science Faction (t/invertebrates lettering).
282 Corbis: Lawson Wood (bl).
284 Getty Images: Louie Psihoyos / Science Faction (t/vertebrates lettering).
284–285 Getty Images: Louie Psihoyos / Science Faction (b/background).
285 DK Images: Natural History Museum, London (cla).
286 Alamy Images: Michael Patrick O'Neill (bl).
286–287 DK Images: Natural History Museum, London.
288 Alamy Images: Danita Delimont (crb). **Getty Images:** George Grall / National Geographic (crb/gar fish).
290 Corbis: Louie Psihoyos / Science Faction (cb).
292 Yale University Peabody Museum Of Natural History: (bl).
296 Bayerische Staatssammlung für Paläontologie und Geologie, München (cb).
298 DK Images: Natural History Museum, London (tr).
299 Alamy Images: Javier Etcheverry (cra/giganotosaurus). **DK Images:** Natural History Museum, London (tl). **Ryan Somma:** (b).
301 DK Images: Robert L. Braun—modelmaker (tl); Natural History Museum, London (bl) (cra). **Getty Images:** Ira Block / National Geographic (tr).
304 Corbis: Paul Vicente / EPA (bl).
308–309 DK Images: Natural History Museum, London; Natural History Museum, London (ca).
310 DK Images: Royal Tyrrell Museum of Palaeontology, Alberta, Canada (cra).
311 Corbis: Louie Psihoyos / Science Faction (cla); Kevin Schafer (br). **DK Images:** Peabody Museum of Natural History, Yale University (tl); Luis Rey—modelmaker (fcra). **Getty Images:** Louie Psihoyos / Science Faction (cra).
312 Getty Images: Ira Block / National Geographic (bl); Spencer Platt (cla).
314 DK Images: Queensland Museum, Brisbane, Australia (clb).
315 DK Images: Royal Tyrrell Museum of Palaeontology, Alberta, Canada (tr); Peter Minister (b).
318 DK Images: Natural History Museum, London (br) (tl). **Getty Images** (cl).
322 DK Images: Royal Tyrrell Museum of Palaeontology, Alberta, Canada (tr/skull) (c).
323 DK Images: Royal Tyrrell Museum of Palaeontology, Alberta, Canada (b).
325 DK Images: American Museum of Natural History (br).
330 DK Images: American Museum of Natural History (ca). **Getty Images:** Louie Psihoyos / Science Faction (cr).
331 DK Images: American Museum of Natural History.
334 DK Images: Royal Tyrrell Museum of Palaeontology, Alberta, Canada (bl/skeleton and skull). **Getty Images:** O.

Louis Mazzatenta / National Geographic (br).
335 DK Images: American Museum of Natural History (c). **Hai-lu You:** (tr).
336 Corbis: Peter Foley / EPA (tr); Joe McDonald (cra). **MENG Jin:** ZHANG Chuang and XING Lida (lillustrators) (ca).
337 Carnegie Museum of Natural History: (crb). **Getty Images:** Nicole Duplaix / National Geographic (cla). **Steve Morton:** Museum Victoria / Monash University (tl).
338–339 Alamy Images: WaterFrame (main).
340–389 Alamy Images: WaterFrame (sidebands).
340–341 Plate Tectonic and Paleogeographic Map C.R. Scotese, © 2007, PALEOMAP Project (www.scotese.com).
340 Alamy Stock Photo: Michelle Gilders (cra). **Corbis:** Momatiuk–Eastcott (tr). **DK Images:** Natural History Museum, London (bc) (fbr). **Getty Images:** Bill Hatcher / National Geographic (cl).
341 Alamy Images: JTB Photo Communications, Inc. (tl). **DK Images:** Natural History Museum, London (clb) (bl). **Getty Images:** Kim Westerskov (cra) (t/plants lettering).
342 Alamy Images: WaterFrame (b/background). **Corbis:** image100 (clb). **Science Photo Library:** Maurice Nimmo (cra); Maria and Bruno Petriglia (ca).
345 Corbis: Frans Lanting (cl). **DK Images:** Natural History Museum, London (bl) (bc).
346 Alamy Images: WaterFrame (t/invertebrates lettering). **SeaPics.com:** James D. Watt (clb).
349 Getty Images: Richard Herrmann / Visuals Unlimited (tr).
350 Alamy Images: WaterFrame (t/vertebrates lettering) (b/background). **Corbis:** Jonathan Blair (cl). **DK Images:** Natural History Museum, London (crb).
351 Corbis: Radius Images (clb); Visuals Unlimited (br).
353 Ray Carson –UF Photography: (crb). **DK Images:** Natural History Museum, London (ca) (br) (clb).
355 Ryan Somma: (br).
356 DK Images: Natural History Museum, London (ca).
357 Corbis: Bettmann (tl). **DK Images:** Natural History Museum, London (cla). **Getty Images:** Kevin Schafer / Visuals Unlimited (br). **Ryan Somma:** (crb).
358 Corbis: Michael & Patricia Fogden (cla). **DK Images:** Natural History Museum, London (cra). **Hessisches Landesmuseum Darmstadt:** Photo: Wolfgang Fuhrmannek. Hessisches Landesmuseum Darmstadt. (tc). **Dr. Kent A. Sundell, owner—Douglas Fossils (www.douglasfossils.com):** (bc).
359 Corbis: Kevin Schafer (tr). **DK Images:** Natural History Museum, London (tl/skulls and foot).
360 Getty Images: Johnny Sundby / America 24-7 (tl).
361 DK Images: Natural History Museum, London (crb). **Getty Images:** Stan Honda / AFP (tc).
362–363 Photolibrary: Tam C. Nguyen (main).
363 DK Images: Natural History Museum, London (crb).
364–365 Photolibrary: Tam C. Nguyen (b/background).
364–365 Plate Tectonic and Paleogeographic Map C.R. Scotese, © 2007, PALEOMAP Project (www.scotese.com).
364 Corbis: (cl); **Getty Images:** Pakawat Thongcharoen / Moment (tr). **DK**

Images: Natural History Museum, London (br).
365 DK Images: Natural History Museum, London (bl) (br). **Getty Images:** Thomas Dressler / Gallo Images (cra/left image); Joseph Sohm—Visions of America / Photodisc (tl); Space Frontiers / Hulton Archive (cra).
366 Corbis: Mike Grandmaison (cra). **Getty Images:** Joseph Sohm—Visions of America / Stockbyte (clb) (t/plants lettering). **Photolibrary:** Tam C. Nguyen (b/background).
367 Corbis: David Muench (tc/taxodium).
370 DK Images: Natural History Museum, London (clb) (br/Palmoxylon).
371 DK Images: Natural History Museum, London (cra/porana) (b/background).
372 Photolibrary: Tam C. Nguyen (t/invertebrates lettering).
373 Getty Images: David Wrobel / Visuals Unlimited (cra).
374 Alamy Images: John T. Fowler (br).
378–379 Alamy Images: Phil Degginger.
380 DK Images: Natural History Museum, London (ca) (cb) (b/background). **Photolibrary:** Tam C. Nguyen (t/vertebrates lettering).
381 Bone Clones, Inc.: (bl). **Corbis:** Louie Psihoyos (cla). **DK Images:** Natural History Museum, London (tr) (clb).
382 DK Images: Natural History Museum, London (ca).
383 Nicholas D. Pyenson: Paleontologists from the San Diego Natural History Museum (cb). **Valley Anatomical Preparations, Inc.:** (bl).
384 Agate Fossil Beds National Monument: (tl). **Mira Images:** Phil Degginger c/o Mira.com (clb). **Ryan Somma:** (tc) (bc).
385 Ashfall Fossil Beds / University of Nebraska State Museum: (tr); Natural History Museum, London (br). **DK Images:** American Museum of Natural History (tl). **Photolibrary:** Steve Turner / Oxford Scientific (OSF) (fbr).
388 DK Images: Natural History Museum, London (br).
389 DK Images: Natural History Museum, London (cl) (bl) (clb) (t).
390–391 Corbis: Micha Pawlitzki/zefa (main).
392–415 Corbis: Micha Pawlitzki / zefa (sidebands).
392–394 Corbis: Micha Pawlitzki / zefa (b/background).
392–393 Plate Tectonic and Paleogeographic Map C.R. Scotese, © 2007, PALEOMAP Project (www.scotese.com).
392 DK Images: Natural History Museum, London (bl/both skulls on the left). **Getty Images:** William Albert Allard / National Geographic (tr); Vilem Bischof / AFP (br); Stocktrek Images (cl).
393 Corbis: Momatiuk–Eastcott (tl). **DK Images:** Natural History Museum, London (bc). **Getty Images:** PhotoLink / Photodisc (cra). **Science Photo Library:** Javier Trueba / MSF (bl).
394 Corbis: Craig Aurness (crb); Wayne Lawler / Ecoscene (cra) (b/background); Micha Pawlitzki / zefa (t/plants lettering). **Science Photo Library:** David Scharf (cl).
395 Alamy Images: Walter H. Hodge / Peter Arnold, Inc. (br).
396 Alamy Images: blickwinkel / Koenig (bc).
397 Photo Biopix.dk: J.C. Schou (cla). **Getty Images:** DEA / RANDOM / De Agostini Picture Library (bl). **Wikimedia Commons:** Kahuroa (cra).
398 Alamy Images: Marek Piotrowski

(tr). **Corbis:** Stuart Westmorland (tc).
399 Science Photo Library: Eye of Science (clb).
400 Ardea: Steve Hopkin (clb) (b/background). **Corbis:** Micha Pawlitzki / zefa (t/invertebrates lettering).
402 The Natural History Museum, London: (cla).
403 Corbis: Ken Wilson / Papilio (bc).
404 Getty Images: David Wrobel / Visuals Unlimited (c).
406 Corbis: Micha Pawlitzki / zefa (vertebrates lettering) (b/background). **Getty Images:** Tom Walker / The Image Bank (crb); American Museum of Natural History (tr).
407 DK Images: Natural History Museum, London (crb) (tc). **Image courtesy of DigiMorph.org:** (bc).
408 DK Images: Natural History Museum, London (c/foot) (cr). **The Natural History Museum, London:** (b).
409 DK Images: Natural History Museum, London (cra) (tr). **Getty Images:** Theo Allofs / Photonica (br).
410 Corbis: Ted Soqui (crb). **DK Images:** Down House / Natural History Museum, London (cra). **photo by David K. Smith:** Arctodus simus specimen from The Mammoth Site, Hot Springs, South Dakota (clb).
411 DK Images: Natural History Museum, London (cr).
413 DK Images: National Museum of Wales (br).
414 DK Images: Natural History Museum, London (t/mastodon bones).
415 DK Images: Rough Guides / Simon Bracken (bl); Natural History Museum, London (tr) (br).
417 Stone Age Institute: Dr. Sileshi Semaw (project director) (ca).
418 The Natural History Museum, London: (tl); **Science Photo Library:** Javier Trueba / MSF (tr); **The Natural History Museum, London:** (fbr); **Science Photo Library:** Javier Trueba / MSF (bl).
419 Bone Clones, Inc.: (cl); **National Geographic Creative:** Robert Clark (clb); **Science Photo Library:** Pascal Goetgheluck (l); **Alamy Stock Photo:** Banana Pancake (bc); **National Geographic Creative:** Robert Clark (clb); **Stefan Fichtel** (cr).

All other images © Dorling Kindersley
For further information see:
www.dkimages.com